FRACTURE AND FATIGUE CONTROL IN STRUCTURES

Applications of Fracture Mechanics

STANLEY T. ROLFE

Ross H. Forney Professor of Engineering
University of Kansas

JOHN M. BARSOM

Associate Research Consultant
United States Steel Research Laboratory

PRENTICE-HALL, INC., *Englewood Cliffs, New Jersey*

Library of Congress Cataloging in Publication Data

ROLFE, STANLEY THEODORE,
Fracture and fatigue control in structures.

Includes bibliographical references and index.
1. Fracture Mechanics. 2. Metals–Fatigue.
I. Barsom, John M., joint author. II. Title.
TA409.R64 620.1'125 76-16001
ISBN 0-13-329953-8

Civil Engineering and Engineering Mechanics Series
N. M. NEWMARK AND W. J. HALL, *Editors*

© 1977 by Prentice-Hall, Inc.
Englewood Cliffs, New Jersey 07632

All rights reserved. No part of this book
may be reproduced in any form or by any means
without permission in writing
from the publisher.

10 9 8 7 6 5 4 3 2

Printed in the United States of America

PRENTICE-HALL INTERNATIONAL, INC., *London*
PRENTICE-HALL OF AUSTRALIA PTY. LIMITED, *Sydney*
PRENTICE-HALL OF CANADA, LTD., *Toronto*
PRENTICE-HALL OF INDIA PRIVATE LIMITED, *New Delhi*
PRENTICE-HALL OF JAPAN, INC., *Tokyo*
PRENTICE-HALL OF SOUTHEAST ASIA PTE. LTD., *Singapore*
WHITEHALL BOOKS LIMITED, *Wellington, New Zealand*

To

Phyllis and Valentina

Contents

Foreword

In his well-known text on "Mathematical Theory of Elasticity", Love inserted brief discussions of several topics of engineering importance for which linear elastic treatment appeared inadequate. One of these topics was rupture. Love noted that various safety factors, ranging from 6 to 12 and based upon ultimate tensile strength, were in common use. He commented that "the conditions of rupture are but vaguely understood." The first edition of Love's treatise was published in 1892. Fifty years later, structural materials had been improved with a corresponding decrease in the size of safety factors. Although Love's comment was still applicable in terms of engineering practice in 1946, it is possible to see in retrospect that most of the ideas needed to formulate the mechanics of fracturing on a sound basis were available. The basic content of modern fracture mechanics was developed in the 1946 to 1966 period. Serious fracture problems supplied adequate motivation and the development effort was natural to that time of intensive technological progress.

Mainly what was needed was a simplifying viewpoint, progressive crack extension, along with recognition of the fact that real structures contain discontinuities. Some discontinuities are prior cracks and others develop into cracks with applications of stress. The general idea is as follows. Suppose a structural component breaks after some general plastic yielding. Clearly a failure of this kind could be traced to a design error which caused inadequate section strength or to the application of an overload. The fracture failures which were difficult to understand are those which occur in a rather brittle manner at stress levels no larger than were expected when the structure was designed. Fractures of this second kind, in a special way, are also due to overloads. If one considers the stress redistribution around a pre-existing crack subjected to tension, it is clear that the region adjacent to the perimeter of the crack is overloaded due to the severe stress concentration and that local plastic strains must occur. If the toughness is limited, the plastic strains at the crack border may be accompanied by crack extension. However, from similitude, the crack border overload increases with crack size. Thus progressive crack extension tends to be self stimulating. Given a prior crack, and a material of limited toughness, the possibility for development of rapid fracturing prior to general yielding is therefore evident.

Analytical fracture mechanics provides methods for characterizing the "overload" at the leading edge of a crack. Experimental fracture mechanics collects information of practical importance relative to fracture toughness, fatigue cracking, and corrosion cracking. By centering attention on the active region involved in progressive fracturing, the collected laboratory data are in a form which can be transferred to the leading edge of a crack in a structural component. Use of fracture mechanics analysis and data has explained many service fracture failures with a satisfactory degree of quantitative accuracy. By studying the possibilities for such fractures in advance, effective fracture control plans have been developed.

Currently the most important task is educational. It must be granted that all aspects of fracture control are not yet understood. However, the information now available is basic, widely applicable, and should be integrated into courses of instruction in strength of materials. The special value of this book is the emphasis on practical use of available information. The basic concepts of fracture mechanics are presented in a direct and simple manner. The descriptions of test methods are clear with regard to the essential experimental details and are accompanied by pertinent illustrative data. The discussions of fracture control are well-balanced. Readers will learn that fracture control with real structures is not a simple task. This should be expected and pertains to other aspects of real structures in equal degree. The book provides helpful fracture control suggestions and a sound viewpoint. Beyond this the engineer must deal with actual problems with such resources as are needed. The adage "experience is the best teacher" does not seem to be altered by the publication of books. However, the present book by two highly respected experts in applications of fracture mechanics provides the required background training. Clearly the book serves its intended purpose and will be of lasting value.

George R. Irwin
University of Maryland
College Park, Maryland

Preface

The field of fracture mechanics has become the primary approach to controlling brittle fracture and fatigue failures in structures. The emphasis in this book is on an introduction to the field of fracture mechanics from an application viewpoint. Once the topic is introduced, applications of fracture mechanics to the fields of fracture and fatigue control in structures are emphasized. We believe that this textbook will serve as an introduction to the field of fracture mechanics for senior or beginning graduate students, but more importantly, it introduces the practicing engineer to a field that is becoming increasingly important. In recent years, structural failures and the desire for increased safety and reliability of structures have led to the development of various fracture criteria for many types of structures, including bridges, airplanes, and nuclear pressure vessels.

The development of fracture-control plans for new and unusual structures such as offshore drilling rigs, floating nuclear power plants, space shuttles, *et cetera*, is becoming more widespread. Each of the topics of fracture criteria and fracture control is developed from an engineering viewpoint, including economic and practical considerations. We believe that this textbook should assist engineers to become aware of the fundamentals of fracture mechanics, and in particular, of their responsibility in controlling brittle fracture and fatigue failures in structures.

Chapter 1 serves as an overview of the problem of fracture and fatigue in structures as well as an introduction to the field of fracture mechanics. Chapter 2 provides the theoretical development of stress-intensity factors, K_I, and Chapter 3 describes the test methods for obtaining *critical* stress intensity factors, K_{Ic} or K_{Id}. Chapter 4 describes the effect of temperature, loading rate, and plate thickness on the fracture toughness of a wide variety of structural materials, primarily structural steels. Chapter 5 describes the relationship between stress, flaw size, and material toughness, with specific design examples. These first five chapters provide the fundamental basis of linear-elastic fracture mechanics.

Because many structural materials have fracture toughness levels outside the range of linear-elastic fracture mechanics at service temperatures and loading rates, correlations with other more common notch toughness tests are widely used, and these are described in Chapter 6.

Chapters 7 through 11 deal with sub-critical crack growth by fatigue, stress corrosion, or corrosion fatigue. It is in these areas that linear-elastic fracture mechanics has perhaps had its greatest field of application. Chapter 7 describes a method of analyzing fatigue-crack initiation from a blunt-notch using fracture mechanics terminology. Chapter 8 describes one of the most widespread uses of fracture mechanics, namely fatigue-crack propagation under constant-amplitude load fluctuation. Chapter 9 describes fatigue-crack propagation under variable-amplitude load fluctuation and the use of the root-mean-square ΔK value, ΔK_{RMS}.

Chapters 10 and 11 introduce the effects of environments on subcritical crack growth, namely stress corrosion cracking and corrosion fatigue respectively.

Chapters 12 and 13 introduce the various factors affecting fracture criteria and present examples of existing fracture criteria currently in use. Chapter 14 develops the principles of fracture-control plans and Chapter 15 presents various examples of existing fracture-control plans for specific classes of structures such as nuclear pressure vessels, aircraft, and bridges, as well as proposed fracture-control plans for welded ship hulls and floating nuclear power-plant platforms.

Chapter 16 introduces the field of elastic-plastic fracture-mechanics as analyzed by crack-opening displacement (COD), R-curve, or J-integral.

Based on the student's background, an introductory course might consist of Chapters 1 through 6, 8, 10, and 12 through 14, whereas some of the more advanced topics are covered in Chapters 7, 9, 11, 15 and 16. After reading Chapters 1, 2, and 5, the practicing engineer who is generally familiar with behavior of structural materials could move to any chapter of particular interest.

The authors wish to express their appreciation to their respective organizations, The University of Kansas and U.S. Steel Research Laboratory, for their support in preparing this book. Most importantly, however, we wish to acknowledge the support of our many colleagues both within and outside our organizations, who have contributed to the development of this book. Finally, the typing done by Betty Lane, Janice Herrington, Linda Smoot, and Thelma Bernadowski as well as the continued encouragement and support of our families is sincerely appreciated.

STAN ROLFE
JOHN BARSOM

1

Overview of the Problem of Fracture and Fatigue in Structures

1.1. Historical

Although the total number of structures that have failed by brittle fracture* is low, brittle fractures have occurred and do occur in structures. The following limited historical review is not meant to be complete but only to illustrate

FIG. 1.1. Photograph of typical brittle fracture surface.

*Brittle fracture is a type of catastrophic failure in structural materials that usually occurs without prior plastic deformation and at extremely high speeds (as high as 7,000 ft/sec). The fracture is usually characterized by a flat fracture surface (cleavage) with little or no shear lips, as shown in Figure 1.1, and at average stress levels below those of general yielding. Brittle fractures are not so common as fatigue, yielding, or buckling failures, but when they do occur they may be more costly in terms of human life and/or property damage.

the fact that brittle fractures can occur in engineering structures such as tanks, pressure vessels, ships, bridges, airplanes, etc.

Shank[1] and Parker[2] have reviewed many structural failures, beginning in the late 1800s when members of the British Iron and Steel Institute reported the mysterious cracking of steel in a brittle manner. In 1886, a 250-ft-high standpipe in Gravesend, Long Island failed by brittle fracture during its hydrostatic acceptance test. During this same period, other catastrophic brittle failures of riveted structures such as gas holders, water tanks, and oil tanks were reported even though the materials used in these structures had met all existing tensile and ductility requirements.

One of the most famous tank failures was that of the Boston molasses tank, which failed in January 1919 while it contained 2,300,000 gal of molasses. Twelve persons were drowned in molasses or died of injuries, 40 others were injured, and several horses were drowned. Houses were damaged, and a portion of the Boston Elevated Railway structure was knocked over. An extensive lawsuit followed, and many well-known engineers and scientists were called to testify. After years of testimony, the court-appointed auditor handed down the decision that the tank failed by overstress. In commenting on the conflicting technical testimony, the auditor stated in his decision, ". . . amid this swirl of polemical scientific waters, it is not strange that the auditor has at times felt that the only rock to which he could safely cling was the obvious fact that at least one half of the scientists must be wrong . . . ." His statement fairly well summarized the state of knowledge among engineers regarding the phenomenon of brittle fracture. At times, it seems that the statement is still true today.

Prior to World War II, several welded vierendeel truss bridges in Europe failed shortly after being put into service. All the bridges were lightly loaded, the temperatures were low, the failures were sudden, and the fractures were brittle. Results of a thorough investigation indicated that most failures were initiated in welds and that many welds were defective (discontinuities were present). The Charpy impact test results showed that most steels were brittle at the service temperature.

However, in spite of these and other brittle failures, it was not until the large number of World War II ship failures that the problem of brittle fracture was fully appreciated by the engineering profession. Of the approximately 5,000 merchant ships built during World War II, over 1,000 had developed cracks of considerable size by 1946. Between 1942 and 1952, more than 200 ships had sustained fractures classified as serious, and at least nine T-2 tankers and seven Liberty ships had broken completely in two as a result of brittle fractures. The majority of fractures in the Liberty ships started at square hatch corners or square cutouts at the top of the sheerstrake. Design changes involving rounding and strengthening of the hatch corners, removing square cutouts in the sheerstrake, and adding riveted crack arresters in various locations led to immediate reductions in the incidence of failures.[3,4]

Most of the fractures in the T-2 tankers originated in defects in the bottom-shell butt welds. The use of crack arresters and improved workmanship reduced the incidence of failures in these vessels.

Studies indicated that in addition to design faults steel quality also was a primary factor that contributed to brittle fracture in welded ship hulls.[5]

Therefore, in 1947, the American Bureau of Shipping introduced restrictions on the chemical composition of steels, and in 1949, Lloyds Register stated that "when the main structure of a ship is intended to be wholly or partially welded, the committee may require parts of primary structural importance to be steel, the properties and process of manufacture of which have been specially approved for this purpose."[6]

In spite of design improvements, the increased use of crack arresters, improvements in quality of workmanship, and restrictions on the chemical composition of ship steels during the late 1940s, brittle fractures still occurred in ships in the early 1950s.[2] Between 1951 and 1953, two comparatively new all-welded cargo ships and a transversely framed welded tanker broke in two. In the winter of 1954, a longitudinally framed welded tanker constructed of improved steel quality using up-to-date concepts of good design and welding quality broke in two.[7]

Since the late 1950s (although the actual number has been low) brittle fractures still have occurred in ships as is indicated by Boyd's description of ten such failures between 1960 and 1965 and a number of unpublished reports of brittle fractures in welded ships since 1965.[8]

The brittle fracture of the 584-ft-long Tank Barge I.O.S. 3301 in 1972,[9] in which the 1-yr-old vessel suddenly broke almost completely in half while in port with calm seas (Figure 1.2), shows that this type of failure continues to be a problem.

In this particular failure, the material had adequate notch toughness as measured by one method of testing (Charpy V-notch) and marginal toughness as measured by another more severe method of testing (dynamic tear). However, the primary cause of failure was established to be an unusually high loading stress caused by improper ballasting. This failure illustrated the fact that human factors can contribute to brittle fractures in structures, including overloads.

In the mid-1950s two Comet aircrafts failed catastrophically while at high altitudes.[10] An exhaustive investigation indicated that the failures initiated from very small fatigue cracks originating from rivet holes near openings in the fuselage. Numerous other failures of aircraft landing gear and rocket motor cases have occurred from undetected defects or from subcritical crack growth either by fatigue or stress-corrosion. The failures of F-111

FIG. 1.2. Photograph of I.O.S. 3301 barge failure.

aircraft were attributed to brittle fractures of members with preexisting flaws. Also in the 1950s, several failures of steam turbines and generator rotors occurred leading to extensive brittle-fracture studies by manufacturers and users of this equipment.

In 1962, the Kings Bridge in Melbourne failed by brittle fracture at a temperature of 40°F. Poor details and fabrication resulted in cracks which were nearly through the flange *prior to* any service loading. Although this bridge failure was studied extensively (and other bridges had failed previously by brittle fracture), the bridge-building industry did not pay particular attention to the possibility of brittle fractures in bridges until the failure of the Point Pleasant Bridge at Point Pleasant, West Virginia. On December 15, 1967, this bridge collapsed without warning, resulting in the loss of 46 lives. Photographs of this eyebar suspension bridge before and after collapse are shown in Figures 1.3 and 1.4.

An extensive investigation of the collapse was conducted by the National Transportation Safety Board (NTSB),[11] and their conclusion was "that the cause of the bridge collapse was the cleavage fracture in the lower limb of the eye of eyebar 330 at joint C13N of the north eyebar suspension chain in the Ohio side span." Because the failure was unique in several ways, numerous investigations of the failure were made.

Extensive use of fracture mechanics was made by Bennett and Mindlin[12] in their metallurgical investigation and they concluded that

1. "The fracture in the lower limb of the eye of eyebar 330 was caused

FIG. 1.3. Photograph of St. Mary's bridge similar to the Pt. Pleasant bridge before collapse.

FIG. 1.4. Photograph of Point Pleasant bridge after collapse.

by the growth of a flaw to a critical size for fracture under normal working stress.

2. The initial flaw was due to stress-corrosion cracking from the surface of the hole in the eye. There is some evidence that hydrogen sulfide was the reagent responsible for the stress-corrosion cracking. (The final report indicates that the initial flaw was due to fatigue, stress-corrosion cracking and/or corrosion fatigue[11]).
3. The composition and heat treatment of the eyebar produced a steel with very low fracture toughness at the failure temperature.
4. The fracture resulted from a combination of factors; in the absence of any of these it probably would not have occurred: (a) the high hardness of the steel which rendered it susceptible to stress-corrosion cracking; (b) the close spacing of the components in the joint which made it impossible to apply paint to the most highly stressed region of the eye, yet provided a crevice in this region where water could collect; (c) the high design load of the eyebar chain, which resulted in a local stress at the inside of the eye greater than the yield strength of the steel; and (d) the low fracture toughness of the steel which permitted the initiation of complete fracture from the slowly propagating stress-corrosion crack when it had reached a depth of only 0.12 in. (3.0 mm) (Figure 1.5)."

FIG. 1.5. Photograph showing origin of failure in Point Pleasant bridge.

Since the time of the Point Pleasant bridge failure, other brittle fractures have occurred in steel bridges as a result of unsatisfactory fabrication methods, design details, or material properties.[13,14] These and other brittle fractures led to an increasing concern about the possibility of brittle fractures in steel bridges and resulted in the AASHTO (American Association of State Highway and Transportation Officials) Material Toughness Requirements being adopted in 1973 (a complete description of these requirements is presented in Chapter 15).

Fracture mechanics has shown that because of the *interrelation among materials, design, and fabrication,* brittle fractures cannot be eliminated in structures merely by using materials with improved notch toughness. The designer still has fundamental responsibility for the overall safety and reliability of his structure. It is the objective of this book to show how fracture mechanics can be used in design to *prevent* brittle fractures and fatigue failures of engineering structures.

As will be described throughout this textbook, the science of *fracture mechanics* can be used to describe *quantitatively* the trade-offs among these three factors (stress, material toughness, and flaw size) so that the designer can determine the relative importance of each of them during *design* rather than during *failure analysis.*

1.2. Notch-Toughness Testing

In addition to the traditional mechanical property tests that measure strength, ductility, modulus of elasticity, etc., there are many tests available to measure some form of notch toughness. *Notch toughness* is defined as the ability of a material to absorb energy (usually when loaded dynamically) in the presence of a flaw, whereas *toughness* of a material is defined as the ability of a smooth member (unnotched) to absorb energy, usually when loaded slowly. Notch toughness is measured with a variety of specimens such as the Charpy V-notch impact specimen, dynamic tear test specimen, K_{Ic}, precracked Charpy, etc., while toughness is usually characterized by the area under a stress-strain curve in a slow tension test. It is the presence of a notch or some other form of stress raiser that makes structural materials susceptible to brittle fracture under certain conditions.

Traditionally, the notch-toughness characteristics of low- and intermediate-strength steels have been described in terms of the transition from brittle to ductile behavior as measured by various types of impact tests. Most structural steels can fail in either a ductile or brittle manner depending on several conditions such as temperature, loading rate, and constraint. Ductile fractures are generally preceded by large amounts of plastic deformation and usually occur at 45° to the direction of the applied stress. Brittle or cleavage fractures generally occur with little plastic deformation and are usually normal to the direction of the principal stress. The transition from one type of fracture behavior to the other generally occurs with changes in service conditions such as the state of stress, temperature, or strain rate.

This transition in fracture behavior can be related schematically to various fracture states, as shown in Figure 1.6. Plane-strain behavior refers to frac-

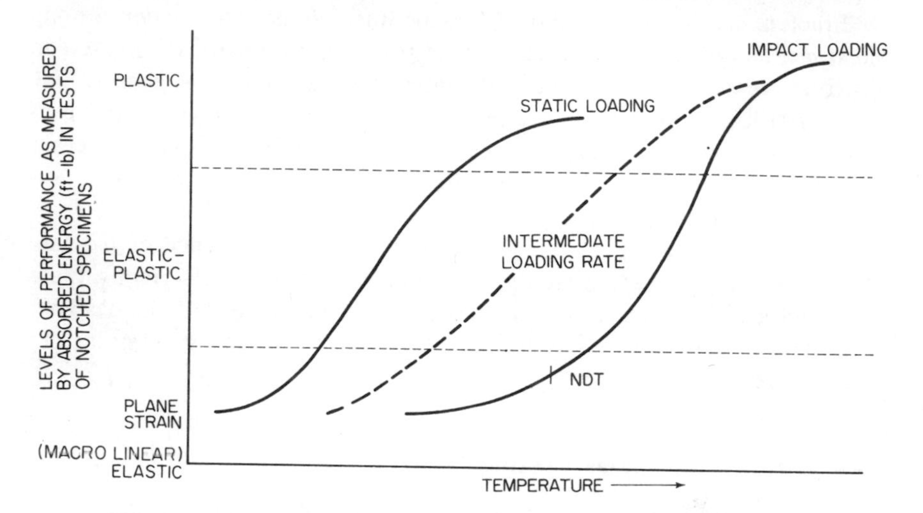

FIG. 1.6. Schematic showing relation between notch-toughness test results and levels of structural performance for various loading rates.

ture under elastic stresses with little or no shear lip development, and it is essentially brittle. Plastic behavior refers to ductile failure under general yielding conditions accompanied usually, but not necessarily, with large shear lips. The transition between these two extremes is the elastic-plastic region, which is also referred to as the mixed-mode region.

For static loading, the transition region occurs at lower temperatures than for impact (or dynamic) loading. Thus, for structures subjected to static loading, the static transition curve should be used to predict the level of performance at the service temperature. For structures subjected to impact or dynamic loading, the impact transition curve should be used to predict the level of performance at the service temperature.

For structures subjected to some intermediate loading rate, an intermediate loading-rate transition curve should be used to predict the level of performance at the service temperature. Because the actual loading rates for many structures are not well defined, the impact loading curve (Figure 1.6) is often used to predict the service performance of structures even though the actual loading may be slow or intermediate. This practice is somewhat conservative and helps to explain why many structures that have low toughness as measured by impact tests have not failed even though their service temperatures are well below an impact transition temperature. As noted in Figure 1.6, a particular notch-toughness value called the nil-ductility transition (NDT) temperature generally defines the upper limits of plane-strain behavior under conditions of impact loading.

One of the fundamental questions to be resolved regarding the interpretation of any particular toughness test for large structures is as follows: What level of material performance should be required for satisfactory performance in a particular structure? That is, as shown schematically in Figure 1.7 for impact loading, one of the following three general levels of material performance could be established at the service temperature for a structural material:

1. Plane-strain behavior—steel 1.
2. Elastic-plastic behavior—steel 2.
3. Fully plastic behavior—steel 3.

Although fully plastic behavior would be a very desirable level of performance for structural materials, it may not be necessary or even economically feasible for many structures. That is, for a large number of structures, a reasonable level of elastic-plastic behavior (steel 2, Figure 1.7) is often satisfactory to prevent initiation of brittle fractures provided the design and

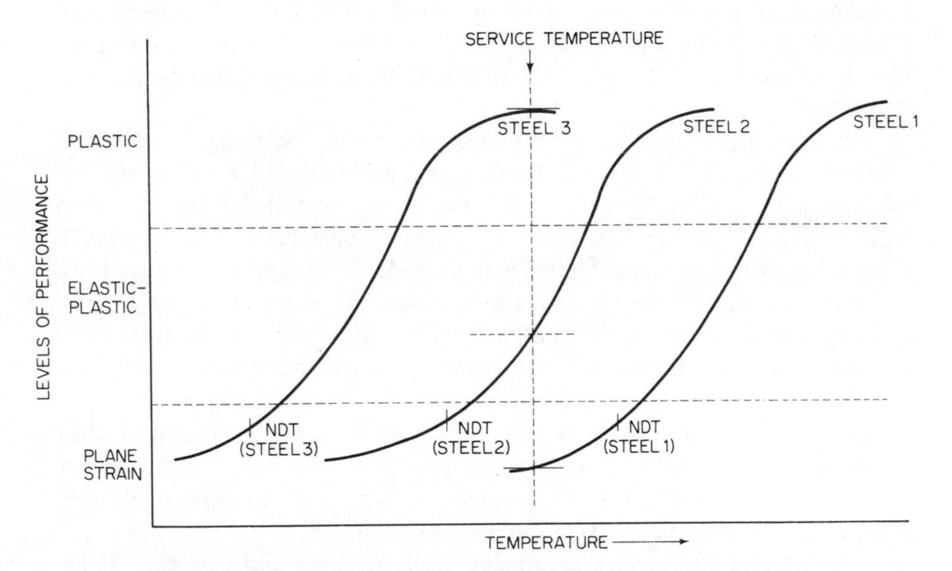

FIG. 1.7. Schematic showing relation between level of performance and transition temperature for 3 arbitrary steels.

fabrication are satisfactory. A more detailed description of fracture criteria and levels of performance is provided in Chapter 12.

Not all structural materials exhibit a brittle-ductile transition. For example, some of the very high-strength structural steels or other structural materials such as aluminum or titanium do not undergo the brittle-ductile transition shown in Figure 1.6. For these materials, temperature has a rather small effect on toughness, as shown in Figure 1.8 for a 250-ksi yield strength steel.

The general purpose of the various kinds of notch-toughness tests is to model the behavior of actual structures so that the laboratory test results can be used to predict service performance. In this sense many different tests have been used to measure the notch toughness of structural materials. These include Charpy V-notch (CVN) impact, drop weight NDT, dynamic tear (DT), wide plate, Battelle drop weight tear test (DWTT), precracked Charpy, as well as many others. A complete description of most of these tests can be found in References 1–7. Generally these notch-toughness tests were developed for specific purposes. For example, the CVN test is widely used as a screening test in alloy development as well as a fabrication and quality control test. In addition, because of correlations with service experience, the CVN test is often used in steel specifications for various structural, marine, and pressure-vessel applications. The NDT test is used to establish the minimum service temperature for various naval and marine applications, whereas the Battelle DWTT test was developed to measure the fracture appearance of line pipe steels as a function of temperature. The precracked Charpy test was developed by introducing a sharp crack into a CVN specimen to model the behavior in actual structures in which sharp cracks may be present.

All these notch-toughness tests generally have one thing in common, however, and that is to produce fracture in steels under carefully controlled laboratory conditions. Hopefully, the results of the test can be correlated with service performance to establish levels of performance, as shown in Figure 1.6, for various materials being considered for specific applications. In fact, the results of the above-mentioned tests have been extremely useful in many structural applications. Structural energy, deformation, and ability to absorb energy, etc., are all important structural parameters, and these tests have served as very good guidelines for the structural engineer.

However, even if correlations are developed for existing structures, they do not necessarily hold for certain designs, new operating conditions, or new materials because the results, which are expressed in terms of energy, fracture appearance, or deformation, cannot always be translated into structural design and engineering parameters such as stress and flaw size. Thus, a much better way to measure notch toughness is with principles of fracture

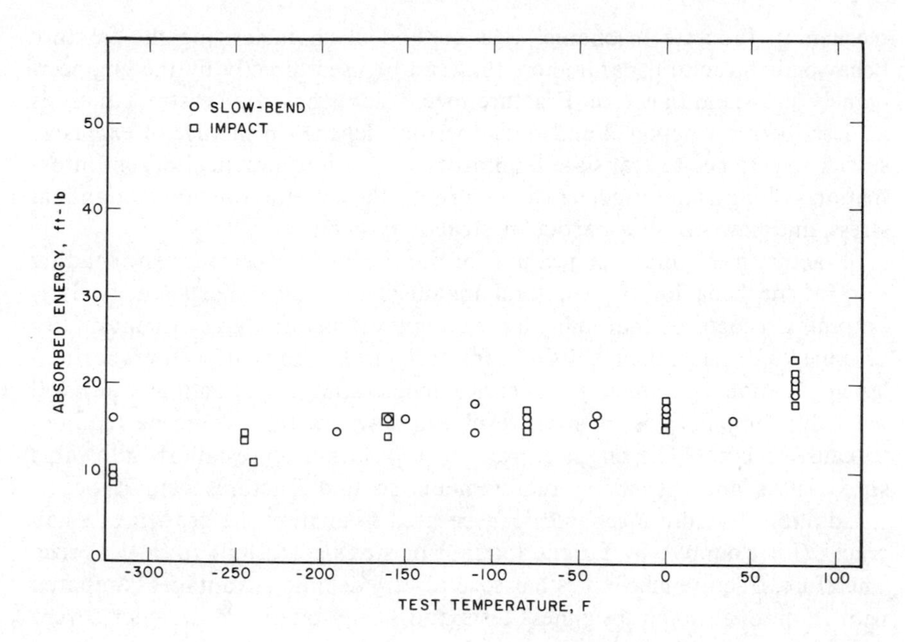

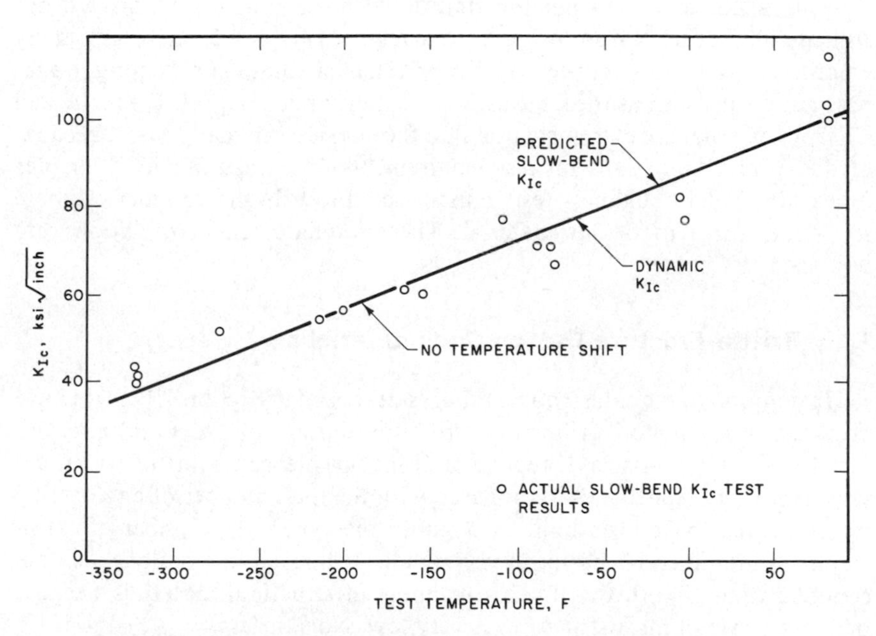

FIG. 1.8. CVN and K_{Ic} test results for 250 ksi yield strength steel showing minor effect of temperature.

mechanics. Fracture mechanics is a method of characterizing the fracture behavior in structural parameters that can be used directly by the engineer, namely stress and flaw size. Fracture mechanics is based on a stress analysis as described in Chapter 2 and thus does not depend on the use of extensive service experience to translate laboratory results into practical design information so long as the engineer can determine the material toughness, nominal stress, and flaw size in a particular structural member.

Fracture mechanics can account for the effect of temperature and loading rate on the behavior of structural members that have sharp cracks. It is becoming recognized that many large complex structures have discontinuities of some kind. Thus, the results of a fracture-mechanics analysis for a particular application (specimen size, service temperature, and loading rate) will yield the combinations of stress level and flaw size that would be required to cause fracture. The engineer can then *quantitatively* establish allowable stress levels and inspection requirements so that fractures cannot occur. In addition, fracture mechanics can be used to analyze the growth of small cracks (for example, by fatigue loading or stress-corrosion) to critical size. Therefore, fracture mechanics has several very definite advantages compared with traditional notch-toughness tests and finally offers the designer a very definitive method of quantitatively designing to prevent brittle fracture in structures.

This is not to imply that the traditional notch-toughness tests are not still useful. In fact, as discussed extensively in Chapter 6, there are many empirical correlations between fracture-mechanics values and existing toughness test results such as the Charpy V-notch, dynamic tear, NDT, precracked Charpy, etc., that are extremely useful to the engineer. In many cases, because of the current limitations on test requirements for measuring K_{Ic} (Chapter 3), existing notch-toughness tests must be used to help the designer estimate K_{Ic} values for a particular material. These estimates and correlations are described in Chapter 6.

1.3. Brittle-Fracture Design Considerations

In addition to the catastrophic failures described in Section 1.1, there have been *numerous* "minor" failures of structures during construction or service that have resulted in delays, repairs, and inconveniences, some of which are very expensive. Nonetheless, compared with the total number of engineering structures that have been built throughout the world, the number of catastrophic brittle fractures has been very small. As a result, the designer seldom concerns himself with the notch toughness of structural materials because the failure rate of most structures is very low. Nonetheless,

1. When designs become more complex,
2. When the use of high-strength thick welded plates becomes more

common compared with the use of lower-strength thin riveted plates,
3. When the choice of construction practices becomes more dependent on minimum cost,
4. When the magnitude of loadings increases, and
5. When actual factors of safety decrease because of more precise computer designs,

the possibility of brittle fractures in large complex structures must be considered, and the designer must become more aware of available methods to prevent brittle failures.

The state-of-the-art *is* that fracture mechanics concepts *are* available that can be used in the design of structures to prevent brittle fractures.

Design codes often include this fact, and in the early 1970s, several design and materials specifications *based on concepts of fracture mechanics* were adopted by various engineering professions. These include

1. ASME Boiler and Pressure Vessel Code Section III—Nuclear Power Plant Components, Appendix G—Protection Against Nonductile Failure.[15]
2. American Association of State Highway and Transportation Officials —Notch Toughness Requirements for Bridge Steels.[16]
3. Air Force Aircraft Designed to Fracture Mechanics Criteria in Addition to the Usual Static Load Limits.[17]

A detailed discussion of these and other specifications or design procedures used in various types of structures is presented in Chapter 15.

The traditional design approach for most structures is generally based on the use of safety factors to limit the maximum calculated stress level to some percentage of either the yield or ultimate stress. It is suggested by Weck[18] that this approach is somewhat outdated.

The factor of safety approach, by itself, does not always give the proper assurance of safety with respect to brittle fracture because large complex structures are not fabricated without some kind of discontinuities. Numerous failure investigations and inspections have shown this to be true. Research by Fisher and Yen[19] has shown that discontinuities exist in practically all structural members, either from manufacture or from the process of fabricating the members by rolling, machining, punching, or welding. The sizes of these discontinuities range from very small microdiscontinuities (<0.01 in.) to several inches long.

In almost every brittle fracture that has occurred in structures, some type of discontinuity was present. These included very small arc strikes in some of the World War II ship failures, very small fatigue cracks (~ 0.07 in.) in the Comet airplane failures, and stress-corrosion or corrosion fatigue cracks (0.12 in.) in the critical eyebar of the Point Pleasant Bridge. Other failures, such as the 260-in.-diameter missile motor case that failed during hydrotest or the F-111 aircraft failure, had somewhat larger, but still undetected, cracks.

Bravenec[20] has reviewed various brittle fractures during fabrication and testing and has shown that cracks have originated from torch-cut edges, mechanical gouges, corrosion pits, weld repairs, severe stress concentrations, etc. Dolan[21] has made the flat statement that "*every structure* contains small flaws whose size and distribution are dependent upon the material and its processing. These may range from nonmetallic inclusions and microvoids to weld defects, grinding cracks, quench cracks, surface laps, etc."

The significant point is that discontinuities or cracks *are* present in many large fabricated structures even though the structure may have been "inspected." Methods of inspection or nondestructive testing are gradually improving, with the result that smaller and smaller discontinuities are becoming detectable. But the fact is that discontinuities are present regardless of whether or not they are discovered. In fact the problem of establishing acceptable levels of discontinuities in welds is becoming somewhat of an economic problem since techniques that minimize the size and distribution of discontinuities are available if the engineer chooses to use them.

Weck[18] has stated that "Some authorities—in the API-ASME Code for instance—have produced porosity charts and rules for permissible sizes of other defects which, in the complete absence of any factual or experimental basis, must have been the result of divine inspiration of the Code makers. What is good enough for the job cannot be established by divine inspiration but only by patient experimental research. Such researches have amply demonstrated that for many jobs far less than absolute perfection is adequate. Whether a given defect is permissible or not depends on the extent to which the defect increases the risk of failure of the structure. It is quite clear that this will vary with the type of structure, its service conditions and the material from which it is constructed."

As described in the following sections and the remainder of this book, fracture mechanics is the best available science that can correctly account *quantitatively* for the factors that influence the true factor of safety or degree of reliability of a structure. Fracture mechanics is not only a quantitative research tool and method of failure analysis, but it has been and should be used in structural design to help determine acceptable stress levels, acceptable discontinuity sizes, and desired material properties for specific service conditions.

1.4. Introduction to Fracture Mechanics

An overwhelming amount of research on brittle fracture in structures of all types has shown that numerous factors (e.g., service temperature, material toughness, design, welding, residual stresses, fatigue, constraint, etc.) can contribute to brittle fractures in large welded structures.[22-29] However, the recent development of fracture mechanics[30-35] has shown that there are three *primary* factors that control the susceptibility of a structure to brittle fracture:

1. MATERIAL TOUGHNESS (K_c, K_{Ic}, K_{Id})
Material toughness can be defined as the ability to carry load or deform plastically in the presence of a notch and can be described in terms of the critical stress-intensity factor under conditions of plane stress (K_c) or plane strain (K_{Ic}) for slow loading and linear elastic behavior. K_{Id} is a measure of the critical material toughness under conditions of maximum constraint (plane strain) and impact or dynamic loading, also for linear elastic behavior. (These terms are described more completely in Chapters 2 and 3.) For elastic-plastic behavior (materials with higher levels of notch toughness than linear elastic behavior) the material toughness is measured in terms of parameters such as R-curve resistance, J_{Ic}, and COD as described in Chapter 16. In addition to metallurgical factors such as composition and heat treatment, the notch toughness of a steel also depends on the application temperature, loading rate, and constraint (state-of-stress) ahead of the notch, as described in Chapter 4.

2. CRACK SIZE (a)
Brittle fractures initiate from discontinuities of various kinds. These discontinuities can vary from extremely small cracks within a weld arc strike (as was the case in the brittle fracture of a T-2 tanker during World War II) to much larger weld or fatigue cracks. Complex welded structures are not fabricated without discontinuities (porosity, lack of fusion, toe cracks, mismatch, etc.), although good fabrication practice and inspection can minimize the original size and number of cracks. Thus, these discontinuities will be present in many welded structures even after all inspections and weld repairs are finished. Furthermore even though only "small" discontinuities may be present initially, these discontinuities can grow by fatigue or stress-corrosion, possibly to a critical size.

3. STRESS LEVEL (σ)
Tensile stresses (nominal, residual, or both) are necessary for brittle fractures to occur. These stresses are determined by conventional stress analysis techniques for particular structures.

These three factors are the primary ones that control the susceptibility of a structure to brittle fracture. All other factors such as temperature, loading rate, stress concentrations, residual stresses, etc., merely affect the above three *primary* factors.

Engineers have known these facts for many years and have reduced the susceptibility of structures to brittle fractures by controlling the above factors in their structures *qualitatively*. That is, good design (e.g., adequate sections, minimum stress concentrations) and fabrication practices (decreased discontinuity size because of proper welding control and inspection), as well as the use of materials with good notch-toughness levels (e.g., as measured with a Charpy V-notch impact test), will minimize and have minimized the probability of brittle fractures in structures. However, the engineer has not had specified design guidelines to evaluate the relative performance and economic trade-offs among design, fabrication, and materials in a *quantitative* manner.

The recent development of fracture mechanics as an applied science has shown that all three of the above factors can be interrelated to predict (or to design against) the susceptibility of various structures to brittle fracture. Fracture mechanics is a method of characterizing fracture behavior in structural parameters familiar to the engineer, namely, stress and crack size. Linear-elastic fracture-mechanics technology is based on an analytical procedure that relates the stress-field magnitude and distribution in the vicinity of a crack tip to the nominal stress applied to the structure, to the size, shape, and orientation of the crack or crack-like discontinuity, and to the material properties. In Figure 1.9 are the equations that describe the elastic-stress field in the vicinity of a crack tip in a body subjected to tensile stresses normal to the plane of the crack (mode I deformation, as will be described in Chapter 2). The stress-field equations show that the distribution of the elastic-stress

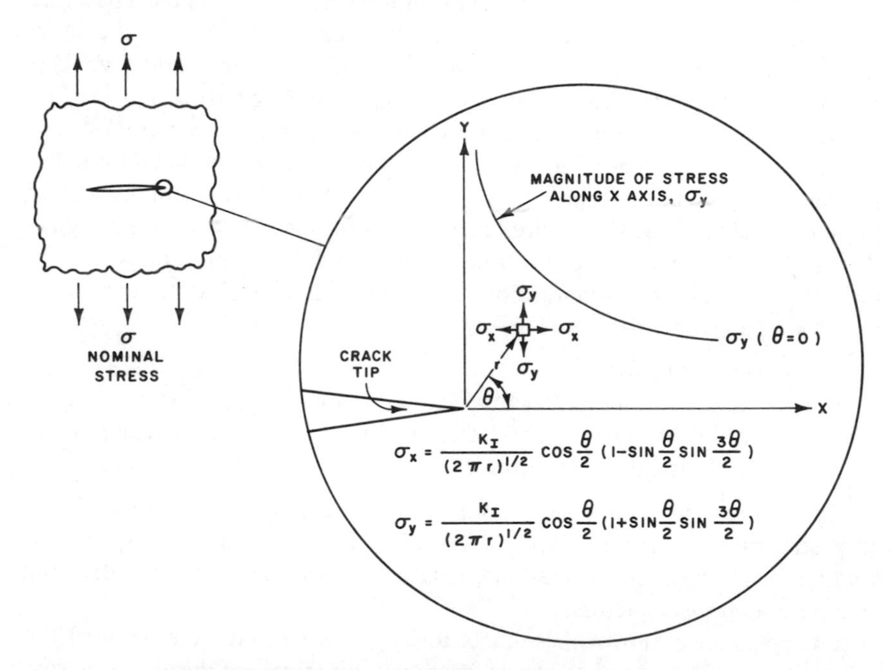

FIG. 1.9. Elastic-stress-field distribution ahead of a crack.

field in the vicinity of the crack tip is invariant in all structural components subjected to this type of deformation and that the magnitude of the elastic-stress field can be described by a single parameter, K_I, designated the stress-intensity factor. Consequently, the applied stress, the crack shape, size, and orientation, and the structural configuration associated with structural components subjected to this type of deformation affect the value of the stress-intensity factor but do not alter the stress-field distribution. Thus it is possible to translate laboratory results into practical design information without the use of extensive service experience or correlations. Relationships between the stress-intensity factor and various body configurations, crack sizes, shapes, orientations, and loading conditions are presented in Chapter 2; however, examples of some of the more widely used stress-flaw size relations are presented in Figure 1.10.

One of the underlying principles of fracture mechanics is that unstable fracture occurs when the stress-intensity factor at the crack tip reaches a

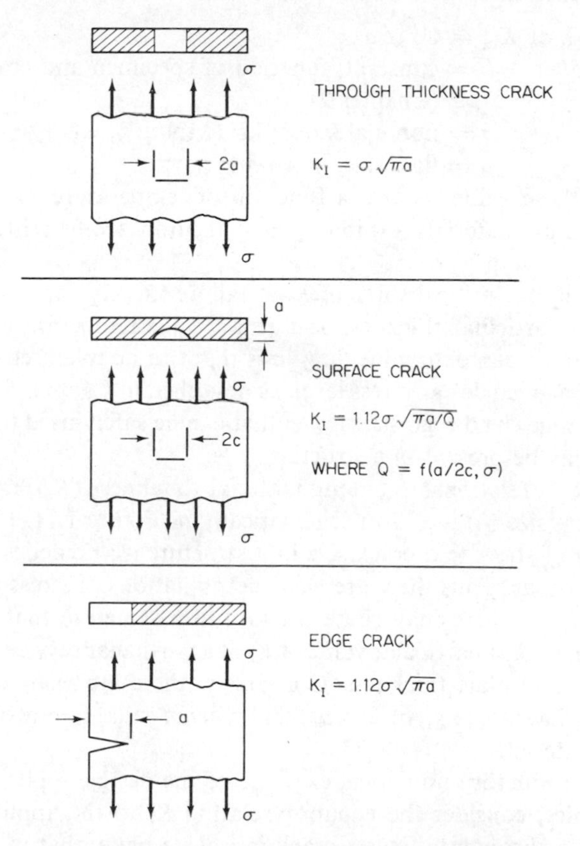

FIG. 1.10. K_I values for various crack geometries.

critical value, K_c. For Mode I deformation and for small crack-tip plastic deformation (plane-strain conditions), the critical-stress-intensity factor for fracture instability is designated K_{Ic}. K_{Ic} represents the inherent ability of a material to withstand a given stress-field intensity at the tip of a crack and to resist progressive tensile crack extension under plane-strain conditions. Thus, K_{Ic} represents the fracture toughness of the material and has units of ksi$\sqrt{\text{in.}}$ (MN/m$^{3/2}$). However, this material-toughness property depends on the particular material, loading rate, and constraint as follows:

K_c = critical-stress-intensity factor for static loading and plane-stress conditions of variable constraint. Thus, this value depends on specimen thickness and geometry, as well as on crack size.

K_{Ic} = critical-stress-intensity factor for static loading and plane-strain conditions of maximum constraint. Thus, this value is a minimum value for thick plates.

K_{Id} = critical-stress-intensity factor for dynamic (impact) loading and plane-strain conditions of maximum constraint,

where K_c, K_{Ic}, or $K_{Id} = C\sigma\sqrt{a}$,

C = constant, function of specimen and crack geometry (Chapter 2),

σ = nominal stress, ksi (MN/m^2),

a = flaw size, in. (mm).

Each of these values is also a function of temperature, particularly for those structural materials exhibiting a transition from brittle to ductile behavior.

By knowing the critical value of K_I at failure (K_c, K_{Ic}, or K_{Id}) for a given material of a particular thickness and at a specific temperature and loading rate, the designer can determine flaw sizes that can be tolerated in structural members for a given design stress level, as described in Chapter 5. Conversely, he can determine the design stress level that can be safely used for an existing crack that may be present in a structure.

This general relationship among material toughness (K_c), nominal stress (σ), and crack size (a) is shown schematically in Figure 1.11. If a particular combination of stress and crack size in a structure (K_I) reaches the K_c level, fracture can occur. Thus there are *many* combinations of stress and flaw size (e.g., σ_f and a_f) which may cause fracture in a structure that is fabricated from a steel having a particular value of K_c at a particular service temperature, loading rate, and plate thickness. Conversely, there are *many* combinations of stress and flaw size (e.g., σ_0 and a_0) that will *not* cause failure of a particular structural material.

As an introductory numerical example of the design application of fracture mechanics, consider the equation relating K_I to the applied stress and flaw size for a through-thickness crack in a wide plate, that is $K_I = \sigma\sqrt{\pi a}$.

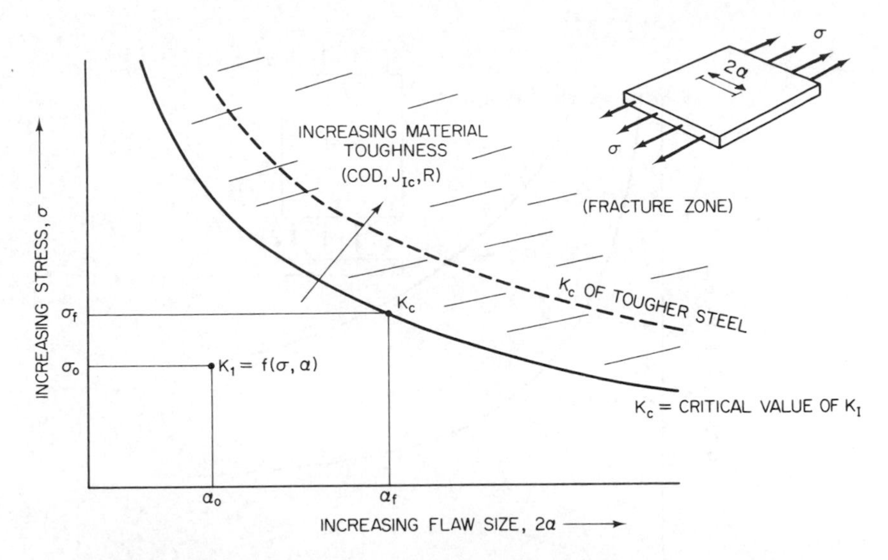

FIG. 1.11. Schematic relation between stress, flaw size, and material toughness.

Assume that laboratory test results show that for a particular structural steel with a yield strength of 80 ksi (552 MN/m²) the K_c is 60 ksi$\sqrt{\text{in.}}$ (66 MN/m$^{3/2}$) at the service temperature, loading rate, and plate thickness used in service. Also assume that the design stress is 20 ksi (138 MN/m²). Substituting $K_I = K_c = 60$ ksi$\sqrt{\text{in.}}$ (66 MN/m$^{3/2}$) into the appropriate equation in Figure 1.12 results in $2a = 5.7$ in. (145 mm). Thus for these conditions the tolerable flaw size would be about 5.7 in. (145 mm). For a design stress of 45 ksi (310 MN/m²), the same material could tolerate a flaw size, $2a$, of only about 1.1 in. (27.9 mm). If residual stresses such as may be due to welding are present so that the total stress in the vicinity of a crack is 80 ksi (552 MN/m²), the tolerable flaw size is reduced considerably. Note from Figure 1.12 that if a tougher steel is used, for example, one with a K_c of 120 ksi$\sqrt{\text{in.}}$ (132 MN/m$^{3/2}$), the tolerable flaw sizes at all stress levels are significantly increased. If the toughness of a steel is sufficiently high, brittle fractures will not occur, and failures under tensile loading can occur only by general plastic yielding, similar to the failure of a tension test specimen. Fortunately, most structural steels have this high level of toughness at service temperatures and loading rates.

A useful analogy for the designer is the relation among applied load (P), nominal stress (σ), and yield stress (σ_{ys}) in an unflawed structural member and among applied load (P), stress intensity (K_I), and critical stress intensity for fracture (K_c, K_{Ic}, or K_{Id}) in a structural member with a flaw. In an unflawed structural member, as the load is increased the nominal stress increases

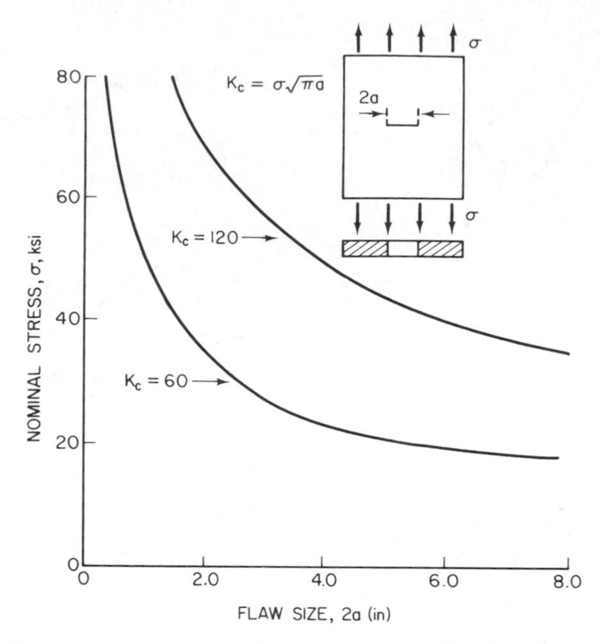

FIG. 1.12. Stress-flaw size relation for through-thickness crack.

until an instability (yielding at σ_{ys}) occurs. As the load is increased in a structural member with a flaw (or as the size of the flaw grows by fatigue or stress-corrosion) the stress intensity, K_I, increases until an instability (fracture at K_c, K_{Ic}, K_{Id}) occurs. Thus the K_I level in a structure should always be kept below the appropriate K_c value in the same manner that the nominal design stress (σ) is kept below the yield strength (σ_{ys}).

Another analogy that may be useful in understanding the fundamental aspects of fracture mechanics is the comparison with the Euler column instability (Figure 1.13). The stress level required to cause instability in a column (buckling) decreases as the L/r ratio increases. Similarly, the stress level required to cause instability (fracture) in a flawed tension member decreases as the flaw size (a) increases. As the stress level in either case approaches the yield strength, both the Euler analysis and the K_c analysis are invalidated because of yielding. To prevent buckling, the actual stress and (L/r) values must be below the Euler curve. To prevent fracture, the actual stress and flaw size, a, must be below the K_c level shown in Figure 1.12. Obviously, using a material with a high level of notch toughness (e.g., a K_c level of 120 ksi$\sqrt{\text{in.}}$ (132 MN/m$^{3/2}$) compared with 60 ksi$\sqrt{\text{in.}}$ (66 MN/m$^{3/2}$) in Figure 1.12) will increase the possible combinations of design stress and flaw size that a structure can tolerate without fracturing.

The critical-stress-intensity factor, K_{Ic}, represents the terminal condi-

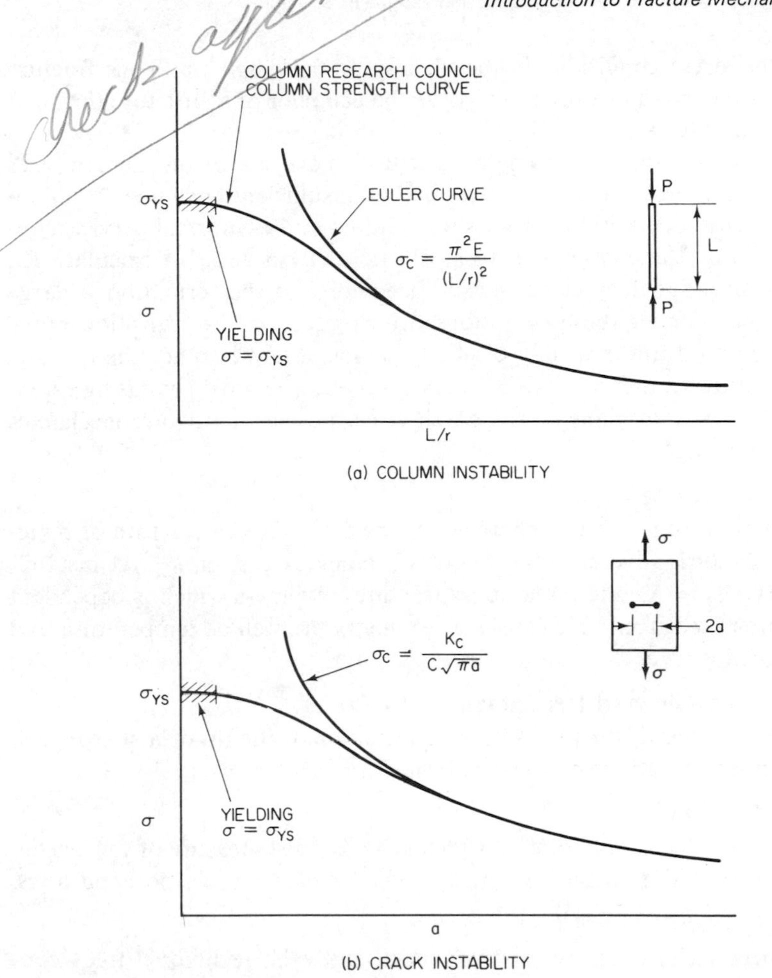

FIG. 1.13. Column instability and crack instability (after Madison and Irwin).

tions in the life of a structural component. The total useful life of the component is determined by the time necessary to initiate a crack and to propagate the crack from subcritical dimensions to the critical size, a_c. Crack initiation and subcritical crack propagation may be caused by cyclic stresses in the absence of an aggressive environment (Chapters 7, 8, and 9), by an aggressive environment under sustained load (Chapter 10), or by the combined effects of cyclic stresses and an aggressive environment (Chapter 11). Because all these modes of subcritical crack propagation are localized phenomena that depend on the boundary conditions at the crack tip, it is logical to expect the rate of subcritical crack propagation to depend on the stress-intensity factor, K_I, which serves as a single-term parameter representative of the stress conditions in the vicinity of the crack tip. Thus fracture mechanics theory can be used to analyze the behavior of a structure throughout its entire life.

Many low-to medium-strength structural materials in the section sizes of interest for most large structures are of insufficient thickness to maintain plane strain conditions under slow loading and at normal service temperatures. For these cases the linear elastic analysis used to calculate K_{Ic} values is invalidated by elastic-plastic behavior and the formation of large plastic zones. Under these conditions, which occur in the transition range between plane strain and fully plastic behavior, analyses other than linear-elastic fracture mechanics (LEFM) must be used. The most promising extensions of LEFM into plane stress as well as elastic-plastic fracture mechanics are the following:

1. *R*-CURVE ANALYSIS
   A procedure used to characterize the resistance to fracture of materials during incremental slow-stable crack extension, K_R. At instability, $K_R = K_c$, the plane stress fracture toughness which is dependent upon specimen thickness and geometry, as well as temperature and loading rate.
2. CRACK-OPENING DISPLACEMENT (COD)
   A measure of the prefracture deformation at the tip of a sharp crack under conditions of inelastic behavior.
3. *J* INTEGRAL
   Path-independent integral which is an average measure of the elastic-plastic stress-strain field ahead of a crack. For elastic conditions, $J_{Ic} = K_{Ic}^2/E(1 - \mu^2)$.

All three techniques are relatively new, and each technique holds considerable promise for specifying material toughness in terms of allowable stress or defect size since they are all three extensions of linear-elastic fracture mechanics. These techniques are described in Chapter 16.

1.5. Fatigue and Stress-Corrosion Crack Growth

Conventional procedures that are used to design structural components subjected to fluctuating loads provide the engineer with a design fatigue curve which characterizes the basic unnotched fatigue properties of the material and a fatigue-strength-reduction factor. The fatigue-strength-reduction factor incorporates the effects of all the different parameters characteristic of the specific structural component that make it more susceptible to fatigue failure than the unnotched specimen. The design fatigue curves are based on the prediction of cyclic life from data on nominal stress (or strain) versus elapsed cycles to failure (*S-N* curves) as determined from laboratory test specimens. Such data represent both the number of cycles required to initiate a crack in the specimen and the number of cycles required to propagate the crack from a subcritical size to a critical dimension which often varies from laboratory to laboratory. The dimension of the critical crack required to cause "failure" in the fatigue specimen also depends on the magnitude of the applied stress and on the test specimen size.

Figure 1.14 is a schematic *S-N* curve divided into an initiation component and a propagation component. The number of cycles corresponding to the endurance limit represents initiation life primarily, whereas the number of cycles expended in crack initiation at a high value of applied stress is negligible. Consequently, *S-N*-type data do not necessarily provide information regarding safe-life predictions in structural components, particularly in components having surface irregularities different from those of the test specimens, and in components containing crack-like discontinuities because the existence of surface irregularities and crack-like discontinuities reduces and may eliminate the crack-initiation portion of the fatigue life of structural components.

Many attempts have been made to characterize the fatigue behavior of

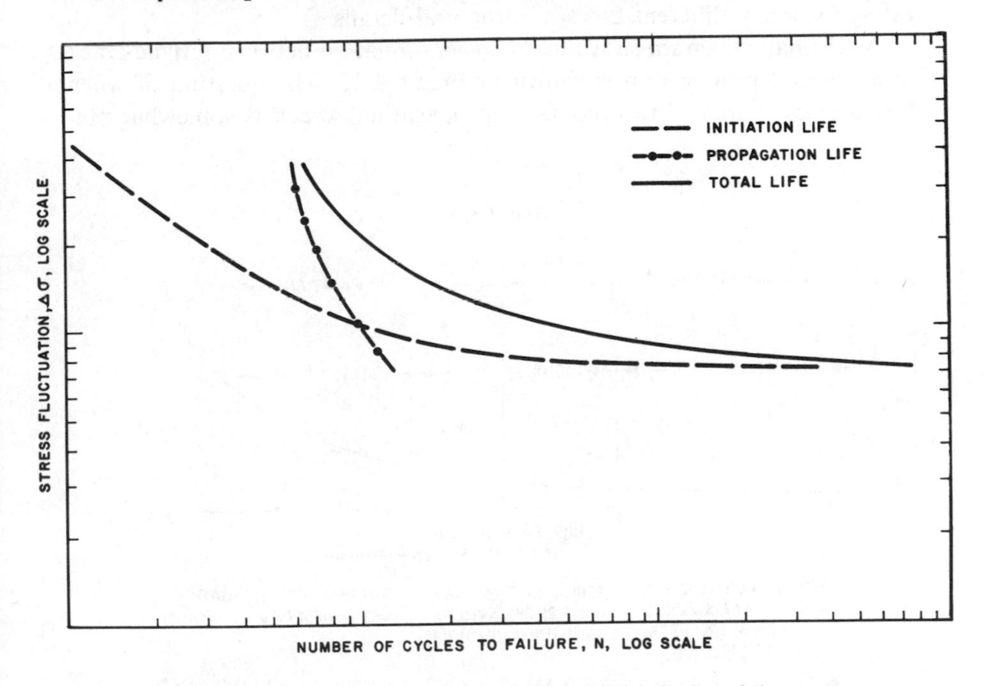

FIG. 1.14. Schematic *S-N* curve divided into "initiation" and "propagation" components.

metals. The results of some of these attempts have proved invaluable in the evaluation and prediction of the fatigue strength of structural components. However, these fatigue-strength evaluation procedures are subject to many limitations, caused primarily by the failure to distinguish adequately between fatigue-crack initiation and fatigue-crack propagation.

As described in Chapter 7, fracture mechanics can be used to determine the initiation life of fatigue cracks from various stress concentration factors. In Chapters 8, 9, and 11 we shall describe the use of fracture mechanics to evaluate the behavior of propagating cracks.

Thus, although *S-N* curves have been widely used to analyze the fatigue behavior of steels and weldments, closer inspection of the overall fatigue process in complex welded structures indicates that a more rational analysis of fatigue behavior may be possible by using concepts of fracture mechanics. Specifically, small (possibly large) fabrication discontinuities are invariably present in welded structures, even though the structure has been "inspected." Accordingly, a conservative approach to designing to prevent fatigue failure is to assume the presence of an initial flaw and analyze the fatigue-crack-growth behavior of the structural member. The size of the initial flaw is obviously highly dependent on the quality of fabrication and inspection. However, such an analysis would minimize the need for expensive fatigue testing for many different types of structural details.

A schematic diagram showing the general relation between fatigue-crack initiation and propagation is shown in Figure 1.15. The question of when does a crack "initiate" to become a "propagating" crack is somewhat phi-

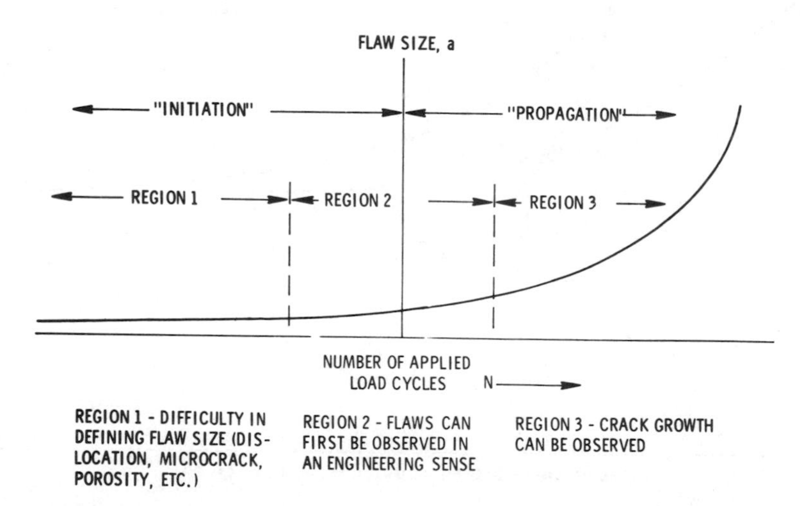

FIG. 1.15. Schematic showing relation between "initiation" life and "propagation" life.

losophical and depends on the level of observation of a crack, i.e., crystal imperfection, dislocation, microcrack, lack of penetration, etc. The fracture-mechanics approach to fatigue is to assume an initial flaw size on the basis of the quality of fabrication or inspection and then to calculate the number of cycles required to grow a crack to the critical size for brittle fracture. Using this approach, inspection requirements can be established logically.

In addition to subcritical crack growth by fatigue, small cracks also can grow by stress-corrosion during the life of structures. Although crack growth by either fatigue or stress-corrosion does not represent catastrophic failure for structures fabricated from materials having reasonable levels of notch toughness, in both mechanisms small cracks can become large enough to cause a decrease in structural efficiency and can thereby require repairs. Furthermore the possibility of both mechanisms operating at once by corrosion fatigue also exists. Thus, a knowledge of the fatigue, corrosion-fatigue, and stress-corrosion behavior of materials is required to establish an overall fracture-control plan that includes inspection requirements.

By testing precracked specimens under static loads in specific environments (such as salt water) and analyzing the results according to fracture-mechanics concepts, a K_I value can be determined, below which subcritical crack propagation does not occur. This threshold value is called K_{Iscc}. The K_{Iscc} value for a particular material and environment is the plane-strain stress-intensity threshold that describes the value below which subcritical cracks (scc) will not propagate and also has units of ksi$\sqrt{\text{in.}}$

For materials that are susceptible to crack growth in a particular environment, the K_{Iscc} value is used as the failure criterion rather than K_{Ic}. Thus, in the schematic relation among material toughness, design stress, and flaw size shown in Figure 1.11, K_{Iscc} would replace K_c as the critical value of K_I. This procedure is described in Chapter 10.

1.6. Fracture Criteria

Criteria selection, i.e., the determination of how much toughness is necessary as well as how much the designer is willing to pay to obtain materials with superior toughness, is the most important part of developing an adequate fracture-control plan and should be based on a very careful study of the particular requirements for a structure. Some of the factors involved in the development of a criterion are

1. A knowledge of the service conditions (temperature, loading, loading rate, etc.) to which the structure will be subjected.
2. The desired level of performance in the structure (plane-strain, elastic-plastic, or plastic).
3. The consequences of structural failure.

Various criteria that have been developed for structural applications are the AASHTO material-toughness requirements, the PVRC recommendations for ferritic materials for nuclear pressure vessels, and Mil Standard 1530 for aircraft. These examples are discussed in Chapters 12–15.

1.7. Fracture-Control Plans

Most engineering structures in existence are performing safely and reliably. The comparatively few service failures in structures indicate that present-day practices governing material properties, design, and fabrication procedures are generally satisfactory. However, the occurrence of infrequent failures indicates that further understanding and possible modifications in present-day practices are needed. The identification of the specific modifications needed requires a thorough study of material properties, design, fabrication, inspection, erection, and service conditions.

Structures are fabricated in various sizes and are subjected to numerous service conditions. Thus, it is very difficult to develop a set of rules that would ensure the safety and reliability of all structures when each structure involves a unique set of operating conditions. The use of data obtained by testing laboratory specimens to predict the behavior of complex structural components may result in approximations or in excessively conservative estimates of the life of such components but does not guarantee a correct prediction of the fracture or fatigue behavior. The safety and reliability of structures and the correct prediction of their overall resistance to failure by fracture or fatigue can be approximated best by using a fracture-control plan. A fracture-control plan is a detailed procedure used

1. To identify all the factors that may contribute to the fracture of a structural detail or to the failure of the entire structure.
2. To assess the contribution of each factor and the synergistic contribution of these factors to the fracture process.
3. To determine the relative efficiency and trade-off of various methods to minimize the probability of fracture.
4. To assign responsibility for each task that must be undertaken to ensure the safety and reliability of the structure.

The development of a fracture-control plan for complex structures is very difficult. Despite the difficulties, attempts to formulate a fracture-control plan for a given application, even if only partly successful, should result in a better understanding of the fracture characteristics of the structure under consideration.

A fracture-control *plan* is a procedure tailored for a given application and cannot be extended indiscriminately to other applications. However, general fracture-control *guidelines* that pertain to classes of structures (such as bridges, ships, pressure vessels, etc.) can be formulated for consideration in the development of a fracture-control plan for a particular structure within any particular class of structures.

The correspondence among fracture-control plans based on crack initiation, crack propagation, and fracture toughness of materials can be readily demonstrated by using fracture-mechanics concepts. The fact that crack initiation, crack propagation, and fracture toughness are functions of the stress-intensity fluctuation, ΔK_I, and of the critical-stress-intensity factor, K_{Ic}—which are in turn related to the applied nominal stress (or stress fluctuation)—demonstrates that a fracture-control plan for various structural applications depends on

1. The fracture toughness, K_{Ic} (or K_c), of the material at the temperature and loading rate representative of the intended application. The fracture toughness can be modified by changing the material used in the structure.
2. The applied stress, stress rate, stress concentration, and stress fluctuation, which can be altered by design changes and by proper fabrication.
3. The initial size of the discontinuity and the size and shape of the critical crack, which can be controlled by design changes, fabrication, and inspection.

The total useful life of structural components is determined by the time necessary to initiate a crack and to propagate the crack from subcritical dimensions to the critical size. The life of the component can be prolonged by extending the crack-initiation life and the subcritical-crack propagation life. Consequently, crack initiation, subcritical crack propagation, and fracture characteristics of structural materials are primary considerations in the formulation of fracture-control guidelines for structures.

Fracture-control guidelines are established in Chapter 14, and fracture-control plans for various classes of structures are presented in Chapter 15.

REFERENCES

1. M. E. Shank, "A Critical Review of Brittle Failure in Carbon Plate Steel Structures Other than Ships," *Ship Structure Committee Report*, *Serial No. SSC-65*, National Academy of Sciences-National Research Council, Washington, D.C., Dec. 1, 1953 (also reprinted as *Welding Research Council Bulletin No. 17*).
2. E. R. Parker, *Brittle Behavior of Engineering Structures*, prepared for the Ship Structure Committee under the general direction of the Committee on Ship Steel–National Academy of Sciences–National Research Council, Wiley, New York, 1957.

3. D. B. Bannerman and R. T. Young, "Some Improvements Resulting from Studies of Welded Ship Failures," *Welding Journal, 25*, No. 3, March 1946.
4. H. G. Acker, "Review of Welded Ship Failures," *Ship Structure Committee Report, Serial No. SSC-63*, National Academy of Sciences–National Research Council, Washington, D.C., Dec. 15, 1953.
5. *Final Report of a Board of Investigation—The Design and Methods of Construction of Welded Steel Merchant Vessels, 15 July, 1946*, GPO, Washington, D.C., 1947.
6. G. M. Boyd and T. W. Bushell, "Hull Structural Steel—The Unification of the Requirements of Seven Classification Societies," *Quarterly Transactions: The Royal Institution of Naval Architects* (*London*), *103*, No. 3, March 1961.
7. J. Turnbull, "Hull Structures," *The Institution of Engineers and Shipbuilders of Scotland, Transactions, 100*, pt. 4, Dec. 1956–1957, pp. 301–316.
8. G. M. Boyd, "Fracture Design Practices for Ship Structures," in *Fracture*, Vol. V: Fracture Design of Structures, edited by H. Libowitz, Academic Press, New York, 1969, pp. 383–470.
9. Marine Casualty Report, "Structural Failure of the Tank Barge I.O.S. 3301 Involving the Motor Vessel Martha R. Ingram on 10 January 1972 Without Loss of Life," *Report No. SDCG/NTSB*, March 74–1.
10. T. Bishop, "Fatigue and the Comet Disasters," *Metal Progress*, May 1955, p. 79.
11. "Collapse of U.S. 35 Highway Bridge, Point Pleasant, West Virginia," *NTSB Report No. NTSB-HAR-71-1*, Dec. 15, 1967.
12. J. A. Bennett and Harold Mindlin, "Metallurgical Aspects of the Failure of the Pt. Pleasant Bridge," *Journal of Testing and Evaluation*, March 1973, pp. 152–161.
13. "State Cites Defective Steel in Bryte Bend Failure," *Engineering News Record, 185*, No. 8, Aug. 20, 1970.
14. "Joint Redesign on Cracked Box Girder Cuts into Record Tied Arch's Beauty," *Engineering News Record, 188*, No. 13, March 30, 1972.
15. *ASME Boiler and Pressure Vessel Code*, Section III, Section G, *ASME*, New York, 1972.
16. *American Association of State Highway and Transportation Officials—Material Specifications*, Association General Offices, Washington, D.C., 1974.
17. *MIL-Std—1530*, "Aircraft Structural Integrity Program, Airplane Requirements," United States Air Force, Wright-Patterson, Dayton, Ohio, Sept. 1, 1972.
18. R. Weck, "A Rational Approach to Standard for Welded Construction," *British Welding Journal*, Nov. 1966.
19. J. W. Fisher and B. T. Yen, "Design, Structural Details, and Discontinuities in Steel, Safety and Reliability of Metal Structures," *ASCE*, Nov. 2, 1972.
20. E. V. Bravenec, "Analysis of Brittle Fractures During Fabrication and Testing," *ASCE*, March 1972.
21. T. J. Dolan, "Preclude Failure: A Philosophy for Materials Selection and Simulated Service Testing," Wm. M. Murray Lecture, SESA Fall Meeting,

Houston, Oct. 1969, published *SESA Journal of Experimental Mechanics*, Jan. 1970.

22. E. R. PARKER, *Brittle Behavior of Engineering Structures*, Wiley, New York, 1957.
23. Welding Research Council, *Control of Steel Construction to Avoid Brittle Failure*, edited by M. E. Shank, M. I. T. Press, Cambridge, Mass., 1957.
24. W. J. HALL, H. KIHARA, W. SOETE, and A. A. WELLS, *Brittle Fracture of Welded Plate*, Prentice-Hall, Englewood Cliffs, N.J., 1967.
25. The Royal Institution of Naval Architects, *Brittle Fracture in Steel Structures*, edited by G. M. Boyd, Butterworth's, London, 1970.
26. C. F. TIPPER, *The Brittle Fracture Story*, Cambridge University Press, New York, 1962.
27. H. LIBOWITZ, ed., *Fracture, An Advanced Treatise*, Vols. I–VII, Academic Press, New York, 1969.
28. The Japan Welding Society, "Cracking and Fracture in Welds," in *Proceedings of the First International Symposium on the Prevention of Cracking in Welded Structures, Tokyo, Nov. 8–10, The Japan Welding Society*, Tokyo, 1971.
29. W. S. PELLINI, "Principles of Fracture—Safe Design," *Welding Journal* (*Welding Research Supplement*), pt. I, March 1971, pp. 91-S–109-S, and pt. II, April 1971, pp. 147-S–162-S.
30. S. T. ROLFE, "Fracture Mechanics in Bridge Design," *Civil Engineering ASCE*, Aug. 1972.
31. American Society for Testing and Materials, "Fracture Toughness Testing and Its Applications," *ASTM Special Technical Publication No. 381*, Philadelphia, 1964.
32. American Society for Testing and Materials, "Plane Strain Crack Toughness Testing of High Strength Metallic Materials," edited by W. F. Brown and J. E. Srawley, *ASTM STP No. 410*, Philadelphia, 1966.
33. American Society for Testing and Materials, "Review of Developments in Plane Strain Fracture Toughness Testing," edited by W. F. Brown, *ASTM STP No. 463*, Philadelphia, 1970.
34. American Society for Testing and Materials, "Fracture Toughness, Proceedings of the 1971 National Symposium on Fracture Mechanics," pt. II, *ASTM STP No. 514*, Philadelphia, 1971.
35. American Society of Civil Engineers, "Safety and Reliability of Metal Structures," ASCE Specialty Conference held in Pittsburgh, Nov. 2–3, published by ASCE, New York, 1972.

2

Stress Analysis for Members with Cracks

2.1. Introduction

The fundamental principle of fracture mechanics is that the stress field ahead of a sharp crack in a structural member can be characterized in terms of a single parameter, K, the stress-intensity factor, that has units of ksi$\sqrt{\text{in.}}$ This parameter, K, is related to both the nominal stress level (σ) in the member and the size of the crack present (a). Thus all structural members, or test specimens, that have flaws can be loaded to various levels of K, analogous to the situation where various unflawed members can be loaded to various stress levels, σ. In this chapter we shall describe the analytical development of the stress-intensity factor, K, for various crack geometries. In the appendix to this chapter we shall describe the original Griffith theory[1-4] that formed the basis of fracture mechanics.

Because failure of most equipment or structural members is caused by the propagation of cracks, an understanding of the magnitude and distribution of the stress field in the vicinity of the crack front is essential to determine the safety and reliability of equipment and structures. Because fracture mechanics is based on a stress analysis, a *quantitative* evaluation of the safety and reliability of structures is possible. Fracture mechanics can be subdivided into two general categories, namely linear-elastic and elastic-plastic. The theory of linear-elastic fracture mechanics is well developed and forms the basis of both categories. It is described in this chapter. Elastic-plastic fracture mechanics is not so well developed and is described in Chapter 16.

2.2. Linear-Elastic Fracture Mechanics

Linear-elastic fracture-mechanics technology is based on an analytical procedure that relates the stress-field magnitude and distribution in the vicinity of a crack tip to the nominal stress applied to the structure, to the size, shape, and orientation of the crack or crack-like discontinuity, and to material properties.

To establish methods of stress analysis for cracks in elastic solids, it is convenient to define three types of relative movements of two crack surfaces.[5] These displacement modes (Figure 2.1) represent the local deformation in an

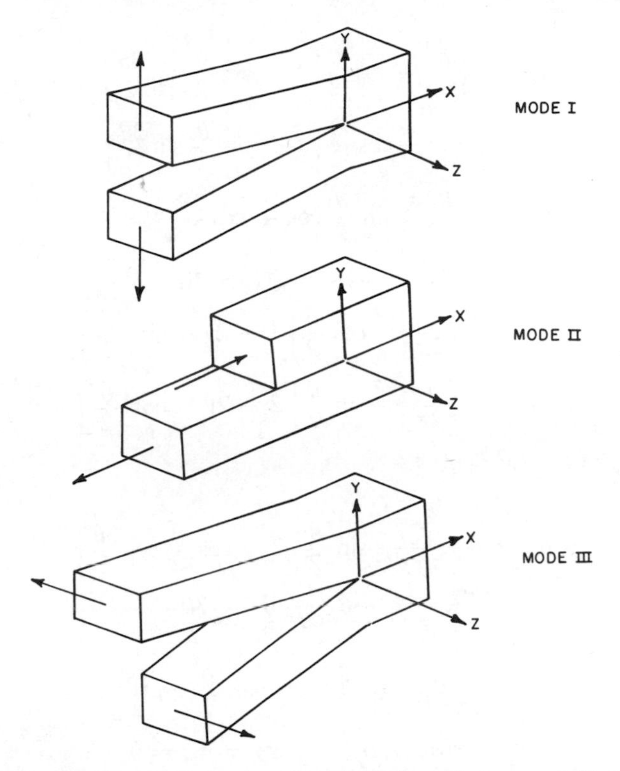

FIG. 2.1. The three basic modes of crack surface displacements (Ref. 5).

infinitesimal element containing a crack front. The opening mode, Mode I, is characterized by local displacements that are symmetric with respect to the *x-y* and *x-z* planes. The two fracture surfaces are displaced perpendicular to each other in opposite directions. Local displacements in the sliding or shear mode, Mode II, are symmetric with respect to the *x-y* plane and skew symmetric with respect to the *x-z* plane. The two fracture surfaces slide over each other in a direction perpendicular to the line of the crack tip. The tearing mode, Mode III, is associated with local displacements that are skew symmetric with respect to both *x-y* and *x-z* planes. The two fracture surfaces slide over each other in a direction that is parallel to the line of the crack front. Each of these modes of deformation corresponds to a basic type of stress field in the vicinity of crack tips. In any problem the deformations at the crack tip can be treated as one or a combination of these local displacement modes. Moreover, the stress field at the crack tip can be treated as one or a combination of the three basic types of stress fields.

By using a method that was developed by Westergaard,[6] Irwin[7] found that the stress and displacement fields in the vicinity of crack tips subjected to the three modes of deformation are given by

Mode I:

$$\begin{aligned}
\sigma_x &= \frac{K_{\mathrm{I}}}{(2\pi r)^{1/2}} \cos\frac{\theta}{2}\left[1 - \sin\frac{\theta}{2}\sin\frac{3\theta}{2}\right] \\
\sigma_y &= \frac{K_{\mathrm{I}}}{(2\pi r)^{1/2}} \cos\frac{\theta}{2}\left[1 + \sin\frac{\theta}{2}\sin\frac{3\theta}{2}\right] \\
\tau_{xy} &= \frac{K_{\mathrm{I}}}{(2\pi r)^{1/2}} \sin\frac{\theta}{2}\cos\frac{\theta}{2}\cos\frac{3\theta}{2} \\
\sigma_z &= \nu(\sigma_x + \sigma_y), \qquad \tau_{xz} = \tau_{yz} = 0 \\
u &= \frac{K_{\mathrm{I}}}{G}\left[\frac{r}{2\pi}\right]^{1/2} \cos\frac{\theta}{2}\left[1 - 2\nu + \sin^2\frac{\theta}{2}\right] \\
v &= \frac{K_{\mathrm{I}}}{G}\left[\frac{r}{2\pi}\right]^{1/2} \sin\frac{\theta}{2}\left[2 - 2\nu - \cos^2\frac{\theta}{2}\right] \\
w &= 0
\end{aligned} \tag{2.1}$$

Mode II:

$$\begin{aligned}
\sigma_x &= -\frac{K_{\mathrm{II}}}{(2\pi r)^{1/2}} \sin\frac{\theta}{2}\left[2 + \cos\frac{\theta}{2}\cos\frac{3\theta}{2}\right] \\
\sigma_y &= \frac{K_{\mathrm{II}}}{(2\pi r)^{1/2}} \sin\frac{\theta}{2}\cos\frac{\theta}{2}\cos\frac{3\theta}{2} \\
\tau_{xy} &= \frac{K_{\mathrm{II}}}{(2\pi r)^{1/2}} \cos\frac{\theta}{2}\left[1 - \sin\frac{\theta}{2}\sin\frac{3\theta}{2}\right] \\
\sigma_z &= \nu(\sigma_x + \sigma_y), \qquad \tau_{xz} = \tau_{yz} = 0 \\
u &= \frac{K_{\mathrm{II}}}{G}\left[\frac{r}{2\pi}\right]^{1/2} \sin\frac{\theta}{2}\left[2 - 2\nu + \cos^2\frac{\theta}{2}\right] \\
v &= \frac{K_{\mathrm{II}}}{G}\left[\frac{r}{2\pi}\right]^{1/2} \cos\frac{\theta}{2}\left[-1 + 2\nu + \sin^2\frac{\theta}{2}\right] \\
w &= 0
\end{aligned} \tag{2.2}$$

Mode III:

$$\begin{aligned}
\tau_{xz} &= -\frac{K_{\mathrm{III}}}{(2\pi r)^{1/2}} \sin\frac{\theta}{2} \\
\tau_{yz} &= \frac{K_{\mathrm{III}}}{(2\pi r)^{1/2}} \cos\frac{\theta}{2} \\
\sigma_x &= \sigma_y = \sigma_z = \tau_{xy} = 0 \\
w &= \frac{K_{\mathrm{III}}}{G}\left[\frac{2r}{\pi}\right]^{1/2} \sin\frac{\theta}{2} \\
u &= v = 0
\end{aligned} \tag{2.3}$$

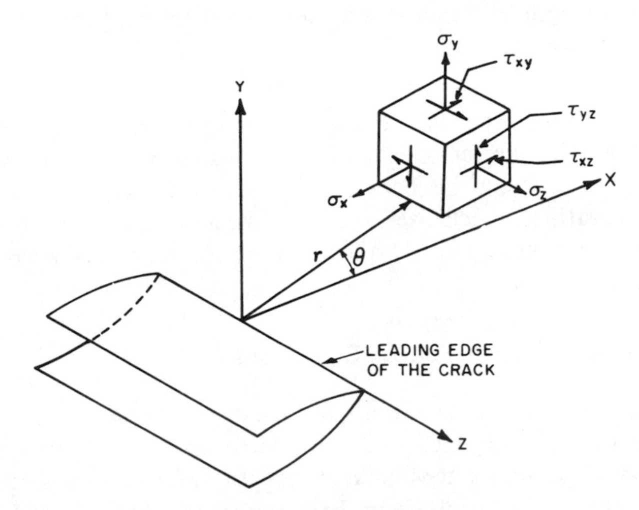

FIG. 2.2. Coordinate system and stress components ahead of a crack tip (Ref. 5).

where the stress components and the coordinates r and θ are shown in Figure 2.2; u, v, and w are the displacements in the x, y, and z directions, respectively; ν is Poisson's ratio; and G is the shear modulus of elasticity.

Equations (2.1) and (2.2) represent the case of plane strain ($w = 0$) and neglect higher-order terms in r. Because higher-order terms in r are neglected, these equations are exact in the limit as r approaches zero and are a good approximation in the region where r is small compared with other x-y planar dimensions. These field equations show that the distribution of the elastic-stress fields and of the deformation fields in the vicinity of the crack tip are invariant in all components subjected to a given mode of deformation and that the magnitude of the elastic-stress field can be described by single-term parameters, K_{I}, K_{II}, and K_{III}, that correspond to Modes I, II, and III, respectively. Consequently, the applied stress, the crack shape and size, and the structural configuration associated with structural components subjected to a given mode of deformation affect the value of the stress-intensity factor but do not alter the stress-field distribution. Dimensional analysis of Equations (2.1), (2.2), and (2.3) indicates that the stress-intensity factor must be linearly related to stress and must be directly related to the square root of a characteristic length. Based on Griffith's original analysis (see the appendix to this chapter for a description of the Griffith theory) of glass members with cracks and the subsequent extension of that work to more ductile materials, the characteristic length is the crack length in a structural member. Consequently, the magnitude of the stress-intensity factor must be directly related to the magnitude of the applied nominal stress, σ, and the square root of the crack length, a. In all cases, the general form of the stress-intensity factor is given by

$$K = f(g) \cdot \sigma \cdot \sqrt{a} \tag{2.4}$$

where $f(g)$ is a parameter that depends on the specimen and crack geometry and has been the subject of extensive investigations and research. Fortunately various relationships between the stress-intensity factor and various body configurations, crack sizes, orientations, and shapes, and loading conditions have been published,[5,8,9] and the more common ones are presented in the following section.

2.3. Stress-Intensity-Factor Equations

Stress-intensity-factor equations for the more common basic specimen geometries that are subjected to Mode I deformation are presented in this section. Extensive stress-intensity-factor equations for various geometries and loading conditions are available in the literature in tabular form.[5,8,9]

2.3.1. *Through-Thickness Crack*

The stress-intensity factor for an infinite plate subjected to uniform tensile stress, σ, and that contains a through-thickness crack of length $2a$ (Figure 2.3) is

$$K_I = \sigma\sqrt{\pi a} \tag{2.5}$$

A tangent-correction factor having the form

$$\left(\frac{2b}{\pi a}\tan\frac{\pi a}{2b}\right)^{1/2} \tag{2.6}$$

is used for plates of finite width, b. Consequently, the stress-intensity factor for a plate of finite width $2b$ that is subjected to uniform tensile stress, σ, and that contains a through-thickness crack of length $2a$ (Figure 2.3) is

$$K_I = \sigma\left[\pi a\left(\frac{2b}{\pi a}\tan\frac{\pi a}{2b}\right)\right]^{1/2} \tag{2.7}$$

The values for the tangent correction factor for various ratios of crack length to plate width are given in Table 2.1.

2.3.2. *Double-Edge Crack*

The stress-intensity factor for double-edge-notched plate that is subjected to uniform tensile stress (Figure 2.4) is given by

$$K_I = 1.12\sigma\sqrt{\pi a} \tag{2.8}$$

where the constant 1.12 represents a free-surface correction factor for edge notches that are perpendicular to the applied tensile stress.

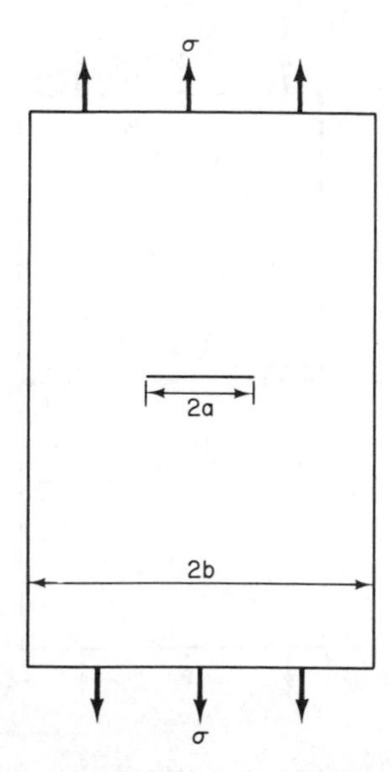

FIG. 2.3. Finite-width plate containing a through-thickness crack.

TABLE 2.1. Correction Factors for a Finite-Width Plate Containing a Through-Thickness Crack (Reference 5)

a/b	$[2b/\pi a \cdot \tan \pi a/2b]^{1/2}$ rad
0.074	1.00
0.207	1.02
0.275	1.03
0.337	1.05
0.410	1.08
0.466	1.11
0.535	1.15
0.592	1.20

The tangent-correction factor, Equation (2.6), can be used to obtain good estimates of the stress-intensity-factor values for double-edge-notched specimens having a finite width, b. However, more exact solutions are available in the literature[5,8,9] that incorporate the effects of specimen length relative to specimen width and the decay of the free-surface-correction factor as the ratio of crack length to specimen width increases.

2.3.3. *Single-Edge Notch*

The stress-intensity-factor equation for a semiinfinite single-edge-notched specimen (Figure 2.5) is given by Equation (2.8). Moreover, the correction

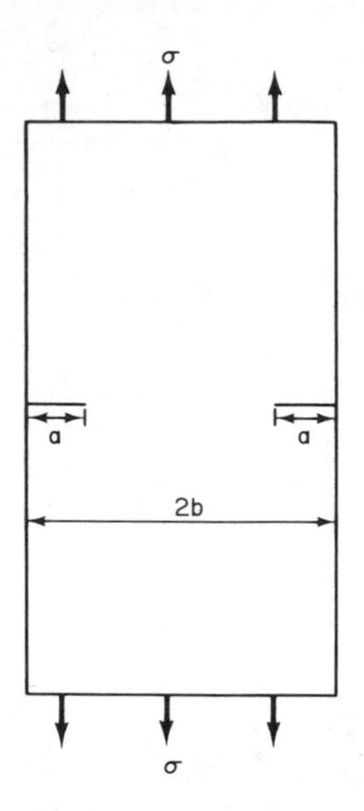

FIG. 2.4. Double-edge-notched plate of finite width.

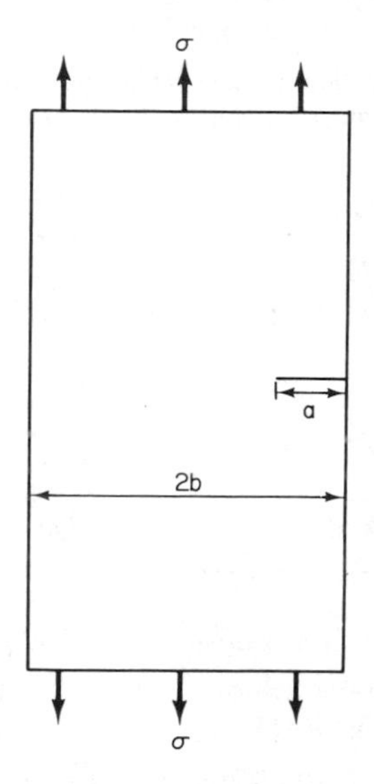

FIG. 2.5. Single-edge-notched plate of finite width.

factors used to calculate the stress-intensity factor for a double-edge-notched specimen of finite width can be used also to calculate the stress-intensity factor for a single-edge-notched specimen of finite width. However, an additional correction factor is necessary to account for bending stresses caused by lack of symmetry in the single-edge-notched specimen.

The stress-intensity-factor equation that incorporates the various correction factors for a single-edge-notched specimen is given by[5,10,11]

$$K_{\mathrm{I}} = \sigma\sqrt{\pi a}\, f\left(\frac{a}{b}\right) \tag{2.9}$$

The values of the function $f(a/b)$ are tabulated in Table 2.2 for various values of the ratio of crack length to specimen thickness.

TABLE 2.2. Correction Factors for a Single-Edge-Notched Plate (Reference 5)

a/b	$f(a/b)$
0.10	1.15
0.20	1.20
0.30	1.29
0.40	1.37
0.50	1.51
0.60	1.68
0.70	1.89
0.80	2.14
0.90	2.46
1.00	2.86

2.3.4. Cracks Emanating from Holes

The stress-intensity factor for cracks emanating from circular or elliptical holes in infinite plates (Figure 2.6[12]) is given by[13]

$$K_{\mathrm{I}} = \sigma\sqrt{\pi a}\, f(\lambda, \delta) \tag{2.10}$$

where $\lambda = a/a_N$ and $\delta = b_N/a_N$ and a, a_N, and b_N and the value of $F(\lambda, \delta)$ are shown in Figure 2.6.

2.3.5. Single Crack in Beam in Bending

The stress-intensity factor for a beam in bending that contains an edge crack (Figure 2.7) is given by

$$K_{\mathrm{I}} = \frac{6M}{(W-a)^{3/2}}\, f\left(\frac{a}{W}\right) \tag{2.11}$$

where the values of $f(a/W)$ are presented in Table 2.3 for various values of the ratio of crack length, a, to beam depth, W.

TABLE 2.3. Stress-Intensity-Factor Coefficients for Notched Beams (Reference 5)

a/W	0.05	0.1	0.2	0.3	0.4	0.5	0.6 (and larger)
$f(a/W)$	0.36	0.49	0.60	0.66	0.69	0.72	0.73

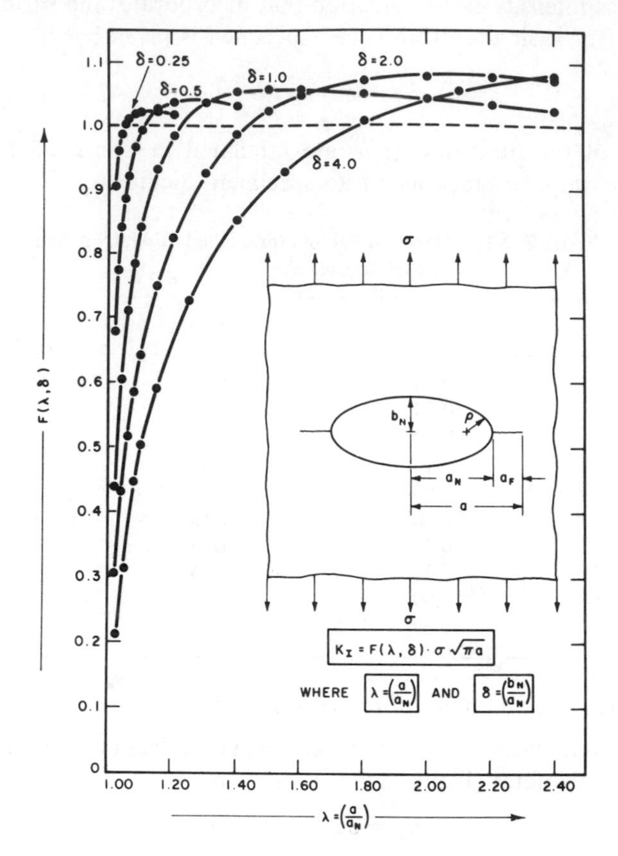

FIG. 2.6. Newman's analysis of stress-intensity-factor coefficients for cracks emanating from elliptical holes contained in an infinite sheet subjected to uniaxial tension.

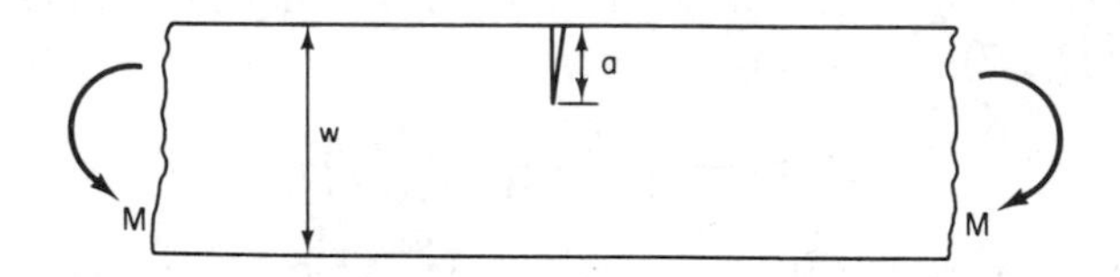

FIG. 2.7. Edge-notched beam in bending.

2.3.6. Elliptical or Circular Crack in Infinite Plate

The stress-intensity factor at any point along the perimeters of elliptical or circular cracks embedded in an infinite body that is subjected to uniform tensile stress (Figure 2.8) is given by[14]

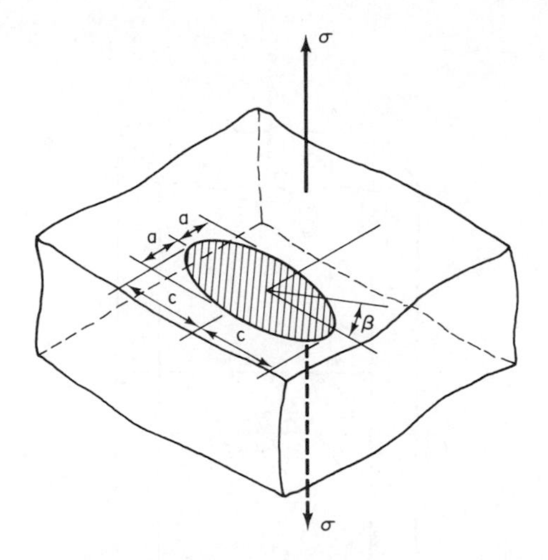

FIG. 2.8. Elliptical crack in an infinite body subjected to uniform tension (Reference 5).

$$K_I = \frac{\sigma\sqrt{\pi a}}{\Phi_0}\left(\sin^2\beta + \frac{a^2}{c^2}\cos^2\beta\right)^{1/4} \tag{2.12}$$

where K_I corresponds to the value of the stress-intensity factor for a point on the perimeter of the crack whose location is defined by the angle β, and Φ_0 is the elliptic integral

$$\Phi_0 = \int_0^{\pi/2}\left[1 - \left(\frac{c^2 - a^2}{c^2}\right)\sin^2\theta\right]^{1/2} d\theta \tag{2.13}$$

For circular cracks ($a = c$), Equation (2.12) reduces to

$$K_I = \frac{2}{\sqrt{\pi}}\sigma\sqrt{a} \tag{2.14}$$

For $c = \infty$ and $\beta = \pi/2$, Equation (2.12) reduces to Equation (2.5) for a through-thickness crack.

2.3.7. Surface Crack

The stress-intensity factor for a part-through thumbnail crack in a plate subjected to uniform tensile stress (Figure 2.9) can be calculated by using Equations (2.12) and (2.13) and a free-surface-correction factor equal to 1.12. The stress-intensity factor for $\beta = \pi/2$ is given by

$$K_I = 1.12\sigma\sqrt{\pi\frac{a}{Q}} \tag{2.15}$$

where $Q = \Phi_0^2$, and Φ_0 is the elliptic integral given by Equation (2.13). Q has been designated a shape factor because its value depends on the values of a

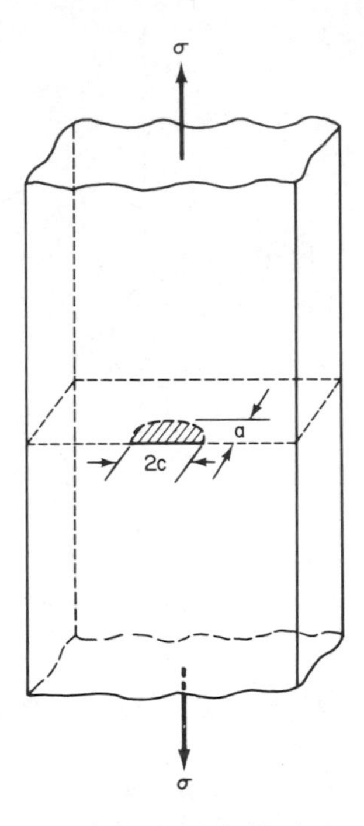

FIG. 2.9. Plate containing a semi-elliptical surface crack.

and c. A graphic representation of this dependence is shown in Figure 2.10 for various crack shapes and values of the ratio of nominal applied stress to yield stress, σ_{ys}. The use of different curves for different values of σ/σ_{ys} is intended to account for the effects of plastic deformation in the vicinity of the crack tip on the stress-intensity-factor value. These effects are discussed in the following section.

2.3.8. Crack in WOL-Type Specimen

The stress-intensity-factor equation for wedge-opening-loading- (WOL) type specimens (Figure 2.11) can be presented in the form[15]

$$K_{\mathrm{I}} = \frac{P}{B\sqrt{a}} f\left(\frac{a}{W}\right) \tag{2.16}$$

where P is the applied load, B is the specimen thickness, and $f(a/W)$ is a measure of the compliance of the specimen. For a constant value of specimen height, H, to specimen width, W, the function $f(a/W)$ can be expressed in a polynomial form as a function of the dimensionless crack length a/W. Two WOL specimen geometries have been used extensively. These geometries are usually referred to as the T-type WOL specimen having $H/W = 0.972$ and the compact tension specimen (CTS) having $H/W = 1.2$. The dimensions for 1-in.-thick specimens for each of these two geometries are shown

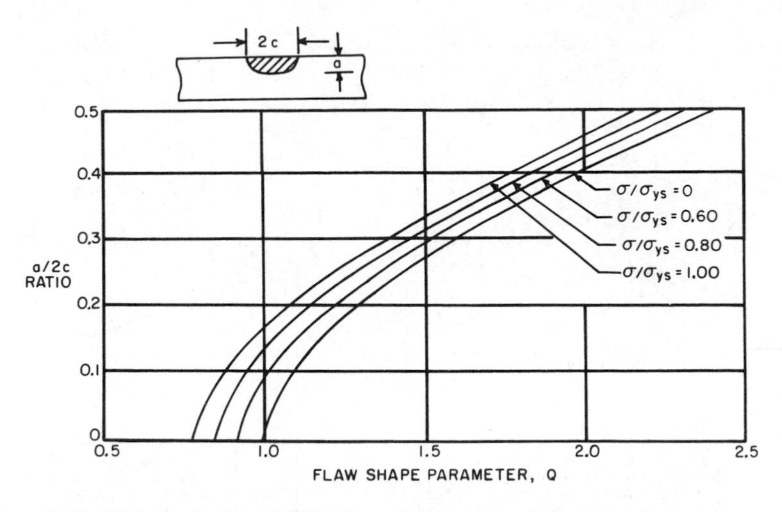

FIG. 2.10. Dependence of the flaw shape parameter on the ratio of depth and crack surface length.

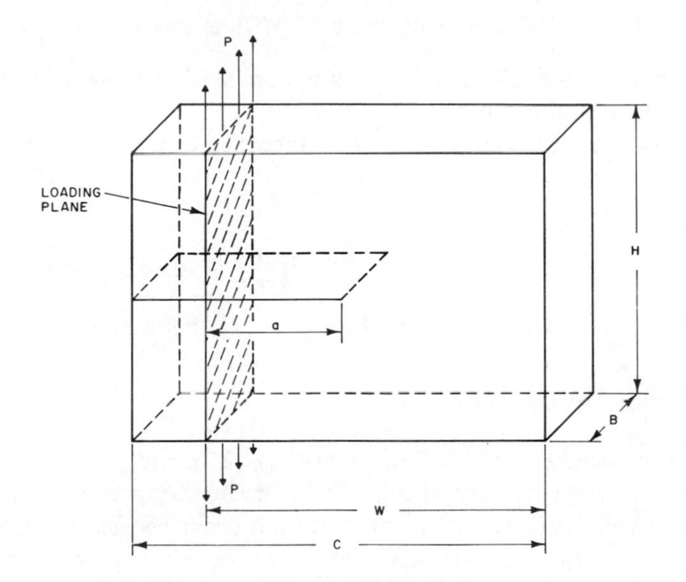

FIG. 2.11. Idealized model of wedge-opening-loading specimen.

in Figures 2.12 and 2.13, respectively. Various sizes of a given WOL specimen geometry are proportional specimens.

The $f(a/W)$ for the T-type WOL specimen can be represented as[15,16]

$$f\left(\frac{a}{W}\right) = \left[30.96\left(\frac{a}{W}\right) - 195.8\left(\frac{a}{W}\right)^2 + 730.6\left(\frac{a}{W}\right)^3 - 1186.3\left(\frac{a}{W}\right)^4 + 754.6\left(\frac{a}{W}\right)^5\right] \qquad (2.17)$$

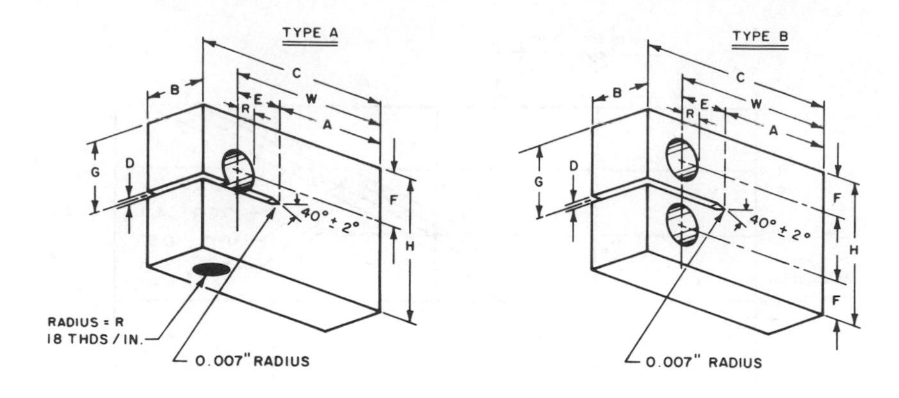

SPEC	B	W	C	A	E	H	G	D	R	F
IT-A	1.000	2.550	3.200	1.783	0.767	2.480	1.240	0.094	0.350	1.000
IT-B	1.000	2.550	3.200	1.783	0.767	2.480	1.240	0.094	0.250	0.650

FIG. 2.12. Two types of 1T WOL specimens.

In the range $0.25 < a/W < 0.75$, the polynomial is accurate to within 0.5% of the experimental compliance.

The $f(a/W)$ for the compact tension specimen can be represented as[15,17]

$$f\left(\frac{a}{W}\right) = \left[0.2960\left(\frac{a}{W}\right) - 1.855\left(\frac{a}{W}\right)^2 + 6.557\left(\frac{a}{W}\right)^3 - 10.17\left(\frac{a}{W}\right)^4 + 6.389\left(\frac{a}{W}\right)^5\right] \tag{2.18}$$

The polynomial is accurate within 0.25% for a/W between 0.3 and 0.7.

2.4. Crack-Tip Deformation

The stress-field equations, Equations (2.1), (2.2), and (2.3), show that the elastic stress in the vicinity of a crack tip where $r \ll a$ can be very large. In reality, such high stress magnitudes do not occur because the material in this region undergoes plastic deformation, thus creating a plastic zone that surrounds the crack tip. Figure 2.14 is a schematic presentation of the change in the distribution of the y component of the stress caused by the localized plastic deformation in the vicinity of the crack tip.

The size of the plastic zone, r_y, can be estimated from the stress-field equations by treating the problem as one of plane stress and setting the y component of stress, σ_y, equal to the yield strength, σ_{ys}, which results in[18]

$$r_y = \frac{1}{2\pi}\left(\frac{K}{\sigma_{ys}}\right)^2 \tag{2.19}$$

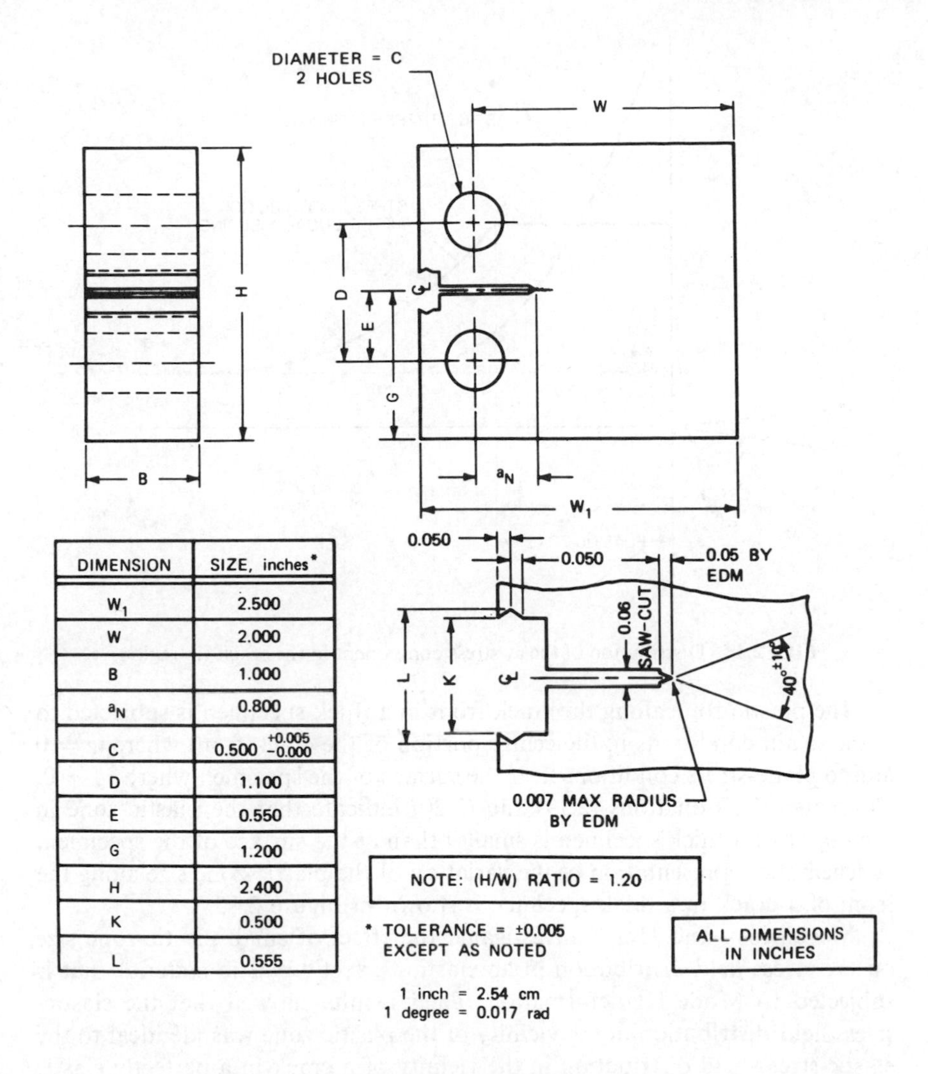

DIMENSION	SIZE, inches*
W_1	2.500
W	2.000
B	1.000
a_N	0.800
C	$0.500\ ^{+0.005}_{-0.000}$
D	1.100
E	0.550
G	1.200
H	2.400
K	0.500
L	0.555

FIG. 2.13. 1-inch-thick compact tension specimen (CTS).

Irwin[19] suggested that the plastic-zone size under plane-strain conditions can be obtained by considering the increase in the tensile stress for plastic yielding caused by plane-strain elastic constraint. Under these conditions, the yield strength is estimated to increase by a factor of $\sqrt{3}$. Consequently the plane-strain plastic-zone size becomes

$$r_y = \frac{1}{6\pi}\left(\frac{K}{\sigma_{ys}}\right)^2 \tag{2.20}$$

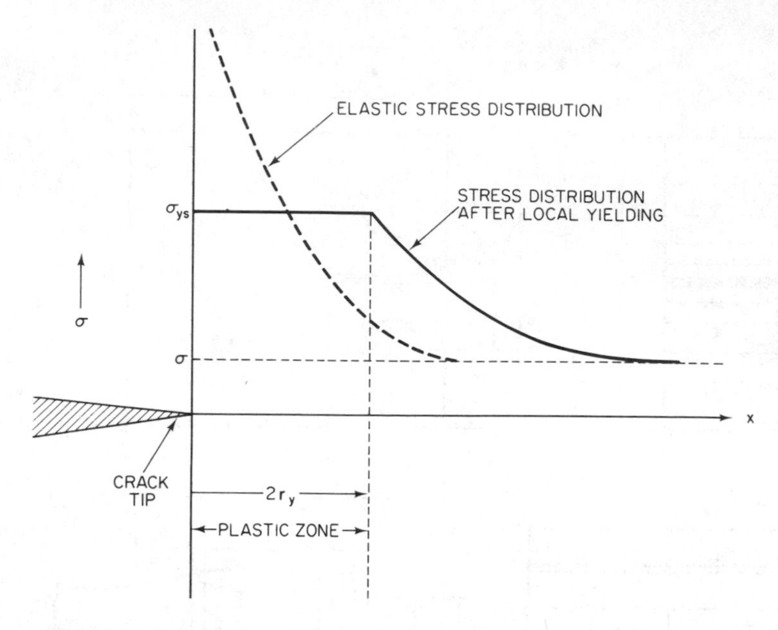

FIG. 2.14. Distribution of the σ_y stress component in the crack-tip region.

The plastic zone along the crack front in a thick specimen is subjected to plane-strain conditions in the center portion of the crack front where $w = 0$ and to plane-stress conditions near the surface of the specimen where $\sigma_z = 0$. Consequently, Equations (2.19) and (2.20) indicate that the plastic zone in the center of a thick specimen is smaller than at the surface of the specimen. A schematic representation of the variation of the plastic-zone size along the front of a crack in a thick specimen is shown in Figure 2.15.

McClintock and Hult[20] investigated the effect of small plastic-zone size on the stress-field distribution in an elastic-perfectly-plastic material that is subjected to Mode III deformation. Their results showed that the elastic-stress-field distribution in the vicinity of the plastic zone was identical to the elastic-stress-field distribution in the vicinity of a crack in a perfectly elastic material whose tip was placed at the center of the plastic zone. Irwin[18] suggested that the effect of small plastic zones corresponds to an apparent increase of the elastic crack length by an increment equal to r_y. This plastic-zone correction factor is valid for small plastic-zone sizes and should not be applied when large plastic deformations occur.

2.5. Superposition of Stress-Intensity Factors

Components that contain cracks may be subjected to one or more different types of Mode I loads such as uniform tensile loads, concentrated tensile loads, or bending loads. The stress-field distributions in the vicinity of

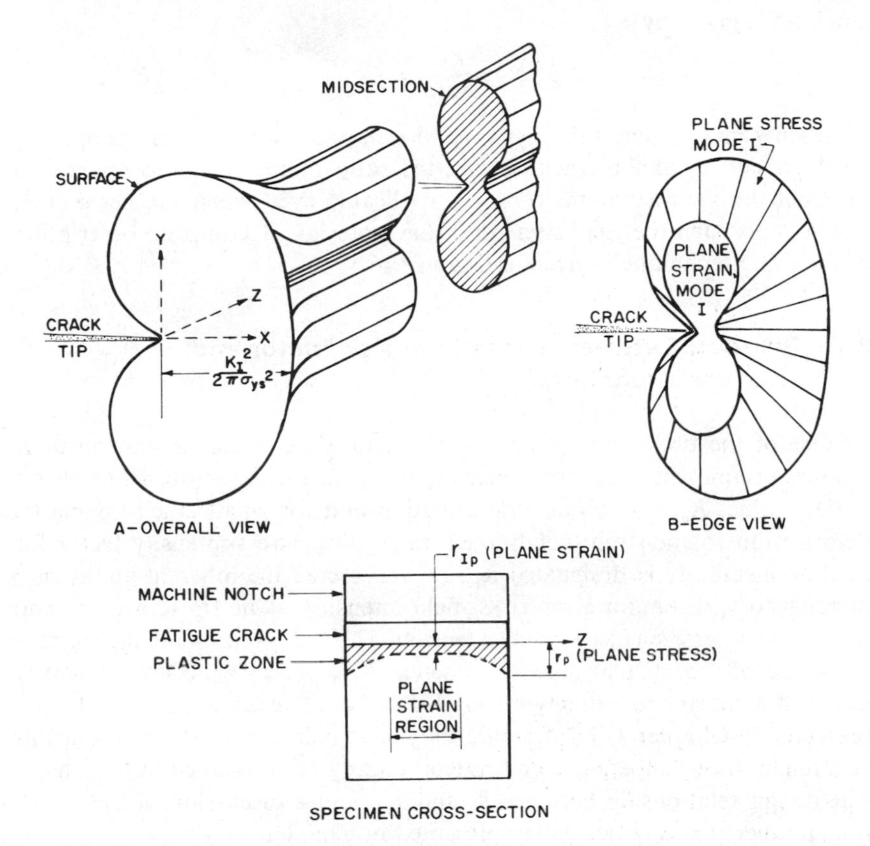

FIG. 2.15. Schematic representation of plastic zone ahead of crack tip.

the crack tip that is subjected to these loads are identical and are represented by Equation (2.1). Consequently, the total stress-intensity factor can be obtained by algebraically adding the stress-intensity factors that correspond to each load. However, some components may be subjected to loads that correspond to various modes of deformation. Because the stress-field distributions in the vicinity of a crack, Equations (2.1), (2.2), and (2.3), are different for different modes of deformation, the stress-intensity factors for different modes of deformation cannot be added. Under these loading conditions, the total energy-release rate, G (see the appendix to this chapter), rather than the stress-intensity factors, can be calculated by algebraically adding the energy-release rate for the various modes of deformation.

2.6. Crack-Opening Displacement (COD)

The tensile stresses applied to a body that contains a crack cause a displacement of the crack surfaces. The crack-opening displacement, δ, at the crack tip is given by[19]

$$\delta = \frac{K_I^2}{E\sigma_{ys}} = \frac{G}{\sigma_{ys}} \tag{2.21}$$

Irwin[19] noted that this linear relationship between the crack-opening displacement, δ, at the crack tip and the ratio of the strain-energy-release rate and the yield strength, G/σ_{ys}, is applicable even when the net section stresses approach the yield strength of the material. A complete description of the COD technique is given in Chapter 16.

2.7. Relation Between Stress-Intensity Factor and Fracture Toughness

One of the underlying principles of fracture mechanics is that unstable fracture occurs when the stress-intensity factor at the crack tip, K, reaches a critical value, K_c. For Mode I deformation and for small crack-tip plastic deformation (plane-strain conditions), the critical stress-intensity factor for fracture instability is designated K_{Ic}. K_{Ic} represents the inherent ability of a material to withstand a given stress-field intensity at the tip of a crack and to resist progressive tensile crack extension. Thus, K_{Ic} represents the fracture toughness of a particular material, whereas K_I represents the stress intensity ahead of a sharp crack in any material. Procedures used to measure K_{Ic} are presented in Chapter 3. The fracture toughness behavior of structural metals as a function of temperature and rate of loading are presented in Chapter 4. The design relationship between K_I and K_{Ic} and a discussion of the use of fracture mechanics in design are presented in Chapter 5.

REFERENCES

1. A. A. Griffith, "The Phenomena of Rupture and Flaw in Solids," *Transactions, Royal Society of London, A-221*, 1920.
2. C. E. Inglis, "Stresses in a Plate Due to the Presence of Cracks and Sharp Corners," *Proceedings, Institute of Naval Architects, 60*, 1913.
3. G. R. Irwin, "Fracture Dynamics," in *Fracturing of Metals*, American Society of Metals, Cleveland, 1948.
4. E. Orowan, "Fracture Strength of Solids," in *Report on Progress in Physics*, Vol. 12, Physical Society of London, 1949.
5. C. P. Paris and G. C. Sih, "Stress Analysis of Cracks," in "Fracture Toughness Testing and Its Applications," *ASTM STP No. 381*, American Society for Testing and Materials, Philadelphia, 1965.
6. H. M. Westergaard, "Bearing Pressures and Cracks," *Transactions, ASME, Journal of Applied Mechanics*, 1939.

7. G. R. Irwin, "Analysis of Stresses and Strains Near the End of a Crack Traversing a Plate," *Transactions, ASME, Journal of Applied Mechanics*, vol 24, 1957.

8. H. Tada, P. C. Paris, and G. R. Irwin, ed., *Stress Analysis of Cracks Handbook*, Del Research Corporation, Hellertown, Pa., 1973.

9. G. C. Sih, *Handbook of Stress-Intensity Factors for Researchers and Engineers*, Institute of Fracture and Solid Mechanics, Lehigh University, Bethlehem, Pa., 1973.

10. O. L. Bowie, "Rectangular Tensile Sheet with Symmetric Edge Cracks," *Transactions, ASME, Journal of Applied Mechanics, 31*, Series E. No. 2, June 1964.

11. B. Gross, J. E. Srawley, and W. F. Brown, Jr., "Stress Intensity Factors for a Single Edge Notch Tension Specimen by Boundary Collocation of a Stress Function," *NASA TN D-2395*, Aug. 1964.

12. S. R. Novak and J. M. Barsom, "AISI Project 168-Toughness Criteria for Structural Steels: Brittle-Fracture (K_{Ic}) Behavior of Cracks Emanating from Notches," presented at the Ninth National Symposium on Fracture Mechanics, University of Pittsburgh, Aug. 25–27, 1975, to be published by American Society for Testing and Materials.

13. J. C. Newman, Jr., "An Improved Method of Collocation for the Stress Analysis of Cracked Plates with Various Shaped Boundaries," *NASA Technical Note, NASA TN D-6376*, Aug. 1971.

14. G. R. Irwin, "The Crack Extension Force for a Part Through Crack in a Plate," *Transactions, ASME, Journal of Applied Mechanics*, vol 29, No. 4, 1962.

15. W. K. Wilson, "Stress Intensity Factors for Compact Specimens Used To Determine Fracture Mechanics Parameters," *Research Report 73-1E7-FMPWR-R1*, Westinghouse Research Laboratories, Pittsburgh, July 27, 1973.

16. W. K. Wilson. "Analytical Determination of Stress Intensity Factors for Manjoine Brittle Fracture Specimen," *Research Report AEC, WERL 0029–3*, Westinghouse Research Laboratories, Pittsburgh, Aug. 26, 1965.

17. W. K. Wilson, "Stress Intensity Factors for Compact Tension Specimen," *Research Memorandum 67-1D6-BTLFR-M1*, Westinghouse Research Laboratories, Pittsburgh, June 12, 1967.

18. G. R. Irwin, "Plastic Zone Near a Crack and Fracture Toughness," 1960 Sagamore Ordnance Materials Conference, Syracuse University, 1961.

19. G. R. Irwin, "Linear Fracture Mechanics, Fracture Transition, and Fracture Control," *Engineering Fracture Mechanics, 1*, No. 2, Aug. 1968.

20. F. A. McClintock and J. Hult, "Elastic-Plastic Stress and Strain Distribution Around Sharp Notches in Repeated Shear," Ninth International Congress of Applied Mechanics, New York, 1956.

Appendix

The Griffith Theory

The first analysis of fracture behavior of components that contain sharp discontinuities was developed by Griffith[1]. The analysis was based on the assumption that incipient fracture in ideally brittle materials occurs when the magnitude of the elastic energy supplied at the crack tip during an incremental increase in crack length is equal to or greater than the magnitude of the elastic energy at the crack tip during an incremental increase in crack length. This energy approach can be presented best by considering the following example. Consider an infinite plate of unit thickness that contains a through-thickness crack of length $2a$ and that is subjected to uniform tensile stress, σ, applied at infinity. The total potential energy of the system, U, may be written as

$$U = U_0 - U_a + U_\gamma \tag{A2.1}$$

where U_0 = elastic energy of the uncracked plate,

U_a = decrease in the elastic energy caused by introducing the crack in the plate,

U_γ = increase in the elastic-surface energy caused by the formation of the crack surfaces.

Griffith used a stress analysis that was developed by Inglis[2] to show that

$$U_a = \frac{\pi\sigma^2 a^2}{E} \tag{A2.2}$$

Moreover, the elastic-surface energy, U_γ, is equal to the product of the elastic-surface energy of the material, γ_e, and the new surface area of the crack:

$$U_\gamma = 2(2a\gamma_e) \tag{A2.3}$$

Consequently, the total elastic energy of the system, U, is

$$U = U_0 - \frac{\pi\sigma^2 a^2}{E} + 4a\gamma_e \tag{A2.4}$$

The equilibrium condition for crack extension is obtained by setting the first derivative of U with respect to crack length, a, equal to zero. The resulting equation can be written as

$$\sigma\sqrt{a} = \left(\frac{2\gamma_e E}{\pi}\right)^{1/2} \tag{A2.5}$$

which indicates that crack extension in ideally brittle materials is governed by the product of the applied nominal stress and the square root of the crack length and by material properties. Because E and γ_e are material properties, the right-hand side of Equation (A2.5) is equal to a constant value that is characteristic of a given ideally brittle material. Consequently, Equation (A2.5) indicates that crack extension in such materials occurs when the product $\sigma\sqrt{a}$ attains a constant critical value. The value of this constant can be determined experimentally by measuring the fracture stress for a large plate that contains a through-thickness crack of known length and that is subjected to remotely applied uniform tensile stress. This value can also be measured by using other specimen geometries, which is what makes this approach to fracture analysis so powerful. Values of this constant for other geometries are discussed in the main body of Chapter 2.

Equation (A2.5) can be rearranged in the form

$$\frac{\pi\sigma^2 a}{E} = 2\gamma_e \tag{A2.6}$$

The left-hand side has been designated the energy-release rate, G, and represents the elastic energy per unit crack surface area that is available for infinitesimal crack extension. The right-hand side of Equation (A2.6) represents the material's resistance to crack extension, R.

In 1948 Irwin[3] suggested that the Griffith fracture criterion for ideally brittle materials could be modified and applied to brittle materials and to metals that exhibit plastic deformation. A similar modification was proposed by Orowan at about the same time. The modification recognized that a material's resistance to crack extension is equal to the sum of the elastic-surface energy and the plastic-strain work, γ_p, accompanying crack extension. Consequently, Equation (A2.6) was modified to

$$\frac{\pi\sigma^2 a}{E} = 2(\gamma_e + \gamma_p) \tag{A2.7}$$

Because the left hand side is the energy-release rate, G, and because $K_I = \sigma\sqrt{\pi a}$ for a through-thickness crack of length $2a$ (Figure 2.3), the following relation exists between G and K_I for plane strain conditions

$$\frac{\pi\sigma^2 a}{E} = G = \frac{K_I^2}{E}(1 - \nu^2) \tag{A2.8}$$

The energy-balance approach to crack extension defines the conditions required for instability of an ideally sharp crack. This approach is not applicable to analysis of stable crack extension such as occurs under cyclic-load fluctuation or under stress-corrosion-cracking conditions. However, the stress-intensity parameter, K, *is* applicable to stable crack extension, and therefore development of linear-elastic fracture-mechanics theory has assisted greatly in improving our understanding of subcritical crack extension and crack instability.

3

Experimental Determination of K_{Ic} and K_{Id}

3.1. General

In Chapter 2 we described various analytical relationships for determining stress-intensity factors in elastic bodies with different shaped cracks. These stress-intensity factors (K_I, K_{II}, or K_{III} for the opening, edge-sliding, or tearing modes of crack extension—as described in Chapter 2) are a function of load, crack size, and geometry. Ideally, a K_I stress-intensity factor associated with a specific crack geometry (e.g., three of the most common geometries are the edge crack, surface crack, and through-thickness crack, as shown in Figure 3.1) can be used to model a particular crack geometry in an actual structure.

These particular stress-intensity values, K_I, are calculated for different load levels in the same general manner as particular stresses, σ, are calculated for different load levels in uncracked members subjected to tension. That is, for a member loaded in tension, the stress is calculated as $\sigma = P/A$ for various loads, P. In an analogous fashion, the stress-intensity is calculated as $K_I = C\sigma\sqrt{a}$ for various nominal stress levels, σ. Note that the calculations for stress-intensity factors, K_I, K_{II}, or K_{III} (analogous to the calculation of stress, σ), are the same for any structural material as long as the general boundary conditions described in Chapter 2 are satisfied.

However, because actual structural materials have certain limiting characteristics (e.g., yielding in ductile materials or fracture in brittle materials), there are limiting values of both σ and K_I, namely σ_{ys} (the yield strength) and K_{Ic}, K_{Id}, or K_c (critical stress-intensity factors).

The critical stress-intensity factor, K_c, at which unstable crack growth occurs for conditions of *static* loading at a particular temperature actually depends on specimen thickness or constraint, as shown in Figure 3.2. The limiting value of K_c for a particular test temperature and slow loading rate is K_{Ic} for plane-strain (maximum constraint) conditions. Under conditions of *dynamic* (or impact) loading, the critical value is defined as K_{Id}.

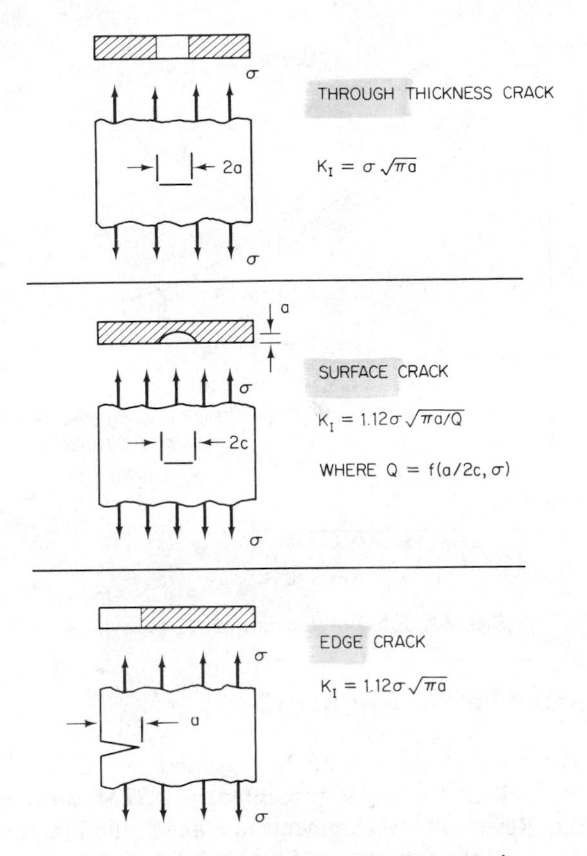

FIG. 3.1. K_I values for various crack geometries.

After considerable study and experimental verification, a special ASTM committee on fracture testing, Committee E-24, prepared a "Recommended Method of Test for Plane Strain Fracture Toughness of Metallic Materials" (ASTM Designation: E-399—74),[1] which is used to measure K_{Ic}. In the following sections we shall describe the background of the K_{Ic} test method and the details of the particular testing procedures.

At present, the only *standardized* ASTM method for measuring critical stress-intensity factors is the K_{Ic} test method. ASTM Committee E-24 on Fracture Testing of Metals is in the process of developing a "Recommended Practice for *R*-Curve Determination,"[2] which measures the resistance to fracture of materials during incremental slow-stable crack extension, and can be used to obtain the plane-stress fracture toughness value, K_c. The *R*-Curve analysis is described in Chapter 16. State-of-the-art techniques for conducting plane-strain critical stress-intensity factor tests for K_{Id} values are presented in this chapter.

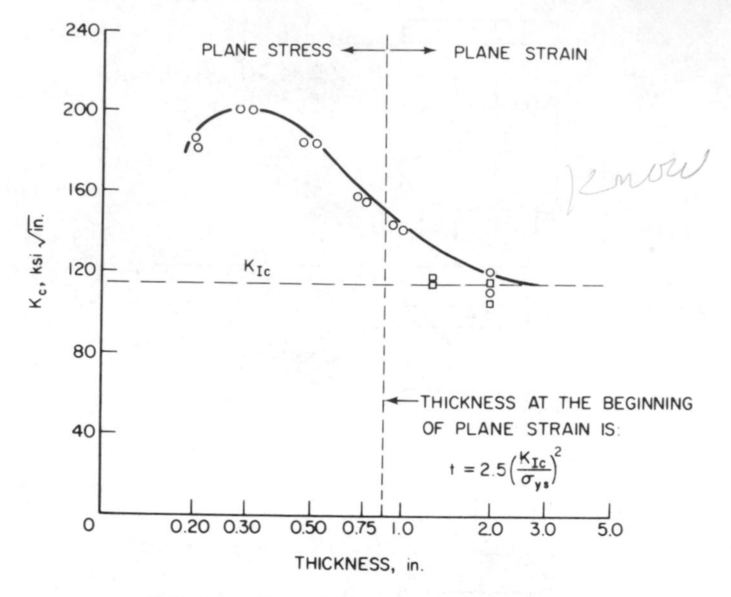

FIG. 3.2. Effect of thickness on K_c behavior.

3.2. Background of K_{Ic} Test Method

A detailed review of the overall background and development of the E-399—74 test method for K_{Ic} is presented in *ASTM Standard Technical Publication 463*,"Review of Developments in Plane Strain Fracture Toughness Testing".[3] In one of the papers in STP 463, "Progress in Fracture Testing of Metallic Materials," J. G. Kaufman[3] stated that early in the development of the test methods it was established that elastic fracture mechanics was the best analysis by which the resistance of materials to unstable crack growth could be described. Although it was realized that most real structural materials do not behave in a purely elastic manner on fracturing, it was hoped that it would be possible with appropriate modification to take into account the finite sizes of structural members and test specimens. If the small crack-tip plasticity could be accounted for, it was hoped that small specimens could be used to recreate and describe the situation of unstable crack growth occurring in a large structure. Thus, the early work of the ASTM committee was directed toward work on the elastic-fracture problem, even though most structural materials fracture in an inelastic mode.

Early in this program it was realized that the critical stress-intensity factor for unstable crack growth (K_c) was a thickness-dependent property as shown by test data for 7075-T6 and T651 sheet and plate materials (Figure 3.3). Over a certain range of thickness for this material (as well as other

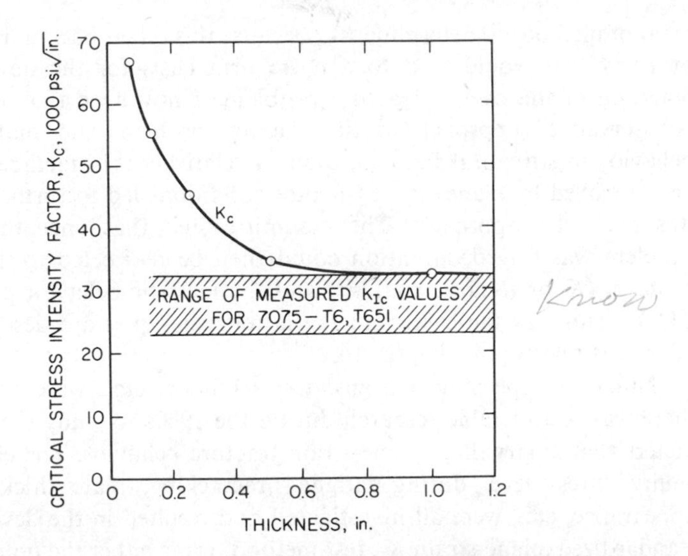

FIG. 3.3. Test data for 7075–T6 & T651 aluminums (Ref. 3.1).

structural materials), the critical combination of load and the crack length at instability, i.e., K_c, decreases with an increase in thickness, reaching a rather constant minimum value (K_{Ic}) when plane-strain conditions are approached. In the fully plane-strain condition, the crack instability leading to complete fracture almost immediately follows crack initiation. It was also observed that the *initial* crack growth, whether stable or unstable, took place at about the same level of stress intensity for a given material regardless of specimen thickness. This initial crack growth was referred to as *pop-in* and led to the belief that a lower level of critical stress-intensity factor associated with plane-strain conditions (K_{Ic}) could be determined reproducibly. Thus, although much of the early work on fracture was for thin sheet specimens, it became clear that not only was thickness an important variable, but because crack toughness decreased with increasing thickness to an apparently constant value (K_{Ic} at a given temperature and loading rate), this value of K_{Ic} might be considered to be a material property in the same manner that yield strength is considered to be a material property. If this were the case, then particular K_{Ic} levels could be established for various structural materials in the same manner that yield strength values are established for various structural materials, as long as the thickness of plate was large enough to establish plane-strain conditions. The K_{Ic} value would also be a minimum value for conditions of maximum structural constraint (for example, stiffeners, intersecting plates, etc.) that might lead to plane-strain conditions even though the individual structural members might be relatively thin. Therefore, emphasis on the early ASTM E-24 fracture committee work shifted to the development of methods to determine K_{Ic}, because this might be a more fundamental property that would lead to a more firm basis for the development and handling of the overall fracture problem. Knowledge that K_{Ic} provided a "conservative" approach to the fracture problem and that the material behavior in structural failures, even in relatively thin members, might well be controlled by plane-strain fracture conditions led to further emphasis on this particular approach. The committee felt that once the plane-strain problem was solved, attention could then be redirected to the thin-section problem (K_c or R-curve analysis) and the inelastic-behavior problem (J_{Ic} or COD). This was the case, and the latter techniques are described in state-of-the-art reviews in Chapter 16.

Numerous specimen designs, test methods, etc., were considered, and there was considerable research during the 1960s to study the various parameters that might affect plane-strain fracture behavior. The effects of notch acuity, stress level during fatigue precracking, plate thickness, fracture appearance, etc., were all investigated and resulted in the development of a standardized, plane-strain K_{Ic} test method, using either the notch-bend specimen or the compact-tension specimen (CTS). The latter specimen was formerly referred to as the wedge-opening-loading (WOL) specimen developed by Manjoine of Westinghouse.

Using the recommended test specimens, which are shown in Figures 3.4 and 3.5, round robin test programs were conducted to establish the fact that the recommended test procedure did indeed give reproducible results. The results of that program indicated that reproducible K_{Ic} values could be obtained by different laboratories within about 15%.

Thus, in the early 1970s, the test procedure was made a standard ASTM test method, E-399—74.[1] In subsequent sections in this chapter we shall describe the details of this test procedure as used to measure K_{Ic} for particular structural materials at a given loading rate and temperature.

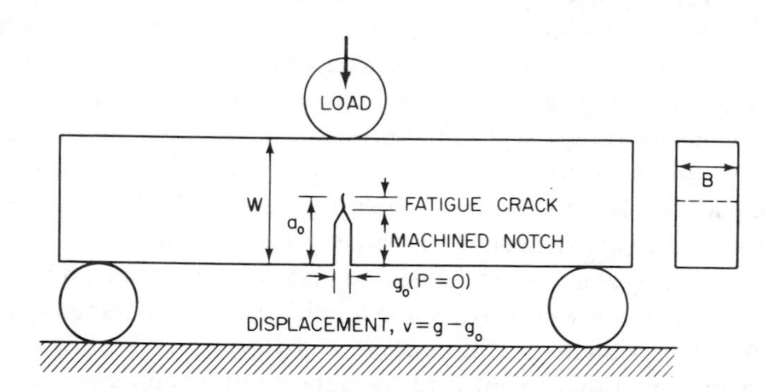

FIG. 3.4. Essentials of bend specimen for K_{Ic} test.

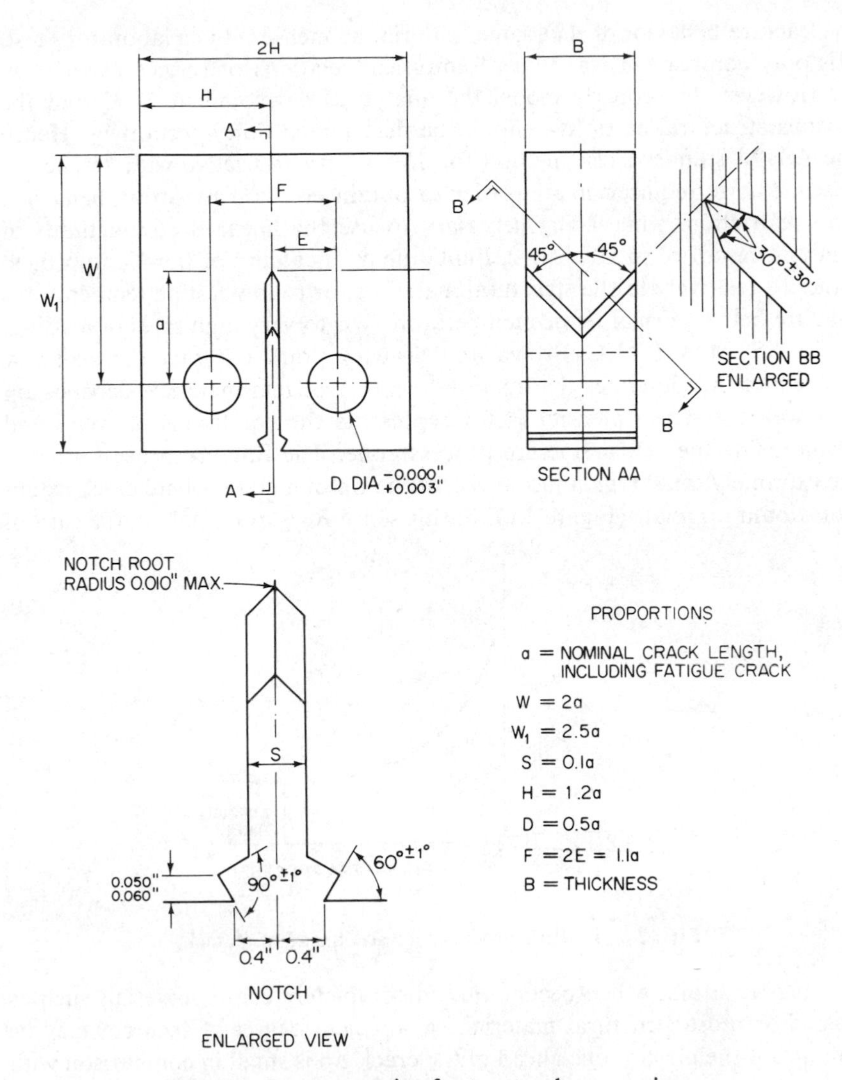

FIG. 3.5. Compact tension fracture toughness specimen.

3.3. Specimen Size Requirements

The dominant advantage of measuring the notch toughness of structural materials in terms of K_{Ic} is that this K_{Ic} value can be compared directly to the various K_I levels calculated for a particular structure of different design loadings. Prior to the development of fracture mechanics, the engineer had no such direct comparison of the fracture resistance of a structural material to the fracture behavior of that same material as measured in a laboratory test. His only comparison was that of empirical relations or service experience.

However, to properly model the analytical development of K_I and the various structural cases, K_{Ic} should be determined fairly accurately. Hence the ASTM standard test method for K_{Ic} is very restrictive with respect to specimen size requirements in order to obtain elastic plane-strain behavior. This restriction, while being necessary to use the linear-elastic methods of analysis described in Chapter 2, limits the applicability of the K_{Ic} approach either to relatively brittle structural materials, or to low testing temperatures that are below normal service temperatures, or to very high rates of loading.

In ASTM STP-410,[4] Brown and Srawley point out that the accuracy with which K_{Ic} describes the fracture behavior of real materials depends on how well the stress-intensity factor represents the conditions of stress and strain inside the actual fracture process zone. The fracture process zone is the extremely small region just ahead of the tip of a crack where crack extension would originate (Figure 3.6). In this sense K_I is exact only in the case of

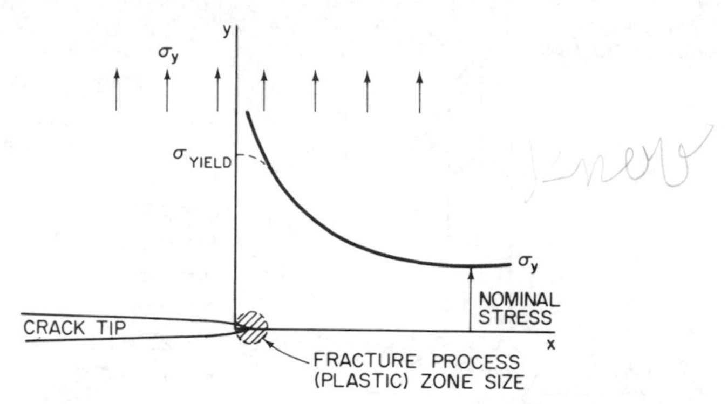

FIG. 3.6. Fracture process zone size ahead of a crack.

zero plastic strain, which occurs only in completely brittle materials such as glass. For most structural materials, a sufficient degree of accuracy may be obtained if the plastic zone ahead of the crack tip is small in comparison with the region around the crack in which the stress-intensity factor yields a satisfactory approximation of the exact elastic-stress field. That is, as shown schematically in Figure 3.7, a small amount of yielding (plastic zone) ahead of the crack tip does not change the overall distribution of the stress field away from the crack front. Any loss in accuracy of the analysis associated with increasing the relative size of the plastic zone is gradual, and it is not possible to prescribe theoretical limits on the applicability of linear elastic fracture mechanics. Obviously, the decision of what is sufficient accuracy depends on the particular application. However, because a *standardized ASTM test method must be reproducible*, the specimen size requirements were

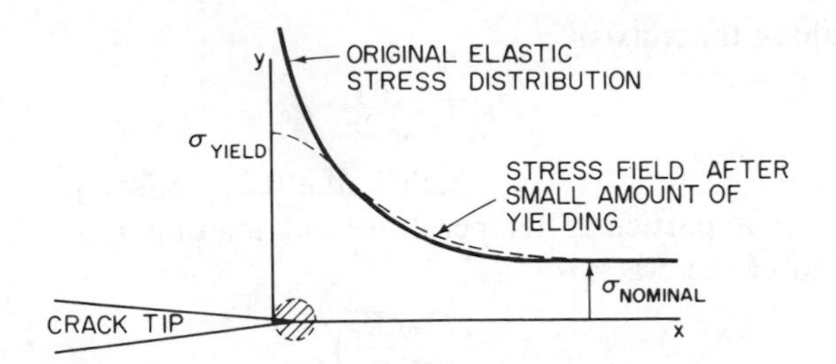

FIG. 3.7. Schematic showing small effect of yielding on nominal stress field.

chosen so that there should be essentially no question regarding this point. As discussed later, there are situations where satisfactory engineering decisions can be made even though the accuracy of the analysis is not as good as specified in the E-399—74 test method.

Figure 3.8 shows the elastic-stress-field distribution ahead of a crack as developed in Chapter 2. The extent of the plastic zone ahead of the crack front can be estimated by using the following expression for stress in the y direction, σ_y:

$$\sigma_y = \frac{K_I}{\sqrt{2\pi r}} \cos\frac{\theta}{2}\left(1 + \sin\frac{\theta}{2}\sin\frac{3\theta}{2}\right)$$

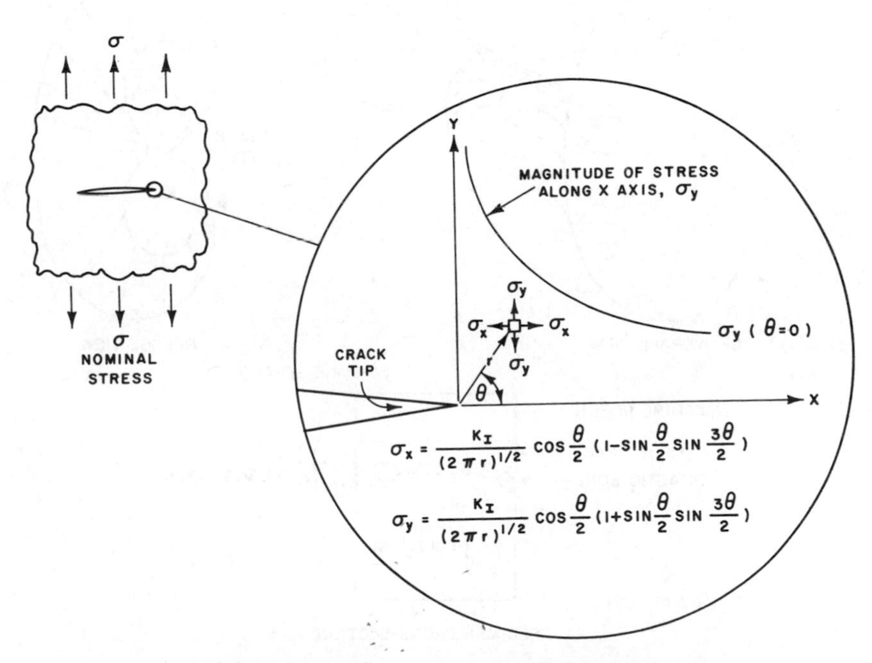

FIG. 3.8. Elastic-stress-field distribution ahead of crack.

for $\theta = 0$ (along the x axis)

$$\sigma_y = \frac{K_I}{\sqrt{2\pi r}}$$

Letting $\sigma_y = \sigma_{\text{yield stress}}$ (σ_{ys}), which is the 0.2% offset yield strength of the material at a particular temperature and loading rate, the extent of yielding ahead of the crack is

$$r_y = \frac{1}{2\pi}\left(\frac{K_I}{\sigma_{ys}}\right)^2$$

At instability, $K_I = K_c$, and the limiting value of r_y, or the plastic zone, is

$$r_y = \frac{1}{2\pi}\left(\frac{K_c}{\sigma_{ys}}\right)^2$$

This value of r_y is estimated to be the plastic-zone radius at instability under plane-stress conditions.

As shown in Figure 3.9, this value is assumed to occur at the surface of

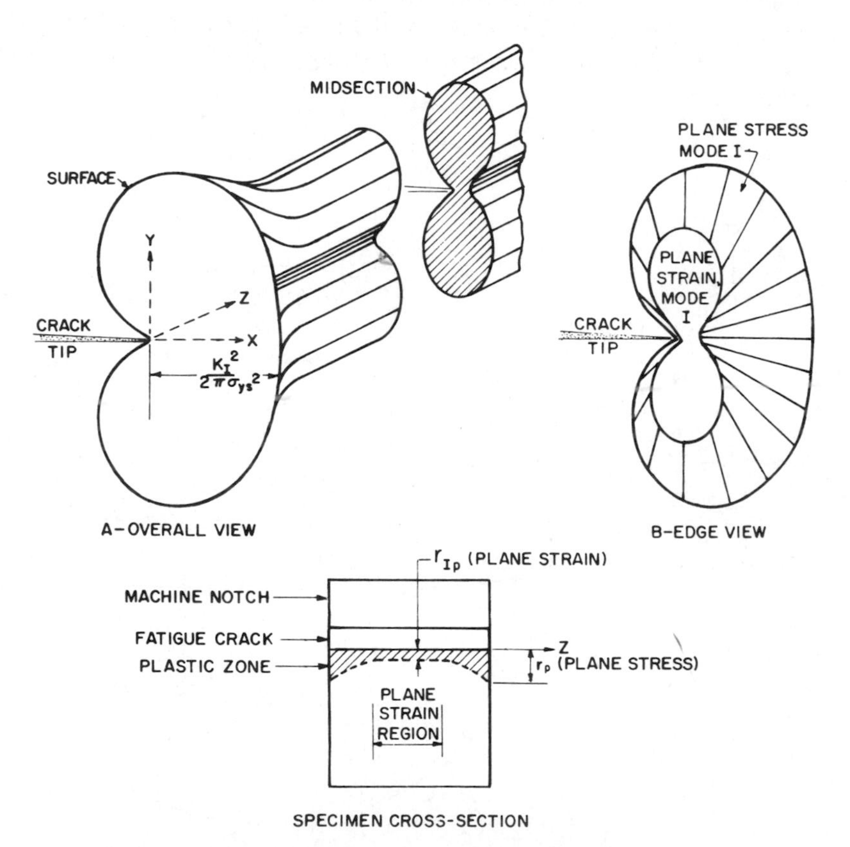

FIG. 3.9. Schematic of plastic zone size.

a plate where the lateral constraint is zero and plane-stress conditions exist. Because of the increase in the tensile stress for plastic yielding under plane strain conditions (Chapter 2), the plastic-zone radius at the center, where the constraint is greater and plane-strain conditions exist, is equal to one third of this value, or

$$r_{y\,(\text{plane strain})} \simeq \frac{1}{6\pi}\left(\frac{K_{Ic}}{\sigma_{ys}}\right)^2.$$

Thus the relative plastic-zone size ahead of a sharp crack is proportional to the $(K_{Ic}/\sigma_{ys})^2$ value of a particular structural material.

In establishing the specimen size requirements for K_{Ic} tests, the specimen dimensions should be large enough compared with the plastic zone, r_y, so that any effects of the plastic zone on the K_I analysis can be neglected. The pertinent dimensions of plate specimens for K_{Ic} testing are crack length (a), thickness (denoted as B in the ASTM standard), and the remaining uncracked ligament length ($W - a$, where W is the overall specimen depth). After considerable experimental work[4], the following minimum specimen size requirements to ensure elastic plane-strain behavior were established:

$$a \geq 2.5\left(\frac{K_{Ic}}{\sigma_{ys}}\right)^2$$

$$B \geq 2.5\left(\frac{K_{Ic}}{\sigma_{ys}}\right)^2$$

$$W \geq 5.0\left(\frac{K_{Ic}}{\sigma_{ys}}\right)^2$$

The following calculation shows that for specimens meeting this requirement, the specimen thickness is approximately 50 times the radius of the plane-strain plastic-zone size:

$$\frac{\text{specimen thickness}}{\text{plastic-zone size}} = \frac{B}{r_y} \simeq \frac{2.5(K_{Ic}/\sigma_{ys})^2}{(1/6\pi)(K_{Ic}/\sigma_{ys})^2} \simeq 2.5(6\pi) \simeq 47$$

Thus, the restriction that the plastic zone be "contained" within an elastic-stress field certainly appears to be satisfied. In fact, during the development of the recommended test method, there was considerable debate about whether or not this requirement was too conservative. As described later, there were test results available that indicated the requirement was too conservative. However, because it was felt that there should be no ambiguity regarding an ASTM standard, Brown and Srawley, in ASTM STP 463,[3] stated that

> "... it has to be accepted that it will frequently not be possible to determine K_{Ic} for a given material in any meaningful sense, usually because the material is not available in sufficient thickness, but sometimes for another reason. The concept of K_{Ic} is fundamentally incompatible with a test method that could be applied

to all available forms of materials, and neglect of this fact can only result in misrepresentation of the relative merits of different materials. The issue, therefore, is whether or not the ASTM Test Method could be made less restrictive with regard to specimen dimensions without significant risk that spurious results would be obtained in some cases. In our opinion there is no convincing evidence that the provisions of ASTM Method E 399-70 T are unduly restrictive; on the contrary, there is reason to suppose that they are not restrictive enough for some materials which are of technological importance. The test method should be applicable to any material that is available in sufficient bulk, and its provisions would be more than enough in other cases, but it is not feasible to discriminate between the more and the less favorable materials. It follows that the discovery of any number of favorable cases does not justify any relaxation of the provisions of the test method, though the discovery of a single unfavorable case may make it necessary to further restrict the test method. There is a further consideration which should not be neglected. Although the direct application of the ASTM Method E 399-70 T is restricted unavoidably to those materials that can be obtained in sufficient bulk, it does serve as a primary reference method for other fracture toughness tests that are based more empirically but less restricted in scope of application. This is an important reason why any margin of error in the provisions should be on the side of restrictiveness rather than uncertainty about the meaning of the results."

Accordingly the preceding specimen size requirements were adopted into ASTM Test Method E-399—74, Standard Method of Test for Plane Strain Fracture Toughness of Metallic Materials.

By adhering to the above test specimen dimensions (a, B, W), the following two essential conditions are satisfied:

1. The test specimen is large enough so that linear-elastic behavior of the material being tested occurs over a large enough stress field so that any effect of the plastic zone ahead of the crack can be neglected, and
2. There is a triaxial tensile stress field present such that the shear stress is very low compared to the maximum normal stress and a plane-strain opening mode behavior would be expected (Mode I, Chapter 2.)

Note that before a K_{Ic} test specimen can even be machined, the K_{Ic} value *to be obtained* must already be known or at least estimated. Three general means of sizing test specimens before the K_{Ic} value is even known are as follows:

1. Overestimate the K_{Ic} value on the basis of experience with similar materials and judgment based on other types of notch-toughness tests. In Chapter 6 we shall describe various empirical correlations with other types of notch-toughness tests that can be used, such as the CVN impact test specimen.

2. Use specimens that have as large a thickness as possible, namely a thickness equal to that of the plates to be used in service.
3. Use the following ratio of yield strength (σ_{ys}) to modulus of elasticity (E) to select a specimen size. However, these estimates are limited to very high-strength structural materials, i.e., steel having yield strengths of at least 150 ksi and aluminums having yield strengths of at least 50 ksi.

σ_{ys}/E	*Minimum Recommended Thickness and Crack Length* *in.*	*mm*
0.0050–0.0057	3.00	75.0
0.0057–0.0062	2.50	63.0
0.0062–0.0065	2.00	50.0
0.0065–0.0068	1.75	44.0
0.0068–0.0071	1.50	38.0
0.0071–0.0075	1.25	32.0
0.0075–0.0080	1.00	25.0
0.0080–0.0085	0.75	20.0
0.0085–0.0100	0.50	12.5
0.0100 or greater	0.25	6.5

Fortunately for the structural designer (because he would like his structure to be built from materials that do *not* exhibit elastic plane-strain behavior) but *unfortunately* for the materials engineer responsible for determining K_{Ic} values, many low- to medium-strength structural materials in the section sizes of interest for most large structures (such as ships, bridges, pressure vessels) are of insufficient thickness to maintain plane-strain conditions under slow loading and at normal service temperatures. In those cases, the linear-elastic analysis used to calculate K_{Ic} values is invalidated by general yielding and the formation of large plastic zones. Under these conditions, which occur in the transition range approaching plane stress (mixed mode) and general yielding, alternative methods must be used for fracture analysis, as described in Chapters 6 and 13–16. Nonetheless, the basis for all subsequent fracture criteria and fracture control rests on a knowledge (or estimate) of how much a material exceeds K_{Ic}-type behavior at the service temperature and loading rate.

3.4. K_{Ic} Test Procedure

The general test procedure to determine K_{Ic} as well as a commentary on the significance of various aspects of that test method are presented in this section. The steps in conducting a K_{Ic} test are as follows:

1. *Determine critical specimen size dimensions as described in Section 3.3.*

$$a = \text{crack depth} \geq 2.5\left(\frac{K_{Ic}}{\sigma_{ys}}\right)^2$$

$$B = \text{specimen thickness} \geq 2.5\left(\frac{K_{Ic}}{\sigma_{ys}}\right)^2$$

$$W = \text{specimen depth} \geq 5.0\left(\frac{K_{Ic}}{\sigma_{ys}}\right)^2$$

Methods to estimate the probable K_{Ic} value were discussed in Section 3.3.

2. *Select a test specimen and prepare shop drawings.* There are two standard specimen designs, namely the slow-bend test specimen and the compact-tension specimen (CTS). Working drawings of these two specimens are presented in terms of a, W, and B in Figures 3.10 and 3.11. The initial

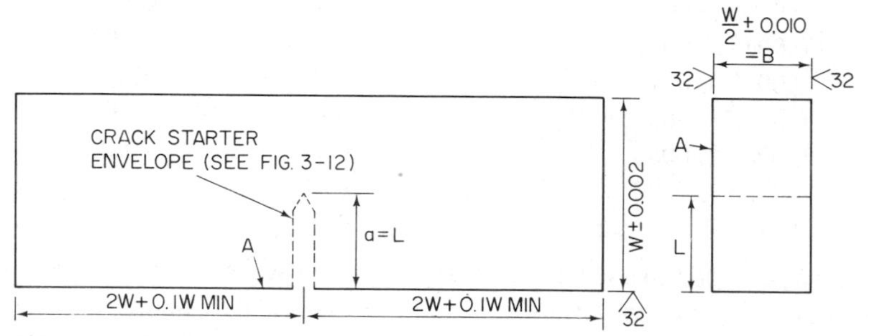

NOTE 1– DIMENSIONS AND TOLERANCES ARE IN INCHES UNLESS OTHERWISE INDICATED.

NOTE 2– "A" SURFACES ARE TO BE PERPENDICULAR TO CENTER LINE OF CRACK-STARTER ENVELOPE WITHIN ±0.005-in. TIR.

NOTE 3– INTEGRAL OR ATTACHABLE EDGES FOR CLIP GAGE ATTACHMENT MAY BE USED (SEE FIG. 3–16)

METRIC EQUIVALENTS

U.S. CUSTOMARY UNITS, in.	METRIC EQUIVALENTS, mm
0.002	0.05
0.005	0.13
0.010	0.25

S=L=4W

FIG. 3.10. Slow-bend specimen.

machined crack length, a, should be $0.45W$ so that the crack can be extended by fatigue to approximately $0.5W$. As stated previously, the K_{Ic} measurement capacity is limited primarily by the thickness of the material available since a and W can generally be made quite large, subject only to machining and testing capabilities. Thus selection of the specimen thickness, B, is usually made first.

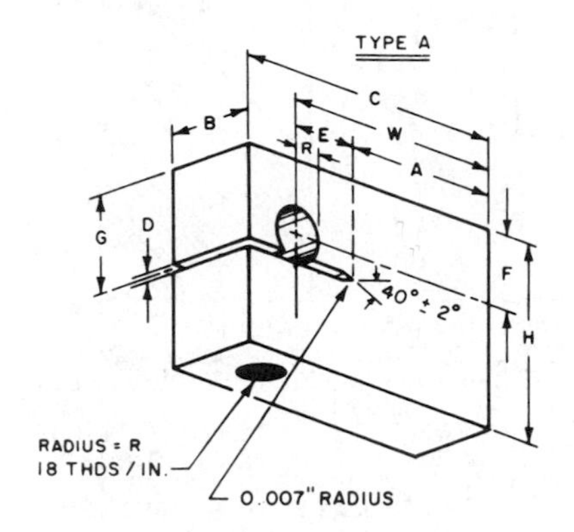

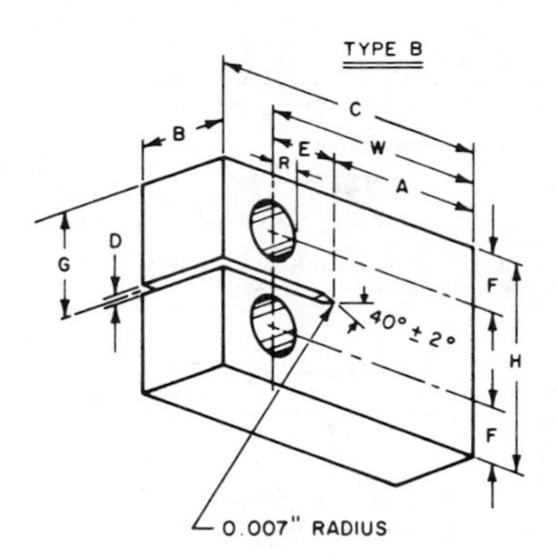

SPEC	B	W	C	A	E	H	G	D	R	F
IT-A	1.000	2.550	3.200	1.783	0.767	2.480	1.240	0.094	0.350	1.000
IT-B	1.000	2.550	3.200	1.783	0.767	2.480	1.240	0.094	0.250	0.650

1 inch = 25.4 mm
1 degree = 0.017 rad

FIG. 3.11. CTS specimen.

3. *Fatigue-crack the test specimen.* The purpose of notching the test specimen is to simulate an ideal plane crack with essentially zero root radius to agree with the assumption made in the K_I analysis. Because a fatigue crack is considered to be the sharpest crack that can be reproduced in the laboratory, the machine notch is extended by fatigue. The fatigue crack should extend at least $0.05W$ ahead of the machined notch to eliminate any effects of the geometry of the machined notch. Examples of notch geometry are presented in Figure 3.12. It has been found that a chevron notch has several advantages compared with a straight crack front. The chevron notch has been found to keep the crack in-plane and to ensure that it extends well beyond the notch root ($0.05W$). Thus the machining operation for the original chevron notch is not so critical as for a straight machined crack front. If a straight machined crack front is used, it is difficult to produce a fatigue

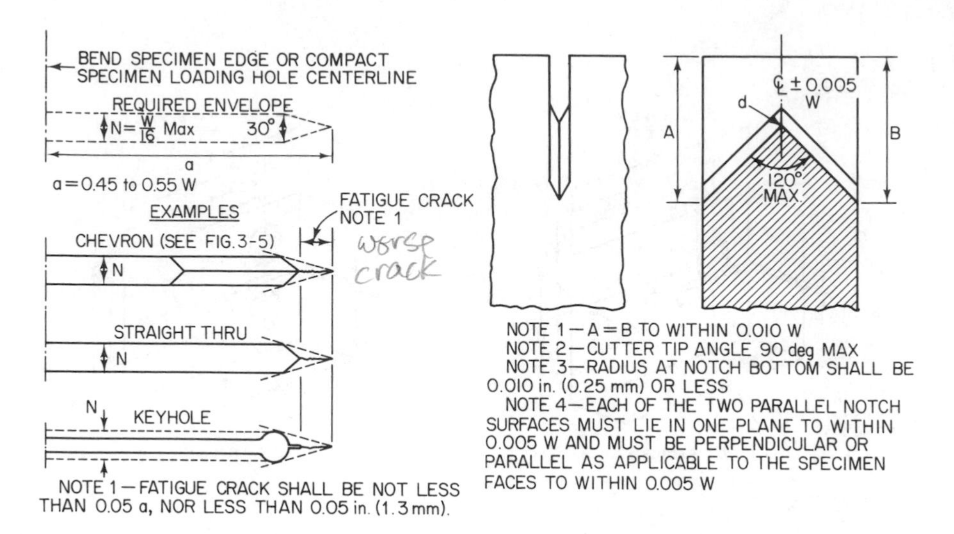

FIG. 3.12. Chevron notch for CTS & SB test specimens.

crack that is uniform unless an extremely sharp uniform machined notch is used, such as might be obtained by electrodischarge machining (EDM).

To ensure that the plastic-zone size during the final fatigue cycle is less than the plastic-zone size during actual K_{Ic} testing, the last 2.5% of the overall length of notch plus fatigue crack is loaded at a maximum stress-intensity level during fatigue $K_{f(max)}$ such that $K_{f(max)}/E \leq 0.002^{1/2}$. $K_{f(max)}$ is the maximum K_I level to which the specimen is subjected during cyclic loading and is calculated using the general expression for K_Q given later. K_Q is a conditional calculation of K_{Ic} based on the actual fracture test results, but the expression for K_Q is a general one for K_I. Figure 3.13 shows the general relation among $K_{f(max)}$, ΔK_f, and K_Q for corresponding values of load.

$K_{f(max)}$ must not exceed 60% of the K_Q value determined from the actual fracture test results. Fatigue cracking should be considered as a special type of critical machining operation because cracks produced at high stress-intensity levels (high values of $K_{f(max)}$) can significantly affect the subsequent fracture test results. Control of the plastic-zone size during fatigue cracking is particularly important when the fatigue cracking is done at room temperature and the actual K_{Ic} test is conducted at lower temperatures. In this case, $K_{f(max)}$ at room temperature must be kept to very low values so that the plastic-zone size corresponding to K_Q at low temperatures is smaller than the plastic-zone size corresponding to $K_{f(max)}$ at room temperature.

Results of a limited study of the effects of stress level during fatigue cracking of air-melted (AM) 18Ni-8Co-3Mo maraging steel are summarized in Table 3.1 and Figure 3.14. The bend test results presented in Figure 3.14 showed that as long as the nominal stress for fatigue cracking was less than

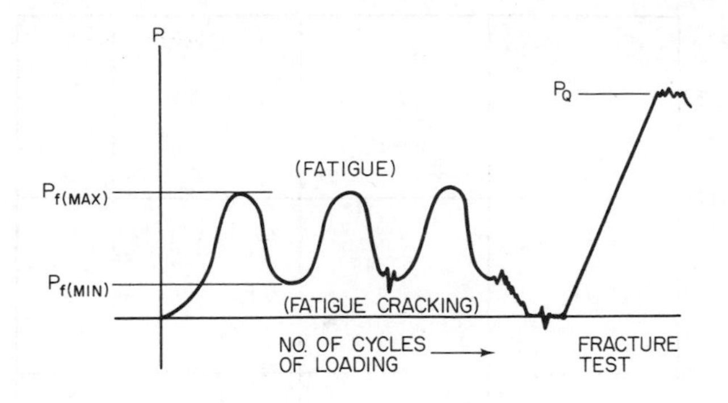

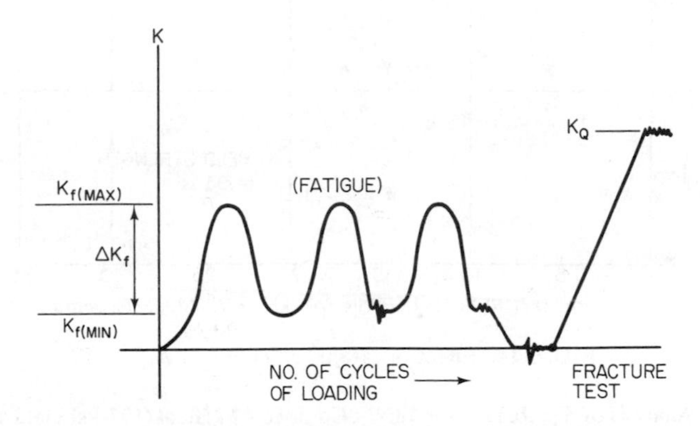

FIG. 3.13. Relation between P_{max}, K_{max}, ΔK fatigue, and K_Q.

about 20% of the yield strength the K_{Ic} values averaged 105 ksi$\sqrt{\text{in.}}$ Stressing at about 30% of the yield strength increased the K_{Ic} value to 120 ksi$\sqrt{\text{in.}}$

In terms of $K_{f(max)}/E$, stressing at 25% of the yield strength corresponded to a $K_{f(max)}/E$ value of about 0.0012, which is less than the specification limit of 0.002. However, values of $K_{f(max)}$ and $K_{f(max)}/K_{Ic}$ presented in Table 3.1 show that even the restriction that $K_{f(max)} \leq 0.6K_Q$ may not be sufficient to ensure that fatigue cracking has no effect on the subsequent K_{Ic} test results.

This behavior is similar to that observed by Brown and Srawley[3] and confirms their statement that fatigue cracking of K_{Ic} specimens should be conducted at the lowest practical stress level.

4. *Obtain test fixtures and displacement gage.* Recommended bend test fixtures for slow-bend testing and tension testing fixtures for compact-tension specimen testing are described in reference 3.1. These fixtures were developed to minimize friction and have been used successfully by numerous laboratories. Other fixtures can be used as long as good alignment is maintained and frictional errors are minimized.

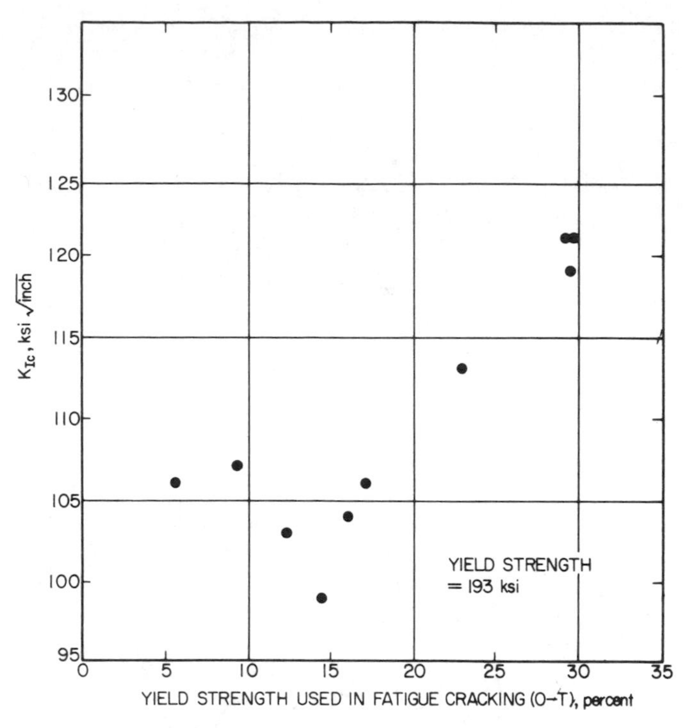

FIG. 3.14. Effect of fatigue cracking on K_{Ic}.

TABLE 3.1. Analysis of K_{Ic} Results for 18Ni-8Co-3Mo AM Steel (193-ksi yield strength)

Specimen Number	*Nominal Stress for Fatigue Cracking (psi)*	*Number of Fatigue Cycles*	K_Q *ksi*$\sqrt{in.}$	*Nominal Stress (ksi)*	$K_{f(max)}$	$\frac{K_{f(max)}}{K_{Ic}}$
18N-2	57,000	8,000	119	118	58	0.55
18NF-8a	57,000	6,400	121	115	60	0.57
18NF-11	57,000	6,700	121	116	60	0.57
18NF-1	33,000	31,900	106	106	33	0.31
18NF-7	31,000	33,600	104	124	36	0.34
18NF-5	24,000	67,400	103	102	24	0.23
18NF-3	28,000	70,800	99	101	27	0.26
18NF-10	11,000	156,000	106	106	11	0.10
18N-3	18,000	123,000	107	108	18	0.17
18NF-12	45,000	15,300	113	90	57	0.54

A key item in the K_{Ic} test is the accurate measurement of some quantity that can be related to the beginning of crack extension from the fatigue crack, i.e., unstable crack growth. The basic measurement selected is the relative displacement of two points located symmetrically on opposite sides of the crack plane, as shown in Figure 3.15. An extremely sensitive and highly linear displacement gage that has no lost motion between the gage and the locating portions on the specimen has been developed and is shown in Figure 3.16.

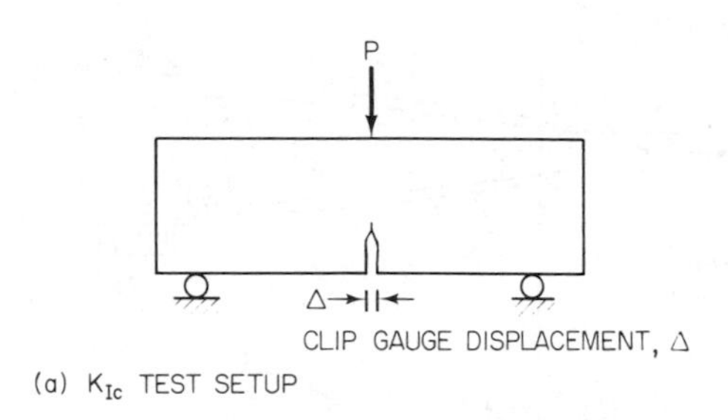

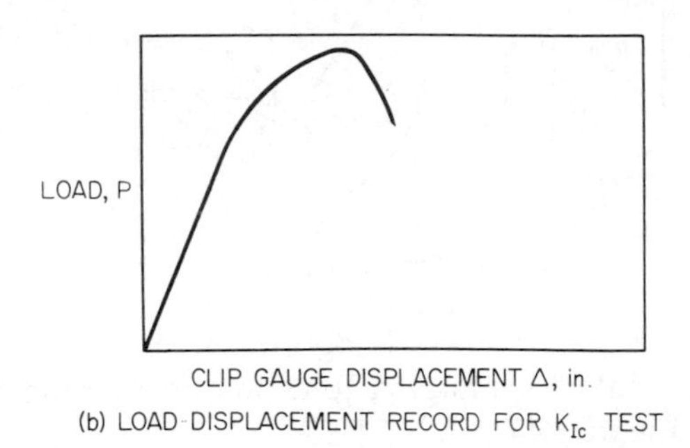

FIG. 3.15. Schematic showing displacement measurement for K_{Ic} test.

5. *Test Procedure*

a. *Bend testing.* The bend test fixture should be set up so that the line of action of the applied load passes midway between the support roll centers for three-point loading and symmetric with respect to the crack plane for four-point loading. The specimen should be located with the notch center line midway between the rolls to within 0.5% of the span and square to the roll axes within 2°. The displacement gage should be seated on the knife-edges to maintain registry between knife-edges and gage grooves. In the case of attachable knife-edges, the gage should be seated before the knife-edge positioning screws are tightened. The specimen should be loaded at a rate such that the rate of increase of stress intensity is within the range 30–150 ksi$\sqrt{\text{in.}}$/min (0.55–2.75 MPa$\sqrt{\text{m}}$/s), corresponding to a loading rate for the 1-in.-thick specimen of between 4,000 and 20,000 lb/min (0.03–0.15

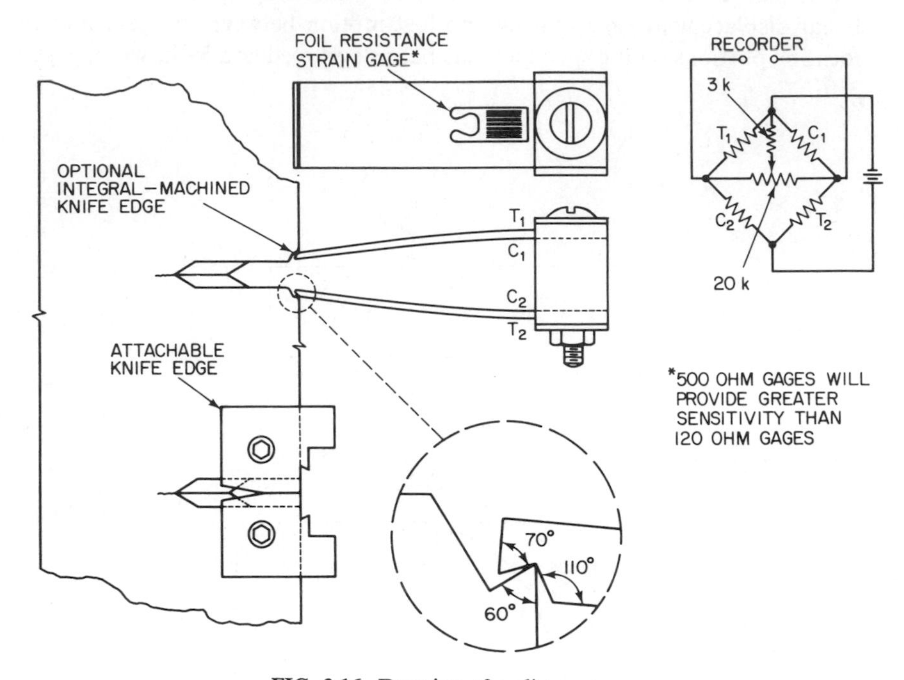

FIG. 3.16. Drawing of a clip gage.

kN/sec). This loading rate is approximately 10^{-5} in./in./sec and is defined as "static" loading. A four-point slow-bend K_{Ic} test setup is shown in Figure 3.17.

b. *Compact-tension testing.* Friction and eccentricity of loading introduced by the clevis itself can be minimized by adherence to the specified tolerances for the specimen clevis and pins given in reference 1. Eccentricity of loading can also result from misalignment external to the clevis or from incorrect positioning of the specimen with respect to the center of the clevis opening. Satisfactory alignment can be obtained by keeping the center line of the upper and lower loading rods coincident within 0.03 in. (0.76 mm) during the test and by centering the specimen with respect to the clevis opening within 0.03 in. (0.76 mm). The displacement gage should be seated in the knife-edges to maintain registry between the knife-edges and the gage groove. In the case of attachable knife-edges, the gage should be seated before the knife-edge positioning screws are tightened. The specimens should be loaded at a rate such that the rate of increase of stress intensity is within the range 30–150 ksi$\sqrt{\text{in.}}$/min, corresponding to a loading rate for the 1-in.-thick specimen between 4,500 and

FIG. 3.17. Four-point slow-bend K_{Ic} test setup.

22,500 lb/min. A photograph showing a typical compact-tension test setup is presented in Figure 3.18.

c. *Test record.* A test record consisting of an autographic plot of the output of the load-sensing transducer versus the output of the displacement gage should be obtained. The initial slope of the linear portion should be between 0.7 and 1.5. It is conventional to plot the load along the vertical axis, as in an ordinary tension test record. Select a combination of load-sensing transducer and autographic recorder so that the maximum load can be determined from the test record with an accuracy of $\pm 1\%$. With any given equipment, the accuracy of readout will be greater the larger the scale of the test record. Continue the test until the specimen can sustain no further increase in load. A schematic test record is presented in Figure 3.19.

d. *Measurements.* Measurements of the specimen dimensions and fracture surfaces and features should be made as required to calculate K_Q (B, S, W, a; Figures 3.10 and 3.11). The crack length, a, should be measured to the nearest 0.5% at the center of the crack front and midway between the center and the end of the crack front on each side (Figure 3.20). Use the average of these three measurements as the crack length in subsequent calculations provided that the length of either surface trace is within 10% of the average crack length.

FIG. 3.18. Compact tension specimen (CTS) K_{Ic} test setup.

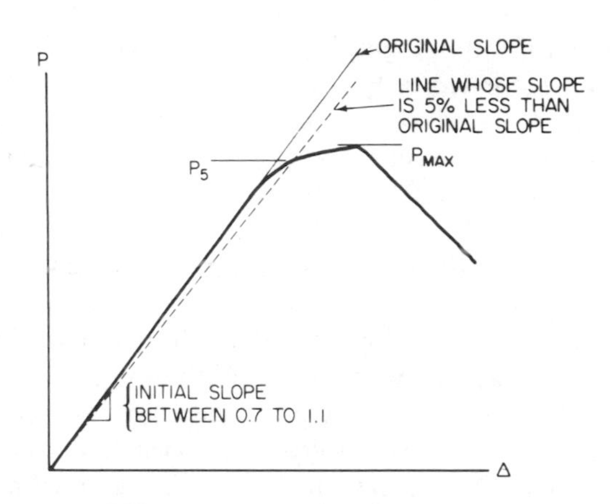

FIG. 3.19. Schematic P-Δ test record.

6. *Analysis of P-Δ records.* If a material exhibited perfectly elastic behavior until fracture, the load-displacement curve would be merely a straight line until fracture. However, even very brittle structural materials exhibit some nonlinear behavior and thus the idealized perfectly elastic load-displacement curve is rarely seen. The principal types of load-displacement curves observed are presented in Figure 3.21, which shows that considerable variation in behavior occurs for different structural materials. This

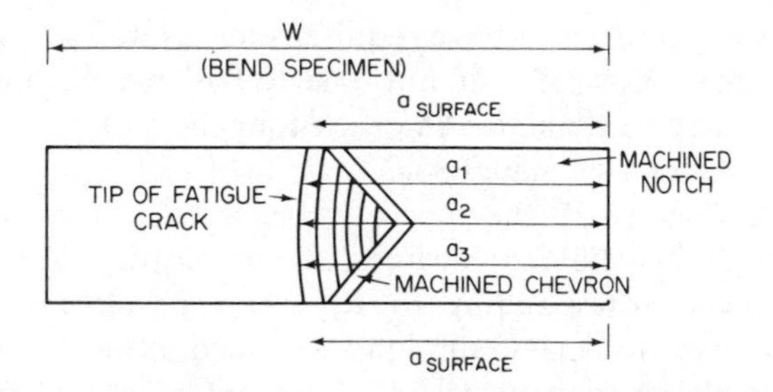

FIG. 3.20. Various crack length measurements to be made.

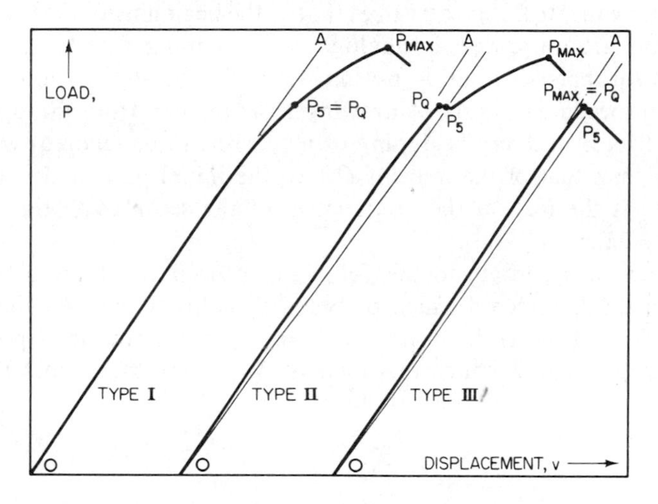

FIG. 3.21. Types of load displacement curves illustrating procedure for determination of K_{Ic}.

is not unexpected, however, because of the wide variety of materials used in different structural applications.

To establish that a valid K_{Ic} has been determined (that is, one that has satisfied the various restrictions to ensure that plane-strain behavior was obtained), it is necessary first to calculate a conditional result, K_Q, which involves a construction on the test record. Then it must be determined whether or not this K_Q value is consistent with the size and yield strength of the specimen according to the general requirements that

$$a \geq 2.5\left(\frac{K_Q}{\sigma_{ys}}\right)^2$$

$$B \geq 2.5\left(\frac{K_Q}{\sigma_{ys}}\right)^2$$

$$W \geq 5.0\left(\frac{K_Q}{\sigma_{ys}}\right)^2$$

If this K_Q value meets the above requirements, as well as other subsequent requirements, then $K_Q = K_{Ic}$. If not, the test is invalid, and whereas the results may be used to *estimate* the crack toughness of a material, they are not valid ASTM standard values.

As noted in Figure 3.21, the general types of P-Δ curves are not perfectly elastic but do exhibit different degrees of nonlinearity. Various criteria to establish the load corresponding to K_{Ic} were considered, such as initial deviation from linearity, maximum load, specified offset, and a secant offset. After considerable experimentation, a 5% secant offset was chosen to define K_{Ic} as the critical stress-intensity factor at which the crack reaches an effective length equal to 2% greater than that at the beginning of the test. Although somewhat arbitrary, this is analogous to defining the 0.2% offset yield strength for materials that do not have a well-defined yield point.

The procedure consists of drawing a secant line from the origin (slight nonlinearity at the very beginning of a record can be ignored) with a slope 5% less than that of the tangent OA to the initial part of the record. The load, P_5, is the load at the intersection of the secant with the test record (Figure 3.22).

Define P_Q according to the following procedure. If the load at every point on the P-Δ record which precedes P_5 is lower than P_5, then P_Q is P_5 (Figure 3.21, Type I). If, however, there is a maximum load preceding P_5 that is larger than P_5, then this load is P_Q (Figure 3.21, types II and III).

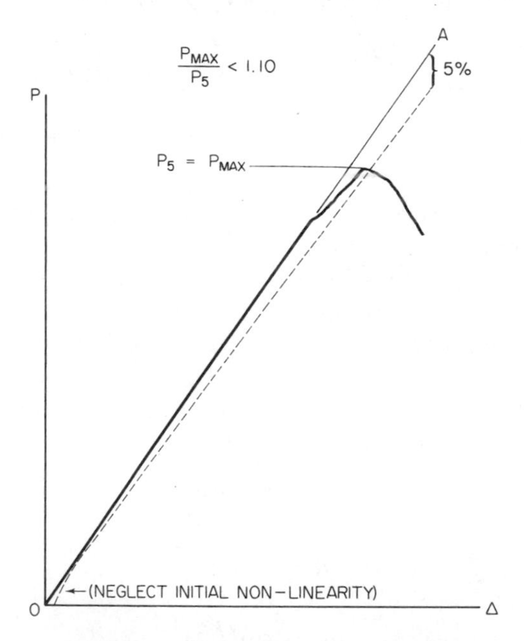

FIG. 3.22. Construction of 5% secant offset.

Thus $P_Q = P_5$ in the example shown in Figure 3.22. If P_{max}/P_Q is greater than 1.10, the test is not a valid test because it is possible that K_Q is not representative of K_{Ic}. For relatively tough structural materials, P_{max} is usually $> 1.10P_5$ as shown in typical P-Δ records for steels that are too tough to exhibit plane-strain behavior (Figure 3.23). For these tough materials, a

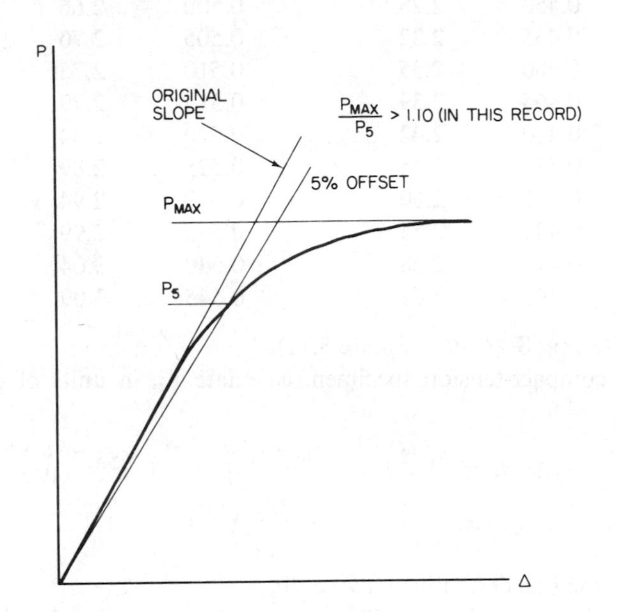

FIG. 3.23. P-Δ **test record for structural material with high level of notch toughness.**

"roundhouse" P-Δ curve is obtained, indicating general-yielding-type behavior rather than plane-strain behavior.

7. *Calculation of conditional* K_{Ic} (K_Q). After determining P_Q for either the bend specimen or the compact-tension specimen, calculate K_Q using the following expression or table:

Bend Specimen (Figure 3.10)

$$K_Q = \frac{P_Q S}{BW^{3/2}}\left[2.9\left(\frac{a}{W}\right)^{1/2} - 4.6\left(\frac{a}{W}\right)^{3/2} + 21.8\left(\frac{a}{W}\right)^{5/2} - 37.6\left(\frac{a}{W}\right)^{7/2} + 38.7\left(\frac{a}{W}\right)^{9/2}\right]$$

where P_Q = load as determined above, lb,
B = thickness of specimen, in.,
S = span length, in.,
W = depth of specimen, in.,
a = crack length as determined above, in.

To facilitate calculation of K_Q, values of the power series given in brackets in the above expression are tabulated in the following table for specific values of a/W for bend specimens:

a/W	$f(a/W)$	a/W	$f(a/W)$
0.450	2.28	0.500	2.66
0.455	2.32	0.505	2.70
0.460	2.35	0.510	2.75
0.465	2.39	0.515	2.79
0.470	2.42	0.520	2.84
0.475	2.45	0.525	2.89
0.480	2.50	0.530	2.94
0.485	2.54	0.535	2.99
0.490	2.58	0.540	3.04
0.495	2.62	0.545	3.09

COMPACT-TENSION SPECIMEN (Figure 3.11)

For the compact-tension specimen, calculate K_Q in units of psi$\sqrt{\text{in.}}$ as follows:

$$K_Q = \frac{P_Q}{BW^{1/2}}\left[29.6\left(\frac{a}{W}\right)^{1/2} - 185.5\left(\frac{a}{W}\right)^{3/2} + 655.7\left(\frac{a}{W}\right)^{5/2} - 1017.0\left(\frac{a}{W}\right)^{7/2} + 638.9\left(\frac{a}{W}\right)^{9/2}\right]$$

where P_Q = load as determined above, lb,
B = thickness of specimen, in.,
W = width of specimen, in.,
a = crack length as determined above, in.

To facilitate calculation of K_Q, values of the power series given in brackets in the above expression are tabulated below for specific values of a/W for compact-tension specimens:

a/W	$f(a/W)$	a/W	$f(a/W)$
0.450	8.34	0.500	9.60
0.455	8.45	0.505	9.75
0.460	8.57	0.510	9.90
0.465	8.69	0.515	10.05
0.470	8.81	0.520	10.21
0.475	8.93	0.525	10.37
0.480	9.06	0.530	10.54
0.485	9.19	0.535	10.71
0.490	9.33	0.540	10.89
0.495	9.46	0.545	11.07
		0.550	11.26

8. *Final Check for* K_{Ic}. Calculate $2.5(K_Q/\sigma_{ys})^2$ where $\sigma_{ys} = 0.2\%$ offset yield strength in tension. If this quantity is less than both the thickness and the crack length of the specimen, then K_Q is equal to K_{Ic}. Otherwise it is necessary to use a larger specimen to determine K_{Ic} in order to satisfy this requirement. The dimensions of the larger specimen can be estimated on the basis of K_Q. If K_Q is invalid, the strength ratio, R, is a useful comparative measure of the toughness of materials when the test specimens tested are of the same size and that size is insufficient to produce a valid K_{Ic}.

For the bend specimen,

$$R = \frac{\text{bending moment at failure}}{\text{bending moment at yielding}}$$

$$R = \frac{(P/2)(S/2)}{\sigma_{ys}[(W-a)^2/6]B} = \frac{6PW}{\sigma_{ys}(W-a)^2B}$$

For the CTS specimen,

$$R = \frac{2P(2W+a)}{B(W-a)^2\sigma_{ys}}$$

3.5. Typical K_{Ic} Test Results

Examples of several types of P-Δ records for materials that exhibit behaviors ranging from almost perfectly elastic to general yielding are presented in Figures 3.24 through 3.27. Comparison of the various materials, specimen sizes, and test results are presented in Table 3.2. Comments on each of these records are as follows:

3.5.1. *Very High-Strength Aluminum* (Figure 3.24)

The P-Δ record for an extremely high-strength aluminum alloy with a yield strength of 71 ksi, Table 3.2, is presented in Figure 3.24.

The test record is the general Type III (Figure 3.21) with essentially linear-elastic behavior to P_{max}, which occurs before P_5. P_5 and P_{max} are essentially identical, and K_{Ic} is 19.8 ksi$\sqrt{\text{in.}}$ The required B is only 0.2 in. [$B = 2.5(K_{Ic}/\sigma_{ys})^2$], and thus the specimen thickness used (1.37 in.) is obviously much greater than necessary. Materials with this low a $(K_{Ic}/\sigma_{ys})^2$ ratio, 0.08, are not widely used in structural applications.

3.5.2. *18Ni Air Melt Maraging Steel* (Figure 3.25)

As shown in Figure 3.25, the P-Δ record for this material is also representative of Type III types (Figure 3.21). The interpretation of the record is straightforward and $K_Q = K_{Ic} = 113$ ksi$\sqrt{\text{in.}}$ The $(K_{Ic}/\sigma_{ys})^2$ ratio of $(\frac{113}{190})^2 = 0.35$ is more representative of high-strength low-toughness structural materials than the value of 0.08 for the structural material presented in Figure 3.24.

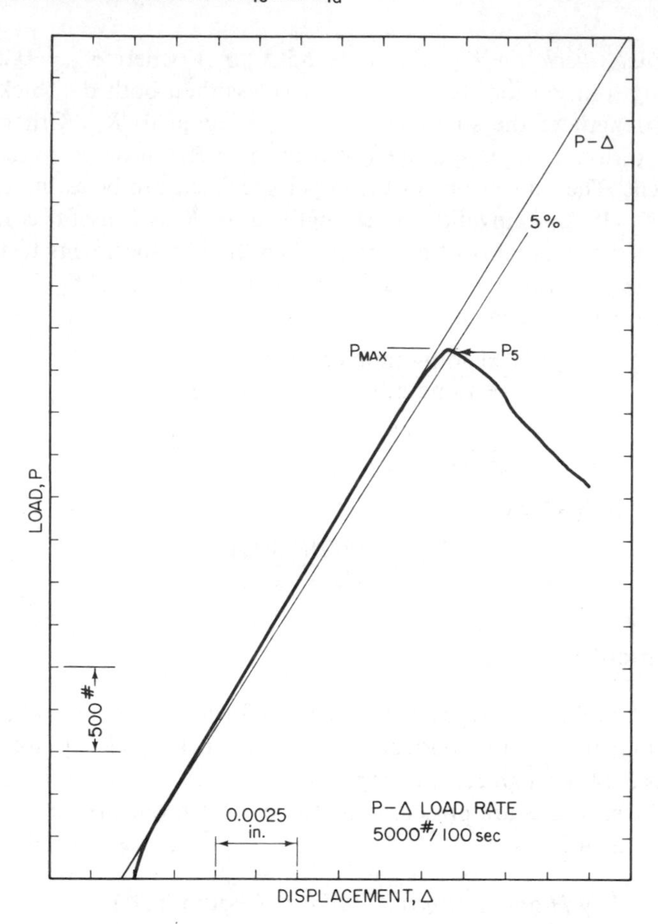

FIG. 3.24. Actual P-Δ test record for a very high strength aluminum.

3.5.3. 12Ni Vacuum Melt Maraging Steel (Figure 3.26)

This test record is representative of Type I materials with relatively high levels of toughness. These types of materials exhibit various amounts of non-linear behavior between P_5 and P_{max}, and in this case P_{max}/P_5 is quite large, equal to 1.45, clearly violating the requirement that $P_{max}/P_5 \leq 1.10$.

It should be noted, however, that from a structural design viewpoint, such behavior is very desirable because considerable yielding occurs before fracture. The fact that the K_{Ic} test procedure is limited to elastic behavior has hampered the widespread application of fracture mechanics, particularly for specification purposes.

Using either a K_{max} calculated from P_{max} or a 1.5 amplification factor applied to K_Q, the probable minimum specimen thickness for plane-strain behavior in this material would be 2–3 in. (Table 3.2). If material of this

TABLE 3.2. Typical Load-Displacement Results for Various Structural Materials

Material and Figure No. for P-Δ Record	σ_{ys} (*ksi*)	σ_T (*ksi*)	K_{Ic} *Test Specimen Dimensions (in.)*			P_5 (*lb*)	P_{max} (*lb*)	P_Q (*lb*)	K_Q ($ksi\sqrt{in.}$)	K_{Ic} ($ksi\sqrt{in.}$)	*Minimum Req'd* $B = 2.5\left(\frac{K_{Ic}}{\sigma_{ys}}\right)^2$ (*in.*)
			B	a	W						
7001-T75 very high-strength aluminum (Figure 3.24)	70.6	80.5	1.37	1.08	2.00	3,140	3,150	3,150	19.8	19.8 valid	0.20
18Ni maraging steel (Figure 3.25)	190.0	196.0	1.24	0.95	3.50	22,950	22,950	22,950	113.0	113.0 valid	0.88
12Ni maraging steel (Figure 3.26)	183.0	191.0	1.00	0.46	3.00	55,000	80,150	55,000	143.0	Invalid	$B_{est} = 1.5(2.5)\left(\frac{K_Q}{\sigma_{ys}}\right)^2 = 2.3$ $B_{est} = 2.5\left(\frac{K_{max}}{\sigma_{ys}}\right)^2 = 3.2$
A517 steel (Figure 3.27)	110.0	121.0	2.00	2.60	6.00	47,800	66,000	47,800	150.0	Invalid	$B_{est} = 1.5(2.5)\left(\frac{K_Q}{\sigma_{ys}}\right)^2 = 7.0$ $B_{est} = 2.5\left(\frac{K_{max}}{\sigma_{ys}}\right)^2 = 8.8$

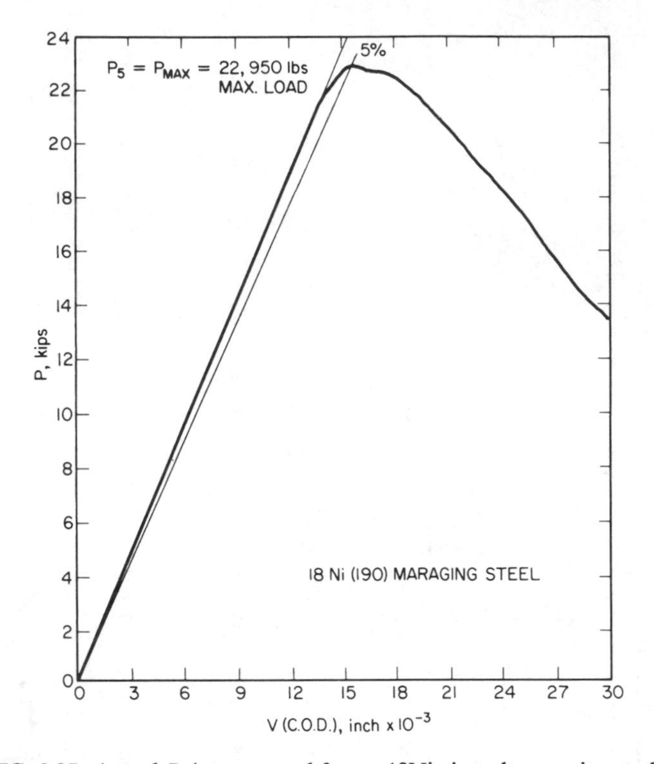

FIG. 3.25. Actual *P*-Δ test record for an 18Ni air melt maraging steel.

thickness were available, additional K_{Ic} tests could be conducted. If not, alternative methods of fracture analysis using either correlations (Chapter 6) or other elastic-plastic fracture mechanics approaches (Chapter 16) would need to be used.

3.5.4. *A517 Structural Steel* (Figure 3.27)

The test record for a 100-ksi yield strength structural steel is shown in Figure 3.27 and indicates another example of inelastic behavior where P_5 is considerably less than P_{max} ($P_{max}/P_5 = 66/47.8 = 1.38$, which is well above the minimum allowable of 1.1). As sometimes happens while conducting K_{Ic} tests, an apparent pop-in occurred (this behavior was verified on a duplicate specimen by interrupting the test and heat-tinting to establish the crack growth by pop-in) at a load of about 56 kips, giving an "apparent" K_{Ic} of 177 ksi$\sqrt{\text{in.}}$ K_Q was 150 ksi$\sqrt{\text{in.}}$, and thus it could be estimated that a lower-bound value of K_{Ic} was in the range 150–177 ksi$\sqrt{\text{in.}}$ It would appear that a specimen thickness of 7–8 in. is necessary to provide sufficient constraint for plane-strain behavior in this steel. Although extremely useful for general engineering estimates of the fracture toughness of this material, such calculations are not standardized.

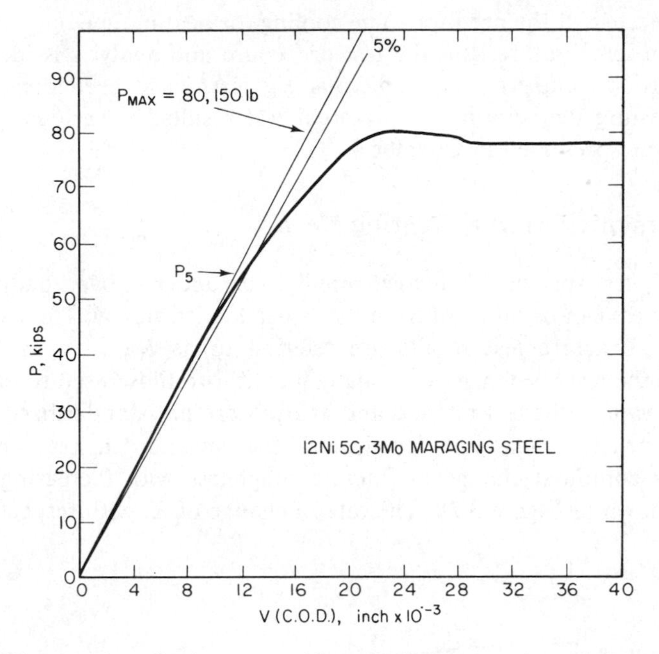

FIG. 3.26. Actual *P*-Δ test record for a 12Ni vacuum melt maraging steel.

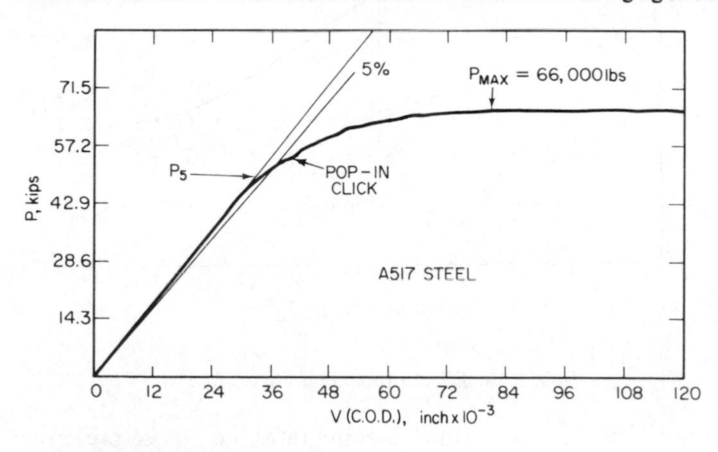

FIG. 3.27. Actual *P*-Δ test record for an A517 structural steel.

3.6. Low-Temperature K_{Ic} Testing

The fracture toughness of many structural materials is a function of temperature as well as loading rate. The effect of temperature on K_{Ic} can be determined by conducting either the slow-bend or compact-tension K_{Ic} test specimens described in the preceding sections at any test temperature following the standard method of testing. Obviously, it is more difficult to conduct the test because of the need for some cooling (or heating) medium compared with room temperature. But the test procedure and analysis is identical to that for room temperature. Numerous K_{Ic} tests have been conducted at various testing temperatures, and actual test results on various structural materials are presented in Chapter 4.

3.7. Dynamic Fracture Testing for K_{Id}

If a K_{Ic} test specimen is loaded rapidly, i.e., under impact loading rates, the toughness can be different from that when loaded slowly. These dynamic, or impact, fracture test results are referred to as K_{Id} (dynamic) values. Although the test specimens are usually identical to those used to obtain the static K_{Ic} values, the test method and analysis are not standardized. In fact, for some structural materials, particularly the low-strength structural steels, there is a continual change in fracture toughness with increasing loading rate, as shown in Figure 3.28. The rate of change of K_I with respect to time,

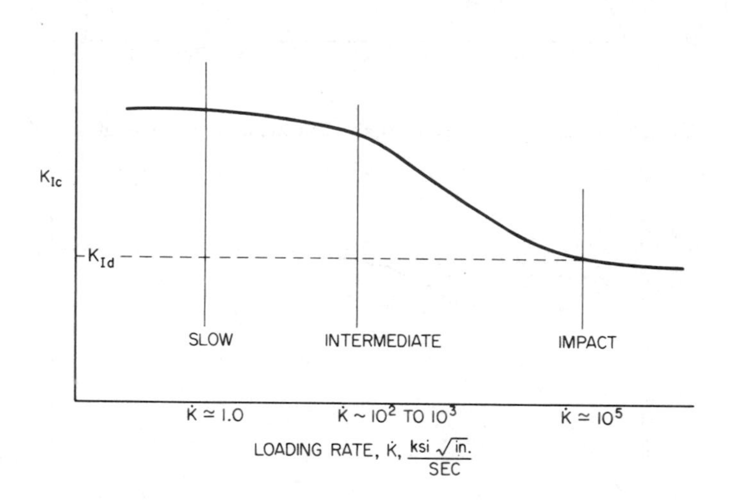

FIG. 3.28. Effect of loading rate on K_{Ic}.

$\dot{K}$, is given in ksi$\sqrt{\text{in.}}$/sec. "Slow" loading rates, i.e., those prescribed in the standard method of K_{Ic} testing, are around 1 ksi$\sqrt{\text{in.}}$/sec, whereas those loading rates generally obtained in K_{Id} testing are around 10^5 ksi$\sqrt{\text{in.}}$/sec.

Because of the difficulty in conducting controlled loading-rate tests, and because of the convenience of using a falling weight as an impact (or dynamic) load, most K_{Id} tests are conducted using a fatigue-cracked slow-bend test specimen loaded in a drop-weight machine. The tests are usually conducted over a range of temperatures and the results compared with a similar series of results obtained from specimens tested over a range of temperatures but at the "slow" standard ASTM loading rate. These dynamic K_{Id} values are considered to be close to the minimum values that can be obtained for materials that are rate-sensitive and thus are useful design values for those structures where either dynamic loading or propagating cracks are present.

Another type of dynamic fracture test, proposed by Crosley and Ripling[5] yields $K_{Ia\,(arrest)}$ values that are considered to be lower-bound values of K_{Id}. K_{Ia} is defined as the stress intensity factor a short time after a run-arrest segment of rapid crack extension. However, this procedure is not well defined, although ASTM has established a committee to develop a testing procedure.

The test specimen most widely used for determining K_{Id} values is a fatigue-cracked three-point bend specimen essentially identical to the K_{Ic} slow-bend specimen (Figure 3.29). The specimen is loaded in three-point bending by a striking tup mounted on a free-falling weight. Either the specimen or the tup (or both) is instrumented to measure applied load as a function of time.

The following dynamic test procedure was developed by Shoemaker and Rolfe[6] at the U.S. Steel Research Laboratory. Another similar procedure was developed by Madison and Irwin[7] at Lehigh. Both methods would be expected to give essentially the same results.

Strain gages are applied to the bend specimen as shown in Figure 3.29. Gage 1 is used as a crack detector to determine the time of crack initiation. The nominal strain at gage 2 is recorded as the crack initiates and is used to

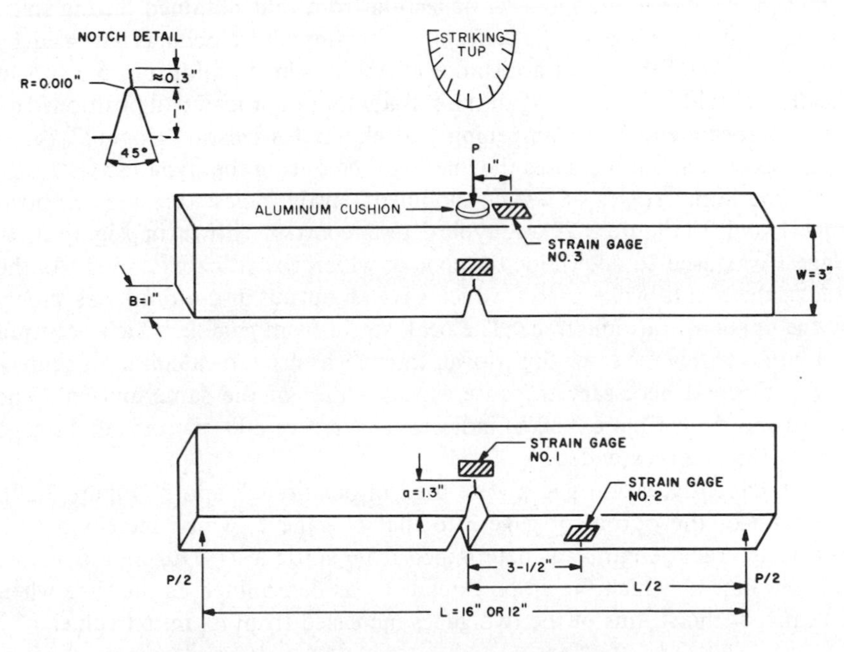

FIG. 3.29. Fatigue-cracked K_{Id} (dynamic K_{Ic}) bend test specimen.

calculate the corresponding nominal elastic stress at the point. Using the elementary strength of materials formula, $\sigma = My/I$, the moment and corresponding equivalent static load necessary to give this stress are calculated. This equivalent static load, which occurs at the time the crack initiates, is used in the following equation to determine the K_{Id} value for the test conditions studied:

$$K_{Id} = f\left(\frac{a}{W}\right)\frac{6Ma^{1/2}}{BW^2}$$

where M = applied bending moment,
a = crack length,
B = specimen thickness,
W = specimen width,
S = span (see Figure 3.29),

$$f(a/W) = A_0 + A_1(a/W) + A_2(a/W)^2 + A_3(a/W)^3 + A_4(a/W)^4.$$

	A_0	A_1	A_2	A_3	A_4
$S/W = 8$	+1.96	−2.75	+13.66	−23.98	+25.22
$S/W = 4$	+1.93	−3.07	+14.53	−25.11	+25.80

To calculate dynamic K_{Ic} values, it is assumed that during the dynamic tests the strain distribution is the same as that obtained from an equivalent static load. This assumption implies that any inertial effects present do not significantly change the mode of deflection from that obtained during static loading. To minimize inertial effects, a low-impact velocity is attained by dropping a 1600-lb weight a distance of about 9 in. In addition, the tup of the drop weight is impacted against a soft aluminum or lead pad positioned on the test specimen. This dampening pad eliminates elastic "ringing" waves in the specimen and increases the loading time during the dynamic test.

In the initial stages of test development, strain gages were used in positions 1 and 2 (Figure 3.29). A typical test record is shown in Figure 3.30. Gage 1 was used to determine the time at which the crack extended. As the crack extended it broke gage 1, which gave an output discontinuity as shown by the upper strain-time trace. The peak strain from gage 2, which occurred at approximately the same time (lower trace), was used to calculate an equivalent static load necessary to cause a peak strain of the same amount. The record shown in Figure 3.30(b) indicates a strain of 200 microin./in. at gage 2 when the crack extended.

Subsequently, strain gages were used in positions 2 and 3 (Figure 3.29). The ratio of the output of gage 3 to that of gage 2, which increased with increasing crack length, was determined from static tests. During a dynamic test, the time at which the crack extended was determined as the time when the ratio of the strains of the two gages increased from its initial value.

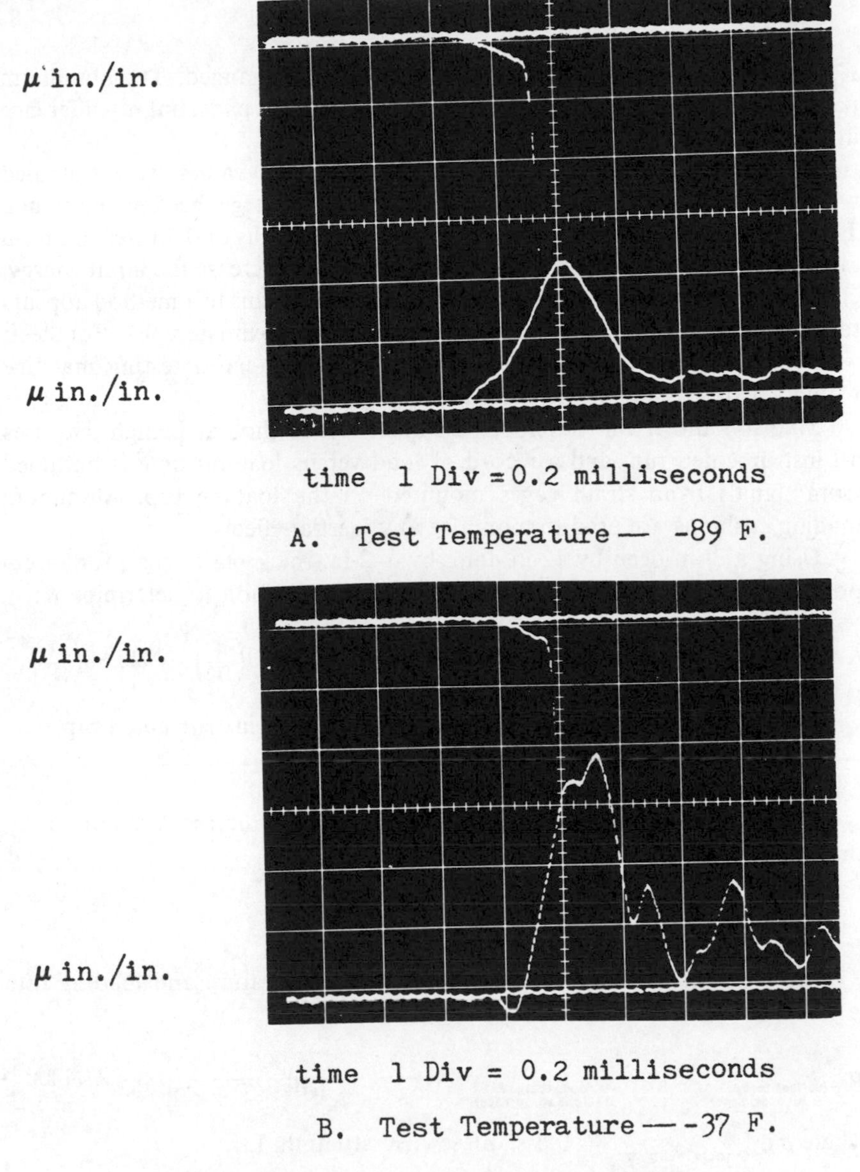

FIG. 3.30. Typical strain-time records for dynamic tests.

During the development of the dynamic testing procedure, two different specimen span lengths (different stiffnesses and natural frequencies), two different sets of gage positions, and different materials for dampening the impact blow were used to check for possible inertial effects. In no instance was it possible to find consistent significant deviations from the mean of the

K_{Id} data as any one of these testing variables was changed. The aluminum pads gave loading times shorter than those for the lead pads, but no difference in K_{Id} was obtained.

As tougher material behavior and increased K_{Id} values were obtained with increased temperature, the dynamic input energy became marginal. This led to strain-time records which had strain plateaus of 0.5-msec duration and greater. When the drop height was increased to increase the input energy, elastic "ringing" waves were encountered. Therefore, this test method appears to be restricted to a maximum K_{Id}/σ_{yd} ratio of approximately 0.7. For steels with high dynamic K_{Id} values, loading with a high-speed testing machine would alleviate many of the experimental difficulties.

Madison and Irwin[7] developed a similar test method at Lehigh that uses an instrumented tup, and a record of load versus loading time is obtained from signals from strain gages mounted on the loading tup. Aluminum loading cushions are used to minimize any inertial effects.

Using a 3-in.-deep by 12-in.-long by $\frac{1}{2}$–2-in.-thick plate (span for three-point bending is 10 in.), they used the following relation to determine K:

$$K = \frac{1.5PS\sqrt{a}}{BW^2}\left[1.93 - 3.12\frac{a}{W} + 14.68\left(\frac{a}{W}\right)^2 - 25.30\left(\frac{a}{W}\right)^3 + 25.90\left(\frac{a}{W}\right)^4\right]$$

where P = fracture load from strain gage records on instrumented tup,
a = effective crack length at fracture,
S = support span = $3.33W$.

The effect of plasticity is estimated by increasing the crack length, a, by the plastic-zone radius, r_y, where

$$r_y \cong \frac{1}{2\pi}\left(\frac{K_{Id}}{\sigma_{yd}}\right)^2$$

The yield stress corresponding to the test temperature and loading rate is estimated by

$$\sigma_{yd\,(\text{@ test temp. \& loading rate})} \cong \left[\sigma_{ys\,(\text{@ room temp. \& static loading rate})} + \frac{174{,}000}{\log(2 + 10^{10}t)(T + 459)} - 27.4\text{ ksi}\right]$$

where $\sigma_{ys\,(\text{@ room temp. \& static loading rate})}$ = 0.2% offset yield strength, ksi,
T = specimen temperature, °F,
t = load rise time for test—time from start of load to fracture, sec.

Static and some elevated-strain-rate yield strength data have been obtained for seven steels from room temperature to −320°F.[8] In Figure 3.31, these values are plotted in terms of the rate-temperature parameter, $T \ln A/\dot{\epsilon}$, suggested by Bennett and Sinclair,[9] where T is the absolute temperature in

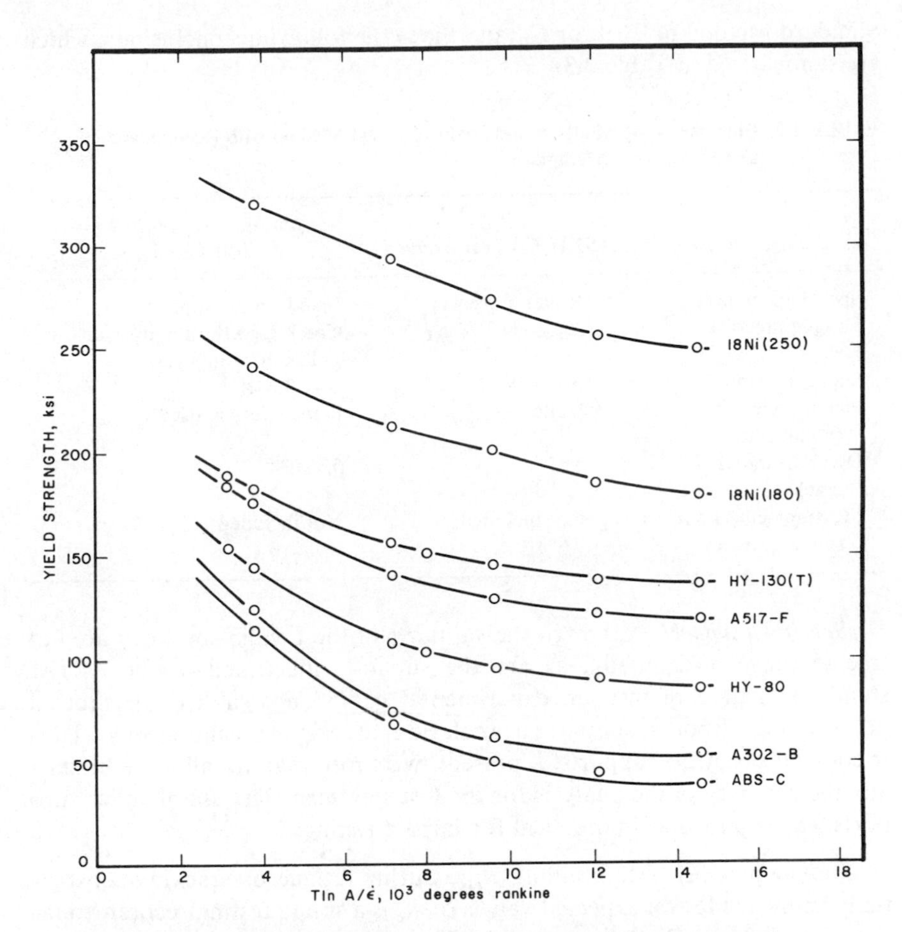

FIG. 3.31. Yield strength for seven steels in terms of the rate-temperature parameter, $T \ln A/\dot{\epsilon}$.

degrees Rankine, A is the frequency factor taken to be a constant of 10^8/sec, $\dot{\epsilon}$ is the strain rate, and ln is the natural logarithm to the base e. This rate-temperature parameter, based on the Arrhenius rate equation, was shown to correlate yield strength behavior of bcc materials for different temperatures and strain rates. This parameter appears to be useful to estimate yield behavior for low-strength alloy steels, for which significant changes in yield strength are observed for changes in temperature and strain rate. Because the effects of rate and temperature on the absolute yield strength decrease with steels of increasing yield strength, this parameter is used to estimate the yield behavior at the test conditions primarily for low-strength steels.

Madison and Irwin[7] compared their testing method with the ASTM

Standard Method of Test for K_{Ic} and made the following conclusions, which are summarized in Table 3.3:

TABLE 3.3. Comparison of Madison and Irwin K_{Id} Test Method with Recommended ASTM K_{Ic} Test Method

Testing Factors	*ASTM K_{Ic} Test Method*	*Madison and Irwin K_{Id} Test Method*
Specimen support	Roller supports	Fixed simple supports
Fatigue precrack	$K_{max} < 60\% K_{Ic}$	Crack growth rate at or below 1×10^{-6} in./cycle
Support span	$S \geq 4W$	$S = 3.33W$
Plastic-zone size (r_y) adjustment	None	Plane stress r_y used
Depth of initial crack	$0.5W$	$0.33W$
Measurement of slow stable growth	Not included	Not included

Specimen support. Although the supports of the Lehigh specimen are not free to move horizontally, as are the supports illustrated in the ASTM standard method of test, an experimental compliance calibration showed no deviation from a simple extrapolation of existing calibrations. Thus friction effects at the supports, if present, were too small to influence materially the accuracy of the analysis for the test specimen. The use of roller supports was regarded as impractical for impact testing.

Fatigue precrack. Maintaining K_{max} during fatigue precracking substantially below the lowest expected values of K_c is a sound testing recommendation from a conservative viewpoint. However, such a limitation of K_{max} resulted in fatigue precracking times which seemed excessive. For the Lehigh specimen, the available data did not indicate an influence of K_{max} on K_c in the Lehigh tests, even though a number of those tests resulted in measured K_c values less than K_{max}. This observation may not apply either to ASTM K_{Ic} testing or to K_c testing of other structural steels. Furthermore, in the case of typical cracks found in large welded structures, "sharpening" of such cracks by low-amplitude fatigue has not been reported and seems most unlikely. Thus the precracking method used herein seems acceptable on practical grounds.

Support span. The span of the Lehigh specimen is the same as the drop-weight tear test developed by the Naval Research Laboratory and Battelle Laboratories, i.e., a specimen which was in common use by the pipeline industry at the time the Lehigh tests were begun. The difference between

3.33W and 4W is not considered to be significant. Indeed, the compliance calibrations were in good agreement.

r_y Adjustment. When the ASTM K_{Ic} testing requirements are met, the plastic-zone radius is small enough that no correction for plasticity is necessary. However, for material thicknesses and testing temperatures indicative of common service conditions for the steels tested by Lehigh University, ASTM-type K_{Ic} measurements are rarely possible. For the tests reported in ref. 7, the plastic-zone size was seldom negligible, and a suitable adjustment was made to account for plasticity, as analyzed in the section on testing procedure. The adjustment consisted of increasing the crack length by the plane-stress radius of the plastic zone computed as

$$r_y = \frac{1}{2\pi}\left(\frac{K_c}{\sigma_{ys}}\right)^2$$

Depth of initial crack. As indicated previously, the K_c testing practice examined in ref. 7 should avoid the region of general yielding. Thus the specimen design should try to maximize the ratio of the K_c value to an appropriate measure of nominal stress relative to net section yielding. In the case of a centrally notched tensile specimen, the average net section tensile stress can be regarded as an appropriate nominal stress of that nature. Studies of that specimen show that maximum ratios of K_c to σ_{NET} are obtained if the crack size is about four-tenths of the specimen width. In the case of the single-edge-notched bend specimen, a corresponding calculation of optimum crack size can be made using, in place of net section stress, the nominal fiber stress for the net ligament beneath the starting crack.

The equation for K can be written in the form

$$K = Y \cdot \sigma_f \sqrt{a}$$

in which σ_f = the nominal fiber stress, ignoring the crack, from the equation

$$\text{bending moment} = \tfrac{1}{6}\sigma_f W^2,$$

$$Y = 1.93 - 3.12(a/W) + 14.68(a/W)^2 - 25.30(a/W)^3 + 25.90(a/W)^4.$$

Using σ_M as the symbol for the corresponding net ligament fiber stress

$$K = \sigma_M\left(1 - \frac{a_1}{W}\right)^2 \sqrt{a}\, Y$$

in which a_1 is the actual crack size and $a = a_1 + r_y$. Trial calculations in a range where r_y is negligible show maximum values of K/σ_M near $a_1 = 0.25W$. At high stress levels any value of a_1/W from 0.2 to 0.3 provides nearly the same maximum ratio of K to σ_M. Note that the ratio of $a_1 = 0.25W$ to half of the net ligament corresponds closely to the optimum ratio of crack length to net section size for the centrally notched tensile specimen. Although it lies slightly beyond the region of maximum K/σ_M values, the choice of $a_1 = W/3$ is appropriate for three reasons: (1) This choice is very close to the high stress level range where K/σ_M has largest values; (2) restriction of the yield pattern to the net ligament is desirable, and assurance of this becomes marginal as a_1/W is reduced below $\frac{1}{3}$; and (3) use of crack depths smaller than the specimen thickness is generally undesirable in K_c testing. Ideally the 2-in.-thick (50-mm) specimens should have lateral dimensions of 6 in. × 24 in. (150 mm × 610 mm) and a starting crack depth of 2 in. (25 mm). This was impractical at the time the Lehigh specimen was being developed.

Ives and Barsom[10] have investigated the instrumentation and procedures used to measure accurately the crack-initiation behavior during K_{Id} tests. They conclude that valid reproducible dynamic results can be obtained when care is taken in the instrumentation of dynamic tests. However, without proper consideration and analysis of the parameters involved in each test prior to actual testing, erroneous results can be obtained. They recommend that crack-opening-displacement data combined with strain-gage measurement on the specimen rather than on the loading tup should be used to calculate K_{Id} data.

Although the K_{Id} test method is yet to be standardized, K_{Id} values are being widely used in analysis, design, and specifications, as discussed throughout this text. Typical K_{Id} results, compared with K_{Ic} test results as a function of temperature, are presented in Chapter 4.

REFERENCES

1. "Standard Method of Test for Plane-Strain Fracture Toughness of Metallic Materials," *ASTM Designation E 399-74*, Part 10, *ASTM Annual Standards.*
2. "Proposed Recommended Practice for R-Curve Determination," *ASTM Annual Book of Standards*, Part 10.
3. W. F. Brown, Jr., Editor, "Review of Developments in Plane Strain Fracture Toughness Testing," *ASTM STP 463*, 1970.
4. W. F. Brown, Jr., and J. E. Srawley, "Plane Strain Crack Toughness Testing of High Strength Metallic Materials," *ASTM STP 410*, 1967.
5. P. B. Crosley and E. J. Ripling, "Plane-Strain Crack Arrest Characterization of Steels," *ASME Paper No. 75-PVP-32*, presented at Second National Congress on Pressure Vessels and Piping, San Francisco, June 23–27, 1975.
6. A. K. Shoemaker and S. T. Rolfe, "Static and Dynamic Low-Temperature K_{Ic} Behavior of Steels," *ASME Trans.*, Sept. 1969.
7. R. B. Madison and G. R. Irwin, "Dynamic K_c Testing of Structural Steel," *Journal of the Structural Division, ASCE, 100*, No. ST 7, Proc. Paper 10653, July 1974, pp. 1331–1349.
8. D. P. Clausing, "Tensile Properties of Eight Constructional Steels Between 70 and −320°F," *Journal of Materials*, ASTM, Vol. 4, No. 2, June 1969.

9. P. E. Bennett and G. M. Sinclair, "Parameter Representation of Low-Temperature Yield Behavior of Body-Centered-Cubic Transition Metals," *Journal of Basic Engineering, Transactions ASME,* Series D, *88,* No. 2, June 1966.
10. K. D. Ives and J. M. Barsom, "Recent Developments in Dynamic Evaluation of K_{Ic} and Analysis of Long Fractures," *Instrument Society of America,* Vol. 12, No. 4, 1973.

4

Effect of Temperature, Loading Rate, and Plate Thickness on Fracture Toughness

4.1. Introduction

In Chapter 3, the ASTM standard test method for determining the critical plane-strain stress-intensity factor, K_{Ic}, was described in detail. Also a nonstandard test method for determining the crack toughness under conditions of impact, or dynamic, loading was described. This dynamic crack toughness was called K_{Id}, the critical plane-strain stress-intensity factor under conditions of dynamic loading. Furthermore, it was stated that these tests are frequently conducted at various temperatures to determine both the "static" and "dynamic" fracture toughness of various structural materials as a function of temperature.

The fact that the inherent fracture toughness of many structural materials increases with increasing test temperature is well known. This increase has been measured using various notch-toughness specimens such as the Charpy V-notch impact specimen, and it is certainly reasonable to expect a similar increase using fracture-mechanics-type test specimens. What is not so widely known is the fact that the same inherent fracture toughness can decrease significantly with increasing loading rate, i.e., comparing K_{Id} test results with K_{Ic} test results. Thus, before the engineer can use fracture-toughness values in design (as described in Chapter 5), the critical fracture-toughness value for the particular service temperature, loading rate, and plate thickness must be known. In this chapter we shall describe the general effects of these three variables on the fracture toughness of various structural materials. Finally, typical test results will be presented for commonly used structural materials.

4.2. Plane-Strain Transition Temperature Behavior

As discussed briefly in Chapter 1, there are numerous types of fracture-toughness specimens used to determine the notch toughness of structural materials. For most structural steels, these specimens [such as the Charpy

V-notch (CVN) impact specimen, the dynamic tear (DT) test specimen, the precracked impact (PCI) specimen, etc.] are tested at several temperatures to determine the notch-toughness behavior in the transition-temperature range. This temperature range is the region throughout which the inherent notch toughness of a material changes from brittle to ductile. This change in behavior is shown schematically in Figure 4.1 for a typical structural steel and

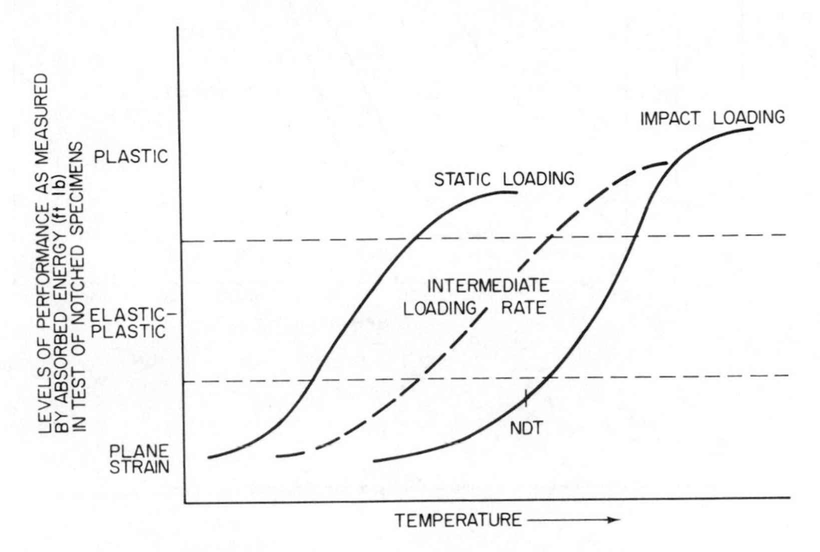

FIG. 4.1. Schematic showing relation between notch-toughness test results and levels of structural performance for various loading rates.

indicates that temperature can have a significant effect on the notch-toughness behavior of structural materials. Examples of CVN, DT, and K_{Ic} test data for an A517 steel are presented in Figure 4.2 and illustrate this marked change in notch toughness with increasing temperature.

This increase in notch toughness with increasing temperature is well known to most design engineers and has led to the *transition-temperature* design approach. In this approach some level of notch toughness is selected as representing an "adequate" level of toughness for the particular structure. In Chapters 12 and 13 we shall describe how an "adequate" level of toughness is established, but one level that has been widely used is the 15-ft-lb CVN impact test value based on the analysis of the World War II ship fractures. The temperature at which the material exhibits this toughness level is required to be at or below the minimum service temperature. Figure 4.3 is a schematic showing two steels with different "15-ft-lb transition temperatures" and their relation to a particular service temperature. In this example, steel A would be considered to be satisfactory because its 15-ft-lb transition temperature

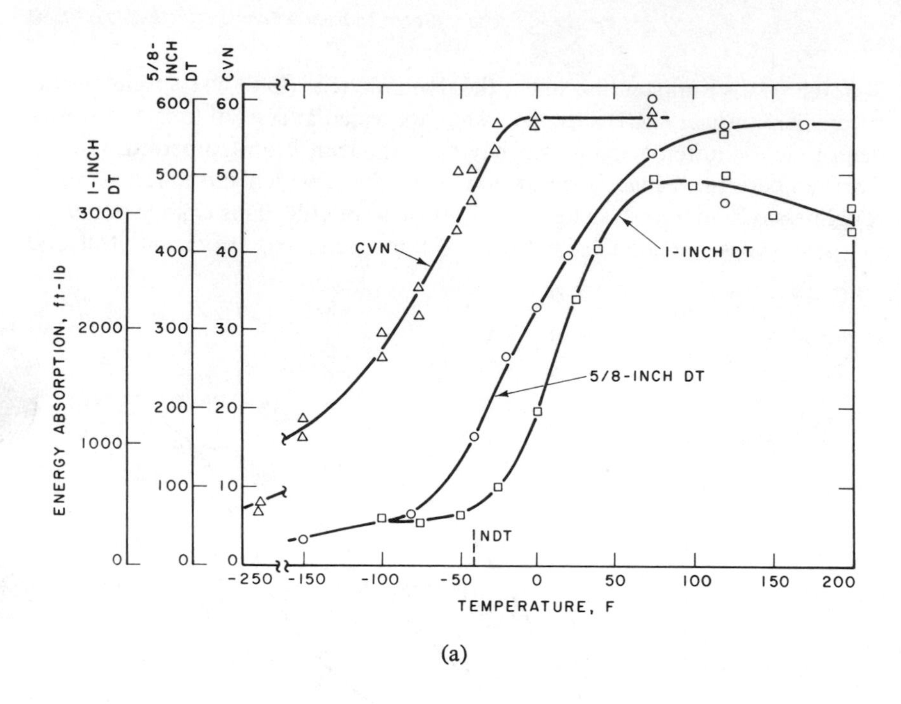

(a)

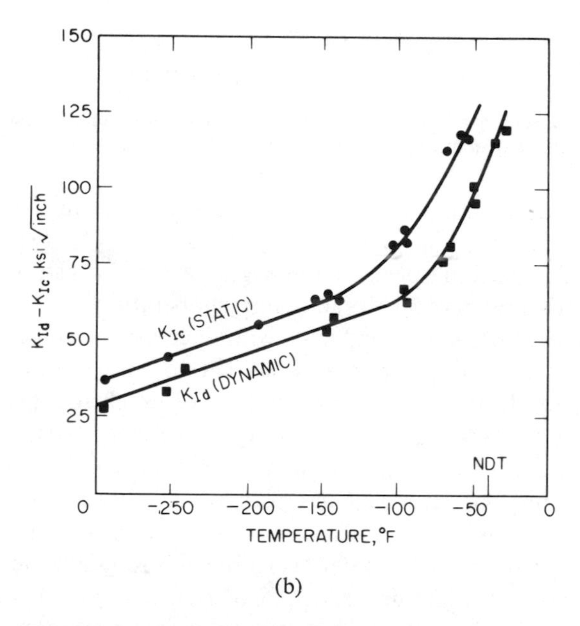

(b)

FIG. 4.2. CVN, DT, and K_{Ic} test results for A517 steel.

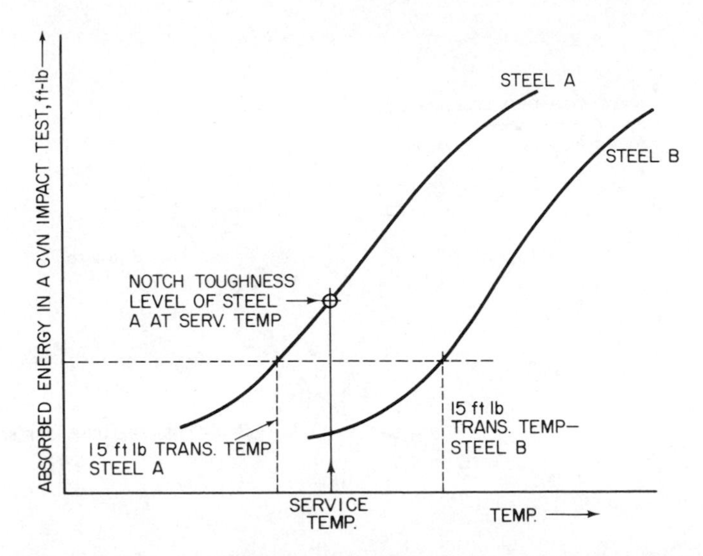

FIG. 4.3. Schematic showing relation between 15-ft-lb transition temperatures and service temperature.

would be below the service temperature, whereas steel B would not meet the criteria. This transition-temperature approach has been widely used by materials engineers to establish various toughness specifications.

In a similar manner, the notch toughness of many structural materials exhibits an inherent change in the plane-strain fracture toughness with change in temperature, as shown in Figure 4.4.[1] The test results presented in Figure 4.4 satisfy the ASTM requirement for minimum specimen thickness in plane-strain crack-toughness testing described in Chapter 3 as given by

$$a \text{ and } B \geq 2.5\left(\frac{K_{Ic}}{\sigma_{ys}}\right)^2$$

where B = specimen thickness,
a = crack length,
σ_{ys} = 0.2% yield strength at the test temperature.

This ASTM requirement can be expressed in terms of Irwin's plane-strain β value[2] as follows:

$$\beta_{Ic} = \frac{1}{B}\left(\frac{K_{Ic}}{\sigma_{ys}}\right)^2 \leq 0.4$$

That is, if β_{Ic} is 0.4, or less, the specimen size is sufficiently large to ensure plane-strain behavior as defined in Chapter 3.

All data points in Figure 4.4 define a single, common function, regardless of whether the points were obtained from tests that satisfied the criterion of $\beta = 0.4$ or less. Because the maximum load at fracture for all data points shown in Figure 4.4 was within the ASTM 5% secant-intercept requirement

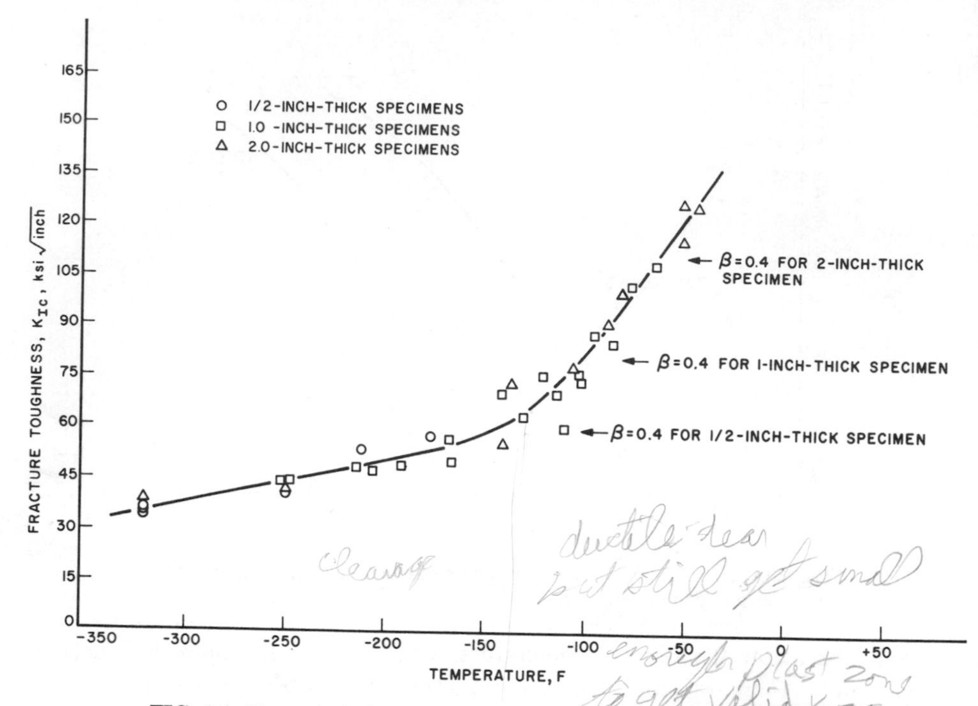

FIG. 4.4. Plane-strain fracture-toughness transition behavior as a function of temperature.

for plane-strain behavior and because all data points define a single, common function, it would appear that all data points are valid K_{Ic} values even though a few points did not satisfy the β criterion. Thus, the curve defined by these data points demonstrates that a K_{Ic} temperature transition exists that is independent of specimen geometry; that is, the rate of increase in K_{Ic} with temperature does not remain constant but increases markedly in a particular temperature range for a particular test specimen and testing conditions. In this case, the change in rate of increase of K_{Ic} occurs in the temperature range $-150°$ to $-100°$F and is associated with the onset of change in the microscopic fracture mode.

Figure 4.5(a) shows the variation of the parameter $(K_{Ic}/\sigma_{ys})^2$, which is proportional to the plastic-zone size, as a function of temperature. Above $-40°$F, the magnitude of the plastic zone appears to increase asymptotically. This behavior indicates that valid K_{Ic} values probably cannot be obtained at temperatures greater than $-40°$F for this particular material regardless of specimen size.

It should be noted that early in the development of fracture mechanics it was thought that there was no plane-strain transition. That is, it was believed that any marked increase in K_{Ic} with temperature such as is shown in Figure 4.5(b) was in reality a loss in through-thickness constraint and actually a

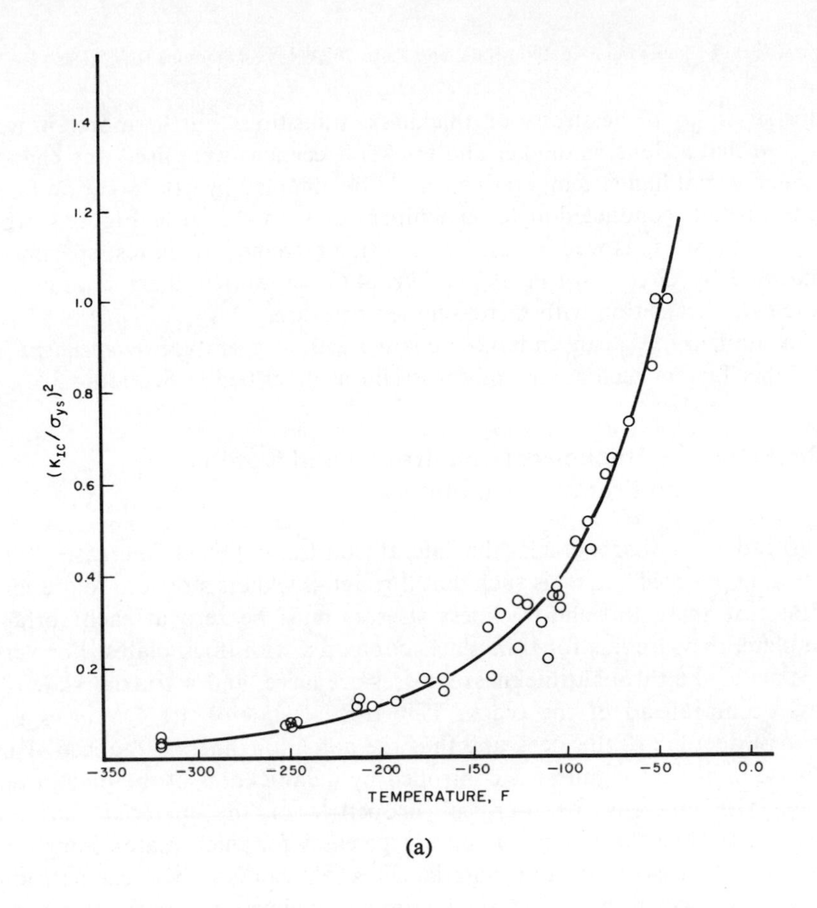

(a)

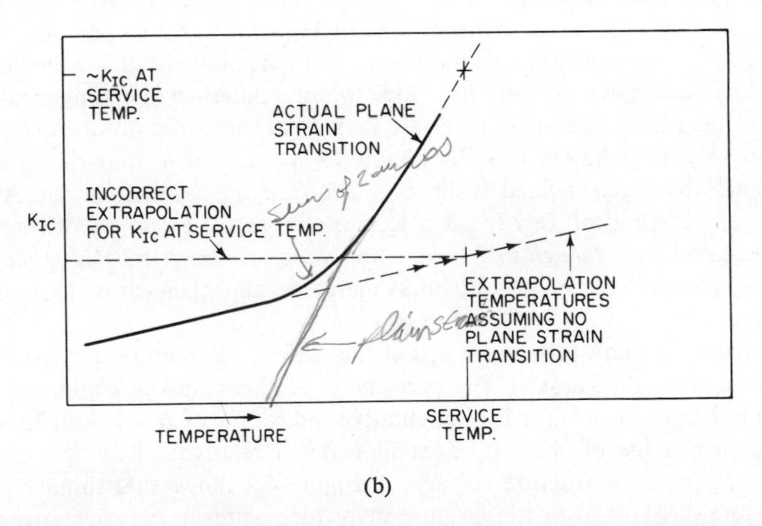

(b)

FIG. 4.5. Effect of plane-strain transition on plastic-zone size and extrapolated K_{Ic}.

plane-strain to plane-stress, or thickness, transition. Furthermore, it was believed that as long as thicker and thicker specimens were used, the change in toughness at higher temperatures could be obtained by extrapolation from K_{Ic} test results conducted at lower temperatures, as shown in Figure 4.5(b). These test results, as well as test results using extremely thick test specimens conducted by Greenburg et al.[3] (Figure 4.6), show that there is indeed a plane-strain transition with increasing temperature.

In summary, K_{Ic} can undergo a change with temperature *independent* of any other factors such as loss of constraint as described in Section 4.3.

4.3. Effect of Thickness (Constraint) and Notch Acuity on Fracture Toughness

Ahead of a sharp crack, the lateral constraint (which increases with increasing plate thickness) is such that through-thickness stresses are present. Because these through-thickness stresses must be zero at each surface of a plate, they are less for thin plates compared with thick plates. For very thick plates the through-thickness stresses are large, and a triaxial state-of-stress occurs ahead of the crack. This triaxial state-of-stress reduces the apparent ductility of the steel, and thus the notch toughness is reduced. This decrease in notch toughness is controlled by the thickness of the plate, even though the inherent metallurgical properties of the material may be unchanged. Thus the notch toughness decreases for thick plates compared with thinner plates of the same material. This behavior is shown schematically in Figure 4.7, which shows that the minimum toughness of a particular material, K_{Ic}, is reached when the thickness of the specimen is large enough so that the state-of-stress is plane strain. In Figure 4.8, actual test results are presented for a high-strength maraging steel that illustrate this behavior.

For thicknesses greater than some value related to the toughness and yield strength of individual materials, maximum constraint occurs and plane-strain (K_{Ic}) behavior results. In Chapter 3 it was shown that this limiting thickness has been defined to be $B \geq 2.5(K_{Ic}/\sigma_{ys})^2$, as given in the ASTM Standard Method of Test for K_{Ic}. Conversely, as the thickness of the plate is decreased, *even though the inherent metallurgical characteristics of the steel are not changed*, the notch toughness increases, and plane-stress (K_c) behavior exists.

Figure 4.9 shows the shear lips at the surface of fracture test specimens with different thicknesses. The percentage of shear lips as compared with the total fracture surface is a qualitative indication of notch toughness. A small percentage of shear-lip area indicates a relative brittle behavior. A comparison of the fracture surfaces in Figure 4.9 shows that thinner plates are more resistant to brittle fracture than thick plates in that the percentage of shear lips is larger for the thinner specimens compared with the thicker ones.

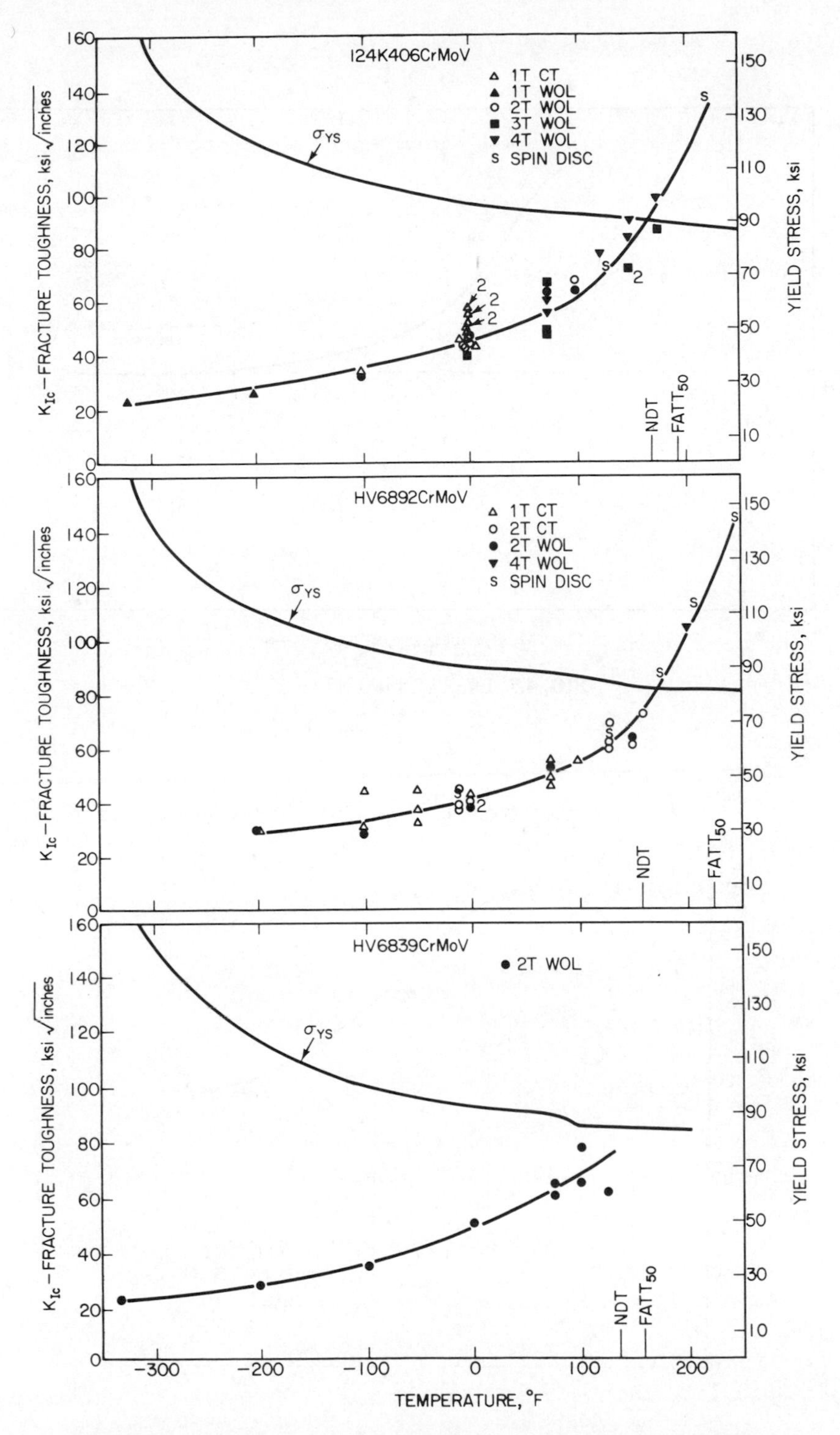

FIG. 4.6. Temperature dependence of the plane-strain fracture toughness (K_{Ic}) of three CrMoV alloy forgings.

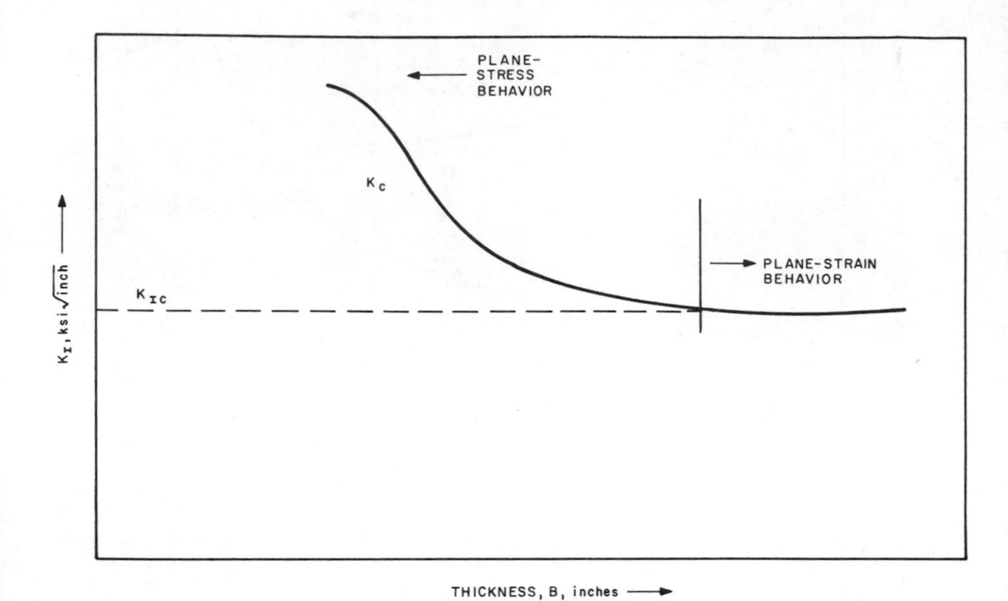

FIG. 4.7. Effect of thickness on K_c.

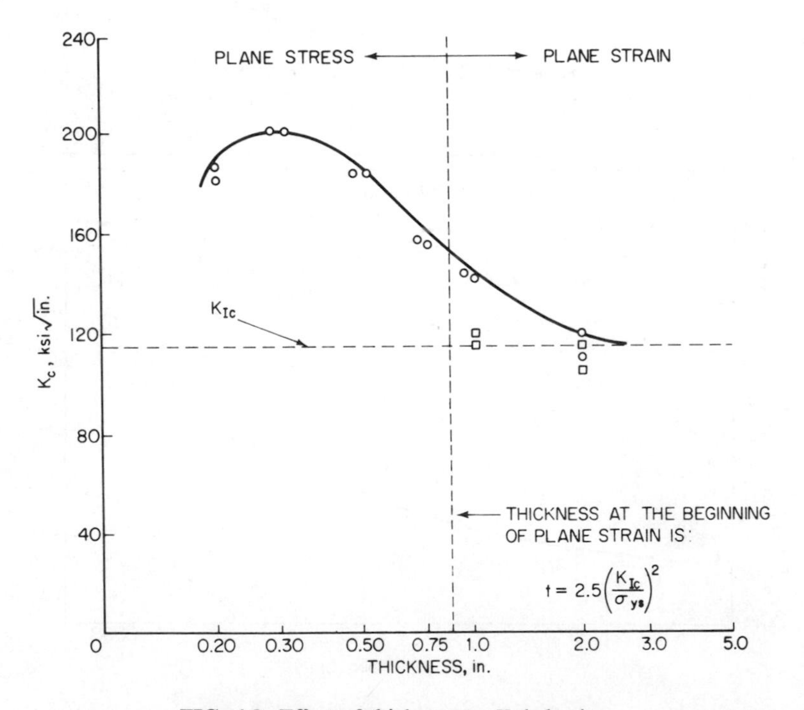

FIG. 4.8. Effect of thickness on K_c behavior.

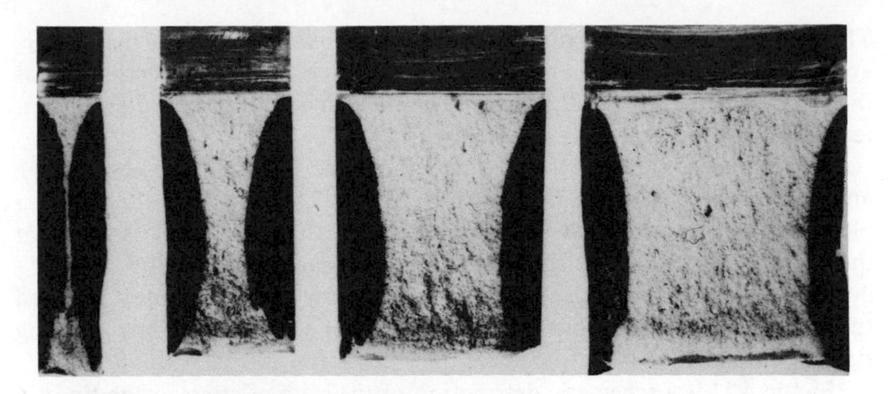

FIG. 4.9. Effect of specimen thickness (2, $1\frac{1}{2}$, 1, and $\frac{1}{2}$ in.) on toughness as determined by size of shear lips.

Pellini[4] has described the physical significance of constraint and plate thickness on crack toughness in terms of plastic flow, as shown in Figure 4.10. This figure shows that the introduction of a circular notch in a bar loaded in tension causes an elevation of the stress-strain, or flow, curve. He describes the plastic flow of the smooth tensile bar as "free" flow, or that normally observed in conventional stress-strain curves, where lateral contraction is not constrained during the initial loading.

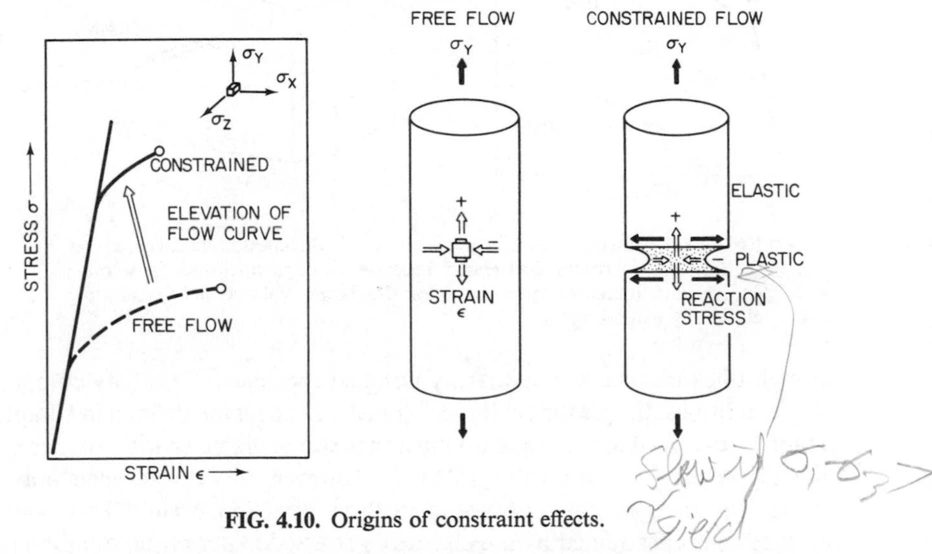

FIG. 4.10. Origins of constraint effects.

In the notched bar, however, the reduced section deforms inelastically while the ends of the specimen are still loaded elastically. Since the amount of elastic contraction (Poisson's ratio) is small compared to the inelastic contraction of the reduced section, a *restriction* to plastic flow is developed.

This restriction is in the nature of a reaction-stress system such that the σ_x and σ_z stresses restrict or constrain the flow in the σ_y (load) direction. Thus the uniaxial stress state of the smooth bar is changed to a triaxial stress system in the notched bar.

For a triaxial state-of-stress, where the three principle stresses, σ_x, σ_y, and σ_z, are equal, there are no shearing stresses. This results in almost complete constraint against plastic flow, thus increasing the elastic stresses at the tip of a crack to extremely high values, compared with the lower "free"-flow stresses in an unrestrained tension specimen. In the case of most notched specimens $\sigma_y > \sigma_x$ or σ_z, but the stresses are not equal. Thus some shearing stresses do occur.

Figure 4.11 is a schematic description of the state-of-stress at the tip of a

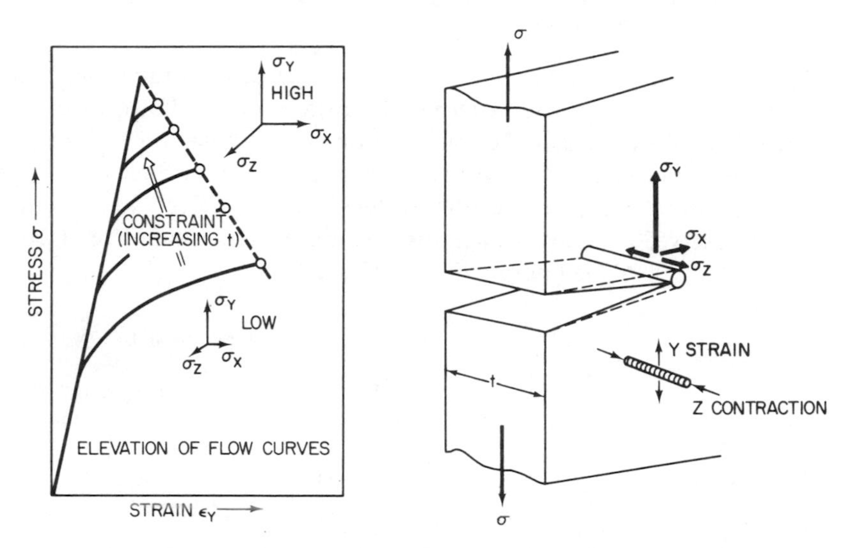

FIG. 4.11. Constraint conditions for through-thickness cracks. Increasing section size increases constraint because through-thickness flow must evolve with increased opposition of the larger volume of surrounding elastically loaded metal.

through-thickness crack in a sharply notched specimen. To satisfy compatibility conditions, the plastic "cylinder" (plastic-zone region defined in Chapter 2) that is developed at the crack tip must increase in diameter with an increase in stress in the y direction due to load. However, this can happen only if through-thickness lateral contraction in the z direction occurs. This lateral contraction is constrained by the elastically stressed material surrounding the "cylinder" and leads to the triaxial state-of-stress which raises the flow stress. Therefore, *as the plate thickness is increased, the constraint increases, and the flow stress curve* is raised, as shown schematically in Figure 4.11.

In summary, the constraint ahead of a sharp crack is increased by increasing the plate thickness. Thus the critical stress intensity, K_c, for a particular structural material tested at a particular temperature and loading rate decreases with increasing specimen thickness (Figures 4.7 and 4.8). Beyond some limiting thickness, maximum constraint is obtained, and the critical stress intensity factor reaches the minimum plane-strain value, K_{Ic}. This maximum constraint occurs when the plate thickness is sufficiently large in a notched specimen of the particular material being tested at a particular test temperature and loading rate. As described previously, the limiting thickness for plane-strain behavior has been established by the ASTM standard test method as

$$B \geq 2.5\left(\frac{K_{Ic}}{\sigma_{ys}}\right)^2$$

For dynamic loading, the limiting thickness would be

$$B \geq 2.5\left(\frac{K_{Id}}{\sigma_{yd}}\right)^2$$

where B = thickness of test specimen,

K_{Ic} = critical plane-strain stress-intensity factor under conditions of static loading as described in ASTM Method E-399 Standard Method of Test for Plane Strain Fracture Toughness of Metallic Materials as described in Chapter 3.

σ_{ys} = static tensile yield strength obtained in "slow" tension test as described in ASTM Test Method E-8—Standard Methods of Tension Testing of Metallic Materials,

K_{Id} = critical plane-strain stress-intensity factor as measured by "dynamic" or "impact" test; the test specimen is similar to a K_{Ic} test specimen but is loaded rapidly as described in Chapter 3; there is no standardized test procedure, as yet.

σ_{yd} = dynamic tensile yield strength obtained in "rapid" tension test at loading rates comparable to those obtained in K_{Id} tests; although extremely difficult to measure, a good engineering approximation based on experimental results of structural steels is

$$\sigma_{yd} = \sigma_{ys} + (20 \text{ to } 30 \text{ ksi})$$

This limiting constraint condition for K_{Ic} or K_{Id} is established for a crack tip of "infinite" sharpness, namely $\rho = 0$. This "infinite" sharpness is obtained by fatigue-cracking the test specimens at low-stress levels, as described in Chapter 3. As a test specimen is loaded, some local plastic flow will occur at the crack tip, and the crack tip will be blunted slightly. For a brittle material, i.e., any structural material tested at a temperature and loading rate where it has very low crack toughness, the degree of crack blunting is very small. Consequently unstable crack *extension* will occur under conditions of continued crack sharpness. In essence, the material fractures under elastic loading and exhibits plane-strain behavior under conditions of maximum constraint.

However, if the inherent toughness of the structural material is such that it is *not* brittle at the particular testing temperature and loading rate (e.g., the start of elastic-plastic behavior), then an increase in plastic deformation at the crack tip occurs and the crack tip is "blunted." As a result, the limit of plane-strain constraint is exceeded. At temperatures above this test temperature, the inherent toughness begins to increase rapidly with increasing test temperature because the effects of the crack blunting and relaxation of plane-strain constraint are synergistic. That is, the crack blunting leads to a relaxation in constraint which causes increased plastic flow, which leads to additional crack blunting. Thus elastic-plastic behavior begins to occur rapidly at increasing test temperatures once this plane-strain constraint (thickness plus notch acuity) is exceeded. Figures 4.12 and 4.13 show examples of the

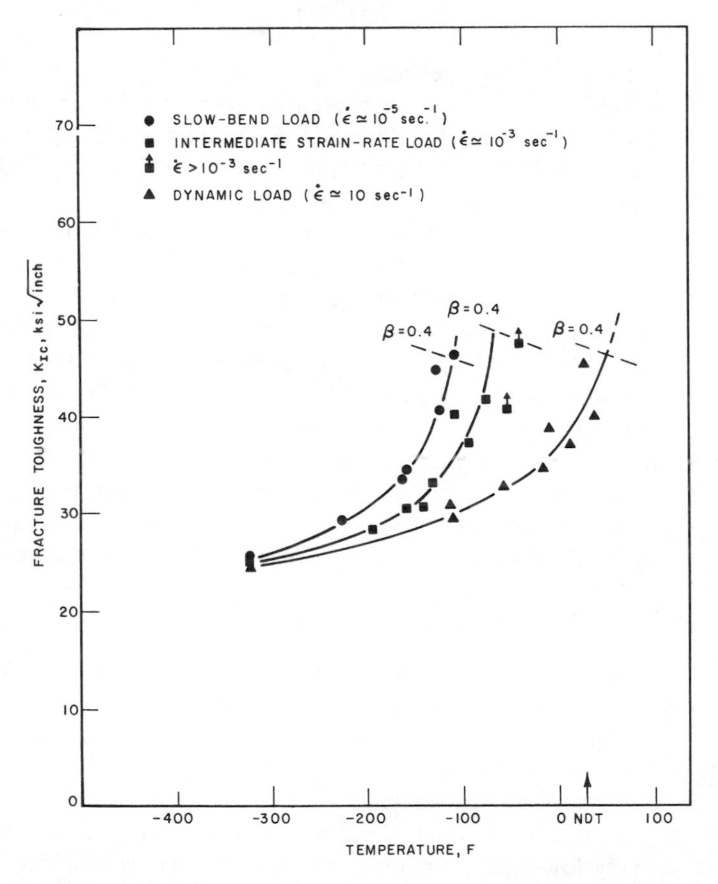

FIG. 4.12. Effect of temperature and strain rate on fracture toughness of A36 steel.

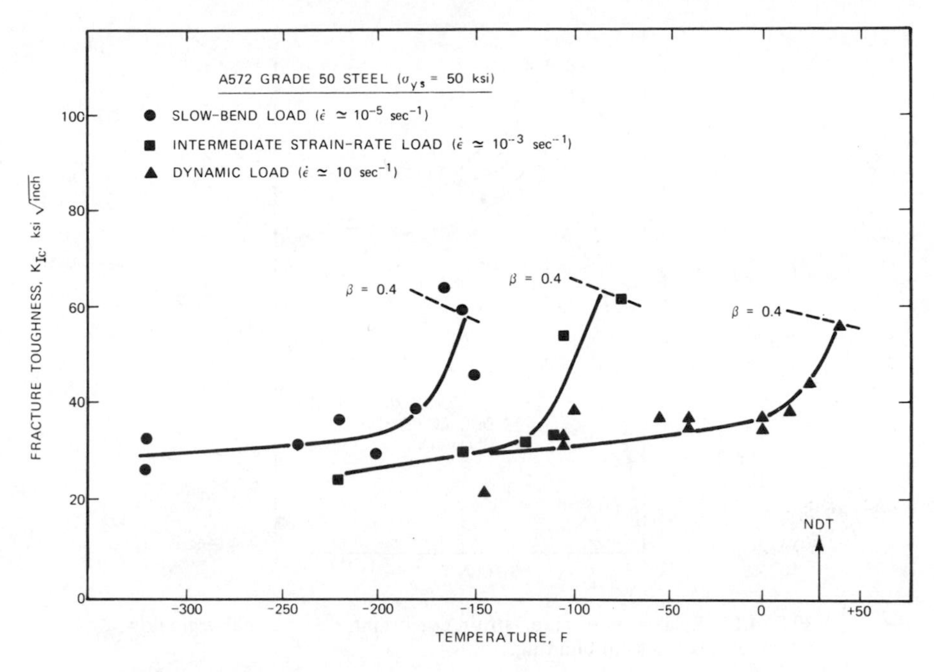

FIG. 4.13. Effect of temperature and strain rate on fracture toughness of A572 grade 50 steel ($\sigma_{ys} = 50$ ksi).

very rapid change in K_{Ic} (after the plane-strain transition) for two structural steels once the limit of plane-strain behavior ($\beta_{Ic} = 0.4$) has been reached.

Pellini[4] has described this behavior in terms of a constraint relation which changes the flow curve as shown schematically in Figure 4.14. The degree of crack-tip blunting establishes the particular flow-stress curve, leading to plane-strain, elastic-plastic, or plastic behavior. For example, the dashed curves *A* and *B* represent flow curves of unconstrained material (as in standard tension tests) leading to plastic or elastic-plastic behavior, depending on the inherent ductility of the structural material. Curve *C* represents the flow curve of a notched fully constrained material leading to failure under conditions of plane strain. If the material is tested at a temperature above the limit of plane strain such that partial crack-tip blunting occurs, a partial relation of constraint occurs, leading to elastic-plastic behavior. If the material is tested at still higher temperatures, considerable crack-tip blunting occurs, and considerable relaxation of constraint occurs, leading to plastic behavior.

Figure 4.15 is a schematic of the metal-grain structure ahead of the crack tip and indicates the microscopic behavior of the particular structural material. The dark line tracings within grains indicate slip on crystal planes, which is necessary to produce deformation. This deformation of individual

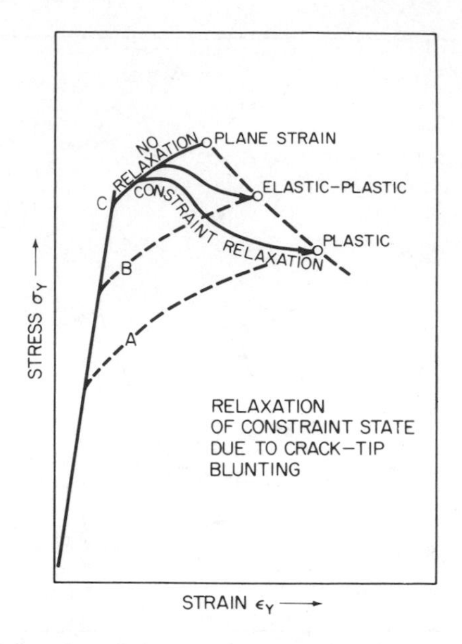

FIG. 4.14. Relaxation of plane-strain constraint, due to metal-grain flow, which causes crack-tip blunting.

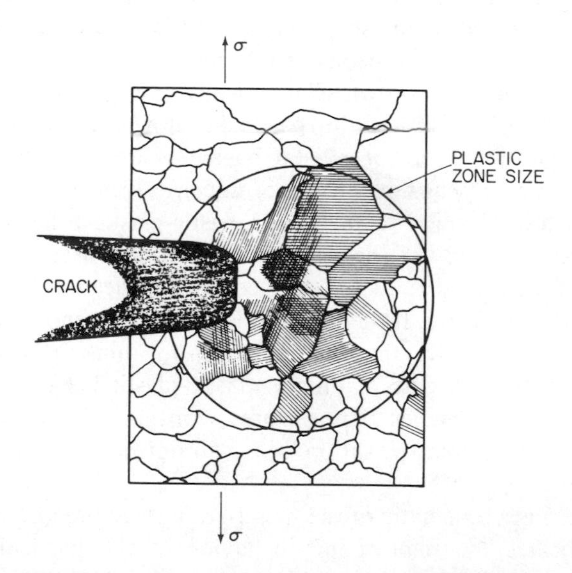

FIG. 4.15. Effect of crack-tip blunting ahead of crack.

grains is necessary to provide for growth of the plastic zone r_y (approximated by the dashed circle) ahead of the crack tip. Depending on the inherent metallurgical structure of the metal, continued loading either increases slip and deformation (elastic-plastic or plastic behavior) or leads to the development of cracks and/or voids (plane-strain behavior).

A material that is brittle at a particular temperature will develop microcracks or voids before the plastic-zone size is very large, resulting in rapid or unstable crack growth, i.e., brittle fracture. As the test temperature is increased for successive test specimens, the metal becomes more ductile and slip occurs before microcracks occur, resulting in larger plastic-zone sizes. This is the beginning of the plane-strain transition where the individual grains begin to undergo large amounts of plasticity (*microscopic plasticity*) but the overall specimen is still in *macroscopic plane strain*.

This transition in behavior from plane-strain to plastic behavior occurs over the region known as the transition-temperature region. This region can be measured with various types of test specimens, but the K_{Ic}, CVN, K_{Id}, and DT test specimens are the most common types used. Note that K_{Ic} or K_{Id} specimens cannot be used to measure the entire range of behavior since they require essentially elastic plane-strain behavior to satisfy the restrictions placed on the analysis described in Chapter 2. Fracture-mechanics-type specimens (J_{Ic}, COD, and R-curve) that *can* be used to obtain quantitative estimates of the toughness in the elastic-plastic region are described in Chapter 16. However, the J_{Ic} and R-curve specimens are research-type specimens that are not standardized, although a tentative ASTM test method has been developed for R-curve testing. The COD-type specimen is standardized and is widely used in the United Kingdom and Japan, although it is not widely used in the United States.

Generally the engineer is required to use the results of auxiliary test specimens, such as the CVN or DT specimens, coupled with empirical correlations as discussed in Chapter 6, in order to describe the various levels of performance quantitatively.

4.4. Effect of Temperature and Loading Rate on K_{Ic} and K_{Id}

In general, the crack toughness of structural materials, particularly steels, increases with increasing temperature and decreasing loading rate. These two general types of behavior are shown schematically in Figures 4.16 and 4.17. Figure 4.16 shows that both K_{Ic} and K_{Id} increase with increasing test temperature but that for any given temperature, the crack toughness measured in an impact test, K_{Id}, generally is lower than the crack toughness measured in a static test, K_{Ic}. Figure 4.17 shows that, at a constant tempera-

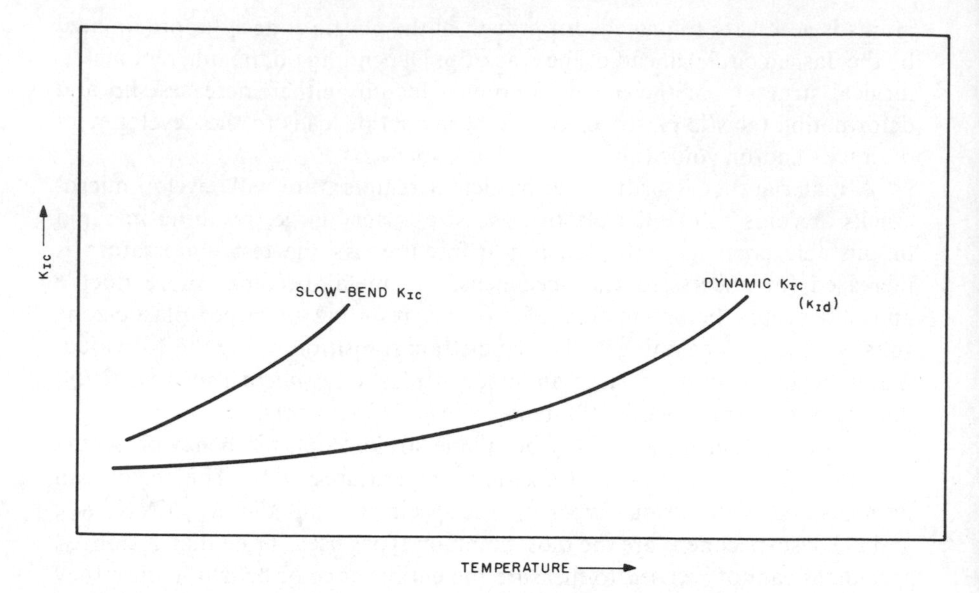

FIG. 4.16. Schematic showing effect of temperature and loading rate on K_{Ic}.

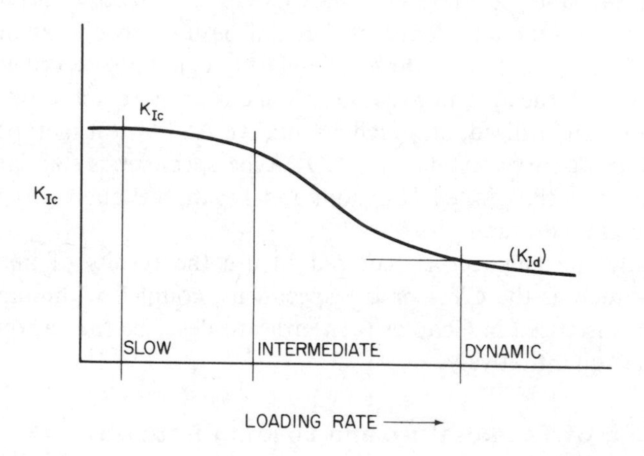

FIG. 4.17. Schematic showing effect of loading rate on K_{Ic}.

ture, crack-toughness tests conducted at higher loading rates generally result in lower toughness values.

This effect of temperature and loading rate is similar to that obtained with Charpy V-notch impact test results, as shown schematically in Figure 4.18. Figure 4.19 shows the general effect of loading rate and temperature on the behavior of K_{Ic} and K_{Id}, for an A36 structural steel ($\sigma_{ys} = 36$ ksi). Figure 4.20 shows the same general effect for impact CVN tests and slow-bend CVN

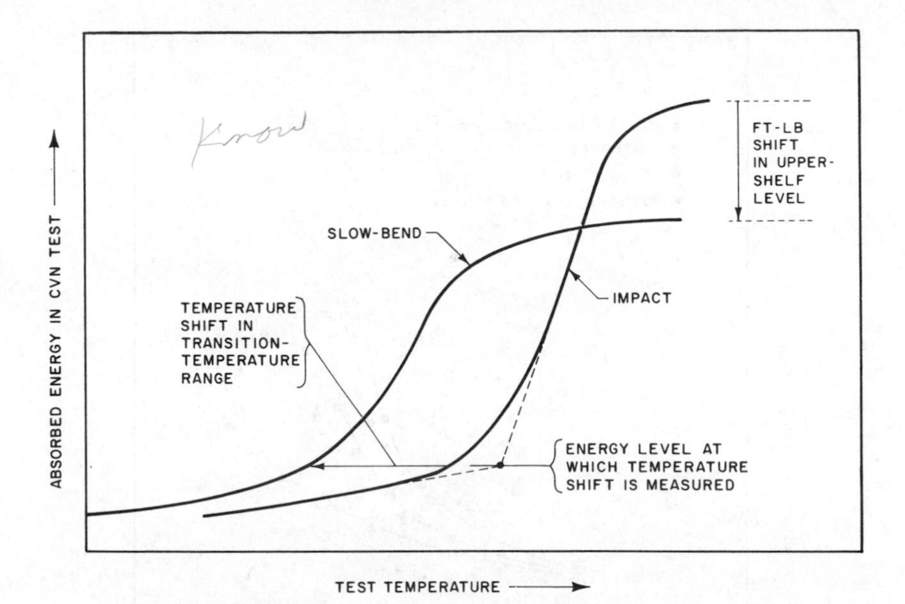

FIG. 4.18. Schematic representation of shift in CVN transition temperature and upper shelf level due to strain rate.

tests of the same A36 structural steel. Thus the transition from brittle to ductile behavior begins at lower temperatures for specimens tested at slow loading rates compared with specimens tested at impact loading rates.

Data obtained for various structural steels demonstrate that a true K_{Ic} temperature transition exists that is independent of geometry. This is the plane-strain transition as was described in Section 4.2. This change in K_{Ic} is not related to a loss in constraint as the test temperature is increased but is an inherent characteristic of many structural materials.

The rate of increase in K_{Ic} with temperature does not remain constant but increases markedly above a given test temperature. Fractographic analyses show that the fracture-toughness transition temperature is associated with the onset of change in the microscopic-fracture mode at the crack tip. At the low end of the transition-temperature range the mode of fracture initiation is cleavage, and at the upper end, the fracture initiation mode is ductile tear. In the transition temperature region, a continuous change in fracture mode occurs.

Thus, there are *two* transitions in fracture behavior with temperature, namely the K_{Ic} temperature transition just described, which is referred to as the plane-strain transition with temperature, and a transition from plane-strain to plane-stress (elastic-plastic) behavior as the constraint at the crack tip decreases. Plane strain generally refers to the *macroscopic* state of stress.

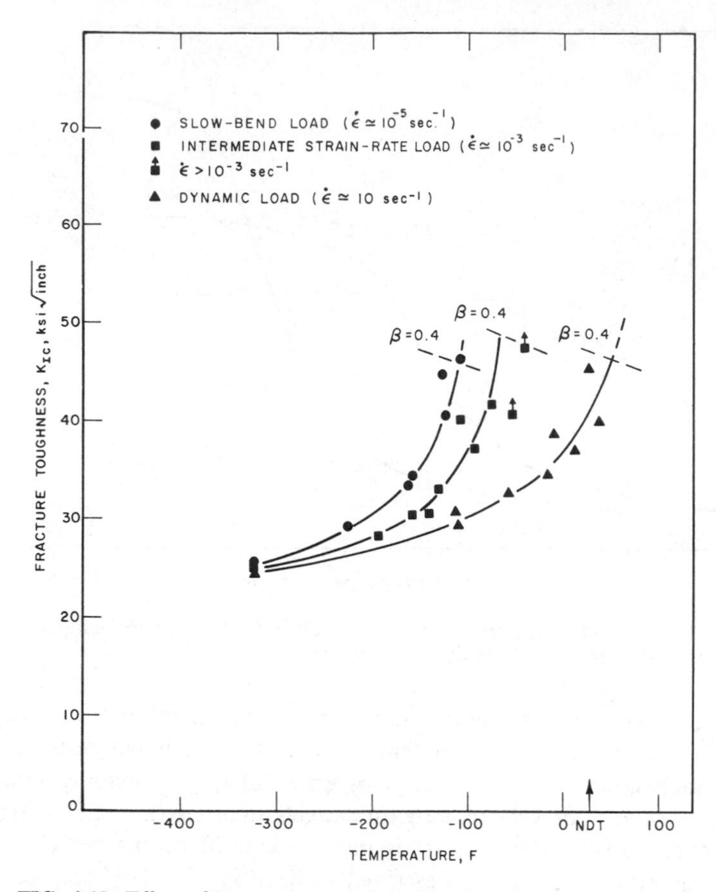

FIG. 4.19. Effect of temperature and strain rate on fracture toughness of A36 steel.

In the plane-strain transition region, the *microscopic* fracture behavior can change with increasing temperature from 100% cleavage or quasi-cleavage to a mixture of quasi-cleavage, tear dimples, and large flat tear areas. Thus, the *microscopic* fracture behavior can change while the *macroscopic* state of stress is still one of plane strain. This change in the *microscopic* fracture behavior leads to the rapid increase in the plane-strain fracture toughness. Figure 4.21 illustrates the *macroscopic* plane-strain transition for 100-ksi yield strength structural steel. Figure 4.22 shows the fracture surfaces of selected specimens in this transition-temperature region and illustrates the *macroscopic* fracture behavior under *macroscopic* plane-strain conditions. Figure 4.23 is a series of fractographs showing the changes in *microscopic* behavior for these same test specimens. At −250°F the fracture surfaces are 100% quasi-cleavage and at +75°F are 100% tear dimples. At −140°F the fractures

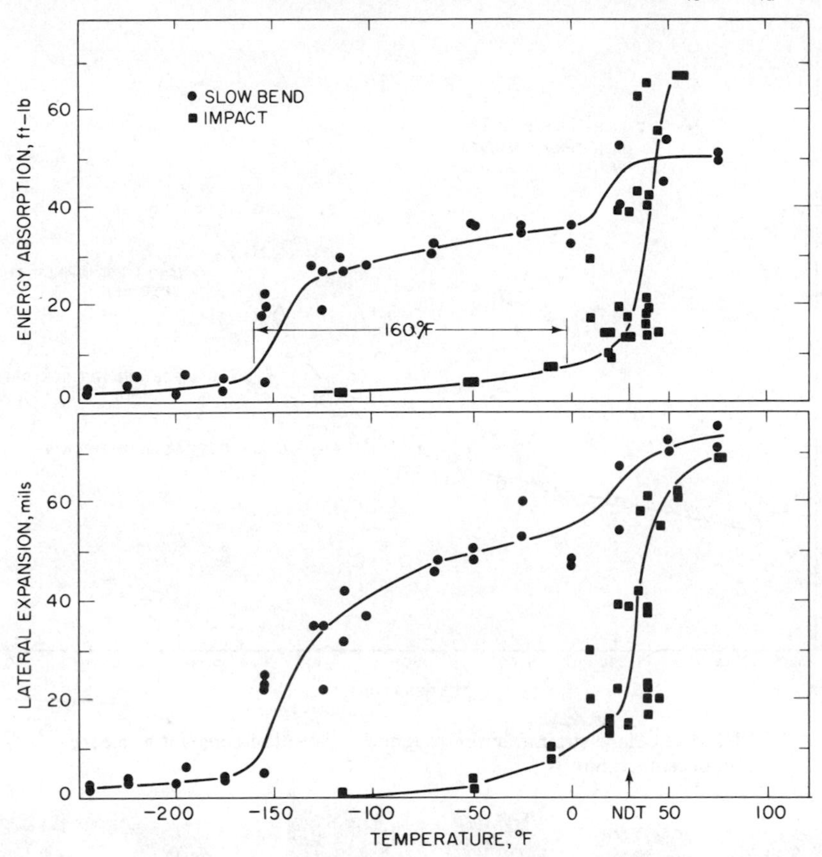

FIG. 4.20. Charpy V-notch energy absorption and lateral expansion for impact and slow-bend tests of standard CVN specimens.

were approximately 90% quasi-cleavage, and at −50°F (approximately the limit of the plane-strain transition at which point the plane-strain *constraint* at the crack tip decreased significantly) the fractures were approximately 90% tear dimples.

Similar behavior exists for CVN specimens,[1] leading to the conclusion that the transition-temperature behavior in K_{Ic} (plane-strain transition) and Charpy tests reflects predominantly a transitional change in the microscopic mode of fracture from quasi-cleavage at very low temperatures to tear dimples at the upper shelf region of the CVN test results.

Under the combined effects of both transitions, i.e., the plane-strain K_{Ic} transition and the transition from plane strain to plane stress, the rate of change of K_c as a function of temperature should be *greater* than the rate shown in Figure 4.21. Data obtained from $\frac{1}{2}$-in.-thick specimens of the same material described in Figure 4.21 substantiate this assumption. The test

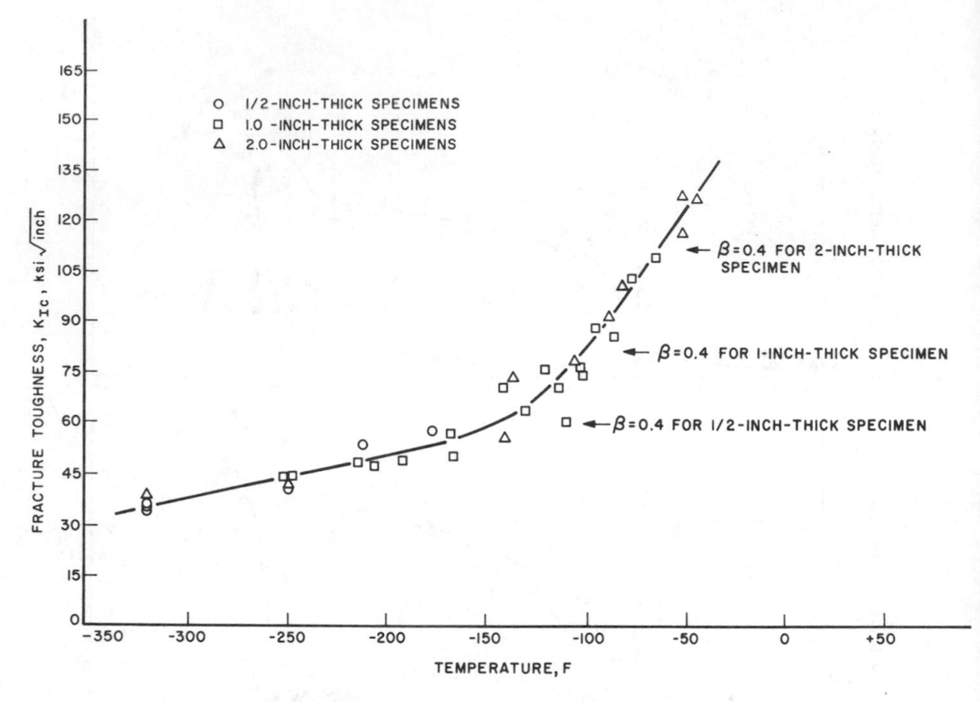

FIG. 4.21. Plane-strain fracture-toughness transition behavior as a function of temperature.

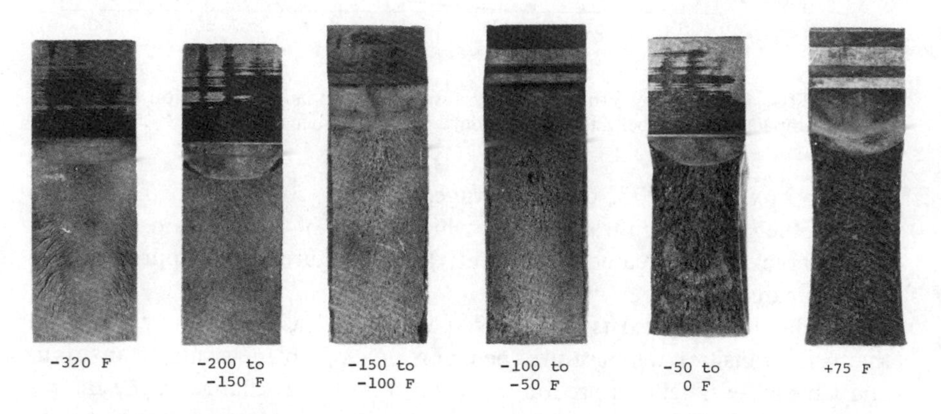

FIG. 4.22. Fracture surfaces of K_{Ic} specimens at various temperatures.

results from these $\frac{1}{2}$-in.-thick specimens are presented in Figure 4.24 and compared with the test results for the 1- and 2-in.-thick valid K_{Ic} test results of Figure 4.21. Approximate values of K_{Ic} were calculated from the plane-stress results obtained on the $\frac{1}{2}$-in.-thick specimens by using Irwin's β relation between K_{Ic} and K_c that is applicable only when the differences

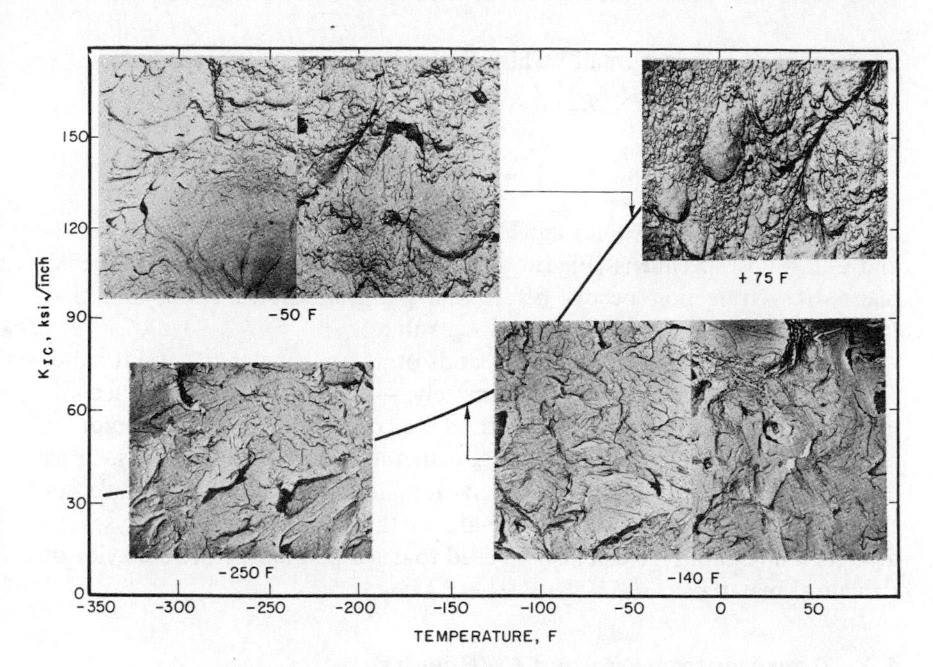

FIG. 4.23. Fractographs of K_{Ic} specimens at various temperatures.

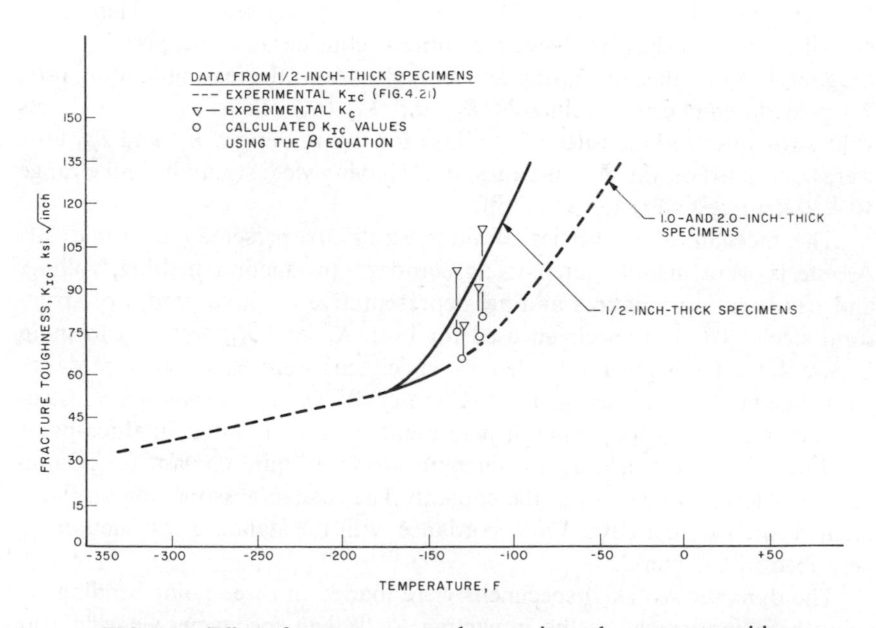

FIG. 4.24. Effect of temperature on plane-strain to plane-stress transition.

between K_c and K_{Ic} are small.[2] This relation is as follows:

$$K_c^2 = K_{Ic}^2[1 + 1.4\beta_{Ic}^2]$$

where

$$\beta_{Ic} = \frac{1}{B}\left(\frac{K_{Ic}}{\sigma_{ys}}\right)^2$$

The calculated K_{Ic} values agree quite well with those obtained from 1- and 2-in.-thick specimens (Figure 4.24), demonstrating the fact that the K_c plane-stress transition occurs before the plane-strain transition and that Irwin's β concept does relate K_c and K_{Ic} values.

Above some temperature that depends on the particular structural material (for this A517 steel it was approximately −40°F), valid K_{Ic} results according to the standard ASTM test method described in Chapter 3 cannot be obtained. For these cases, auxiliary test methods and correlations (such as described in Chapter 6) must be used to estimate K_{Ic} or K_{Id}. For high levels of elastic-plastic behavior, state-of-the-art methods of fracture analyses as described in Chapter 16 can also be used to describe the fracture behavior of structural materials.

4.5. Representative K_{Ic} and K_{Id} Results for Structural Steels

Low-strength structural steels (e.g., steels with yield strengths less than 140 ksi) generally are temperature- and loading-rate-sensitive. That is, as described previously, these steels exhibit a considerable increase in crack toughness with either increasing test temperature or decreasing loading rate. To provide engineering values of K_{Ic} and K_{Id} (as well as crack-toughness values for intermediate rates of loading) for use in design, K_{Ic} and K_{Id} tests were conducted on various structural steels having yield strengths in the range 40–250 ksi by Shoemaker and Rolfe.[5]

The mechanical properties of these steels are presented in Table 4.1. All steels were manufactured using standard production melting, rolling, and heat-treating practices and are representative of these grades of structural steels. The test specimen used for both K_{Ic} and K_{Id} tests is shown in Figure 4.25. The statically loaded K_{Ic} specimens were tested in accordance with the standard K_{Ic} test method (Chapter 3). The specimens were instrumented with a crack-opening-displacement gage and tested in three-point bending. They were held at test temperature in a liquid coolant for 20 min and then tested while still in the coolant. The load–crack-opening-displacement records were analyzed in accordance with the standard test method as described in Chapter 3.

The dynamic K_{Ic} (K_{Id}) specimens were loaded in three-point bending by using a falling weight as the impacting load. The specimens were held at

TABLE 4.1. Longitudinal Static Room-Temperature Mechanical Properties of Seven Structural Steels

				Elongation in 1 in. (%)		*Reduction of Area (%)*		*Charpy V-Notch Energy Absorption (ft-lb)*		
Steel	*Plate Thickness (in.)*	*Yield Strength, 0.2% Offset (ksi)*	*Tensile Strength (ksi)*	*At Maximum Load*	*At Fracture*	*At Maximum Load*	*At Fracture*	*+80°F*	*0°F*	*−80°F*
ABS-C	1	39	63	24.0	36.0	26.0	66.8	87	19	3
A302-B	1	56	88	11.0	26.0	17.4	67.0	61	21	7
HY-80	1	84	99	11.0	25.0	13.4	74.9	116	113	113
A517-F	1	118	129	8.0	19.0	11.8	65.4	47	32	14
HY-130	1	137	143	8.0	20.0	10.8	70.9	98	106	99
18Ni(180)	1	180	189	3.0	14.0	6.6	66.0	55	51	47
18Ni(250)	1	246	253	2.0	10.5	4.0	51.8	21	14	14

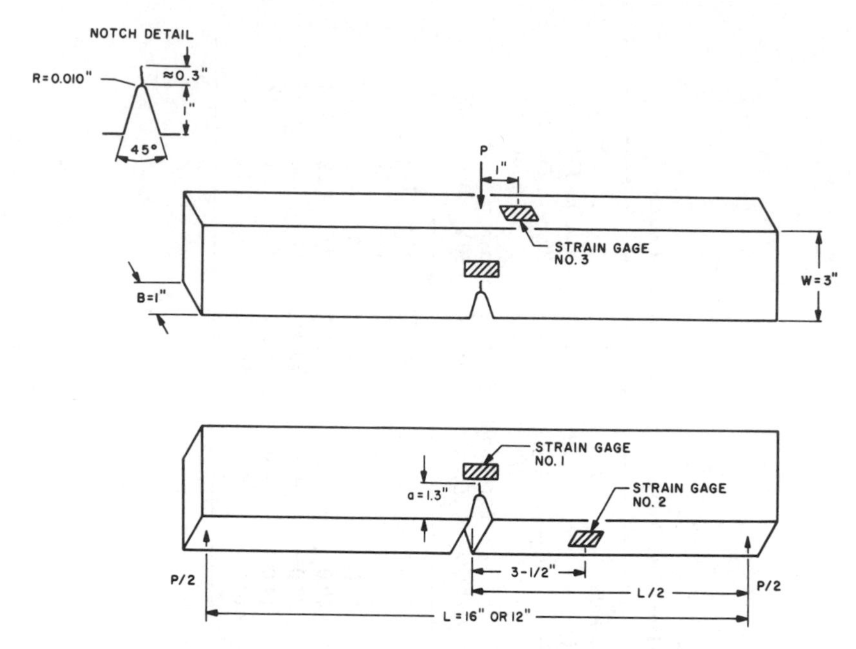

FIG. 4.25. Fatigue-cracked bend specimen.

temperature in a liquid coolant for 20 min and then removed for the test. The time between the removal of the specimen from the bath and the testing was such that the maximum temperature increase was 10°F.

The quasi-static analysis described in Chapter 3 was used to determine the dynamic K_{Ic} values. The output from strain gage 2, shown in Figure 4.25, was determined as the crack extended. An equivalent static load necessary to produce this strain was used to determine the dynamic K_{Ic} values for the test conditions. Because the specimens were dynamically loaded in a drop-weight machine, inertial and "ringing" effects were minimized by using low impact velocities and inserting a dampening material such as soft aluminum or lead between the impacting tup and the specimen. The outputs from gages 1 and 3 (Figure 4.25) were used to determine the time at which the crack extension initiated. A complete description of this procedure is presented in Reference 5.

In addition to the tests at static and dynamic strain rates, several tests were conducted at an intermediate strain rate on three steels: ABS-C, A302-B, and A517. The tests were conducted at the highest speed attainable on a conventional closed-loop hydraulic testing machine.

For all bend tests, the plane-strain critical stress-intensity factor (K_{Ic}) was calculated from the following equation:

$$K_I = \frac{Y \times 6M(a)^{1/2}}{BW^2}$$

where M = applied bending moment,
a = crack length,
B = specimen thickness,
W = specimen width,
$Y = F(a/W)$ as described in Chapter 3.

The validity of the K_{Ic} values was determined according to the present ASTM test method (Chapter 3). The ratio of K_{Ic}/σ_{ys} was always evaluated at the testing temperature and strain rate. For these dynamic tests, strain rates were calculated for a point on the elastic-plastic boundary according to the following equation:

$$\dot{\epsilon} = \frac{2\sigma_{ys}}{tE}$$

where σ_{ys} = yield strength for the test temperature and strain rate,
t = loading time for the test,
E = elastic modulus of the material.

The dynamic K_{Ic} tests (falling weight) were conducted at crack-tip strain rates varying between 10 and 40/sec ($\dot{K}$ between 10^5 and 3×10^5 ksi$\sqrt{\text{in.}}$/sec), and the intermediate strain-rate tests were conducted at strain rates of approximately 10^{-1}/sec ($\dot{K} \approx 3 \times 10^3$ ksi$\sqrt{\text{in.}}$/sec). The static K_{Ic} tests were conducted at strain rates of approximately 10^{-5}/sec ($\dot{K} \approx 3 \times 10^{-2}$ ksi$\sqrt{\text{in.}}$/sec).

The critical stress-intensity factors, K_{Ic} or K_{Id}, for different test temperatures and strain rates are shown in Figures 4.26 through 4.32 for the ABS-C, A302-B, HY-80, A517-F, HY-130, 18Ni(180), and 18Ni(250) steels, respectively. Also shown are the NDT temperatures determined from the drop-weight tests and the 50% fracture-appearance transition-temperature ($FATT_{50}$) values determined from Charpy V-notch impact test results.

The thickness requirement for plane-strain behavior is determined by the ratio of plastic-zone size and specimen thickness. This ratio is defined as

$$\beta = \frac{(K_{Ic}/\sigma_{ys})^2}{B}$$

where K_{Ic} = measured critical stress-intensity factor,
σ_{ys} = yield strength,
B = specimen thickness.

At present, the ASTM standard method of testing requires that β be equal to or less than 0.4 to ensure that crack extension occurs under plane-strain conditions. In Figures 4.26 through 4.32 the dashed lines represent constant values of 0.4 for either static or dynamic loading of the 1-in.-thick plates. Data points falling below these dashed lines are valid K_{Ic} values. The "K_{Ic}"

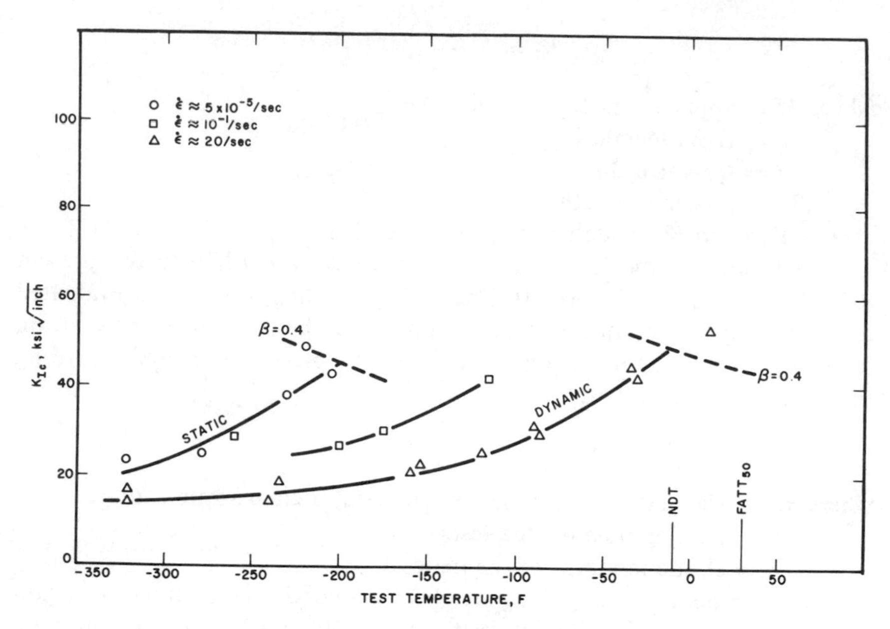

FIG. 4.26. Effect of temperature and strain rate on crack toughness of ABS-C steel.

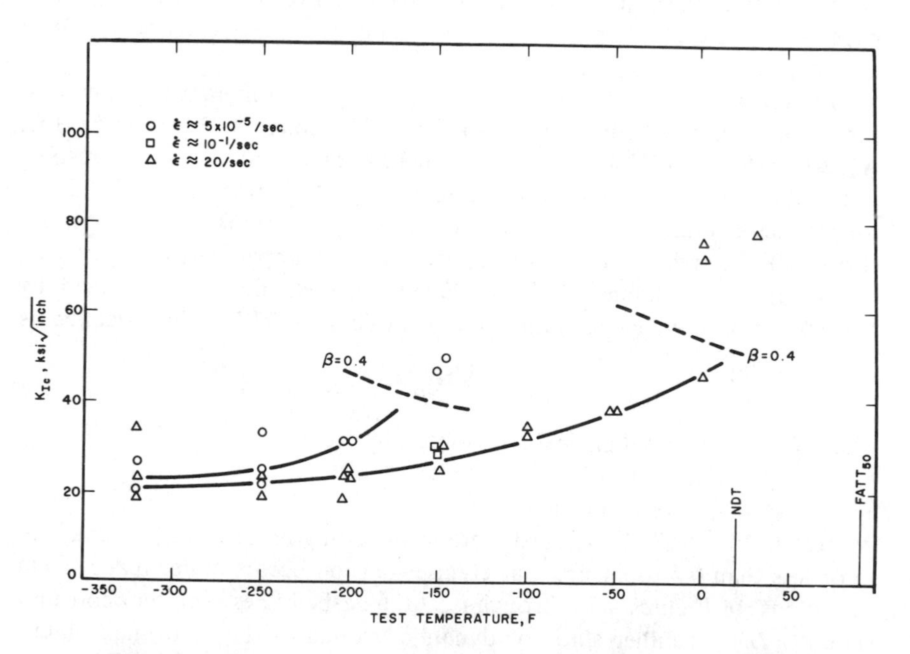

FIG. 4.27. Effect of temperature and strain rate on crack toughness of A302-B steel.

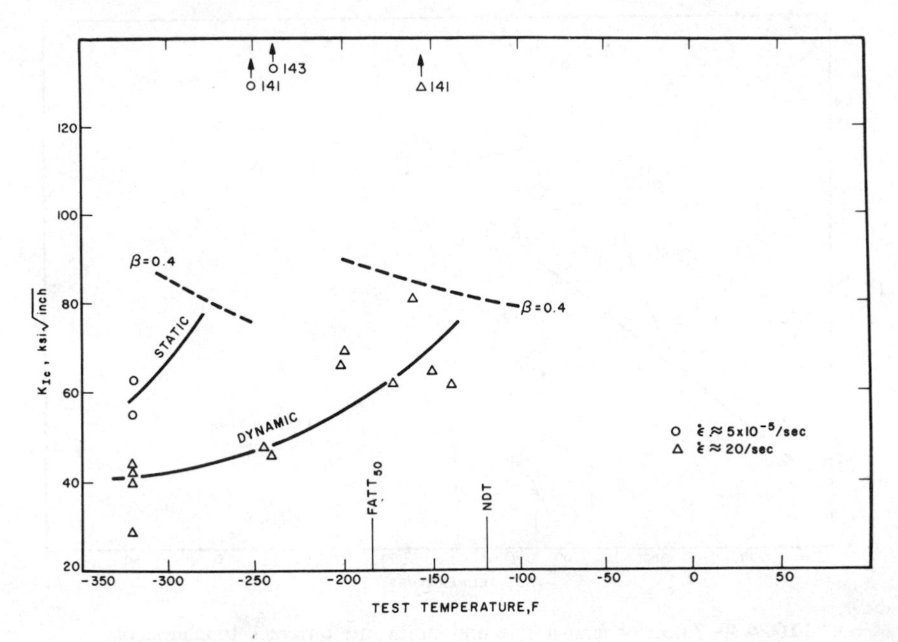

FIG. 4.28. Effect of temperature and strain rate on crack toughness of HY-80 steel.

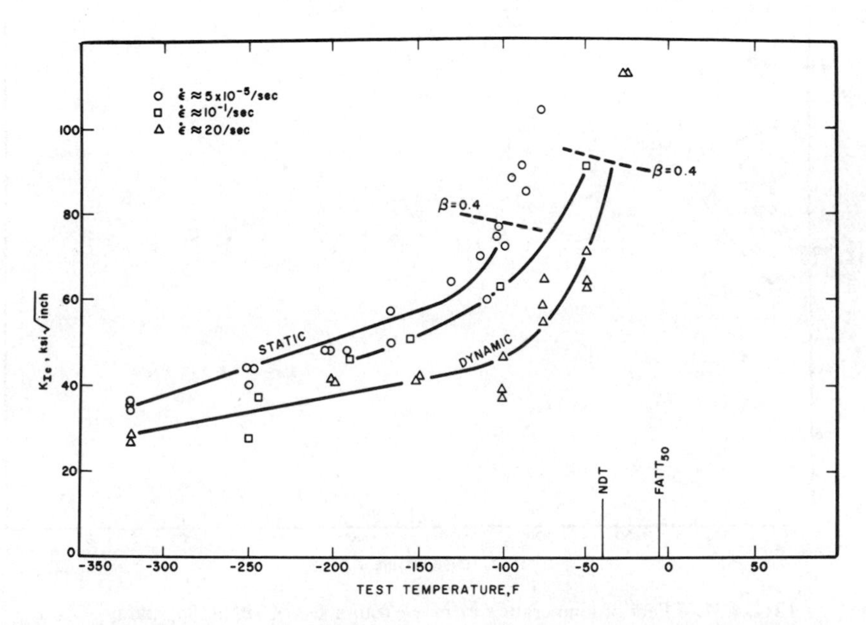

FIG. 4.29. Effect of temperature and strain rate on crack toughness of A517-F steel.

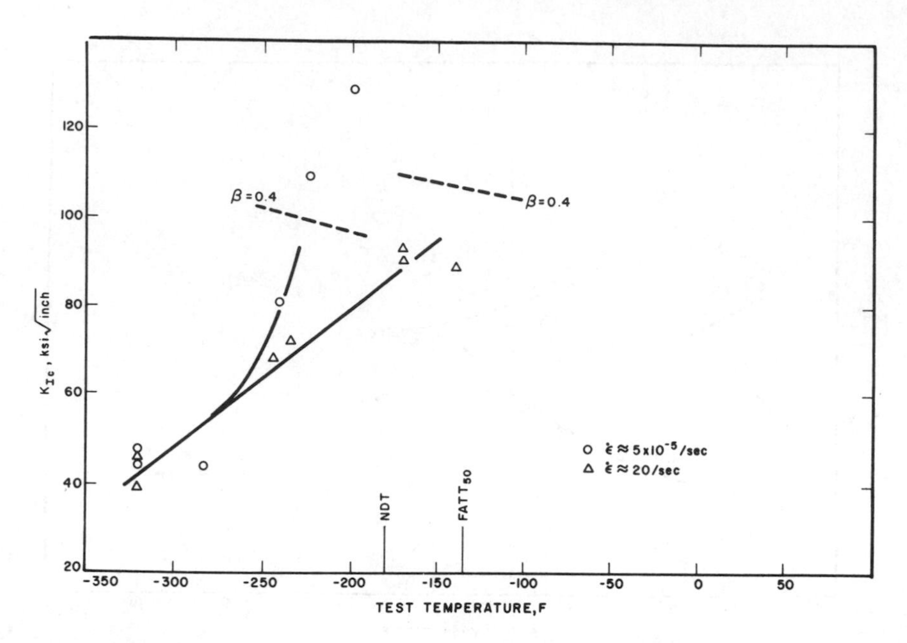

FIG. 4.30. Effect of temperature and strain rate on crack toughness of HY-130 steel.

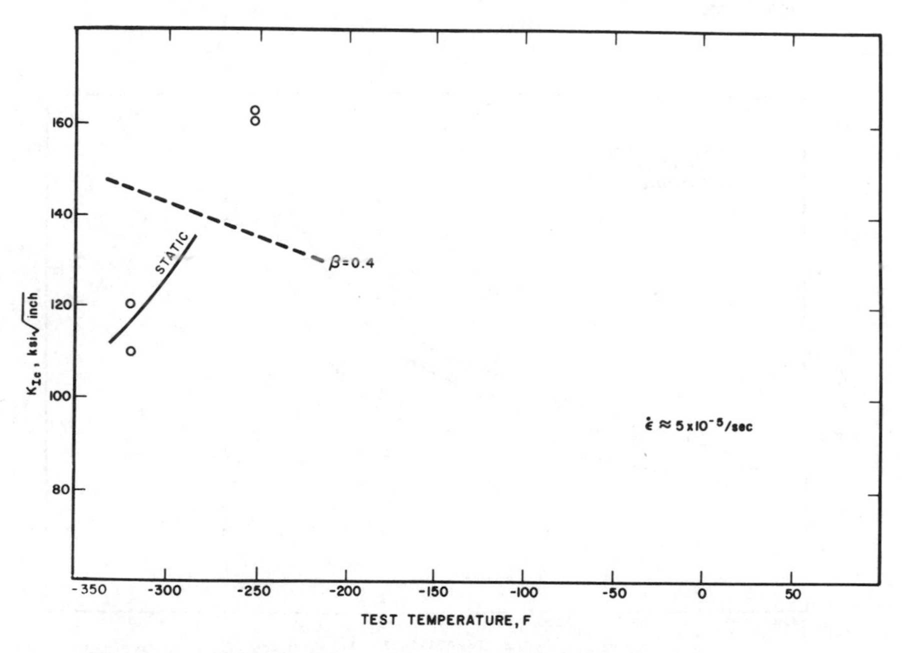

FIG. 4.31. Effect of temperature on crack toughness of 18Ni(180) maraging steel.

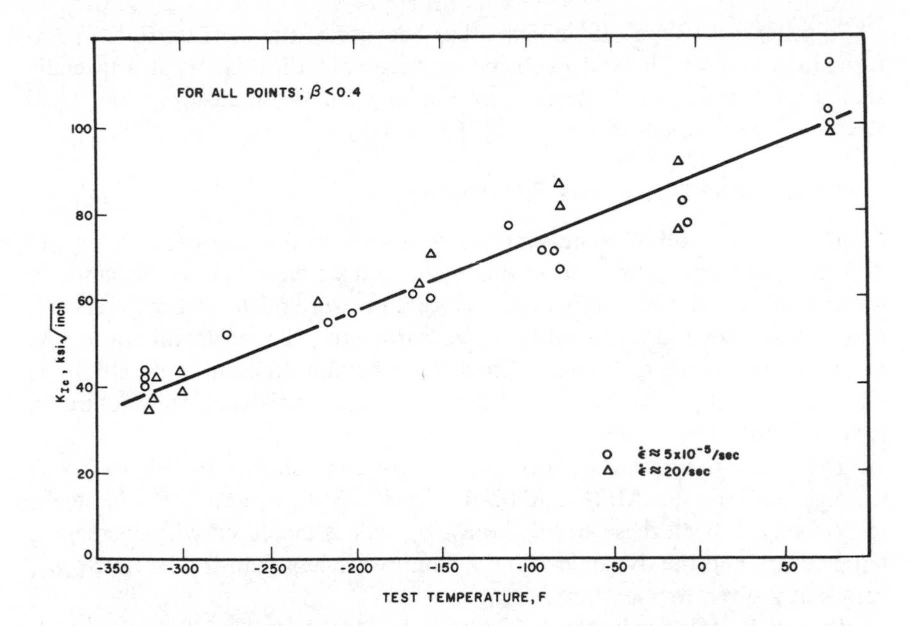

FIG. 4.32. Effect of temperature and strain rate on crack toughness of 18Ni(250) maraging steel.

values measured above these lines might be affected by crack-tip conditions in which the stress state is changing from plane strain to plane stress. Hence, these values cannot be considered valid by current ASTM standards, although they appear to be extensions of the valid results and are useful approximations for engineering purposes.

For the strain rates used in this investigation, the K_{Ic} values for the ABS-C, A302-B, HY-80, A517-F, and HY-130 steels (Figures 4.26 through 4.30) decreased with increasing loading rate for a constant testing temperature. Furthermore, the K_{Ic} values obtained at the intermediate loading rate were always between those obtained by static and by dynamic loading. The most significant effect of increased loading rate on these five steels was the increase in the threshold temperature below which plane-strain behavior occurred. This increased range of K_{Ic} behavior was due to both an increase in yield strength and a reduction in K_{Ic} caused by the increase in strain rate.

Only a few static crack-toughness (K_{Ic}) values were obtained for the 18Ni(180) maraging steel (Figure 4.31). Meaningful dynamic test records could not be obtained because elastic “ringing” was encountered at the high loads necessary to obtain the high K_{Id} values. No significant difference between static and dynamic K_{Ic} values was observed for the 18Ni(250) maraging steel (Figure 4.32).

In general, these results indicate that because of the increase in temperature range over which valid K_{Ic} behavior occurs, loading rate is an especially significant variable in K_{Ic} testing, particularly for those steels having yield strengths less than about 140 ksi.

4.5.1. Crack-Toughness Performance

The best method of comparing the resistance to fracture of steels having different yield strengths is to evaluate the crack-toughness performance in terms of K_{Ic}/σ_{ys}. Because this ratio is both a measure of the critical crack-tip plastic-zone size and a critical flaw size parameter, the larger this ratio, the better the resistance to fracture. Thus steels having different yield strengths but the same K_{Ic}/σ_{ys} ratio should have the same resistance to fracture in terms of critical flaw size.

The crack-toughness performance, K_{Ic}/σ_{ys}, is shown in Figures 4.33 through 4.37 for the ABS-C, A302-B, HY-80, A517-F, and HY-130 steels, respectively. For all these steels, the K_{Ic}/σ_{ys} values increased with increasing temperature and the dynamic K_{Id}/σ_{yd} values were less than the static K_{Ic}/σ_{ys} values at a given temperature.

K_{Ic} and K_{Id} test results for A533 Grade B, Class 1, and A508 steels, which are used in pressure-retaining components of nuclear reactors, are presented

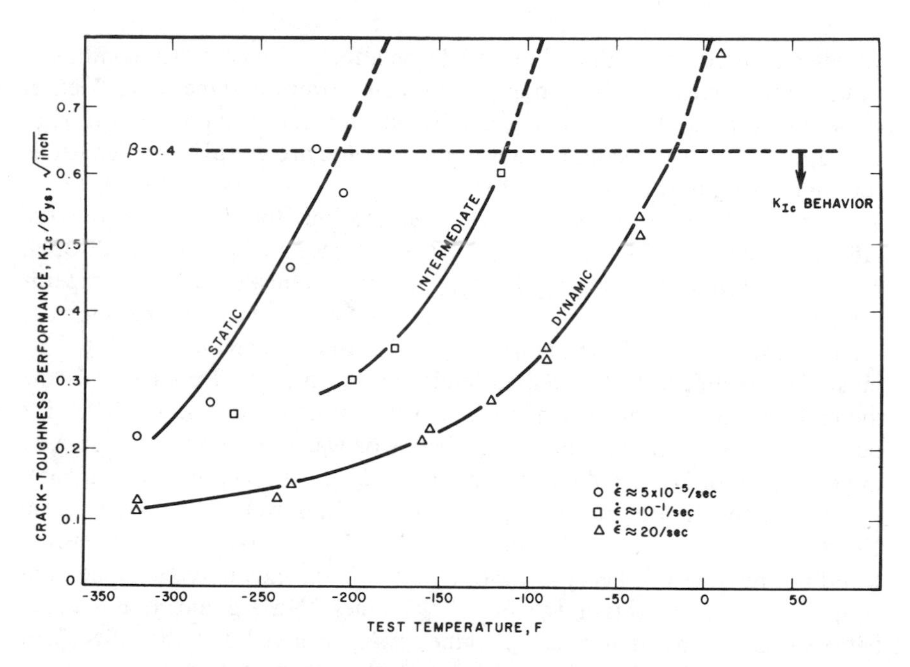

FIG. 4.33. Crack-toughness performance for ABS-C steel.

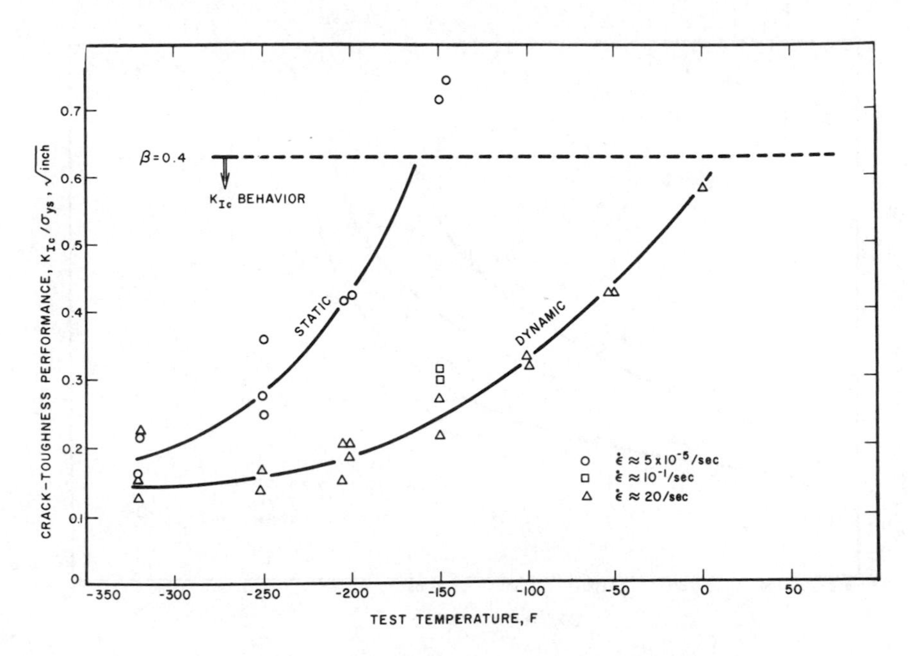

FIG. 4.34. Crack-toughness performance for A302-B steel.

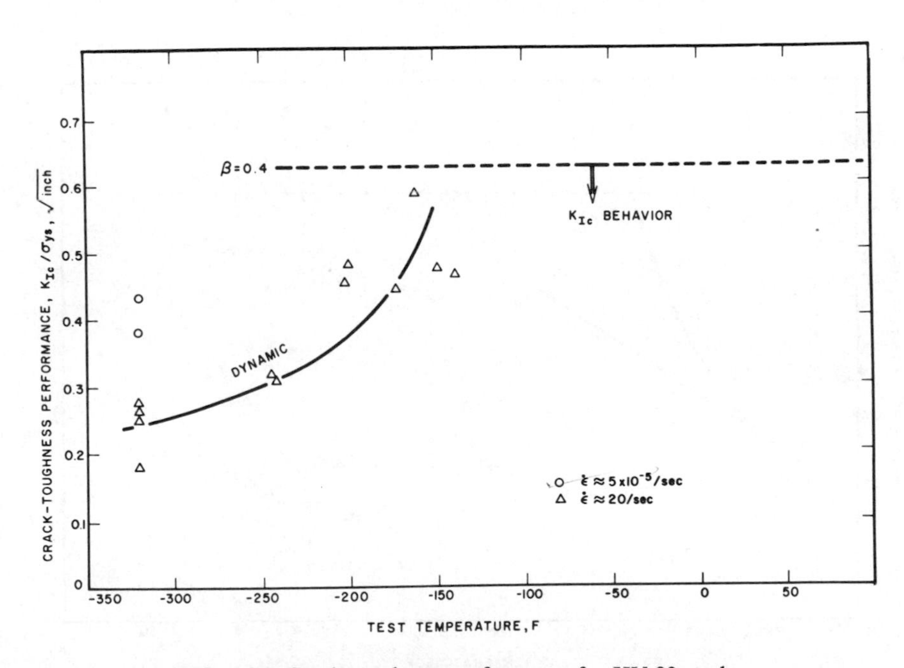

FIG. 4.35. Crack-toughness performance for HY-80 steel.

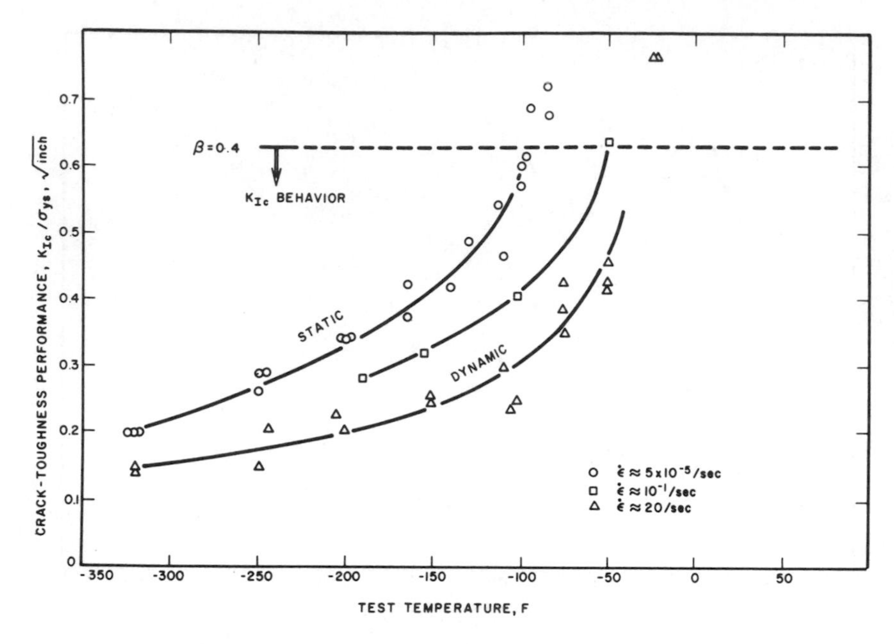

FIG. 4.36. Crack-toughness performance for A517-F steel.

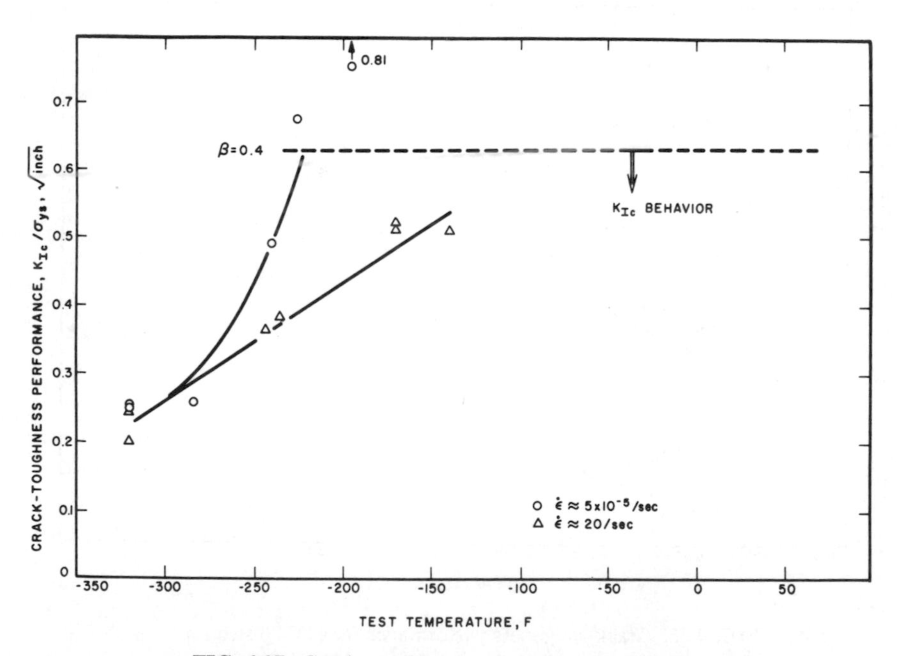

FIG. 4.37. Crack-toughness performance for HY-130 steel.

$$K_{IR} = 26.777 + 1.223 \exp\{0.0145[T - (NDT - 160)]\}$$

RT_{NDT}	K_{IR} (ksi$\sqrt{in.}$)
NDT − 160	28.00
NDT − 140	28.41
NDT − 120	28.96
NDT − 100	29.7
NDT − 80	30.68
NDT − 60	31.99
NDT − 40	33.74
NDT − 20	36.08
NDT	39.21
NDT + 20	43.39
NDT + 40	48.97
NDT + 60	56.44
NDT + 80	66.41
NDT + 100	79.73
NDT + 120	97.54
NDT + 140	121.33
NDT + 160	153.13
NDT + 180	195.61

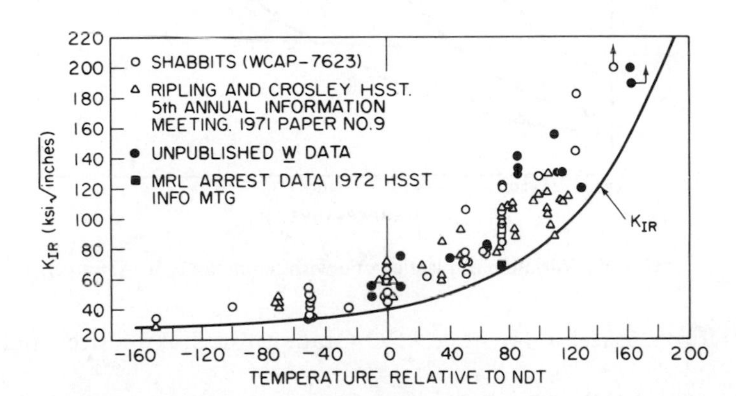

FIG. 4.38. Derivation of curve of reference stress-intensity factor (K_{IR}).

in Figure 4.38. These test results are referenced to the NDT temperature. The lower-bound curve, called K_{IR}, represents a very conservative assumption to the critical dynamic stress-intensity (K_{Id}) versus temperature properties of this class of steels. As part of the development of a fracture-control plan for nuclear pressure vessels, an equation for this lower bound was developed, and is shown in Figure 4.38. The fracture-control plan is described in Section 15.2.

Similar K_{Ic} and K_{Id} test results for various structural steels used in bridges and buildings are presented in Figures 4.39 to 4.43. These results include test

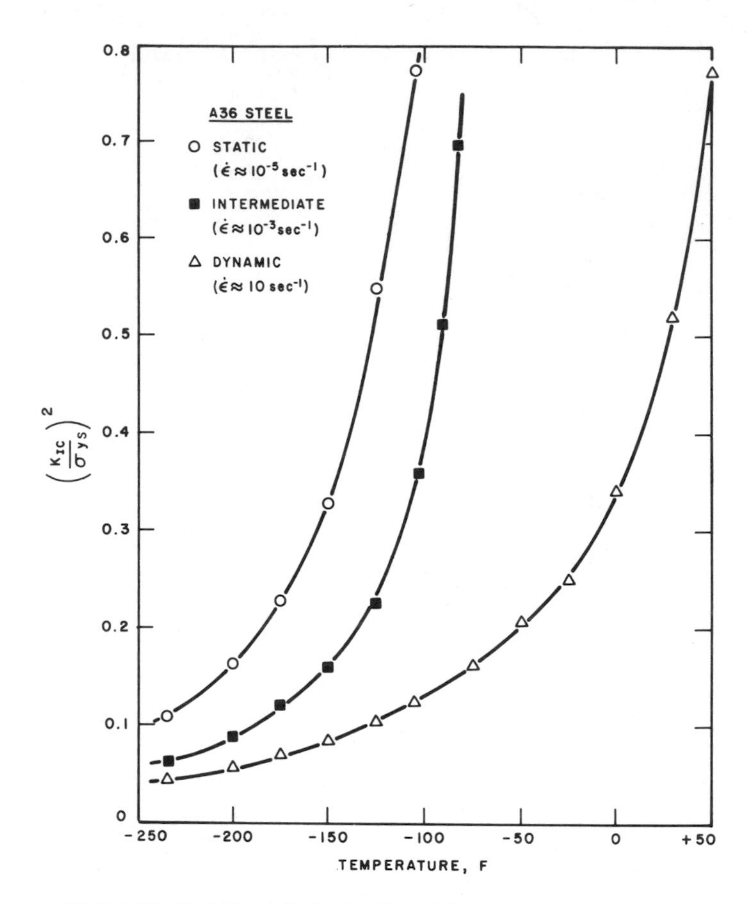

FIG. 4.39. Variation of plastic-zone with temperature in A36 steel.

values for A36 steel, A441 steel, A572 Grade 50 steel, A588 steel, and A517 steel.

K_c, K_{Ic}, and K_{Id} test results for bridge steels have been determined by Roberts et al.[6] as part of a program on fracture toughness of bridge steels sponsored by the Department of Transportation.

All the above results as well as those of Roberts et al.[6] show that these structural materials first undergo a gradual increase in toughness with increasing temperature, referred to as the plane-strain transition. Then, as the inherent toughness increases, and the constraint at the crack tip decreases, these materials undergo a very rapid increase in toughness as they begin to exhibit elastic-plastic behavior. At this point, the K_I analysis is no longer valid because of the excessive yielding at the crack tip, and neither K_{Ic} nor K_{Id} values can be determined. Although this type of behavior is very desirable in structural materials, it makes the application of fracture mechanics in materials testing very difficult.

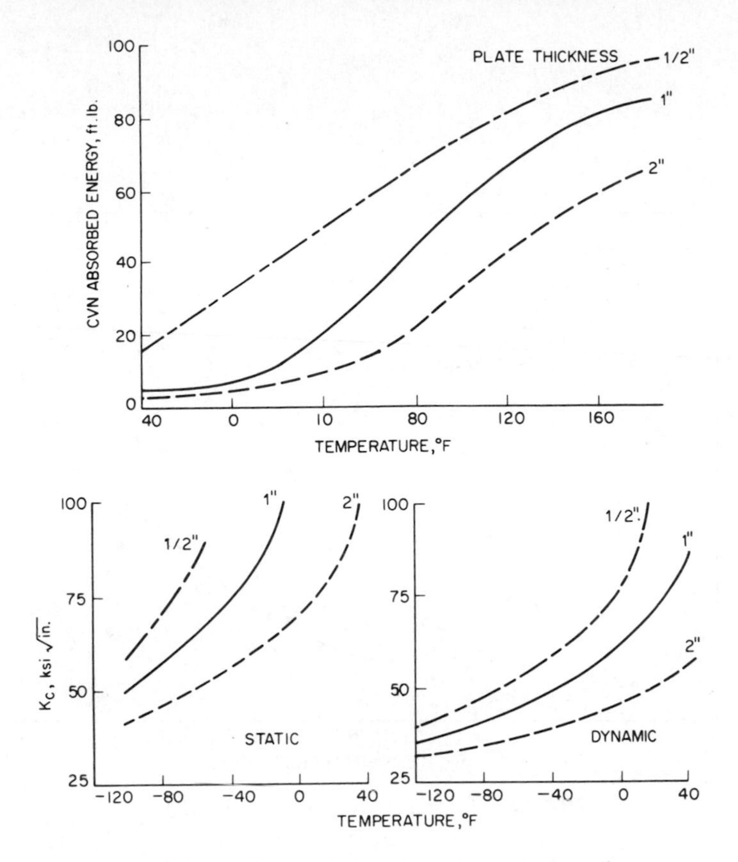

FIG. 4.40. CVN and K_c test results for A441 steel.

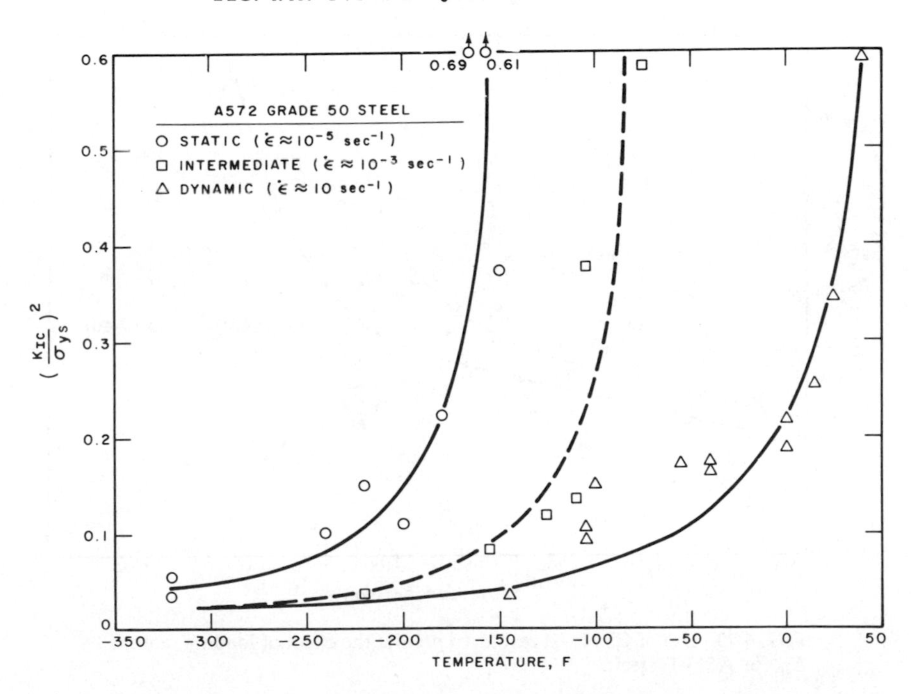

FIG. 4.41. Variation of plastic-zone size with temperature in A572 Grade 50 steel.

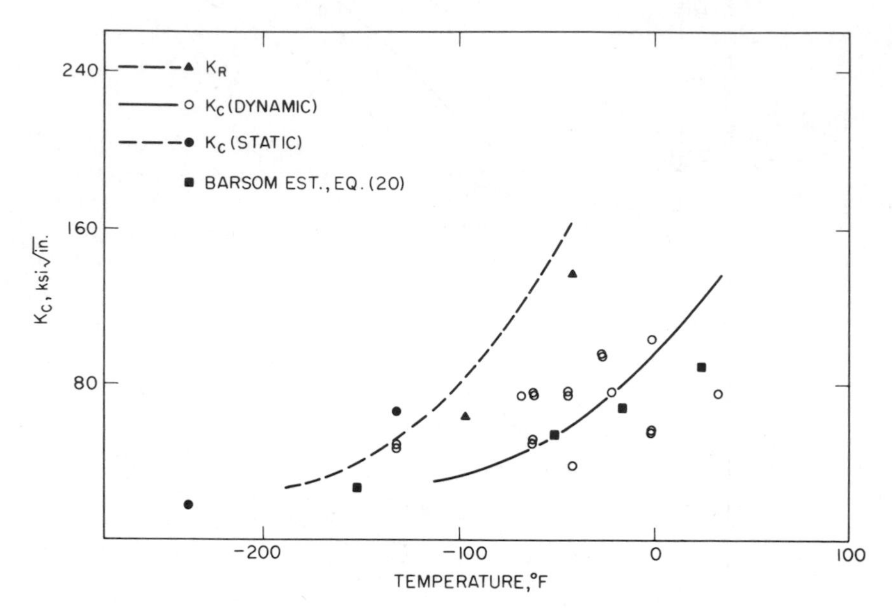

FIG. 4.42. K_c data for A588B 1-in. material.

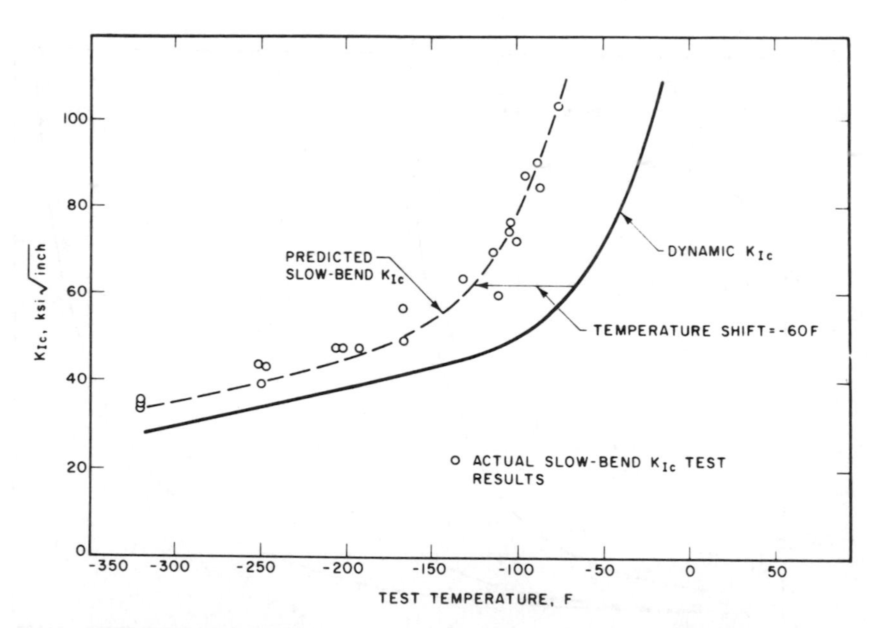

FIG. 4.43. Use of CVN test results to predict the effect of loading rate on K_{Ic} for A517-F steel.

4.6. Loading-Rate Shift for Structural Steels

The K_{Ic} and K_{Id} test results presented in Section 4.5 demonstrated the general effect of loading rate on K_{Ic}. A similar effect of loading rate exists for CVN specimens tested in three-point slow-bend and standard impact loading. The general effect of a slow loading rate (compared with standard impact loading rates for CVN specimens) is to shift the CVN curve to the left and to lower the upper shelf values. This behavior was shown schematically in Figure 4.18.

For low-strength steels the rate of change of absorbed energy as a function of temperature is greater in the impact test than in the slow-bend tests. Thus the magnitude of the temperature shift caused by high-strain-rate testing should be measured, at the same energy level, from the onset of the dynamic temperature transition to the onset of the transition on the static curve as shown in Figure 4.18. This onset of the dynamic temperature transition is defined arbitrarily by the intersection of tangent lines drawn from the lower shelf level and the transition region (Figure 4.18). The loading-rate shift has been verified by Roberts et al.[6] using the 15-ft-lb level so the presence of the shift does not depend on the particular means used to measure it. However, the onset of dynamic transition seems to be the best reference point from which to measure strain-rate effects because this point is located in the energy-absorption region where a change in the microscopic mode of fracture starts to occur at the initial crack front for both static and dynamic testing.[1] Also, because the onset of the static temperature transition occurs at a lower temperature than that marking the onset of the dynamic temperature transition and because the static upper energy-absorption shelf is usually of lower magnitude than that measured in the dynamic test, the static and dynamic energy-absorption curves usually intersect. Thus, measurements of the temperature shift at temperatures above that defined by the onset of dynamic temperature transition may underestimate the magnitude of the shift. Below this reference temperature, the slopes with respect to the temperature axis of both the static and the dynamic CVN energy become very small; consequently, it is difficult to measure the magnitude of the shift between the two curves in the lower shelf region.

Slow-bend and impact CVN test results for nine steels having yield strengths in the range 40–250 ksi are presented in Figures 4.44 through 4.52. The shifts in the transition temperature are related to yield strength, as shown in Figure 4.53. These results are similar to those observed by Roberts et al.[6]

The magnitude of the temperature shift between slow-bend loading and impact loading in steels of various yield strengths has been related to the room-temperature yield strength of the steel and can be approximated by[7]

$$T_{\text{shift}} = 215 - 1.5\sigma_{ys} \qquad \text{for } 36 \text{ ksi} < \sigma_{ys} < 140 \text{ ksi } (248 \text{ to } 965 \text{ MN/m}^2)$$

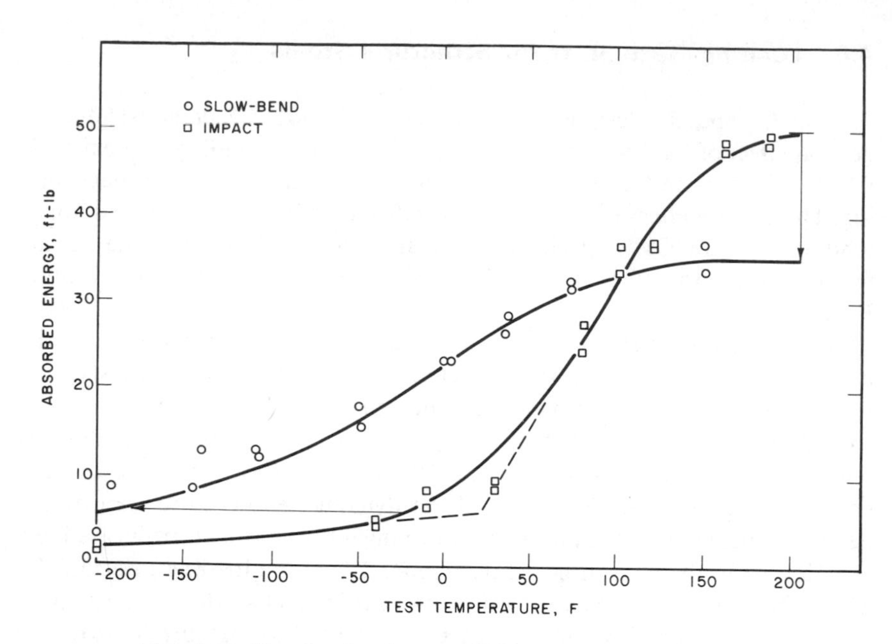

FIG. 4.44. Slow-bend and impact CVN test results for A36 steel.

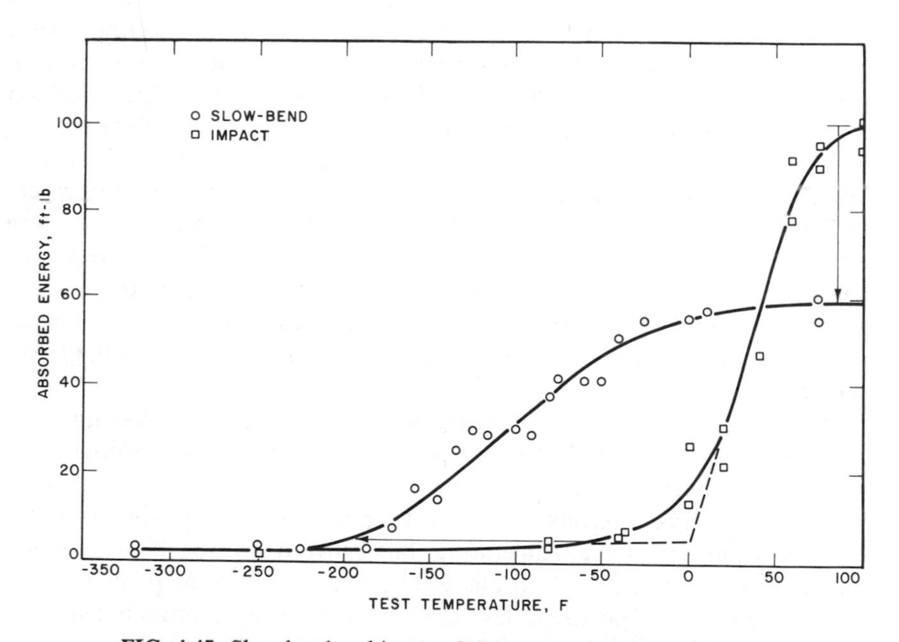

FIG. 4.45. Slow-bend and impact CVN test results for ABS-C steel.

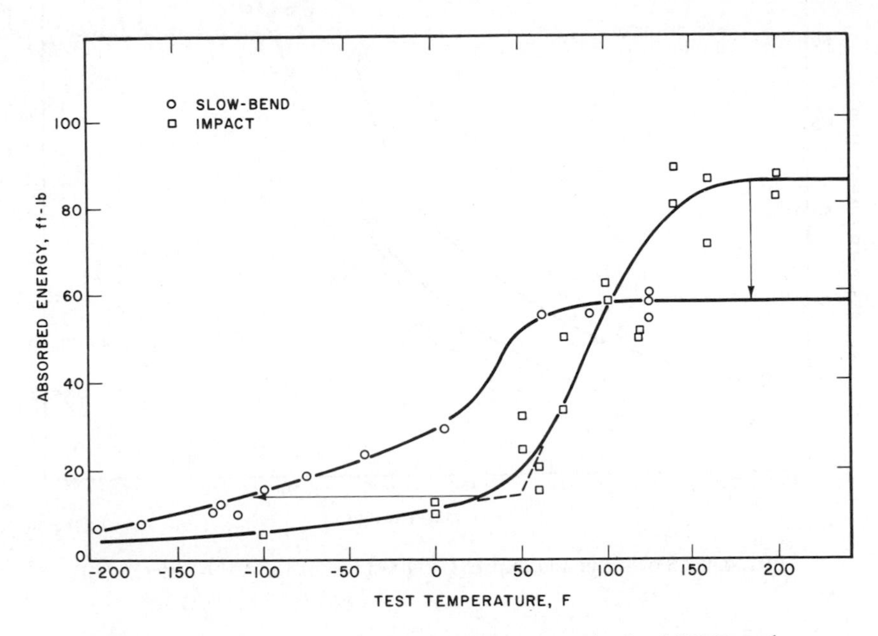

FIG. 4.46. Slow-bend and impact CVN test results for A302-B steel.

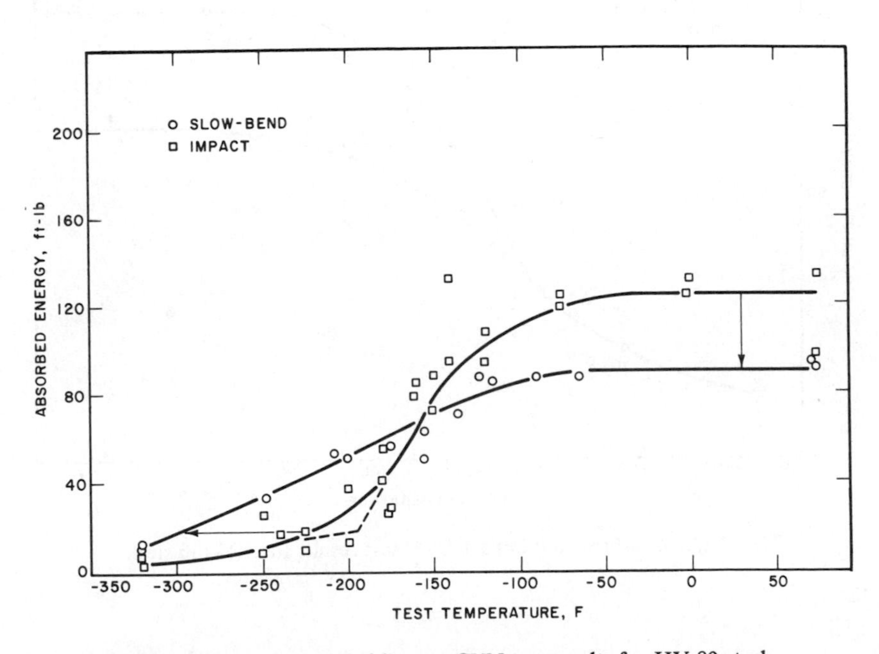

FIG. 4.47. Slow-bend and impact CVN test results for HY-80 steel.

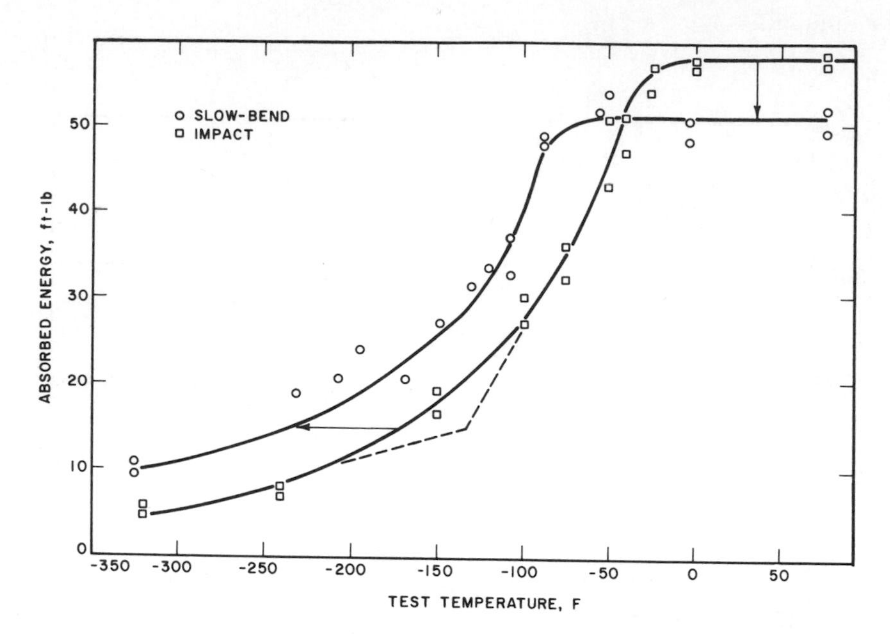

FIG. 4.48. Slow-bend and impact CVN test results for A517-F steel.

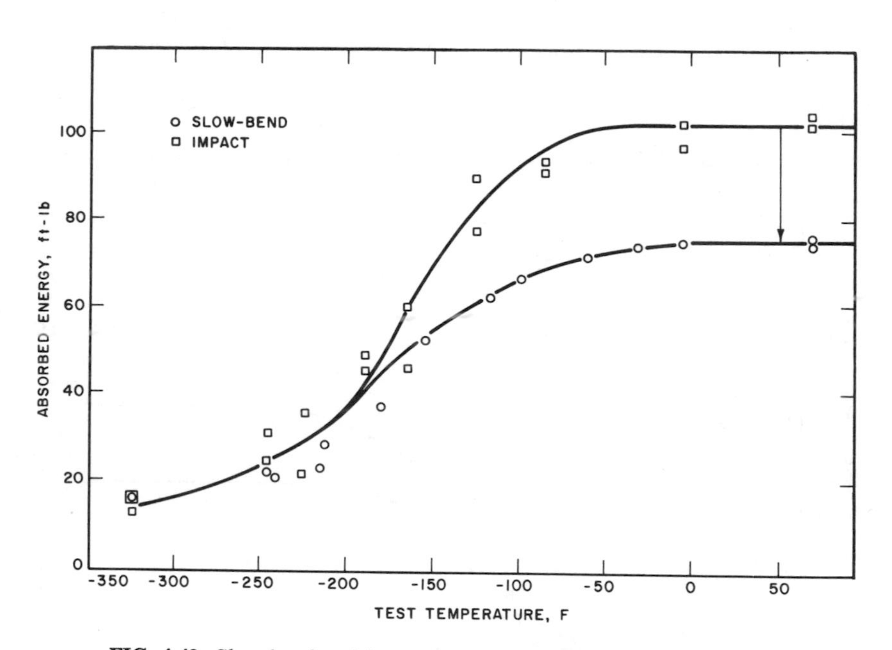

FIG. 4.49. Slow-bend and impact CVN test results for HY-130 steel.

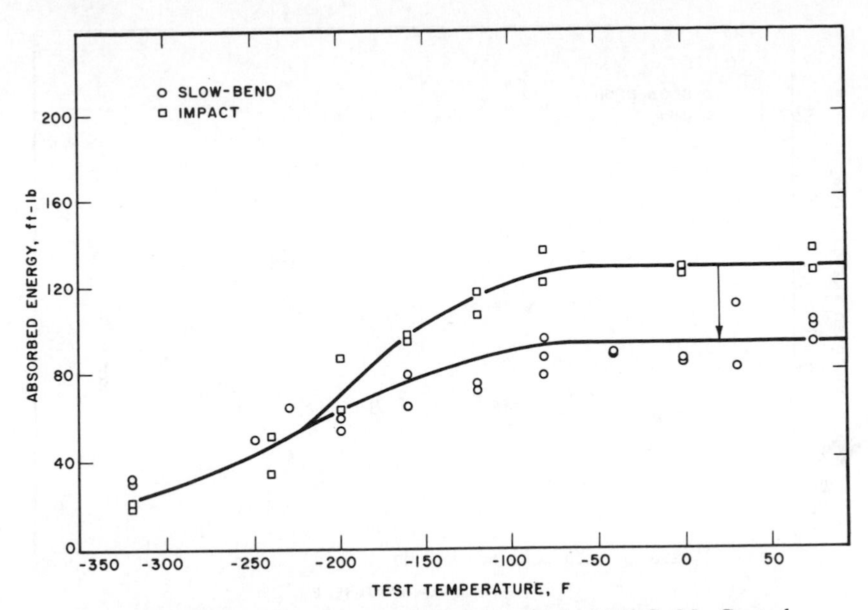

FIG. 4.50. Slow-bend and impact CVN test results for 10Ni-Cr-Mo-Co steel.

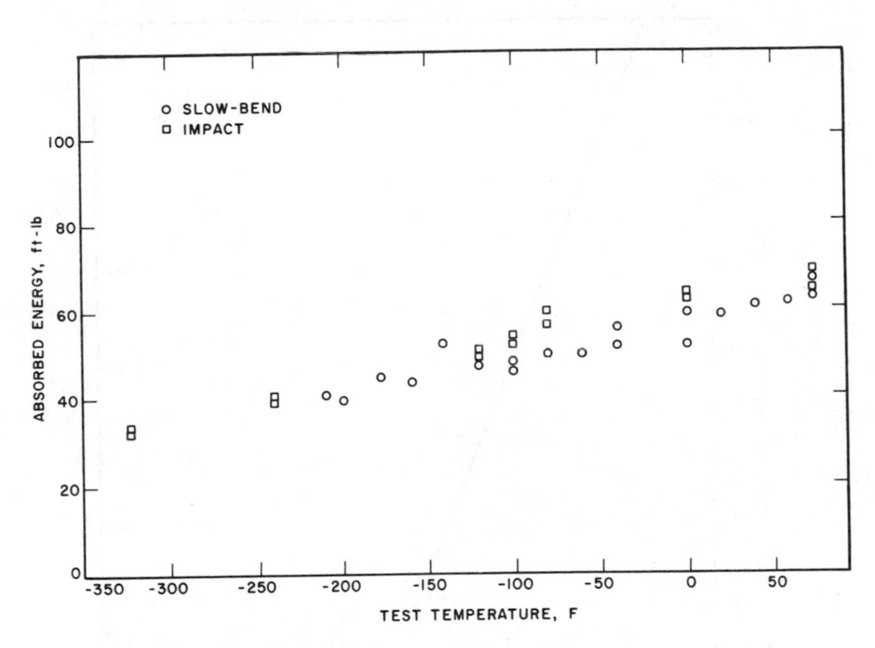

FIG. 4.51. Slow-bend and impact CVN test results for 18Ni(180) steel.

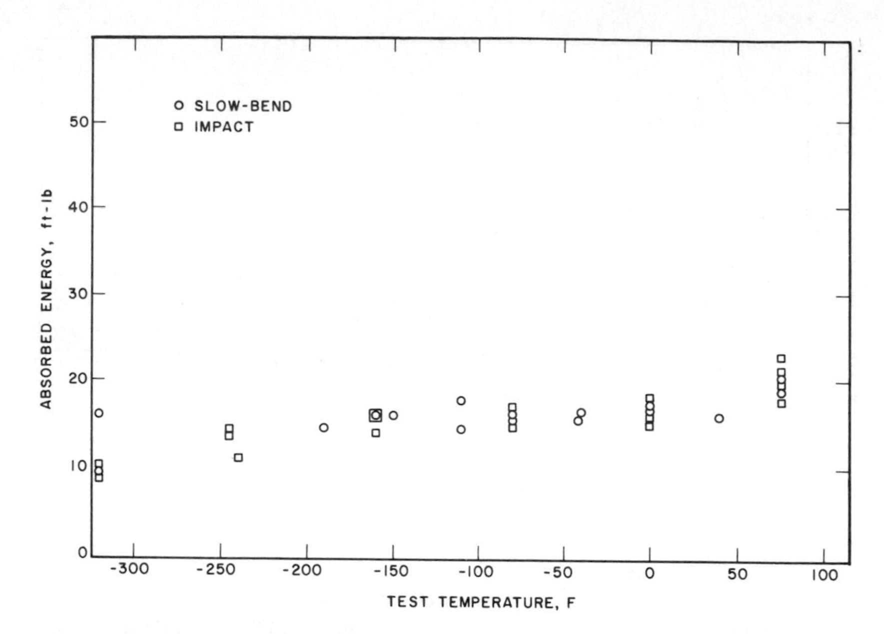

FIG. 4.52. Slow-bend and impact CVN test results for 18Ni(250) steel.

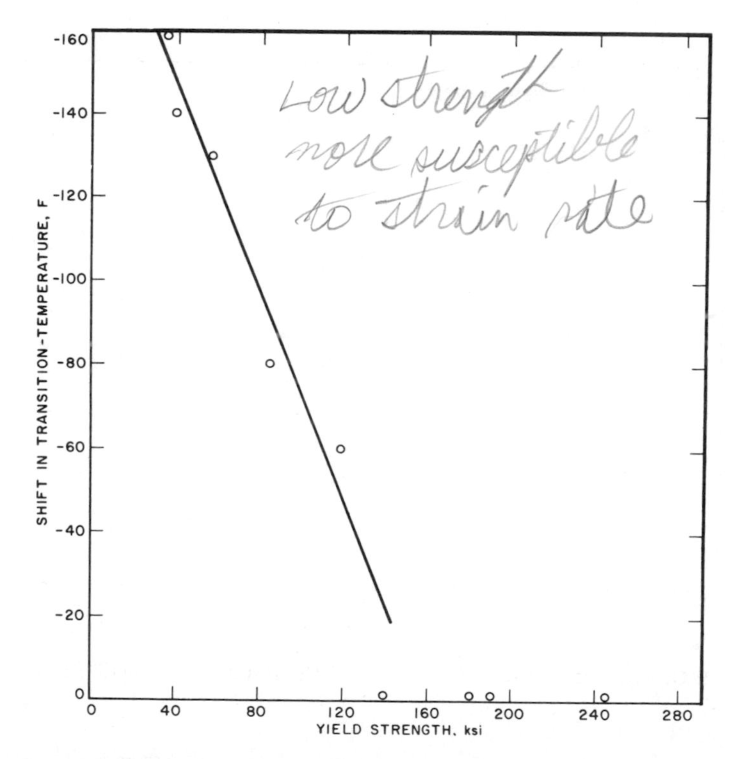

FIG. 4.53. Effect of yield strength on shift in transition temperature between impact and slow-bend CVN tests.

and

$$T_{shift} = 0.0 \qquad \text{for } \sigma_{ys} > 140 \text{ ksi}$$

where T_{shift} = absolute magnitude of the shift in the transition temperature between slow-bend loading and impact loading, °F,

σ_{ys} = room-temperature yield strength, ksi.

Because of this shift, increasing the loading rate can decrease the fracture-toughness value at a particular temperature for steels having yield strengths less than 140 ksi. The change in fracture-toughness values for loading rates varying from slow-bend to impact rates is particularly important to those applications where the actual loading rates are slow or intermediate. As discussed in Chapters 12 and 13, many types of structures fall into this classification.

For example, for a design stress of about 20 ksi (138 MN/m²), a K_{Ic} of 40 ksi$\sqrt{\text{in.}}$ (44 MN/m$^{3/2}$) for a given steel would correspond to the tolerance of a through-thickness flaw of approximately 3 in. (76 mm). If a structure were loaded statically, this size flaw could be tolerated at extremely low temperatures. If the structure were loaded dynamically, however, the temperature at which this size flaw could cause failure would be much higher. That is, using the test results for an ABS-C ship steel presented in Fig. 4.26 as an example, the static K_{Ic} of 40 ksi$\sqrt{\text{in.}}$ occurs at −225°F, whereas the dynamic K_{Id} of 40 ksi$\sqrt{\text{in.}}$ occurs at −50°F. Similar observations could be made for other steels as shown in Figs. 4.27 to 4.30. Note that this difference between the static and dynamic results decreases with increasing yield strength, as indicated in the above equation for T_{shift}.

If the loading rates of structures are closer to those of slow-bend loading than to impact loading, a considerable difference in the behavior of these structures would be expected. Thus, not only should the effects of temperature and constraint (plate thickness) on structural steels be established, but more important, *the maximum loading rates that will occur in the actual structure being analyzed under operating conditions should be established.* As discussed in Chapter 15, the loading-rate shift has been used to establish the AASHTO material-toughness requirements for bridge steels.[7]

The results presented in Figure 4.53 show that the greatest shifts in transition temperature occurred for the low-strength steels and decreased with increasing yield strength up to strength levels of about 140 ksi. Above about 140 ksi, no shift in transition temperature was observed.

A similar shift in K_{Ic} behavior as a function of loading rate has been observed by Shoemaker and Rolfe.[5] To determine whether the shift in CVN test values is the same as the shift in K_{Ic} test results, the dynamic K_{Ic} test results were shifted by an amount equal to the CVN transition-temperature shifts and compared with the actual slow-bend K_{Ic} test results (Figures 4.54 through 4.59). In general, the measured values agreed quite well with the

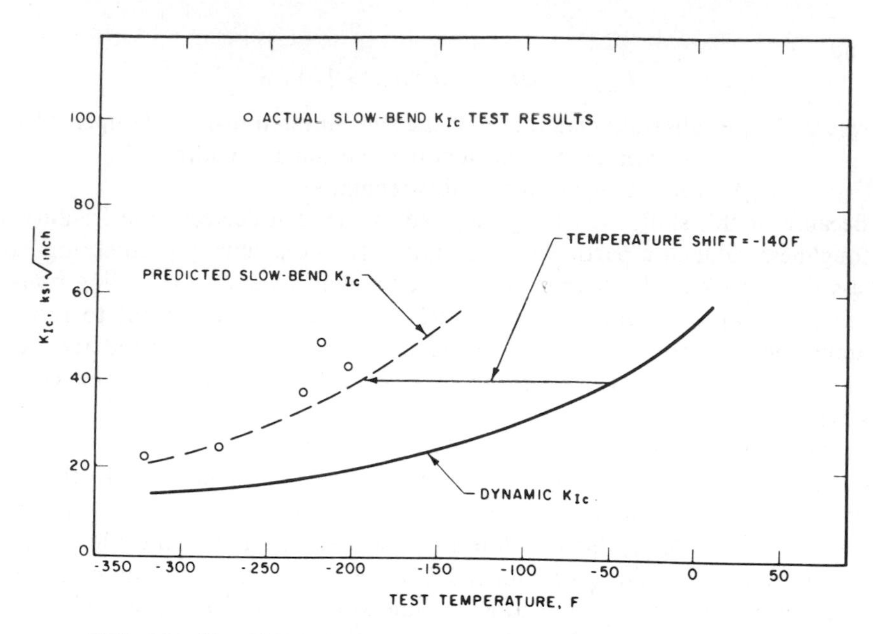

FIG. 4.54. Use of CVN test results to predict the effect of loading rate on K_{Ic} for ABS-C steel.

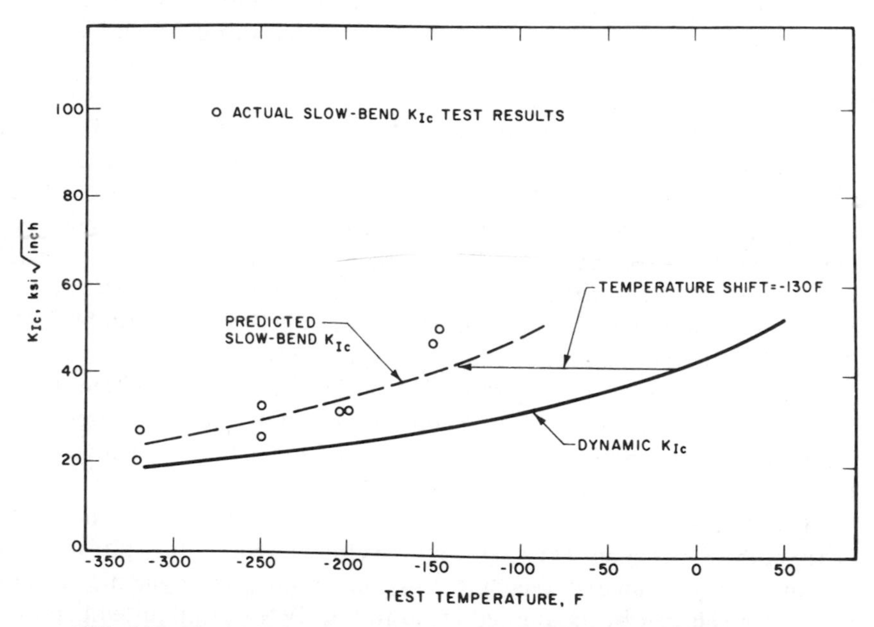

FIG. 4.55. Use of CVN test results to predict the effect of loading rate on K_{Ic} for A302-B steel.

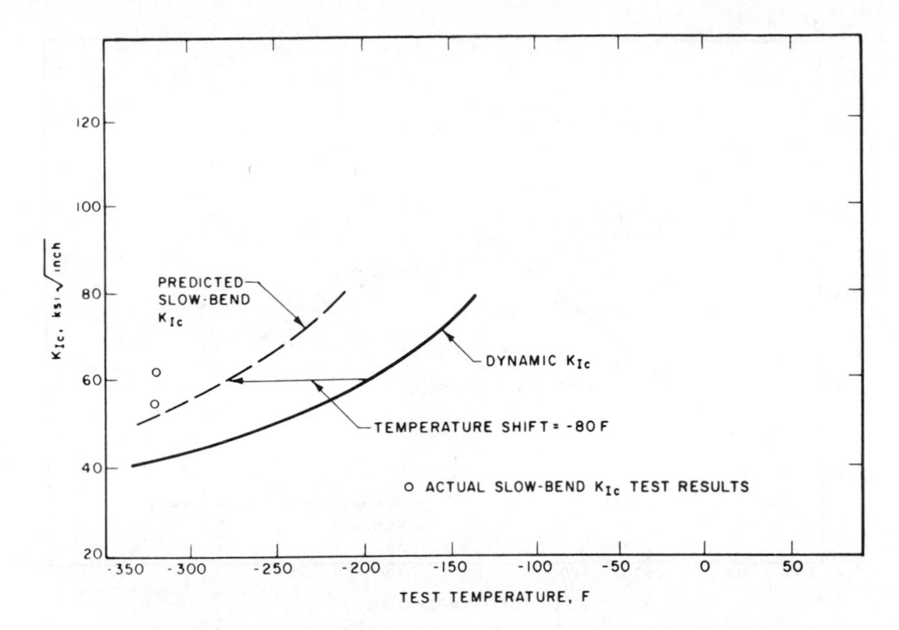

FIG. 4.56. Use of CVN test results to predict the effect of loading rate on K_{Ic} for HY-80 steel.

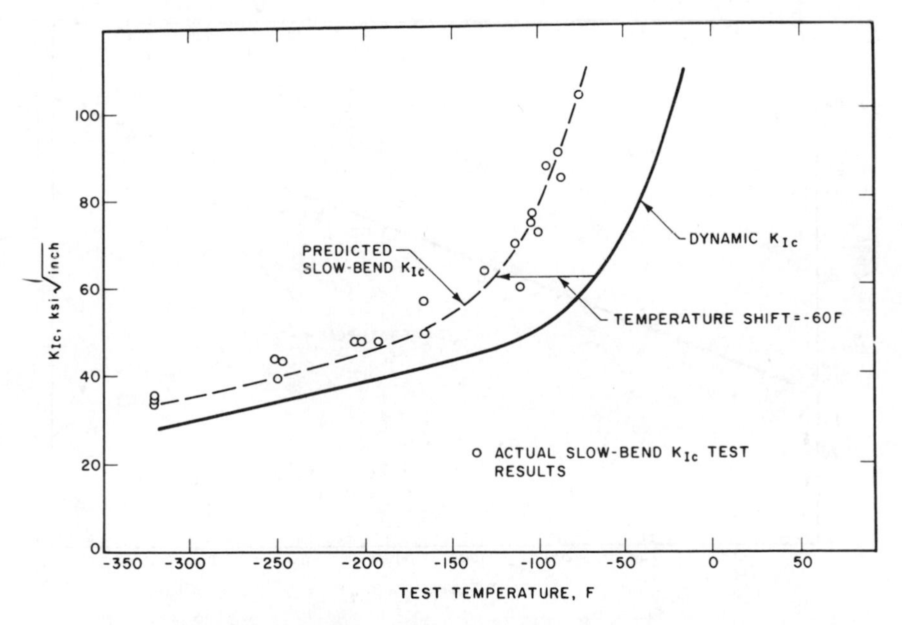

FIG. 4.57. Use of CVN test results to predict the effect of loading rate on K_{Ic} for A517-F steel.

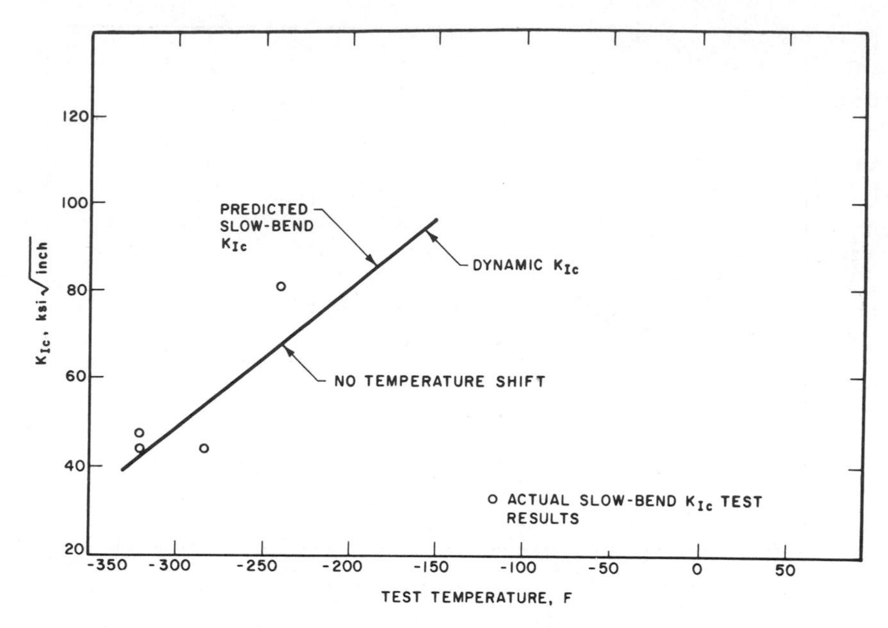

FIG. 4.58. Use of CVN test results to predict the effect of loading rate on K_{Ic} for HY-130 steel.

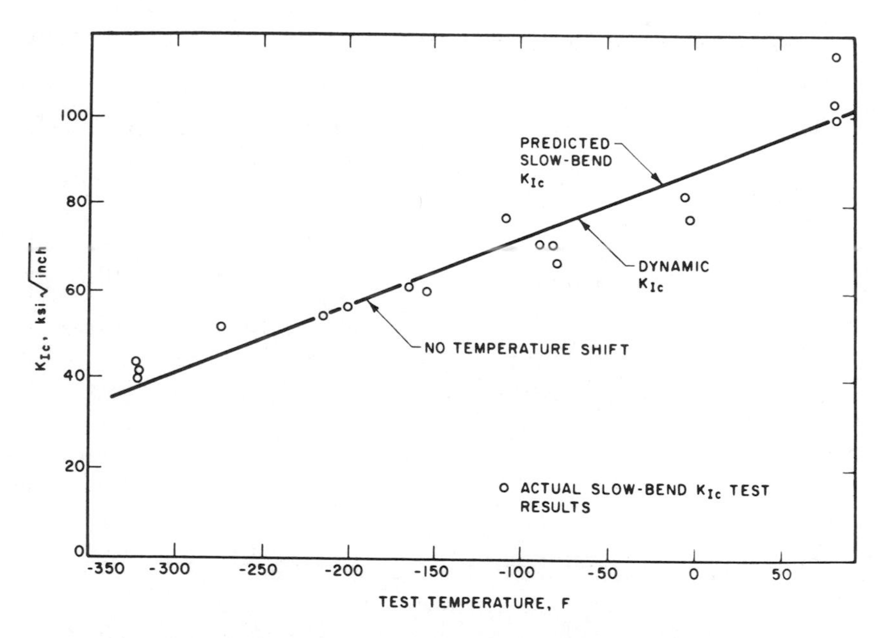

FIG. 4.59. Use of CVN test results to predict the effect of loading rate on K_{Ic} for 18Ni(250) maraging steel.

predicted values. It should be emphasized that a prediction can be made either from slow-bend K_{Ic} values to dynamic values (K_{Id}) or vice versa. Because the dynamic K_{Ic} curves (K_{Id}) were better defined, the results shown in Figures 4.54 through 4.59 were from dynamic K_{Id} values to slow-bend K_{Ic} values. Because dynamic K_{Id} tests are extremely difficult to conduct and analyze, the presently developed prediction procedure should be quite useful in obtaining a first-order approximation of the effects of loading rate on the K_{Ic} behavior of steels by adjusting experimentally obtained slow-bend K_{Ic} values.

REFERENCES

1. J. M. Barsom and S. T. Rolfe, "K_{Ic} Transition-Temperature Behavior of A517-F Steel," *Journal of Engineering Fracture Mechanics, 2*, 1971, p. 341.
2. G. R. Irwin, "Crack-Toughness Testing of Strain-Rate Sensitive Materials," *Journal of Engineering for Power, Transactions of the ASME, Series A, 86*, 1964, p. 444.
3. H. D. Greenberg, E. T. Wessel, and W. H. Pryle, "Fracture Toughness of Turbine-Generator Rotor Forgings," *Engineering Fracture Mechanics, 1*, 1970, pp. 653–674.
4. W. S. Pellini, "Design Options for Selection of Fracture Control Procedures in the Modernization of Codes, Rules, and Standards," in *Proceedings: Joint United States–Japan Symposium on Application of Pressure Component Codes, Tokyo, Japan*, March 13–15, 1973.
5. A. K. Shoemaker and S. T. Rolfe, "The Static and Dynamic Low-Temperature Crack-Toughness Performance of Seven Structural Steels," *Engineering Fracture Mechanics 2*, 1971, pp. 319–339.
6. R. Roberts, G. R. Irwin, G. V. Krishna, and B. T. Yen, "Fracture Toughness of Bridge Steels—Phase II Final Report," *Fritz Engineering Laboratory Report No. 379.2*, Lehigh University, Bethlehem, Pa. June 1974.
7. J. M. Barsom, "Development of the AASHTO Fracture–Toughness Requirements for Bridge Steels," *Journal of Engineering Fracture Mechanics, 7*, 1975, p. 605.

5

Fracture-Mechanics Design

5.1. Introduction

Design is a term used in different ways by engineers. For the structural engineer, the term usually refers to the synthesis of various disciplines (statics, strength of materials, structural analysis, matrix algebra, computers, etc.) to create a structure that is proportioned and then detailed into its final shape using a set of scale drawings. When the word design is used in this sense, designing to prevent brittle fracture usually refers to using an appropriate stress as well as to the elimination (as much as possible) of those structural details that act as stress raisers and that can be potential fracture-initiation sites, e.g., weld joints, mismatch, holes, intersecting plates, etc. Unfortunately, large complex structures (either welded or bolted) cannot be designed (or fabricated) without these discontinuities, although good design and fabrication practices can minimize the size and number of these discontinuities.

On the other hand, the materials engineer or metallurgist uses the word design in a different context, namely that of material selection and the selection of an appropriate design stress level at the particular service temperature and loading rate that the structure will see. Thus from this viewpoint designing to prevent brittle fracture refers to "proper" material selection as well as selection of the proper allowable stress.

Both of these approaches to design are valid, but they should not be confused. The first (and traditional) definition assumes that the designer *starts* with a given material and design stress level (often specified by codes) and involves the process of detailing and proportioning members to carry the given loads without exceeding the allowable stress. The allowable stress is usually a certain percentage of the yield strength for tension members and a certain percentage of the buckling stress for compression members. Both of these "allowable" design stress levels assume "perfect" fabrication in that the structures are usually assumed to have no crack-like discontinuities, defects or cracks. It is realized that stress concentrations or discontinuities will be present, but the designer assumes that his structural materials will yield locally and redistribute the load in the vicinity of these stress concentrations or discontinuities.

The second use of the term design refers to selection of materials and allowable stress levels based on the more appropriate realization of the fact that discontinuities in large complex structures may be present or may initiate under cyclic loads or stress-corrosion cracking and that some level of notch toughness is desirable. This aspect of design has recently been made more quantitative by the development of fracture mechanics as an engineering science.

To provide a safe, fracture-resistant structure, both of these design approaches should be followed. The designer must properly proportion his structure to prevent failure by either tensile overload or compressive instability *and* by unstable crack growth by brittle fracture. Historically, most design criteria have been established to prevent yielding (either in tension or compression) and to prevent buckling. The counterpart of unstable failure by buckling is brittle fracture. Historically the yielding and buckling modes of failure have received considerable attention because analytical design procedures were available to use in various theories of failure. The Euler buckling analysis and the maximum shearing stress theory of failure are among the most widely used analytical design principles to prevent failure by general yielding or buckling. The recent development of fracture mechanics as an analytical design tool finally "fills the gap" in the designers' various techniques for safe design of structures.

There are numerous textbooks that describe the various design techniques to prevent either general yielding or buckling by proper proportioning of the various structural members, and the reader is referred to the large number of design textbooks in his particular field of structures (bridges, ships, pressure vessels, aircraft, etc.) for these well-established design procedures.

This textbook deals primarily with the less well-known definition of design, namely the selection of materials and appropriate design stress levels for fracture-resistant structures. This design approach recognizes the fact that discontinuities may be present in large complex structures. It is true that the incidence of brittle fractures (as well as fatigue cracks) can be *reduced* by good detailing practices and minimizing discontinuities. Examples of design details to optimize fatigue strength in structures are presented in Reference 1.

The remainder of this chapter deals with a description of how fracture mechanics can aid the engineer in the initial selection of

1. Materials
2. Design stress levels, and
3. Tolerable crack sizes for quality control or inspection

for the fracture-resistant design of *any* large complex structure, e.g., bridges, ships, pressure vessels, aircraft, earthmoving equipment, etc. Thus, this textbook should be used in conjunction with textbooks that describe the more traditional methods of designing to prevent either the buckling or yielding modes of failure to ensure that all possible failure modes are considered.

Regardless of the type of structure to be considered, fracture mechanics design assumes that the engineer has established the following general information:

1. Type and overall dimensions of structure (bridge, pressure vessel, etc.).
2. General size of tension members (length, diameter, etc.).
3. Additional performance criteria (e.g. minimum weight, least cost, maximum resistance to fracture, specified design life, loading rate, operating temperature, etc.).
4. Stress and stress range, where crack growth can occur (as discussed in Chapters 7–11).

With this basic information, the designer can incorporate K_{Ic} or K_{Id} values at the service temperature and loading rate in the "design" of a fracture-resistant structure. The fatigue life can be estimated, as well.

If the designer knows or can measure the critical value of notch toughness (K_{Ic} or K_{Id}) at the service conditions (i.e., at the temperature and loading rate of that particular service condition), the philosophy of design using fracture mechanics is straightforward. It generally follows either a fail-safe or safe-life principle, both of which have been used extensively in design. Fail-safe design assumes that if a member fails, the overall structure is still safe from catastrophic fracture. Conversely, safe-life principles assume that for the particular service loading the structure will last the entire design life of the structure and failure will not occur while the structure is in service. The particular merits of these two principles are discussed in Chapter 12. However, both principles recognize the fact that discontinuities may be present in large complex structures or that they may initiate and grow during the service life of the structure. The structural designer, therefore, should not limit his analysis to the traditional approaches of design whereby factors of safety or indices of reliability for unflawed structures are used but rather should incorporate the fact that discontinuities can be present in his structure.

As with any new science the applications are rare not only because of the newness, but in the particular case of fracture mechanics because of the difficulties in obtaining reliable K_{Ic} or K_{Id} values for various structural materials. That is, most of the values presented in Chapter 4 and those found throughout the literature are at temperatures less than the service temperature or at loading rates greater than the service loading rate. In one sense this is very fortunate because it indicates that the behavior of most structural materials appears to be nonplane strain at service temperatures, and this is very desirable from either the safe-life or fail-safe philosophy of design. However,

1. When designs become more complex,
2. When the use of high-strength thick welded structural materials becomes more common compared with the use of lower-strength thin riveted plates,
3. When the fabrication and construction become more complex,
4. When the magnitude of loading increases, and
5. When actual factors of safety decrease because of the use of computer design,

the probability of brittle fracture in large complex structures increases.

In this chapter, design procedures using fracture-mechanics concepts will be limited to design considerations related to the terminal failure *after* possible crack extension had occurred by fatigue (Chapters 7, 8, and 9), stress-corrosion (Chapter 10), or corrosion fatigue (Chapter 11). This particular chapter forms the basis for subsequent development of fracture-control plans where these additional considerations are important.

5.2. General Fracture-Mechanics Design Procedure for Terminal Failure

The critical stress-intensity factor for a particular material at a given temperature and loading rate is related to the nominal stress and flaw size as follows:

$$K_{Ic} \quad \text{or} \quad K_{Id} = C\sigma\sqrt{a} \tag{5.1}$$

where K_{Ic} or K_{Id} = crack toughness of a material, ksi$\sqrt{\text{in.}}$, tested at a particular temperature and loading rate,

C = constant, function of crack geometry, Chapter 2,

σ = nominal applied stress, ksi,

a = flaw size as a critical dimension for a particular crack geometry, in.

Thus, the maximum flaw size a structural member can tolerate at a particular stress level is

$$a = \left[\frac{K_{Ic} \text{ or } K_{Id}}{C \cdot \sigma}\right]^2 \tag{5.2}$$

Accordingly, the engineer can analyze the safety of a structure against failure by brittle fracture in the following manner:

1. Determine the values of K_{Ic} or K_{Id} and σ_{ys} at the service temperature and loading rate for the materials being considered for use in the structure. Note that for a complete analysis of welded structures, values for the weldment should be used also.

2. Select the most probable type of flaw that will exist in the member being analyzed and the corresponding K_I equation. Figure 5.1 shows the fracture-mechanics models that describe three of the more common types of flaws occurring in structural members. Complex-shaped flaws can often be approximated by one of these models. Additional equations to analyze other crack geometries were described in Chapter 2.

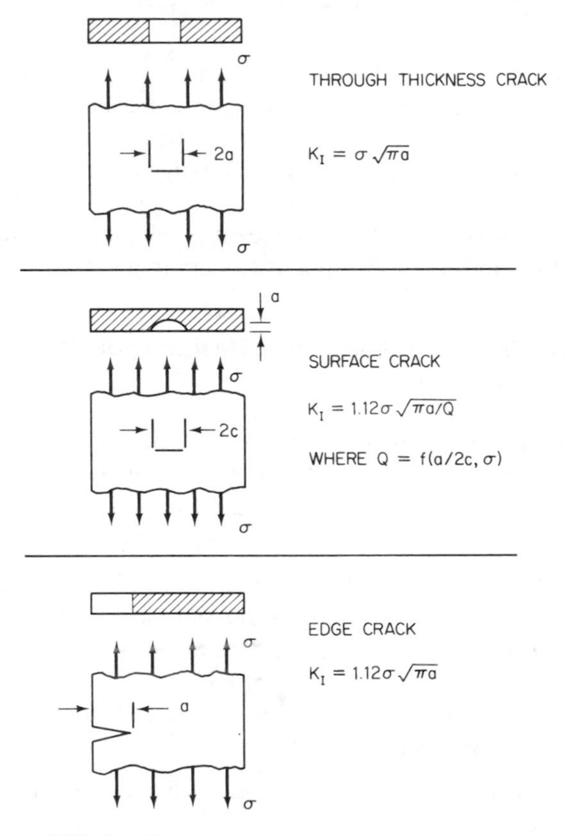

FIG. 5.1. K_I values for various crack geometries.

3. Determine the stress–flaw-size relation at various possible design stress levels using the appropriate K_I expression and the appropriate K_{Ic} or K_{Id} value.

As an example, the relation between stress, flaw size, and stress intensity factor, K_I, for the through-thickness crack geometry shown in Figure 5.1 is

$$K_I = \sqrt{\pi}\,\sigma\sqrt{a} \tag{5.3}$$

Assume that the material being analyzed has a K_{Ic} value at the service temperature of 50 ksi$\sqrt{\text{in.}}$ and a yield strength of 100 ksi. Substituting $K_I = K_{Ic}$, the possible combinations of stress and critical flaw size at failure are described by

$$K_{Ic} = 50 \text{ ksi}\sqrt{\text{in.}} = \sqrt{\pi}\,\sigma\sqrt{a} \tag{5.4}$$

Using this equation, values of the critical crack size for various stress levels are calculated as follows:

σ (*ksi*)	*a* (*in.*)
10	7.96
20	1.99
30	0.88
40	0.50
50	0.32
60	0.22
70	0.16
80	0.12
90	0.10
100	0.08

These results are plotted in Figure 5.2. The curve labeled K_{Ic} is the locus of points at which unstable crack growth (fracture) will occur. Thus, if the nominal design stress level is 30 ksi, then the maximum tolerable flaw size, or *critical crack size*, is 0.88 in. Conversely, if the nominal design stress level

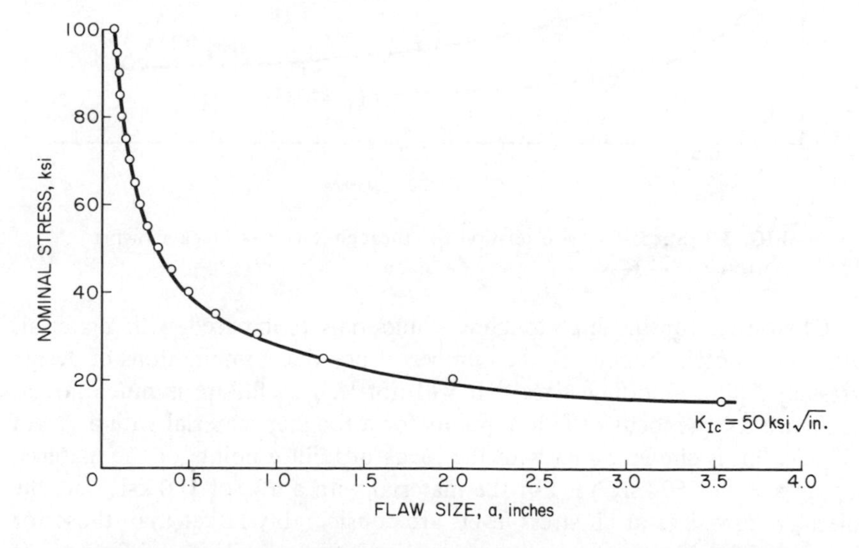

FIG. 5.2. Stress–flaw-size relation for through-thickness crack in material having K_{Ic} = 50 ksi$\sqrt{\text{in.}}$

is 60 ksi, then the crack size is 0.22 in. It can be seen that there is no single critical crack size for a particular structural material (at a given temperature and loading rate) but rather a "semiinfinite" number of critical crack sizes, depending on the nominal design stress level. Figure 5.3 shows the locus of values of stress and flaw size for a "design K_I" of $K_{Ic}/2$, i.e., a factor of safety of 2 against fracture that is based on the critical stress-intensity factor using the traditional definition of factor of safety. In this case a structure having a flaw size of 0.22 in. loaded to a design stress of 30 ksi has a factor of safety of 2 against fracture. Similarly the $K_I = 25$ ksi$\sqrt{\text{in.}}$ curve is the locus of all stress levels and corresponding flaw sizes with a factor of safety of 2 against fracture. The factor of safety also may be based on stress alone or on crack size alone rather than on the critical stress-intensity factor.

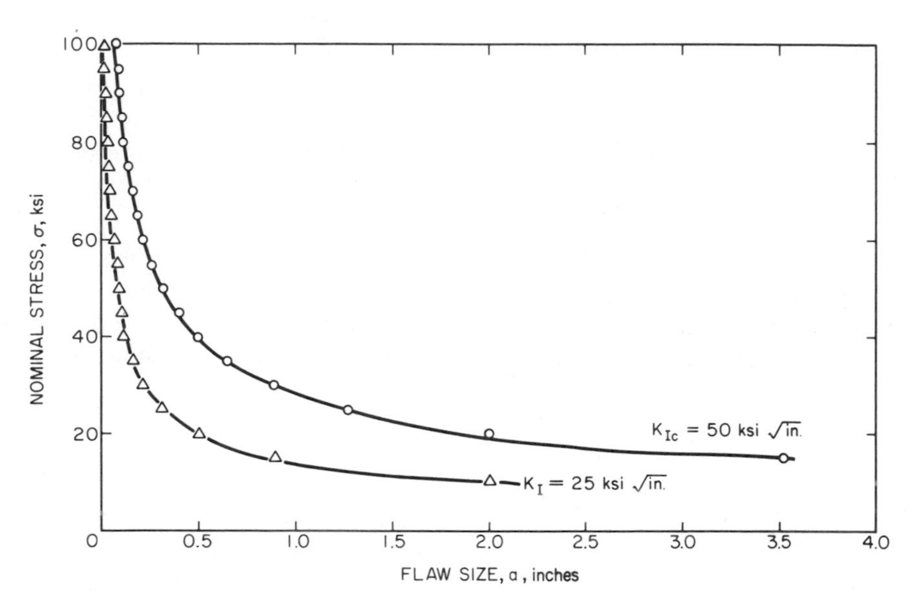

FIG. 5.3. Stress–flaw-size relation for through-thickness crack showing $K_{I(design)} = 25$ ksi$\sqrt{\text{in.}}$

Obviously, for the high-toughness materials (compared with materials with lower notch toughness), the number of possible combinations of design stress and allowable flaw sizes that will not lead to failure is much larger. In Figure 5.4 the locus of failure points for a tougher material with a K_{Ic} of 100 ksi$\sqrt{\text{in.}}$ is shown along with the locus of failure points of the material having a K_{Ic} of 50 ksi$\sqrt{\text{in.}}$ For the material with a K_{Ic} of 100 ksi$\sqrt{\text{in.}}$, the tolerable flaw sizes at all stress levels are considerably larger than those for the material with the lower toughness, and the possibility of fracture is reduced considerably.

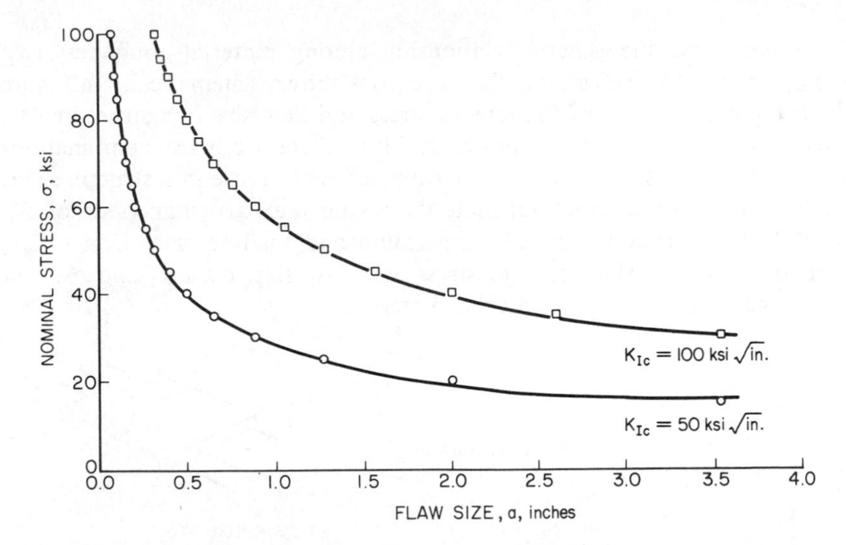

FIG. 5.4. Stress–flaw-size relation for through-thickness crack showing effect of higher K_{Ic} (100 ksi$\sqrt{\text{in.}}$).

Thus it can be seen that to minimize the possibility of brittle fracture in a given structure, the designer has three *primary* factors that can be controlled:

1. Material toughness at the particular service temperature and loading rate (K_{Ic} or K_{Id}), ksi$\sqrt{\text{in.}}$
2. Nominal stress level (σ), ksi.
3. Flaw size present in the structure (a), in.

All three of these factors affect the possibility of a brittle fracture occurring in structures. All other factors such as temperature, loading rate, residual stresses, etc., merely affect the above three primary factors. Design engineers have known these facts for many years and have reduced the susceptibility of their structures to brittle fractures by applying these concepts to their structures qualitatively. That is, the traditional use of "good design" practice, i.e., by using appropriate stress levels and minimizing discontinuities, has led to the reduction of brittle fractures in many structures. In addition, the use of good fabrication practices, that is, decreasing the flaw size by proper welding control and inspection, as well as the use of materials with good notch-toughness levels (e.g., as measured with a Charpy V-notch impact test), has minimized the probability of brittle fractures in structures. This has been the traditional design approach to reducing brittle fractures in structures and has been used reasonably successfully. However, fracture mechanics offers the engineer the possibility of a more *quantitative* approach to designing to prevent brittle fractures in structures.

In summary, the general relationship among material toughness (K_{Ic} or K_{Id}), nominal stress (σ), and flaw size (a) is shown schematically in Figure 5.5. If the particular combinations of stress and flaw size in a structure (K_I) reach the K_{Ic} level, fracture can occur. Thus, there are many combinations of stress and flaw size, σ_f and a_f, that may cause fracture in a structure that is fabricated from a structural material having the particular value of K_{Ic} (or K_{Id}) at a particular service temperature and loading rate. Conversely, there are many combinations of stress and flaw size, e.g., σ_0 and a_0, that will not cause failure of a particular material.

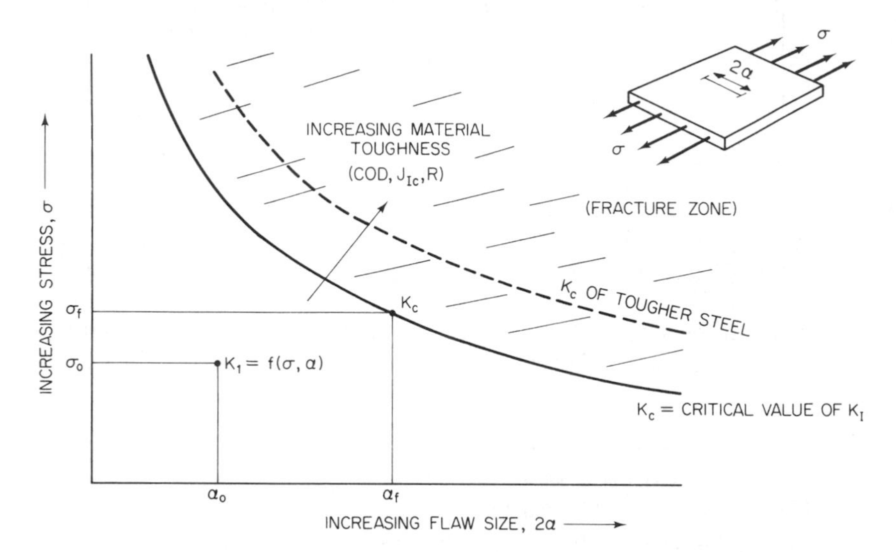

FIG. 5.5. Schematic relation among stress, flaw size, and material toughness.

A useful analogy for the designer is the relation among applied load (P), nominal stress (σ), and yield stress (σ_{ys}) in an unflawed structural member and among applied load (P), stress intensity (K_I), and critical stress intensity for fracture (K_c, K_{Ic}, or K_{Id}) in a structural member with a flaw. In an unflawed structural member, as the load is increased, the nominal stress increases until an instability (yielding at σ_{ys}) occurs. As the load is increased in a structural member with a flaw (or as the size of the flaw grows by fatigue or stress corrosion), the stress intensity, K_I, is increased until an instability (fracture at K_c, K_{Ic}, K_{Id}) occurs. Thus the K_I level in a structure should always be kept below the appropriate critical value (K_c, K_{Ic}, or K_{Id}) in the same manner that the nominal design stress (σ) is kept below the yield strength (σ_{ys}).

Another analogy that may be useful in understanding the fundamental aspects of fracture mechanics is the comparison with the Euler column instability (Figure 5.6). The stress level required to cause instability in a column

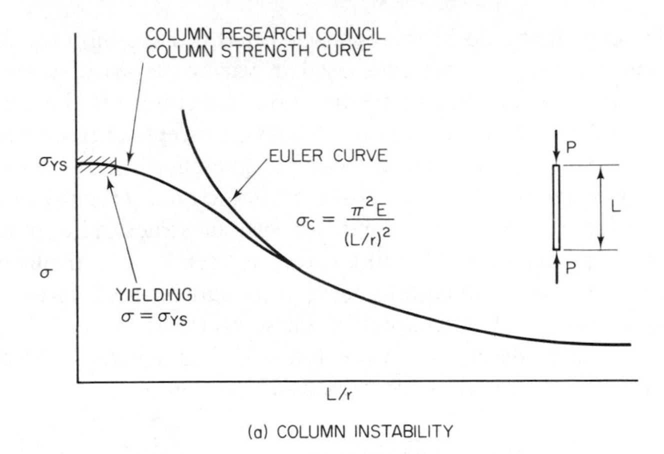

(a) COLUMN INSTABILITY

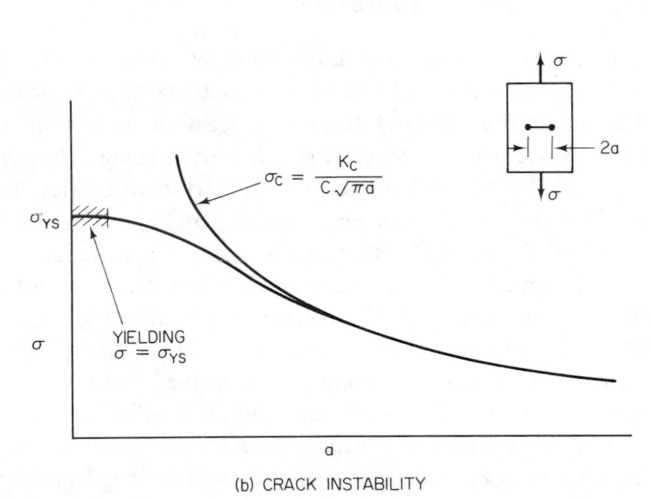

(b) CRACK INSTABILITY

FIG. 5.6. Column instability and crack instability.

(buckling) decreases as the L/r ratio increases. Similarly, the stress level required to cause instability (fracture) in a flawed tension member decreases as the flaw size (a) increases. As the stress level in either case approaches the yield strength, both the Euler analysis and the K_{Ic} analysis are invalidated because of yielding. To prevent buckling, the actual stress and L/r values must be below the Euler curve. To prevent fracture, the actual stress and flaw size, a, must be below the K_{Ic} line shown in Figure 5.6. Obviously, using a material with a high level of notch toughness (e.g., a K_{Ic} level of 100 ksi$\sqrt{\text{in.}}$ compared with 50 ksi$\sqrt{\text{in.}}$ as shown in Figure 5.4) will increase the possible combinations of design stress and flaw size that a structure can tolerate without fracturing.

At this point, it should be reemphasized that (fortunately) the K_{Ic} or K_{Id} levels of many structural materials used in various types of structures are so high that these values cannot be measured directly using the existing test methods described in Chapter 3. Thus although concepts of fracture mechanics can be used to develop fracture-control guidelines and stress–flaw size trade-offs for design use, the state-of-the-art is such that in many cases actual K_{Ic} or K_{Id} values cannot be measured for various structural applications at the service temperatures and loading rates. In these cases various empirical correlations with other notch-toughness tests can be used to approximate K_{Ic} or K_{Id}, as described in Chapter 6. These approximations of K_{Ic} or K_{Id} can then be used as described above, but with less accuracy. Also elastic–plastic analyses can be used as described in Chapter 16.

5.3. Design Selection of Materials

Current methods of design and fabrication of large complex structures are such that engineers expect these structures to be able to tolerate yield stress loading in tension without failing, at least in local regions around points of stress concentration. Since it is generally accepted that yield stress loading may occur in the vicinity of structural discontinuities, the critical crack size, a, is proportional to $(K_{Ic}/\sigma_{ys})^2$ or $(K_{Id}/\sigma_{yd})^2$.

Thus either the K_{Ic}/σ_{ys} or K_{Id}/σ_{yd} ratio becomes a good index for measuring the relative toughness of structural materials. Because for most structural applications it is desirable that the structure tolerate large flaws without fracture, the use of materials with as high a K_{Ic}/σ_{ys} (or K_{Id}/σ_{yd}) as is possible *consistent with economic considerations* is a desirable condition.

The question becomes, *How high must this ratio of K_{Ic}/σ_{ys} or K_{Id}/σ_{yd} be to ensure satisfactory performance in large complex structures, where complete initial inspection for cracks and continuous monitoring of crack growth throughout the life of a structure are not always possible, practical, or economical.*

No simple answer exists because the answer is obviously dependent on the type of structure, frequency of inspection, access for inspection, quality of fabrication, design life of the structure, consequences of failure for a structural member, redundancy of load path, probability of overload, fabrication and material costs, etc., etc., etc. However, fracture mechanics does provide an engineering approach to rationally evaluate this question. Basic assumptions are that flaws do exist in structures, yield stress loading is possible in parts of the structure, and plane-strain conditions can exist. Under these very severe conditions, the K_{Ic}/σ_{ys} or K_{Id}/σ_{yd} ratio for materials used in a particular structure is one of the primary controlling design parameters that can be used to define the relative safety of a structure against brittle fracture.

As an example of the use of the K_{Ic}/σ_{ys} ratio as a material selection parameter, the behavior of a wide plate with a through-thickness center crack of length $2a$ (Figure 5.1) is analyzed for materials having assumed levels of strength and notch toughness. Table 5.1 presents assumed values of K_{Ic} for various steels having yield strengths that range from 40 to 260 ksi. For each of these steels, the critical crack size, $2a$, is calculated for four design stress levels, i.e., $100\% \sigma_{ys}$, $75\% \sigma_{ys}$, $50\% \sigma_{ys}$, and $25\% \sigma_{ys}$. It should be emphasized that there is no single unique critical stress-intensity factor for any one steel at a given test temperature and rate of loading because K_{Ic} or K_{Id} for a given steel depends on the thermomechanical history, i.e., heat treatment, rolling, etc.

TABLE 5.1. Values of Critical Crack Size as a Function of Yield Strength and Notch Toughness

		Critical Flaw Size, 2a (in.) (actual design stress level, ksi, is shown in parenthesis)			
σ_{ys} *(ksi)*	*Assumed* K_{Ic} *Values* *(ksi$\sqrt{in.}$)*	$\sigma = 100\% \sigma_{ys}$	$\sigma = 75\% \sigma_{ys}$	$\sigma = 50\% \sigma_{ys}$	$\sigma = 25\% \sigma_{ys}$
260	80	0.06(260)	0.11(195)	0.24(130)	0.96(65)
220	110	0.16(220)	0.28(165)	0.64(110)	2.55(55)
180	140	0.39(180)	0.68(135)	1.54(90)	6.16(45)
180	220	0.95(180)	1.69(135)	3.80(90)	15.22(45)
140	260	2.20(140)	3.90(105)	8.78(70)	35.13(35)
110	170	1.52(110)	2.70(82.5)	6.08(55)	24.33(27.5)
80	200	3.98(80)	7.07(60)	15.92(40)	63.66(20)
40	100	3.98(40)	7.07(30)	15.92(20)	63.66(10)

Through-Thickness Crack in a Wide Plate (Figure 5.1)

$$K_{Ic} = \sqrt{\pi}\, \sigma_{\text{design}} \sqrt{a}$$

$$\therefore \quad a = \frac{1}{\pi}\left(\frac{K_{Ic}}{\sigma_{\text{design}}}\right)^2$$

$$2a = \frac{2}{\pi}\left(\frac{K_{Ic}}{\sigma_{\text{design}}}\right)^2$$

The results presented in Table 5.1 demonstrate the influence of strength level, notch toughness, and design stress level on the critical crack size in a wide plate. For example, the critical flaw size for the 260-ksi yield strength steel loaded to 50% of the yield strength (design stress of 130 ksi) is 0.24 in. If a design stress of 130 ksi were required for a particular structure, it would be preferable to use a lower-strength, higher-toughness material (e.g., the 180-ksi yield strength material) at a design stress of $75\% \sigma_{ys}$ (135 ksi). For either of the two 180-ksi yield strength steels analyzed in Table 5.1, the critical crack sizes are much larger than 0.24 in. e.g., 0.68 in. for the material with a K_{Ic} of 140 ksi$\sqrt{\text{in.}}$ or 1.69 in. for the material with a K_{Ic} of 220 ksi$\sqrt{\text{in.}}$

Obviously the tougher of these two materials would be a more fracture-resistant structural material, but it probably would be a more expensive material also. This point illustrates one of the basic differences in fracture-resistant design compared with other more traditional modes of design in that economics is always present. That is, structural materials that are extremely notch tough at service temperatures and loading rates are available. However, because the cost of these materials generally increases with their ability to perform satisfactorily under more severe operating conditions, the engineer usually does not want to specify more notch toughness than is required for the particular application. Thus the problem of fracture-resistant design is really one of optimizing structural performance consistent with economic considerations.

Further analysis of Table 5.1 indicates that the traditional method of selecting a design stress as some percentage of the yield strength does not always give the same degree of safety and reliability against fracture as it is presumed to give for yielding. For example, assume that the design stress for the two steels having yield strengths of 220 ksi and 110 ksi is 50% σ_{ys} or 110 ksi and 55 ksi, respectively. For the 220-ksi steel, the critical crack size is 0.64 in., whereas for the 110-ksi yield strength steel, the critical crack size is 6.08 in. If the design stress for the lower yield strength steel were increased to 100% σ_{ys} (110 ksi), the critical crack size would be 1.52 in. This is still significantly greater than the 0.64-in. size for the 220-ksi yield strength steel with a design stress of 50% σ_{ys}.

For the lower-strength steels shown in Table 5.1 that have higher values of K_{Ic}/σ_{ys}, the critical crack sizes become extremely large, indicating that fracture-mechanics theory no longer applies in these cases. This is analogous to saying that for very low L/r ratios the calculated Euler buckling stress is well above the yield stress and the Euler buckling analysis no longer applies.

In summary, there may be situations where the designer should specify a *lower* yield strength material at a *higher* design stress level (as a percentage of the yield strength) to actually *improve* the overall safety and reliability of a structure from a fracture-resistant design viewpoint. Good design dictates that the engineer design to prevent failure of his structure against *all* possible modes of failure, including fracture.

5.4. Design Analysis of Failure of a 260-Inch-Diameter Motor Case

An excellent example of the fact that specifying a percentage of yield strength is not always the best method of establishing the design stress level is the failure of a 260-in.-diameter motor case during hydrotest.[2,3] The motor case failed during hydrotest at a pressure of 542 psi, which was about 56%

of proof pressure. The motor case was constructed of 250 Grade maraging steel plate joined primarily by submerged arc automatic welding. An investigation of the failure was conducted by a committee composed of members from industry and government.

Gerberich[3] summarized the failure as follows:

Although this motor case was designed to withstand proof pressures of 960 psi, it failed during hydrotest at a pressure of 542 psi. In this 240 ksi yield strength material, the failure occurred at a very low membrane stress of 100 ksi. The fracture was both premature and brittle with crack velocities approaching 500 feet/second. The result of this 65-ft-high chamber literally flying apart is shown in Figure 5.7. Postfailure examination revealed that the fracture had originated in an area of two defects that had probably been produced by manual gas-tungsten arc weld repairs. After the crack initiated it branched into multiple cracks as shown in Figure 5.8, leading to complete catastrophic failure of the motor case.

The real lesson in this failure is the lack of design knowledge that went into the material selection. First, the chamber was 0.73 in. thick, which put it into the plane-strain regime for the high-strength material being considered. Grade 250 maraging steel which had a yield strength of about 240 ksi was chosen. For plane-strain conditions this was not a particularly good choice since the base metal had a plane-strain fracture toughness, K_{Ic}, of only 79.6 ksi$\sqrt{\text{in.}}$ At the design stress of 160 ksi, a critical defect only 0.08 in. in depth could have caused catastrophic failure. Still, the chamber manufacturer thought that defects of this size could be detected. In retrospect, this was not a very judicious decision. Furthermore, the error was compounded by the fact that welding this material produced an even lower K_{Ic} value ranging from 39.4 to 78.0 ksi$\sqrt{\text{in.}}$, the toughness level depending on the location of the flaw in the weld.

A postfailure analysis was run on several types of flaw configurations and weld positions which indicated the K_{Ic} value to range from 38.8 to 83.1 ksi$\sqrt{\text{in.}}$ with the average being 55.0 ksi$\sqrt{\text{in.}}$ As the exact value of the fracture toughness at the failure origin in the chamber is not known, this average value is used as an estimate of K_{Ic}. Postfailure examination of the fracture origin indicated the responsible defect had an irregular banana shape that could best be approximated by an internal ellipse that was 0.22 in. in depth and 1.4 in. long. The critical dimension was the in-depth value of 0.22 in. since the crack would first propagate through the thickness from this dimension. To calculate the critical defect size from fracture-mechanics principles, a solution for an internal ellipse gives

$$K_I = \sigma(a)^{1/2} f\left(\frac{a}{c}\right) \tag{5.5}$$

where a is the half-crack depth, c is the half-crack length, and $f(a/c)$ is related to the complete elliptical integral of the second kind (Chapter 2). However, as $a/c \longrightarrow 0$, $f(a/c) \longrightarrow (\pi)^{1/2}$, and this equation becomes

$$K_I = \sigma(\pi a)^{1/2} \tag{5.6}$$

For the shape of flaw under consideration, the difference between these two equations is only 4%, and so for the sake of simplicity the second equation is utilized. Based on the fracture toughness of 55.0 ksi$\sqrt{\text{in.}}$, the second equation was utilized to make a plot of membrane stress versus defect size for crack instability. This is shown in Figure 5.9 as the curve for Grade 250 maraging steel.

An intercept of the 100-ksi failure stress for the chamber predicts a critical defect size, $2a$, of 0.2 in., which is very close to the observed value also indicated in Figure 5.9. Besides the Grade 250 data, there is also shown here a curve for Grade 200 maraging steel. Although this material has a yield strength that is about 10% lower than Grade 250, its toughness is about double with K_{Ic} about

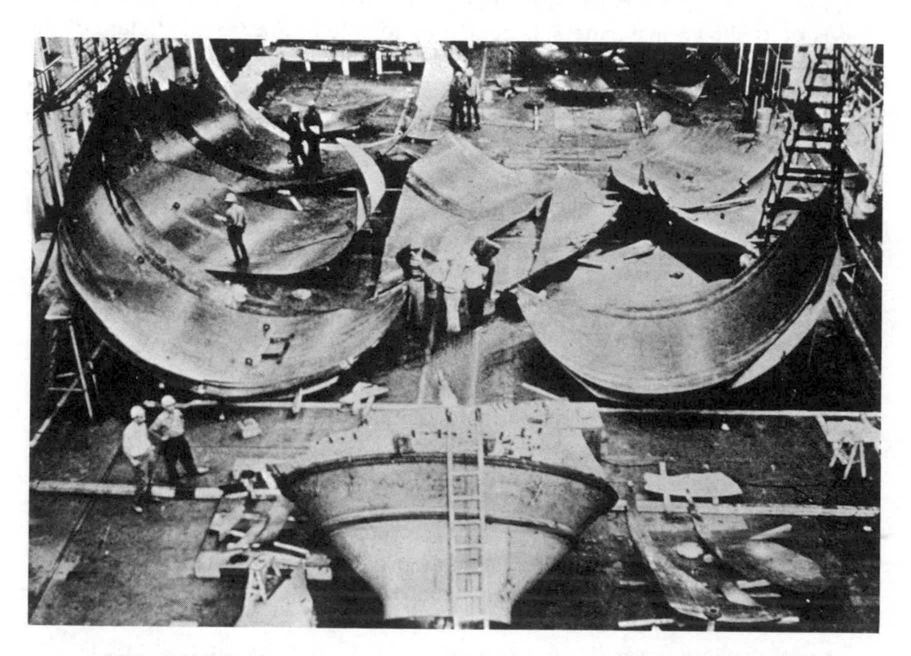

FIG. 5.7. Failed motor case with pieces laid out in approximately the proper relation to each other. (Courtesy of J.E. Srawley, NASA Lewis Research Center.)

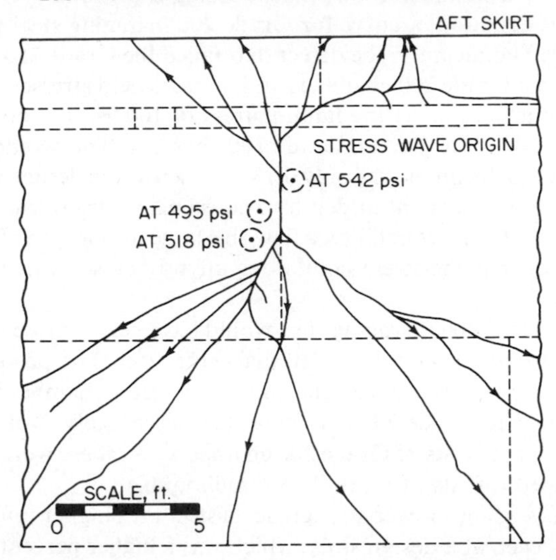

FIG. 5.8. Map of fracture path about failure origin; dotted lines indicate welds (Reference 2).

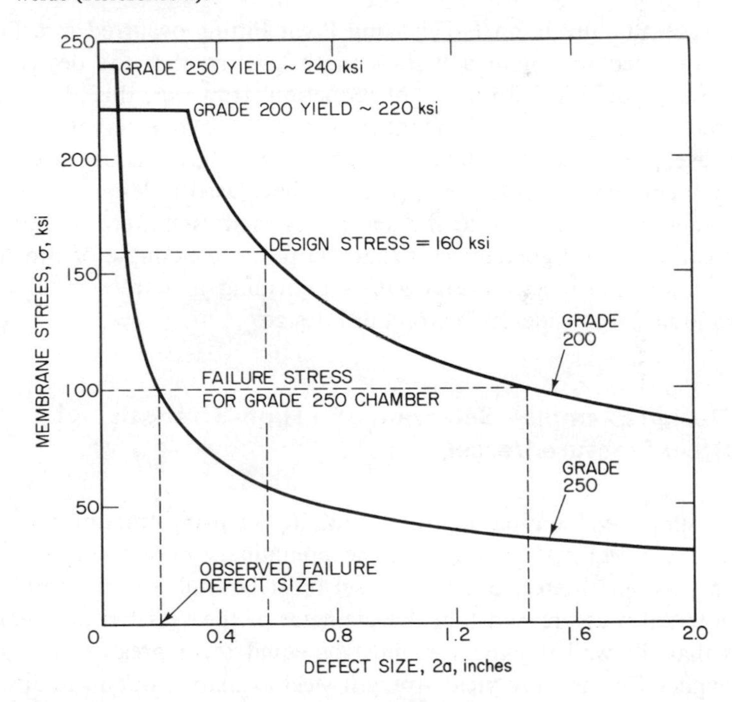

FIG. 5.9. Design curves for critical defect in 260-in. motor case (Reference 3).

150 ksi$\sqrt{\text{in.}}$ for a member 0.7 in. thick. Using the 150-ksi$\sqrt{\text{in.}}$ value for K_{Ic} and the above equation, a curve for Grade 200 maraging steel was constructed for Figure 5.9. Significantly, the defect that failed the Grade 250 chamber would not have failed a Grade 200 chamber, and, in fact, yield stresses could have been reached without failure. At the failure stress of 100 ksi, it would have taken a flaw 1.42 in. in depth to burst a Grade 200 chamber. This would not have been possible since the thickness was only 0.73 in. Even at the design stress of 160 ksi, it would take a flaw 0.56 in. in depth to cause plane-strain fracture, and this is a very large flaw. In all probability, a flaw this large in this type of material would be arrested under plane-stress conditions anyway as soon as it grew through the thickness.

Thus, the Grade 200 maraging steel would have been a more reliable material for this chamber than the Grade 250 maraging steel. The proof of this is that there was a competition to make the 260-in.-diameter chamber for NASA, and the competitor used Grade 200 maraging steel successfully. Not only were there two successful proof tests of Grade 200 chambers, but there were also two firings which developed thrusts of more than 6 million lb.

In summary, using a lower-strength level steel with higher toughness (a larger K_{Ic}/σ_{ys} ratio), even at a design stress which was a higher percentage of the yield strength, would have been better.

The recent failure in an F-111 Wing Pivot Fitting occurred in a D6-AC steel heat-treated to obtain a high-strength level so that the design stress level of $\frac{2}{3}\sigma_{ult}$ would lead to a weight reduction. However, this also led to a lower fracture-toughness level so that the *actual* degree of safety was lowered. In this case, it also seems preferable to have used a heat treatment that gave a higher toughness (but a lower σ_{ult}) and to have used a design stress somewhat higher than $\frac{2}{3}\sigma_{ult}$ so that the *actual* design stresses were the same but the critical crack size greater. This example plus the example of the 260-in. missile motor case illustrate that *both* the yielding and fracture modes of failure should be considered in structural design.

5.5. Design Example—Selection of a High-Strength Steel Pressure Vessel

As a simplified example of the desirability of using fracture-mechanics principles to *select* materials during the preliminary design stages, assume that a high-strength steel pressure vessel must be built to withstand 5,000 psi of internal pressure and that the diameter of the vessel is nominally 30 in. and that the wall thickness, t, must be equal to or greater than 0.5 in. The designer can use any yield strength steel available, but in addition to satisfactory performance, cost and weight of the vessel are important factors that must be considered. The steels available for use in the vessel are shown in Table 5.2 along with their yield strengths and assumed K_{Ic} values at the service temperature and loading rate.

TABLE 5.2. Yield Strength and Crack-Toughness Values of Steels Used in Example Problem

Steel	*Yield Strength,* σ_{ys} *(ksi)*	*Assumed* K_{Ic} *Values (ksi$\sqrt{in.}$)*
A	260	80
B	220	110
C	180	140
D	180	220
E	140	260
F	110	170

As a first step, the designer should estimate the maximum possible flaw size that can exist in the vessel wall, based on fabrication and inspection considerations. In this particular design example, assume that we want to prevent failures caused by a surface flaw of depth 0.5 in. and an $a/2c$ ratio of 0.25 is possible, as shown in Figure 5.10. This size flaw may be due to

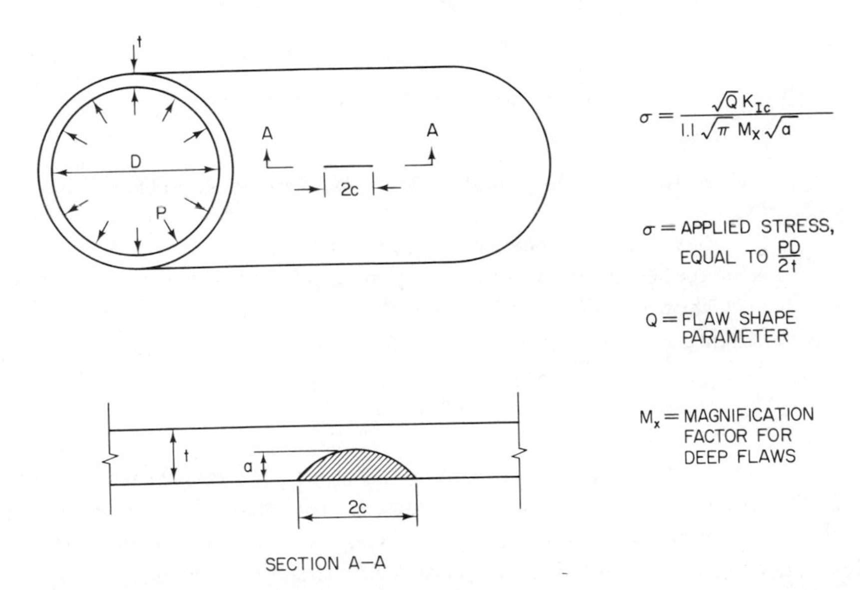

FIG. 5.10. Section of pressure vessel with surface flaw for example problem.

improper fabrication or crack growth by fatigue or stress-corrosion. Obviously, the decision regarding the maximum possible flaw size will depend on many factors related to the design, fabrication, and inspection for the particular structure, but the designer must use the best information available and make the decision. Too often, this step in the design process has been overlooked or, at best, casually handled by assuming that the fabrication will be inspected and "all injurious flaws removed," to quote one specification.

The general relation among K_{Ic}, σ, and a for a surface flaw as developed in Chapter 2 is

$$K_{Ic} = 1.1\sqrt{\pi}M_k\sigma\sqrt{\frac{a}{Q}}$$

where M_k = magnification factor for deep flaws (similar to the tangent correction factor in Chapter 2), assumed in this example to vary linearly between 1.0 and 1.6 as a/t (crack depth/vessel thickness) varies from 0.5 to 1.0 (Figure 5.11); for a/t values less than 0.5, $M_k \simeq 1.0$,

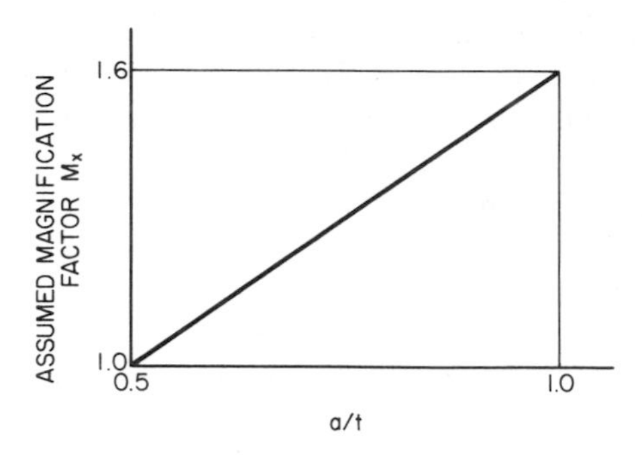

FIG. 5.11. Assumed magnification factor, M_k, for example problem.

Q = flaw shape parameter as shown in Figure 5.12,

σ = applied hoop stress, ksi, equal to $pD/2t$ for the vessel section shown in Figure 5.10.

Rearranging the above equation yields

$$\sigma = \frac{\sqrt{Q}K_{Ic}}{1.1\sqrt{\pi}M_k\sqrt{a}}$$

For $a = 0.5$, and the different values of K_{Ic} for the steels shown in Table 5.2, the calculated *allowable* stress values, σ, for each of the steels being studied are found by (1) calculating the design stress based on the fracture resistance needed for an 0.5-in.-deep crack and then (2) calculating the vessel thickness required at that stress level. Because M_k and Q are functions of the design stress, an iterative procedure must be used.

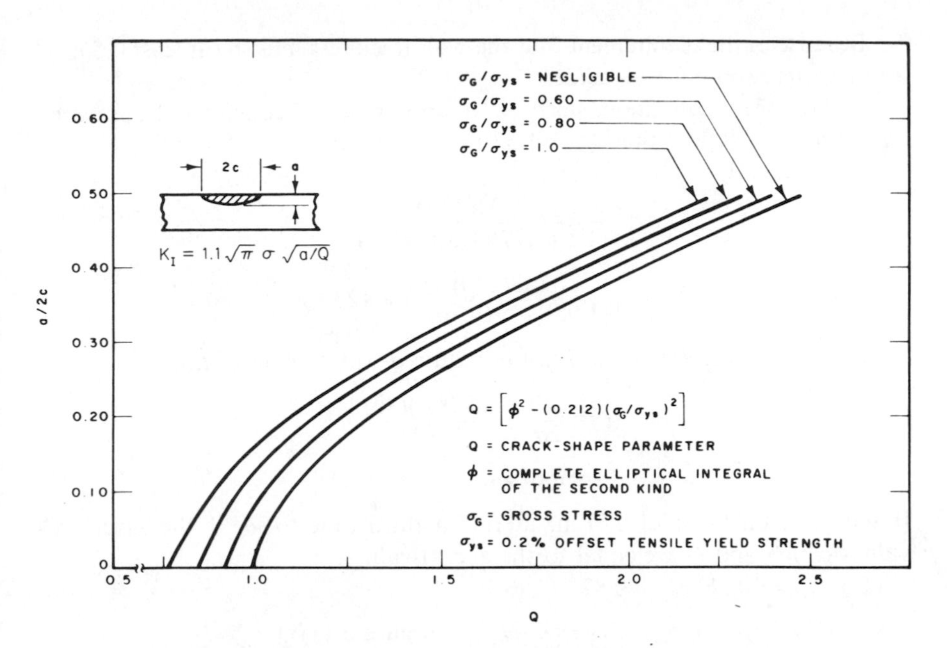

FIG. 5.12. Crack shape parameter, Q, for surface flaw.

A step-by-step calculation of the value for steel D (Table 5.2) ($\sigma_{ys} = 180$ ksi) is as follows: Given

$$\sigma = \frac{\sqrt{Q}\,K_{Ic}}{1.1\sqrt{\pi}\,M_k\sqrt{a}}$$

for steel D

1. $K_{Ic} = 220$ ksi$\sqrt{\text{in.}}$
2. $a = 0.5$ in.
3. Assume that $\sigma/\sigma_{ys} = .55$ and thus $Q = 1.4$ (Figure 5.12).
4. Assume that $M_k = 1.0$ (Figure 5.11) for the first trial.

Thus

$$\sigma = \frac{\sqrt{1.4}(220)}{1.1(1.77)(1.0)\sqrt{0.5}}$$
$$= 189 \text{ ksi}$$

Since this stress is greater than the yield stress, use $\sigma = \sigma_{ys} = 180$ ksi and solve for the wall thickness, t, required to contain the design pressure of 5,000 psi:

$$t = \frac{pD}{2\sigma}$$
$$= \frac{(5{,}000)(30)}{2(180)} = 0.42 \text{ in.}$$

Because of the requirement that the wall thickness must be at least 0.5 in., another iteration must be made.

For the second iteration, since $t \simeq a$, and $M_k \simeq 1.5$ (Figure 5.11), assume that $\sigma/\sigma_{ys} = 0.8$, and thus $Q = 1.33$ (Figure 5.12).

Consequently

$$\sigma = \frac{\sqrt{1.33}(220)}{(1.1)(1.77)(1.5)(\sqrt{0.5})}$$

$$= \frac{1.153(220)}{(1.95)(1.5)(0.707)} = 123 \text{ ksi}$$

For a design stress of 123 ksi, the required wall thickness, t, is

$$t = \frac{pD}{2\sigma} = \frac{(5{,}000)(30)}{2(123)}$$

$$= 0.61 \text{ in.}$$

Based on a thickness of 0.61 in., iterate a third time to see if the assumed values of M_k and Q are equal to the actual values.

For $a/t = 0.5/0.61 = 0.82$,

$$M_k = 1.49 \qquad \text{(Figure 5.11)}$$

For $\sigma/\sigma_{ys} = \frac{123}{180} = 0.68$,

$$Q = 1.36 \qquad \text{(Figure 5.12)}$$

Using $M_k = 1.49$ and $Q = 1.36$, recalculate the design stress:

$$\sigma = \frac{\sqrt{1.36}(220)}{(1.1)(1.77)(1.49)(\sqrt{0.5})}$$

$$= 125 \text{ ksi}$$

For a maximum design stress of 125 ksi, the required wall thickness, t, is

$$t = \frac{pD}{2\sigma} = \frac{(5{,}000)(30)}{2(125)}$$

$$= 0.60 \text{ in.}$$

Because the calculated thickness is essentially equal to the assumed thickness, a final iteration should show that $t_{\text{assumed}} = t_{\text{calculated}}$.

As the fourth iteration, for $\sigma = 125$ ksi and $t = 0.60$ in., assume that

1. $\sigma/\sigma_{ys} = \frac{125}{180} = 0.69$, and therefore $Q = 1.36$.
2. $a/t = 0.5/0.6$, and therefore $M_k = 1.50$.

Thus,

$$\sigma = \frac{\sqrt{1.36}(220)}{(1.1)(1.77)(1.50)(\sqrt{0.5})}$$

$$= 124 \text{ ksi}$$

For a design stress of 124 ksi, the required wall thickness, t, is

$$t = \frac{pD}{2\sigma} = \frac{(5{,}000)(30)}{2(125)}$$

$$= 0.60 \text{ in.}$$

This agrees with the initial value of thickness of 0.60 in. for the assumed value, and thus further trials are not required. Note that the convergence is fairly rapid and that the procedure can easily be programmed for a computer.

Wall thicknesses for the remaining steels in this example are calculated in a similar manner and are presented in Table 5.3. These results show that to withstand an internal pressure of 5,000 psi in a 30-in.-diameter vessel having an 0.5-in.-deep surface flaw the design stresses and wall thicknesses that should be used to give the same resistance to fracture vary considerably for the steels investigated. For example, the allowable design stress level for the 260-ksi yield strength steel is only 70 ksi, and the required wall thickness is 1.07 in., whereas for a lower-strength, tougher steel having a 180-ksi yield strength (and a K_{Ic} of 140 ksi$\sqrt{\text{in.}}$) the design stress is 100 ksi and the required wall thickness is 0.75 in. If an even tougher, 180-ksi yield strength steel is selected, i.e., steel D with a K_{Ic} value of 220 ksi$\sqrt{\text{in.}}$, the design stress can be increased to 124 ksi and the wall thickness decreased to 0.60 in.

TABLE 5.3. Calculated Wall Thickness for $K_I = K_{Ic}$, Example Problem

Steel	*Yield Strength, σ_{ys} (ksi)*	*Assumed K_{Ic} Values (ksi$\sqrt{in.}$)*	*Design Stress, σ_D (ksi)*	$\frac{\sigma_D}{\sigma_{ys}}$	M_k*	Q†	*Wall Thickness, t (in.)*‡
A	260	80	70	0.27	1.0	1.45	1.07
B	220	110	86	0.40	1.1	1.43	0.87
C	180	140	100	0.56	1.2	1.42	0.75
D	180	220	124	0.72	1.43	1.36	0.60
E	140	250	140	1.0	1.51	1.25	0.54
F	110	170	109	0.99	1.26	1.25	0.69

*Magnification factor (Figure 5.11).
†Crack shape parameter (Figure 5.12).
‡Thickness calculated assuming that $p = 5{,}000$ psi and $D = 30$ in.

Because each of the vessels in this example is designed on the basis of *equivalent resistance to fracture in the presence of an 0.5-in.-deep surface flaw*, there would be an obvious savings in weight of the vessel by using a lower-strength, tougher steel compared with the 260-ksi yield strength steel. That is, the weight is proportional to the wall thickness, and the required wall thicknesses for the lower-strength, tougher steels is *less* than for the higher-strength less tough steels.

To show that the highest strength steel may *not* yield the least weight or most economical vessel, an approximate estimate of the weight per foot and assumed cost per foot of the vessel (neglecting the costs of forming, fabrication, etc.) are presented in Table 5.4. The weights per foot of vessel were calculated by estimating the volume of material as the cross-sectional area (πDt) times a 12-in. length and then multiplying by the density of steel, 0.283 lb/in.

TABLE 5.4 Comparison of Weight and Cost of 12-In.-Long Section for Example Problem

Steel	Yield Strength, σ_{ys} (ksi)	Assumed K_{Ic} Values (ksi $\sqrt{in.}$)	Design Stress, σ_D (ksi)	$\frac{\sigma_D}{\sigma_{ys}}$	M_k*	Q†	Wall Thickness, t (in.)‡	Wt. of 12 in. Section (lb)	Assumed§ $/lb Steel	Assumed§ $/ft Vessel
A	260	80	70	0.27	1.0	1.47	1.07	342	1.40	480
B	220	110	86	0.40	1.1	1.43	0.87	275	1.40	385
C	180	140	100	0.56	1.2	1.42	0.75	240	1.00	240
D	180	220	124	0.72	1.43	1.36	0.60	186	1.20	225
E	140	260	140	1.0	1.51	1.25	0.54	173	0.50	85
F	110	170	109	0.99	1.25	1.25	0.69	221	0.15	35

*Magnification factor.
†Crack shape parameter.
‡Thickness calculated assuming that $p = 5{,}000$ psi and $D = 30$ in.
§Estimated values; for examples only.

In this example, as soon as an 0.5-in. crack developed, either by an initial fabrication defect or by subcritical crack growth by fatigue or stress corrosion (Chapters 7–11), failure of the vessel would occur. Thus, the factor of safety or weight of a structure is not necessarily related to the yield strength of the structural material, as the results in Table 5.4 show. In fact the vessel fabricated out of the highest-strength steel actually weighs the most.

A more typical design procedure would be to establish a factor of safety against failure by either the fracture mode or the yielding mode, but preferably against *both* modes of failure. Pressure vessels are usually designed to a given percentage of either the yield strength or tensile strength thus having a factor of safety that is based on stress. A similar factor of safety against fracture should also be established.

Additional design situations for the above example will be analyzed as follows:

Case I: Present design philosophy where "perfect" fabrication is assumed and therefore the flaw size, a, is negligible. Assume a factor of safety of 2.0 against yielding or, $\sigma_{design} = \sigma_{ys}/2$.

Case II: Fracture mechanics design analysis where the maximum possible flaw size present, a, is assumed to be 0.5 in. and a *design stress intensity* of $K_I = K_{Ic}/2$ is used to provide a factor of safety of 2.0 against fracture. The design stress used in this case is thus based on resistance to fracture.

Case I. This design procedure is direct since

$$\sigma_{\text{design}} = \frac{\sigma_{ys}}{2} \quad \text{and} \quad t = \frac{pD}{2\sigma_{\text{design}}}$$

Knowing t for each of the six steels studied in this example, the estimated weight per foot is

$$A \simeq \pi D t$$

$$\text{volume/ft} \simeq A \times 12$$

$$\text{weight/ft} = \text{volume/ft} \times \text{density}$$

$$\text{weight/ft} = \text{volume/ft} \times 0.283 \text{ lb/in.}^3 \qquad \text{(for steel)}$$

$$\text{cost/ft} = \text{weight/ft} \times \text{cost/lb}$$

A typical calculation for steel D with $\sigma_{ys} = 180$ ksi is as follows:

$$\sigma_{\text{design}} = \frac{\sigma_{ys}}{2} = \frac{180}{2} = 90 \text{ ksi}$$

$$t = \frac{pD}{2\sigma_{\text{design}}} = \frac{(5{,}000)(30)}{2(90{,}000)} = 0.83 \text{ in.}$$

$$A \simeq \pi D t = (3.14)(30)(0.83) = 78.2 \text{ in.}^2$$

$$\text{volume/ft} \simeq A \times 12 = 78.2(12) \simeq 938 \text{ in.}^3$$

$$\text{weight/ft} = \text{volume/ft} \times \text{density} = 938 \times 0.283 = 265 \text{ lb/ft}$$

$$\text{cost/ft} = \text{weight/ft} \times \text{cost/ft} = 265 \times (1.20) = \$318/\text{ft}$$

The values for the other steels are calculated in a similar manner and are presented in Table 5.5. These results show the direct effect of reducing weight

TABLE 5.5. Comparison of Results for $\sigma_{\text{design}} = \sigma_{ys}/2$, Example Problem

Steel	*Yield Strength* σ_{ys} *(ksi)*	σ_D *(ksi)*	*t (in.)*	*Weight (lb/ft)*	*Cost ($/ft)*
A	260	130	0.58	185	259
B	220	110	0.68	219	306
C	180	90	0.83	265	265
D	180	90	0.83	265	318
E	140	70	1.07	344	172
F	110	55	1.36	437	66

by using a higher-strength steel. The cost per foot is not necessarily directly related to yield strength because of the numerous factors involved in pricing of structural materials. However, weight is decreased directly by increasing the yield strength, but if fracture is possible, the overall safety and reliability of the vessel may be decreased, as will be illustrated in the following examples. No K_I calculations are made in case I because it is assumed that there are *no* cracks.

Case II. This analysis is similar to the original analysis of this example in which $K_I = K_{Ic}$ (Table 5.4). However, an assumed factor of safety of 2.0 against fracture is obtained by selecting $K_{I(design)} = K_{Ic}/2$. Thus

$$\sigma_{design} = \frac{\sqrt{Q}(K_{Ic}/2)}{1.1\sqrt{\pi}\, M_k\sqrt{a}}$$

and

$$t = \frac{pD}{2\sigma_{design}}$$

M_k and Q must be assumed and several trials conducted to calculate each value of thickness. The calculations are similar to those shown previously (except that $K_I = K_{Ic}/2$), and the final results are summarized in Table 5.6.

TABLE 5.6. Comparison of Results for $K_I = K_{Ic}/2$, Example Problem

Steel	*Yield Strength* σ_{ys} *(ksi)*	σ_D *(ksi)*	K_{Ic} *(ksi$\sqrt{in.}$)*	K_I *(ksi$\sqrt{in.}$)*	t *(in.)*	*Weight (lb/ft)*	*Cost ($/ft)*
A	260	35	80	40	2.14	687	962
B	220	48	110	55	1.56	501	702
C	180	61	140	70	1.23	394	394
D	180	86	220	110	0.87	280	336
E	140	95	260	130	0.79	253	127
F	110	72	170	85	1.04	337	50

These results show the significant effect of toughness on the allowable design stress and illustrate dramatically that high-strength materials with low values of notch toughness do not necessarily yield the least weight vessel when fracture is a possible mode of failure. When cost is the primary criterion, the advantage of the lower-strength, tougher materials based on the assumed prices is obvious.

General Analysis of Cases I and II

The calculated thicknesses tabulated in Tables 5.5 and 5.6 are plotted as a function of yield strength in Figure 5.13. Comparison of the required thicknesses shows that for the two lowest-strength steels, yielding is the most likely mode of failure and the factor of safety against fracture will be greater

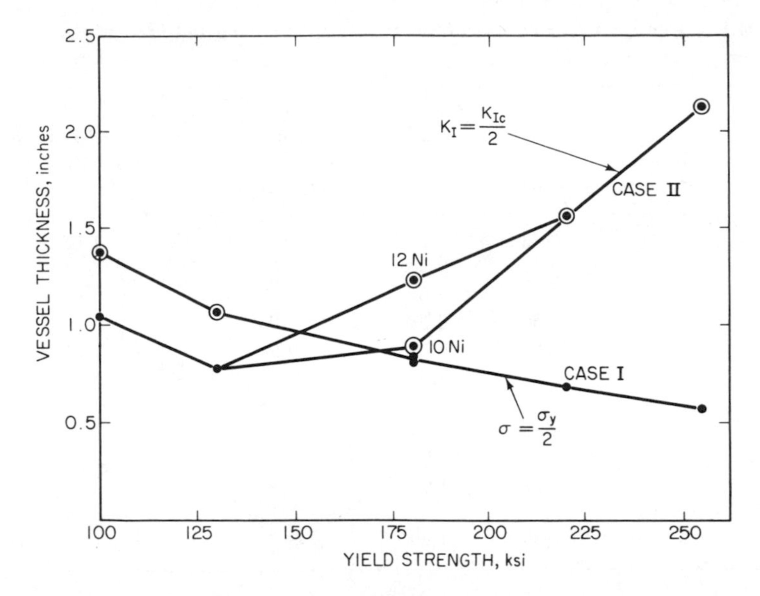

FIG. 5.13. Comparison of vessel thicknesses required for factor of safety of 2.0 against either yielding or fracture in example problem.

than 2.0. Specifically for steel F, the required vessel thickness is 1.36 in., and the design stress is 55 ksi ($\sigma_{ys}/2$); thus, the factor of safety against yielding is 2.0. The corresponding K_I value is 63 ksi$\sqrt{\text{in.}}$, which gives a factor of safety against fracture of $\frac{170}{63} = 2.7$. However, to have a factor of safety of at least 2 against *both* modes of failure, a wall thickness of 1.36 in. is required.

Conversely, the required thickness for the highest-strength steel is 2.14 in. based on a factor of safety against fracture of 2.0. However, the corresponding design stress value is 35 ksi, which gives a factor of safety against yielding of $\frac{260}{35} = 7.42$.

The factors of safety against yielding or fracture are tabulated in Table 5.7 and illustrate the necessity of considering all possible modes of failure prior to selecting a final geometry (thickness) as well as selecting a particular material.

As a last step in this example, the required thicknesses for a factor of safety of at least 2.0 against *both* yielding and fracture are listed in Table 5.8. In addition, the corresponding weight per foot and cost per foot are tabulated for each material. Analysis of the results shows that steel D would be the optimum selection on the basis of minimum weight and steel F would be the least expensive steel. Note that for both cases the factors of safety against yielding and fracture are 2.0 or greater. Thus, the vessels are compared on an equivalent performance basis and show the advantage of using fracture mechanics during the material selection process.

TABLE 5.7. Comparison of Factors of Safety Against Yielding and Fracture for Example Problem

Steel	*Yield Strength,* σ_{ys} *(ksi)*	*t (in.)*	*Factor of Safety Against Yielding*	*Factor of Safety Against Fracture*	*Required Thickness to Satisfy Both Criteria (in.)*
A	260	0.58	2.0	0.37	2.14
		2.14	7.42	2.0	
B	220	0.68	2.0	0.67	1.56
		1.56	4.58	2.0	
C	180	0.83	2.0	0.96	1.23
		1.23	2.95	2.0	
D	180	0.83	2.0	1.51	0.87
		0.87	2.09	2.0	
E	140	1.07	2.0	3.21	1.07
		0.79	1.07	2.0	
F	110	1.36	2.0	2.70	1.36
		1.04	1.29	2.0	

TABLE 5.8 Weight and Cost of Steel for Factor of Safety of 2.0 or Greater Against Both Yielding and Fracture, Example Problem

Steel	*Yield Strength* σ_{ys} *(ksi)*	*t (in.)*	*Weight (lb/ft)*	*Cost ($/ft)*
A	260	2.14	687	962
B	220	1.56	501	702
C	180	1.23	394	394
D	180	0.87	280	336
E	140	1.07	344	172
F	110	1.36	437	66

REFERENCES

1. J. W. Fisher, *Guide to 1974 AASHTO Fatigue Specifications, American Iron and Steel Construction*, New York, 1974.
2. J. E. Srawley and J. B. Esgar, "Investigation of Hydrotest Failure of Thiokol Chemical Corporation 260-Inch-Diameter SL-1 Motor Case," *NASA TMX-1194*, Cleveland, Jan. 1966.
3. W. W. Gerberich, "Fracture Mechanics Approach to Design-Application" presented in a *Short Course on Offshore Structures*, Berkeley, Calif., 1967.

6

Correlations Between K_{Ic} or K_{Id} and Other Fracture-Toughness Tests

6.1. General

In previous chapters we have introduced the concepts and analyses of fracture mechanics; described the test methods used to measure the critical material parameters, K_{Ic} and K_{Id}; demonstrated the general effects of temperature and loading rate on these parameters; and then showed how engineers can use fracture mechanics in design to prevent brittle fractures in structures.

These topics form the basis of fracture mechanics and are well founded on theory, and there is reasonable agreement among most engineers and scientists regarding their correctness even though the field is relatively new. Furthermore, there is general agreement regarding the application of fracture mechanics to those structural materials whose inherent toughness is such that they satisfy the various specimen size and constraint requirements established in Chapters 2 and 3.

However, it is a well known fact that the inherent fracture toughness of most structural materials is such that, using current standard test methods, neither K_{Ic} nor K_{Id} values can be measured at the service temperatures and loading rates for most large complex structures. Research is in progress to develop analytical and experimental techniques, as will be described in Chapter 16, that may result in standardized ASTM test methods.

Therefore, to be able to use fracture mechanics effectively throughout all phases of design, fabrication, quality control, and inspection, the engineer must rely on auxiliary test methods and empirical correlations to obtain estimates of K_{Ic} or K_{Id} for most structural materials. In fact, many recently developed structural codes and specifications that have material-toughness requirements are based on fracture-mechanics principles, but the *specific* toughness tests specified for material purchase or quality control are in terms of *auxiliary* test specimens such as the CVN impact test specimen.

For example, the American Association of State Highway and Transportation Officials (AASHTO) material requirements for bridge steels[1] are based on concepts of fracture mechanics but are specified in terms of Charpy V-notch impact test results. Toughness requirements for thick-walled nuclear pressure-vessel steels are based on minimum dynamic toughness values, K_{Id} (actually K_{IR} for critical reference values[2]). However, the actual material-toughness requirements for steels used in these pressure vessels are specified using NDT (nil-ductility transition) values and CVN impact values using lateral expansion measurements. Proposed toughness requirements for welded ship hulls were developed using K_{Id}/σ_{yd} values, but the proposed material procurement values are in terms of NDT and DT (dynamic tear) values. Thus empirical correlations as well as engineering judgment and experience are used to *translate* fracture-mechanics guidelines or controls into actual material-toughness specifications.

The cost of machining, fatigue precracking, and testing a K_{Ic} or K_{Id} specimen and the size requirements necessary to ensure valid K_{Ic} or K_{Id} test results render the K_{Ic} or K_{Id} test impractical as a quality-control test. Consequently the need exists to correlate K_{Ic} or K_{Id} data with notch-toughness test results obtained with smaller and less costly specimens. Actually, as described in Chapter 15, even though the various fracture-control plans in existence were all developed using concepts of fracture mechanics, few of these plans use either K_{Ic} or K_{Id} values for material specification or quality control. CVN, NDT, and DT specimens are used primarily.

For years, many specimens which are smaller and less costly than K_{Ic} specimens have been used to measure the notch toughness of steels. Each of these specimens has limitations and shortcomings. However, correlations between K_{Ic} and K_{Id} data and various notch-toughness test results can be developed if the limitations of the specimen have been thoroughly investigated and understood.

In the next section we shall describe briefly those test methods for which various correlations with either K_{Ic} or K_{Id} have been developed. In the remainder of this chapter we shall then describe some of the most widely used correlations between these various auxiliary test methods and K_{Ic} or K_{Id}. Correlations with elastic-plastic fracture-mechanics test specimens such as crack-opening displacement (COD) or J_{Ic} are discussed in Chapter 16.

6.2. Other Fracture-Toughness Test Specimens

6.2.1. General

Throughout the years, there have been many fracture-toughness test specimens developed throughout the world. However, only those that are both widely used and that have been correlated with either K_{Ic} or K_{Id} values are discussed in this chapter. For a more complete discussion of the various fracture test specimens that have been developed, the reader is referred to other, more research-oriented texts or technical papers[3-10] and to Chapter 16 for a discussion of COD, *R*-curve, and J integral.

6.2.2. CVN Impact or Slow-Bend Test Specimens

Prior to the development of fracture mechanics, the Charpy V-notch impact fracture-toughness test specimen was the one most widely used to determine the toughness behavior of structural materials. In fact, even today it is widely used throughout the world, not only as a general reference specimen but in many actual toughness specifications. The specimen is a standard ASTM specimen (E-23—Standard Methods for Notched Bar Impact Testing of Metallic Materials) and is shown in Figure 6.1.

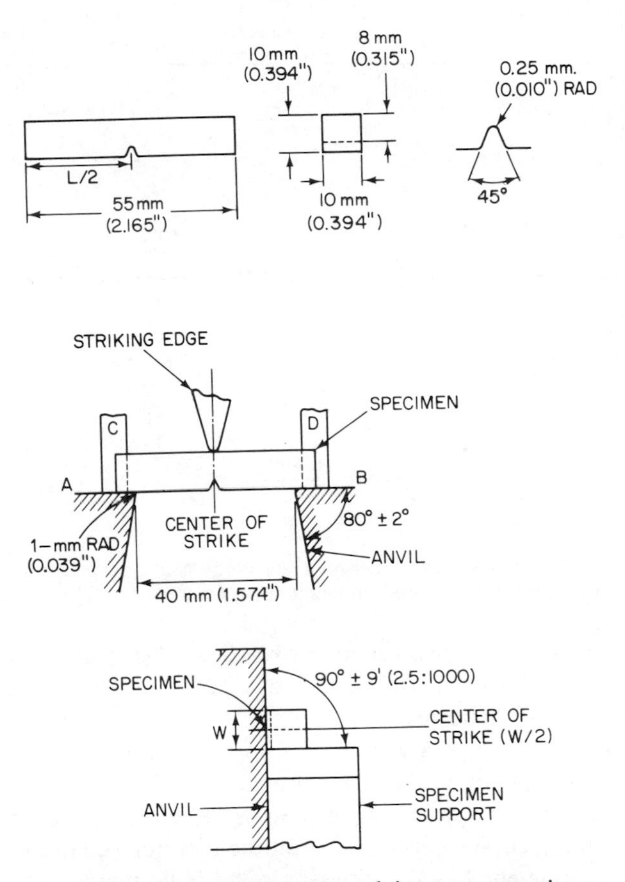

FIG. 6.1. Standard Charpy V-notch impact test specimen.

Using the CVN impact specimen, notch toughness is measured in terms of

1. Energy absorbed, ft-lb, or
2. Lateral expansion (equivalent to notch contraction), mils, or
3. Fracture appearance, percent shear.

Examples of these three measurements for a low-strength structural steel are presented in Figure 6.2. Note that the absorbed energy and the lateral expan-

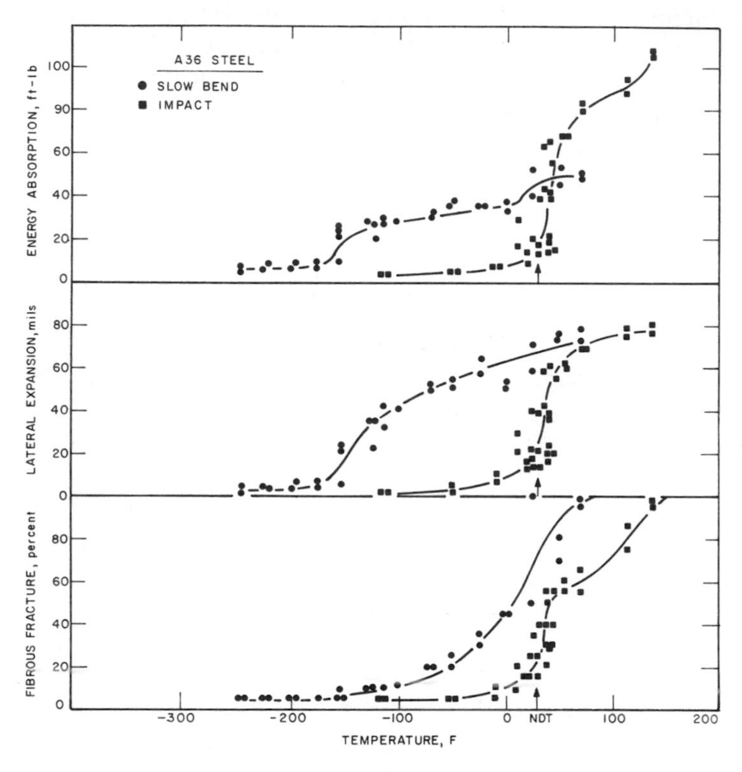

FIG. 6.2. Charpy V-notch energy absorption, lateral expansion, and fibrous fracture for impact and slow-bend test of standard CVN specimens.

sion measurements undergo a change over the same general temperature range. The percent fibrous fracture does not always conform to this behavior, especially in the slow-bend tests.

Examples of CVN impact curves of absorbed energy versus temperature for various structural materials are presented in Figure 6.3. It should be re-emphasized that there is no single unique curve for any single composition but that these curves depend also on the thermomechanical processing history. Approximately 15–20 specimens of a single material are tested at

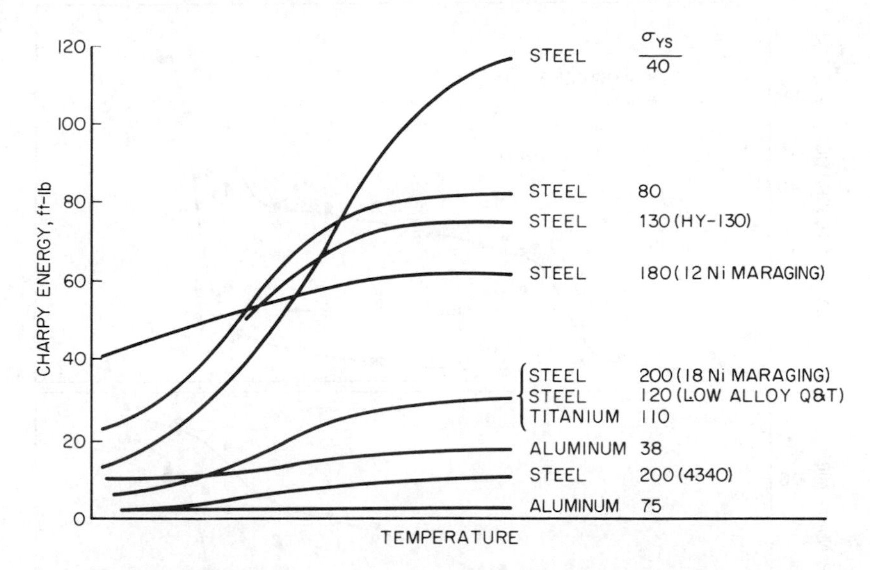

FIG. 6.3. Typical Charpy V-notch energy versus temperature behavior of various materials.

various temperatures, and a smooth curve is fitted to the results. Because of the relative small changes in behavior with temperature of the very high-strength steels, the titaniums, and aluminums, the CVN impact specimen is rarely used to evaluate the toughness behavior of these materials.

Although there are some valid criticisms of the CVN impact test when compared with K_{Ic} or K_{Id} tests, e.g., it has a blunt notch (root radius of 0.01 in.), small size, and does not differentiate between initiation and propagation energies, it is still widely used because it is very fast to conduct, inexpensive, and simple to use. Furthermore it has many years of correlation with service performance, and, as noted later, there are several empirical correlations between CVN impact test results and K_{Ic} or K_{Id}.

The CVN slow-bend specimen is identical to the CVN impact test specimen shown in Figure 6.1. The quantities measured are also identical, and the only difference is the loading rate. The CVN slow-bend specimen is loaded in three-point bending at a loading rate comparable to that used in conducting standard tension tests. The area under the load-displacement curve is measured to determine the total energy to fracture. Typical results are compared with CVN impact results in Figure 6.4 for a low-strength structural steel. The effect of the slower loading rate on a strain-rate-sensitive material is to cause the start of the transition in behavior from brittle to ductile behavior to occur at a lower temperature, compared with impact results.

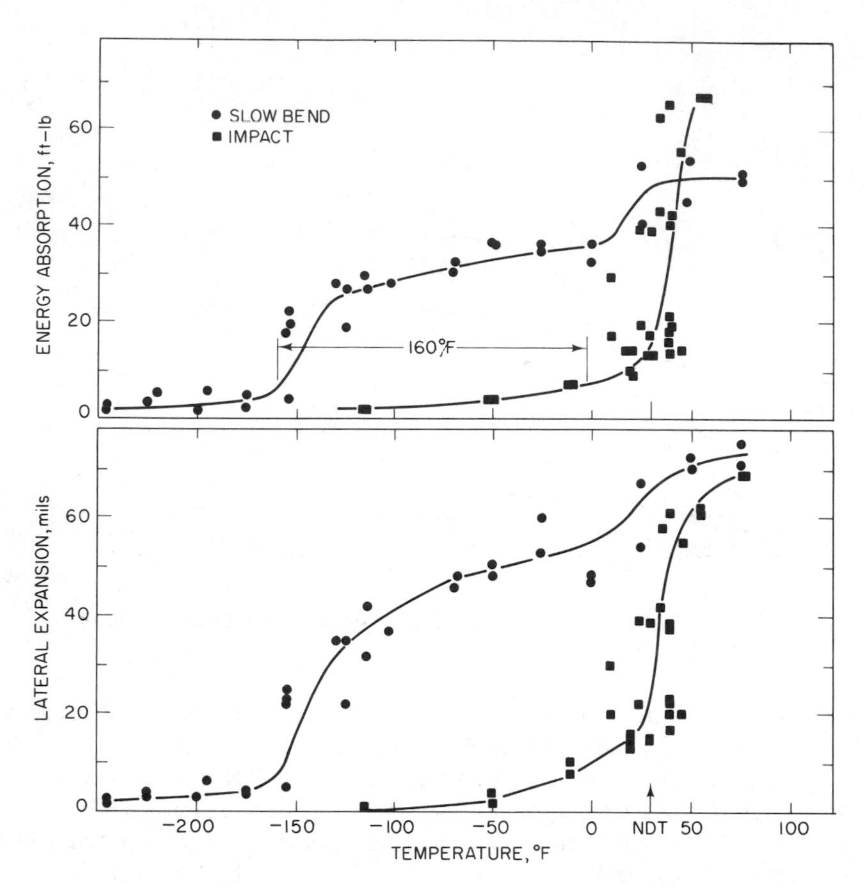

FIG. 6.4. Charpy V-notch energy absorption and lateral expansion for impact and slow-bend tests of standard CVN specimens.

6.2.3. Precracked CVN Impact or Slow-Bend Test Specimens

The *precracked* Charpy V-notch specimen was introduced to eliminate the machined notch in the standard Charpy V-notch specimen. The notch in a regular CVN specimen is precracked under cyclic loading so that the actual test is conducted on a CVN specimen that has a sharp crack, rather than the 0.01-in. notch of the standard CVN specimen.

The specimens can be tested either slowly or under impact conditions, in the same manner as the regular CVN test specimen is tested.

By instrumenting a standard CVN impact testing machine with strain gages whose output is recorded with an oscilloscope, measurements of fracture load can be obtained and values of K_{Id} are estimated. It should be noted, however, that the specimen size used (0.4- × 0.4-in. cross section) rarely meets the size requirements for K_{Id} values presented in Chapter 3. Considerable research is in progress to analyze the load-deflection data (obtained from either slow-bend or impact test results) to develop meaningful fracture-toughness values that can be used to analyze the performance of structures.[11]

6.2.4. Nil-Ductility Transition (NDT) Temperature Test Specimen

The NDT test specimen is a standardized ASTM test specimen and is described in ASTM E208—Standard Method for Conducting Drop Weight Test to Determine Nil-Ductility Transition Temperature of Ferritic Steels. Standard dimensions of the specimen and details of the crack starter are presented in Figure 6.5. The crack starter is a brittle weld in which a saw cut

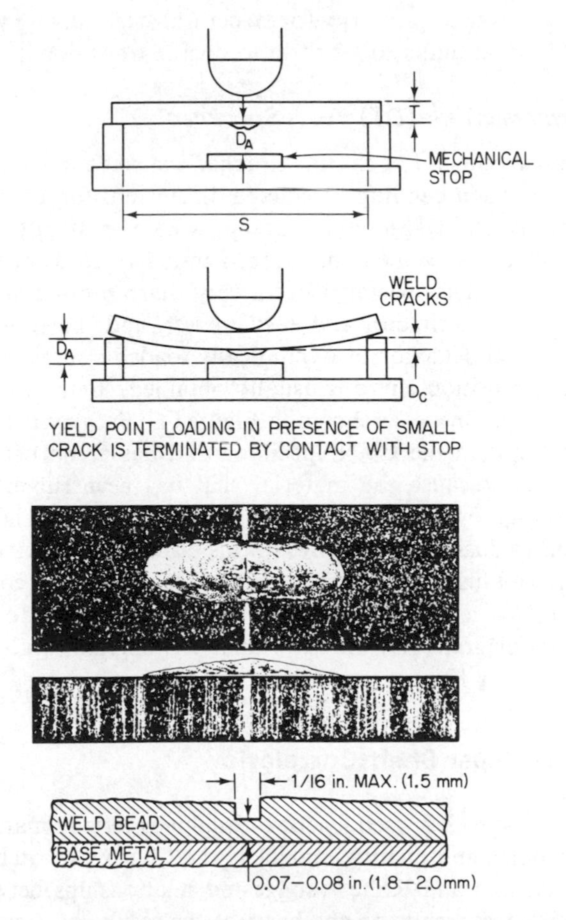

FIG. 6.5. Nil-ductility transition (NDT) temperature test specimen.

is made to localize the fracture in the center of the weld bead. The specimen is loaded in three-point bending by a falling weight at various temperatures. As the weight first hits the specimen, the brittle weld bead cracks, creating a small very sharp semicircular surface crack. As the weight continues to fall, the material being tested is loaded dynamically in the presence of the small weld crack. A mechanical stop is placed under the center of the specimen to limit the amount of deformation to which the specimen is subjected. The NDT temperature is defined as the highest temperature at which a standard specimen breaks in a brittle manner; that is, the material is said to have "nil ductility." Above the NDT temperature, the material has sufficient ductility to deflect to the mechanical stop before fracturing, i.e., deflect inelastically. Thus, a series of about six to eight specimens are tested at various temperatures to establish the NDT temperature of a material, usually within $\pm 10°F$.

The NDT test is used primarily for structural steels having yield strengths less than 140 ksi that undergo a brittle to ductile transition.

6.2.5. *Dynamic Tear (DT) Test Specimen*

The DT test specimen is a sharply notched specimen impact loaded along one edge in three-point bending. Specimen dimensions for a tentative ASTM standard $\frac{5}{8}$-in.-thick test specimen[12] are presented in Figure 6.6, although similar full-thickness specimens have been tested to study the behavior of structural materials. The specimen has a very sharp pressed notch, is standardized ($\frac{5}{8}$-in.-thick specimen), and easy to test, and the results have been related to structural behavior of dynamically loaded structures by Pellini.[13]

A complete transition curve is usually obtained, similar to that obtained for CVN impact specimens, as shown in Figure 6.7. Because the DT specimen is slightly thicker than the CVN specimen and has a sharper notch (additional constraint), because the material that has been subjected to plastic deformation caused by the pressed knife loses some of its original ductility, and because of residual tensile stresses in this plastically deformed region, the transition behavior usually occurs at higher temperatures compared with impact CVN or K_{Id} test results. This test specimen is used to establish the fracture analysis diagram and the ratio analysis diagram, both of which are described in Chapter 12.

6.3. K_{Ic}-CVN Upper Shelf Correlation

One of the most widely used fracture-toughness tests in material development, specifications, and quality control is the Charpy V-notch impact test. Accordingly, Barsom and Rolfe[14] suggested relationships between K_{Ic} and upper shelf CVN test results on the basis of the results of various investigations by Clausing,[15] Holloman,[16] and Gross[17]. Clausing[15] showed that the state of stress at fracture initiation in the CVN impact specimen is plane strain, which is the state of stress in a thick K_{Ic} specimen. Holloman[16] has shown that for the dimensions used in the CVN specimen the maximum possible lateral stress is obtained, indicating a condition approaching maxi-

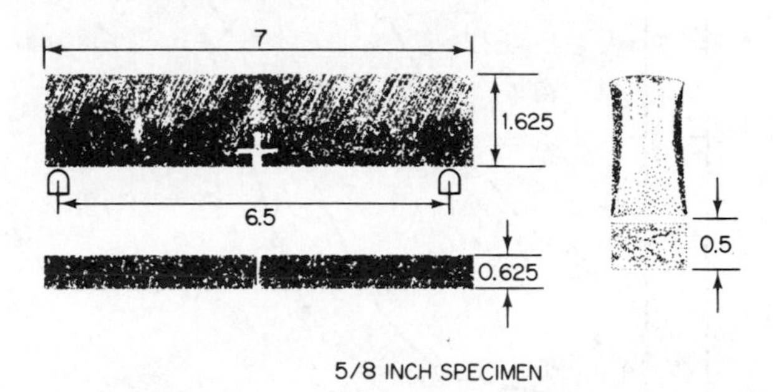

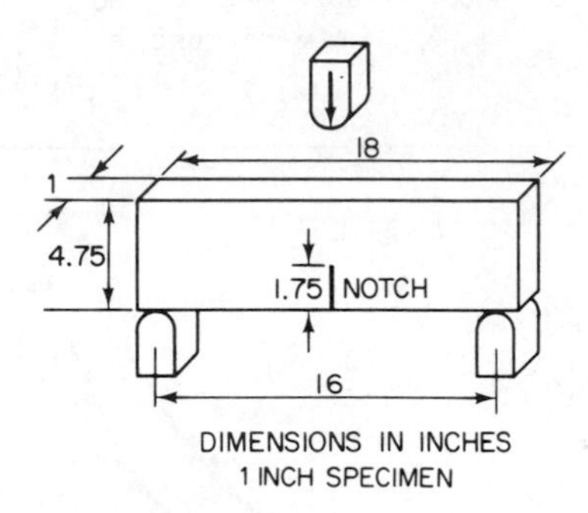

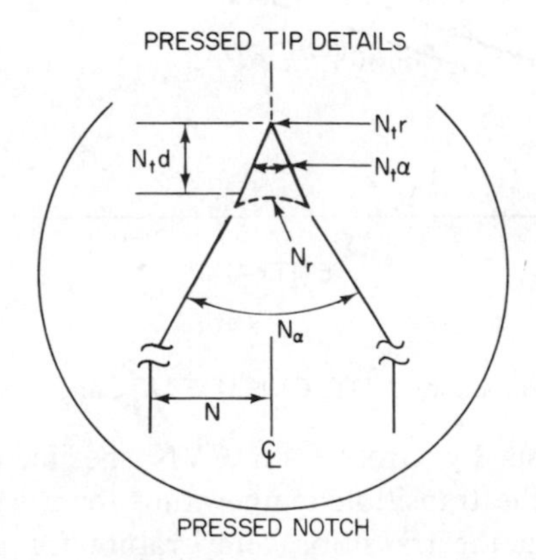

FIG. 6.6. Dynamic tear (DT) test specimen.

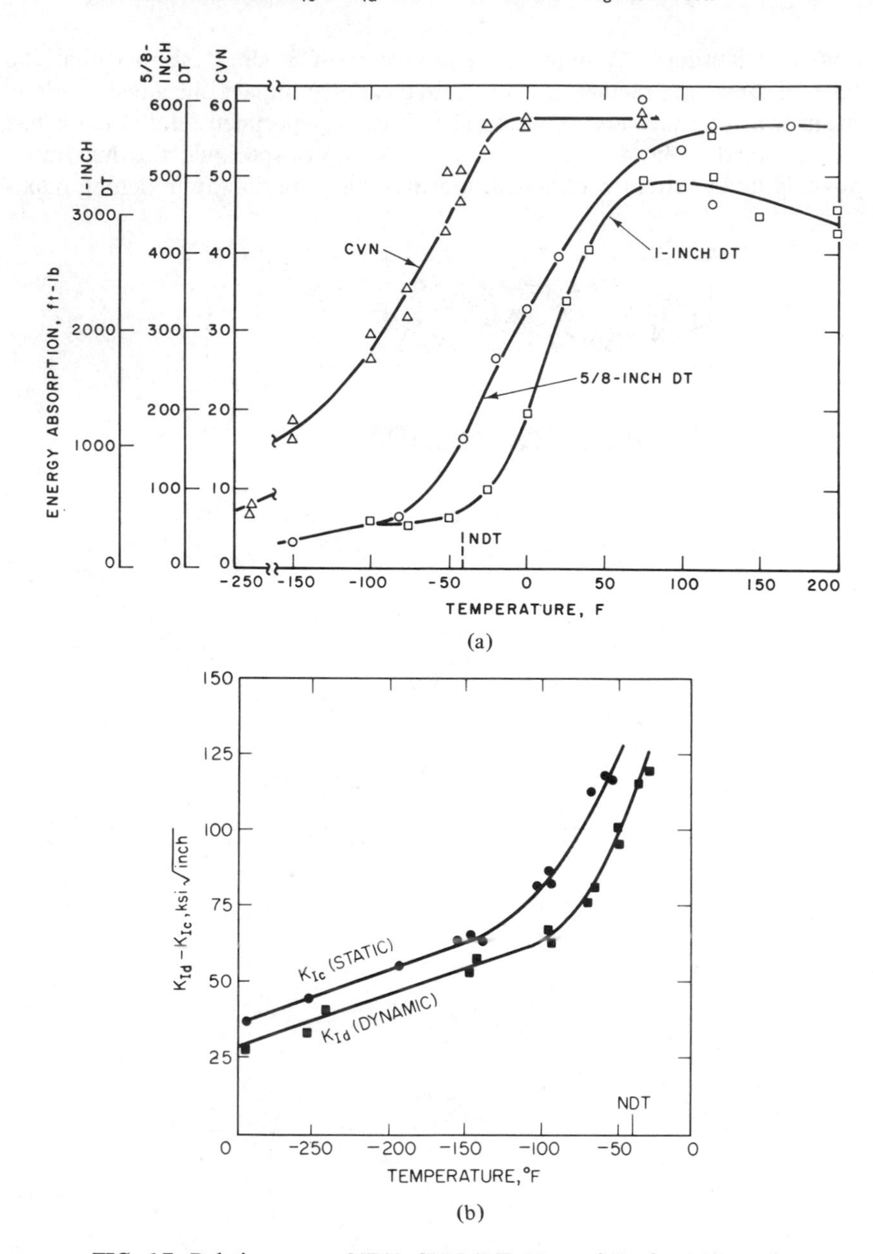

(a)

(b)

FIG. 6.7. Relation among NDT, CVN, DT, K_{Ic}, and K_{Id} for A517 steel.

mum constraint. Tests by Gross[17] on CVN specimens of various thicknesses showed that the transition temperature for a standard CVN specimen is identical with the transition temperature for a CVN specimen of twice the standard width, substantiating the observation that the standard

CVN test specimen has considerable constraint at the notch root. These observations, combined with the conclusion of Barsom and Rolfe[14] that the effect of temperature and rate of loading on CVN and K_{Ic} values is the same, suggested that it should be possible to establish empirical correlations between K_{Ic} and CVN test results. The recent development of the J integral concept,[18,19] as discussed in Chapter 16, suggests a possible theoretical basis for such empirical correlations.

The upper shelf K_{Ic}-CVN correlation shown in Figure 6.8 was developed empirically[14,20] from results obtained on 11 steels having yield strengths in

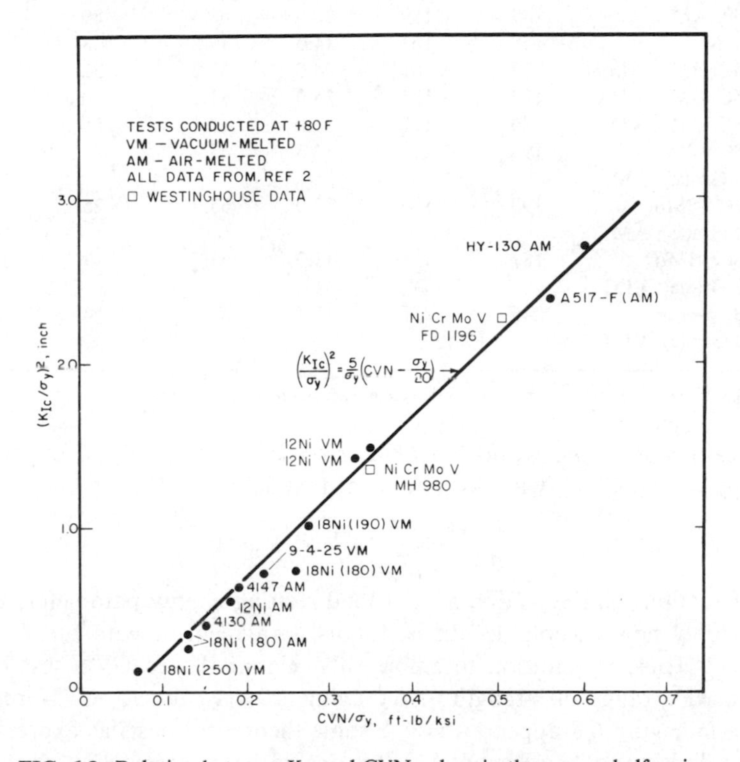

FIG. 6.8. Relation between K_{Ic} and CVN values in the upper shelf region.

the range 110–246 ksi (Table 6.1). The K_{Ic} values for these steels ranged from 87 to 246 ksi$\sqrt{\text{in.}}$, and the CVN impact values ranged from 16 to 89 ft-lb.

At the upper shelf, the effects of loading rate and notch acuity are not so critical as in the transition-temperature region. Thus, the differences in the K_{Ic} and CVN test specimens (namely, loading rate and notch acuity) are not that significant, and a reasonable correlation would be expected.

Furthermore, in a discussion of a paper on the J integral as a fracture criterion[21] P. C. Paris states that "This paper (J integral) finally explains the

TABLE 6.1. Longitudinal Mechanical Properties of Steels Investigated for Room-Temperature K_{Ic}-CVN Correlation

*Steel and Melting Practice**	*Yield Strength, 0.2% Offset (ksi)*	*Tensile Strength (ksi)*	*Elongation in 1 In. (%)*	*Reduction of Area (%)*	*Charpy V-Notch Energy Absorption at +80°F (ft-lb)*	K_{Ic} *(ksi√in.)*
A517-F, AM	110	121	20.0	66.0	62	170
4147, AM	137	154	15.0	49.0	26	109
HY-130, AM	149	159	20.0	68.4	89	246
4130, AM	158	167	14.0	49.2	23	100
12Ni-5Cr-3Mo, AM	175	181	14.0	62.2	32	130
12Ni-5Cr-3Mo, VIM	183	191	15.0	61.2	60	220
12Ni-5Cr-3Mo, VIM	186	192	17.0	67.1	65	226
18Ni-8Co-3Mo (200 Grade), AM	193	200	12.5	48.4	25	105
18Ni-8Co-3Mo (200 Grade), AM	190	196	12.0	53.7	25	112
18Ni-8Co-3Mo (190 Grade), VIM	187	195	15.0	65.7	49	160
18Ni-8Co-3Mo (250 Grade), VIM	246	257	11.5	53.9	16	87

*AM signifies electric-furnace air-melted; VIM signifies vacuum-induction-melted.

reasonableness of the Rolfe–Novak–Barsom correlation[14,20] of upper shelf Charpy values, CVN, with K_{Ic} numbers; that is,

$$\left(\frac{K_{Ic}}{\sigma_{ys}}\right)^2 = \frac{5}{\sigma_{ys}}\left[\text{CVN} - \frac{\sigma_{ys}}{20}\right] \tag{6.1}$$

This equation relating CVN, a limit-load-relating energy parameter, to K_{Ic} is not only now acceptable but is, for us, in agreement with the J failure criteria." Thus, in addition to empirically relating K_{Ic} to CVN test results over a wide range of strength and toughness levels, the K_{Ic}-CVN relation shown in Figure 6.8 appears to have some theoretical basis as expressed in the development of the J integral.[21] It is one of the more widely used correlations for K_{Ic} and has been substantiated by additional tests by other investigators.[22]

Because K_{Ic} is a static test, and the CVN impact test is an impact test, the relationship presented in Equation (6.1) is limited to steels having yield strengths greater than 100 ksi. However, because this correlation is an upper shelf correlation, where the slow bend and impact CVN values are constant for steels of various yield strengths, Equation (6.1) may be applicable to steels having yield strengths < 100 ksi. Because the upper shelf CVN impact results are higher than the upper shelf CVN slow bend results, substituting σ_{yd} into Equation (6.1) may give a better correlation for steels having a yield strength < 100 ksi.

In the above equation

K_{Ic} = critical plane-strain stress-intensity factor at slow loading rates, ksi$\sqrt{\text{in.}}$

σ_{ys} = 0.2% offset yield strength at the upper shelf temperature (or room temperature), ksi.

CVN = standard Charpy V-notch impact test value at upper shelf, ft-lb.

6.4. K_{Ic}-CVN Correlation in the Transition-Temperature Region

To establish correlations between K_{Ic} and CVN test results in the transition-temperature region, the effects of both notch acuity and loading rate should be considered. K_{Ic} values and CVN values in the transition-temperature region can be correlated (1) when the test results for slow-bend K_{Ic} specimens are related to the test results for slow-bend fatigue-cracked CVN specimens and (2) when the test results for dynamic K_{Ic} specimens are related to the test results for dynamic fatigue-cracked CVN impact specimens (Figure 6.9). The correspondence between K_{Ic} and CVN energy-absorption

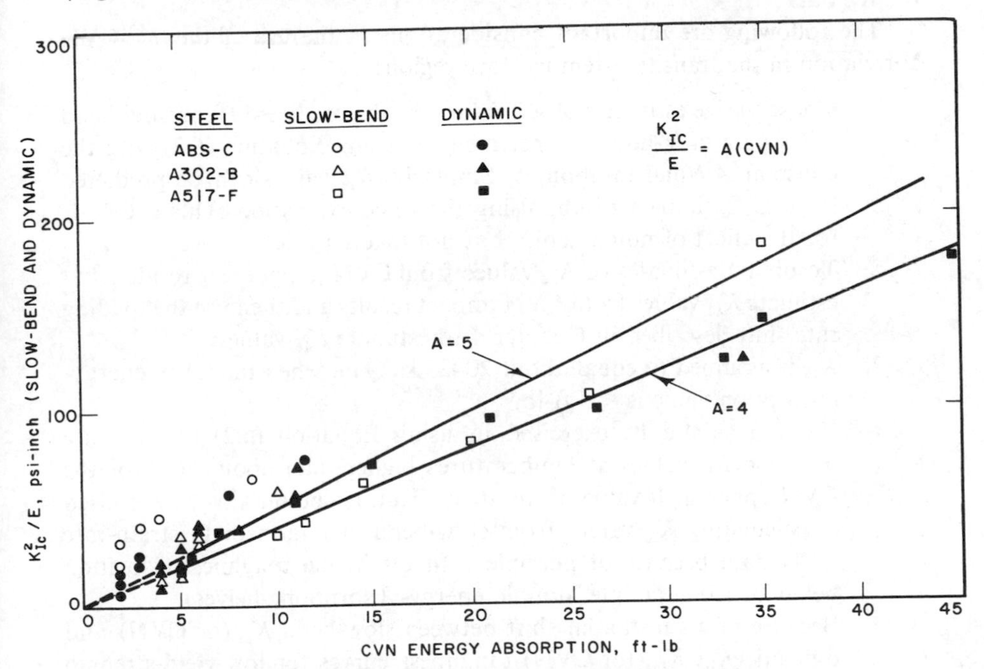

FIG. 6.9. Correlation between K_{Ic} and CVN test results for slow-bend and dynamic loading.

values obtained at a particular test temperature and at the same strain rate for both K_{Ic} and CVN can be approximated by[14,23,24]

$$\frac{K_{Ic}^2}{E} = A(\text{CVN}) \tag{6.2}$$

where E = Young's modulus,
A = constant of proportionality,
and K_{Ic} and CVN are tested at the same temperature and strain rate.

The constant of proportionality, A, incorporates the effects of specimen size as well as notch acuity. Thus by changing the value of A in Equation (6.2) it is possible to correlate K_{Ic} data and CVN energy-absorption values obtained by testing *V-notched* specimens. The constant of proportionality, A, for K_{Ic} and slow-bend *V-notched* Charpy specimens, and K_{Id} and impact V-notched Charpy specimens (as well as at intermediate loading rates) for low-strength structural steels was determined to be equal to about 5.

This equation suggests that the relationship between slow-bend K_{Ic} and slow-bend CVN test results is the same as the relationship between impact K_{Ic} (i.e., K_{Id}) and impact CVN test results. This observation is not unexpected because it was shown in Chapter 4 that a particular change in loading rate causes an equal shift along the temperature axis for both the CVN data and the K_{Ic} data.

The following are important considerations in the use of the K_{Ic}-CVN correlation in the transition-temperature region:

1. Conservative estimates of K_{Ic} values can be predicted from slow-bend CVN data of V-notched specimens by using Equation (6.2) with the constant A equal to about 5. Similarly, K_{Id} values can be predicted from CVN impact results using the same expression. This accounts for the effect of notch acuity but not loading rate.
2. To obtain estimates of K_{Ic} values from CVN impact test results, first estimate K_{Id} values from CVN impact results and then use the loading rate shift described in Chapter 4 to estimate K_{Ic} values.
3. K_{Id} is assumed to equal about 20–25 ksi$\sqrt{\text{in.}}$ when the CVN energy-absorption value is <5 ft-lb.
4. Caution should be exercised in using Equation (6.2) to calculate dynamic K_{Ic} values at temperatures higher than about 50% of the CVN upper shelf value of the steel. Caution should also be exercised in calculating K_{Ic} values from slow-bend or intermediate strain-rate CVN data because of possible artifacts in the toughness-transition behavior (for example, double energy-absorption shelves).
5. Because of a substantial shift between slow-bend K_{Ic} (or CVN) and dynamic K_{Ic}, K_{Id}, (or CVN) toughness curves for low-yield-strength steels (Chapter 4), equations relating these values in the toughness transition zone should not be used for steels having yield strengths less

than 100 ksi (689.5 MN/m²). For example, K_{Ic} values for a 70-ksi (483-MN/m²) yield strength steel would undergo a substantial increase (that is, a toughness transition) within a given temperature range, whereas the impact CVN values within and well above this range remain at a constant value equal to about 5 ft-lb (6.8 J). This behavior is caused by the 100°F (56°C) temperature shift between the slow-bend and impact curves for a steel having 70-ksi yield strength. Therefore a correlation between a slow-bend K_{Ic} value that has undergone a toughness transition and a constant impact CVN energy-absorption value should not be used.

As an example, the K_{Id} at the NDT temperature for an A572 Grade 50 steel is about 58 ksi$\sqrt{\text{in.}}$ (63.8 MN/m$^{3/2}$). The corresponding CVN energy-absorption value calculated by using Equation (6.2) is 22.5 ft-lb (30.6 J). This CVN value agrees very well with the average value of the experimental CVN data as shown in Figure 6.10.[23]

The K_{Id} data presented in Figures 6.11 and 6.12[23,24] have been used in conjunction with Equation (6.2) and the constant of proportionality $A = 5$ to predict the impact CVN energy-absorption behavior of A36 and A572 Grade 50 steels. The resulting impact CVN curves are shown in Figures 6.13 and 6.14[23,24] superimposed on the experimentally determined impact CVN

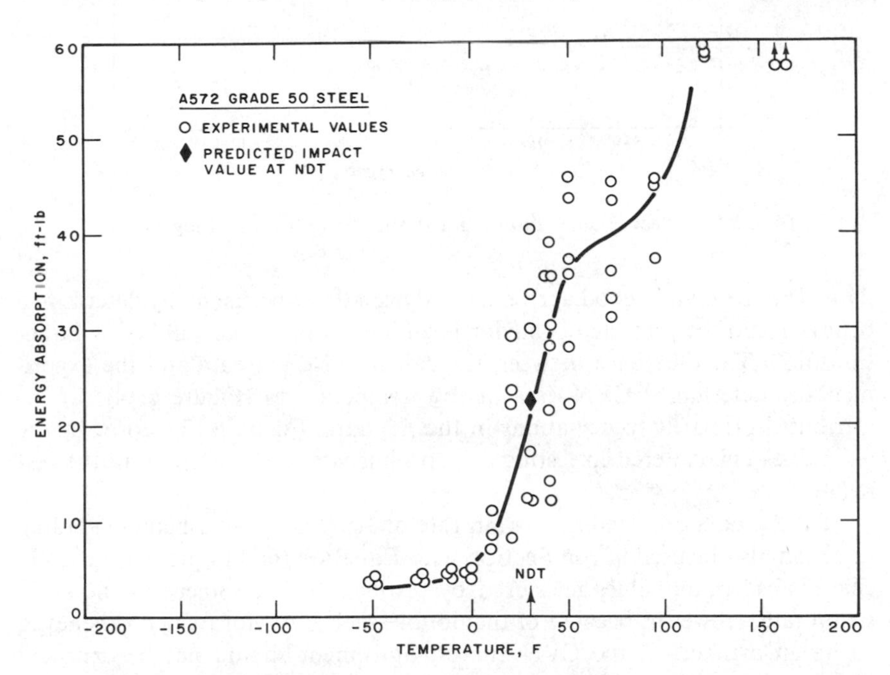

FIG. 6.10. Charpy V-notch energy absorption for impact tests of standard CVN specimens.

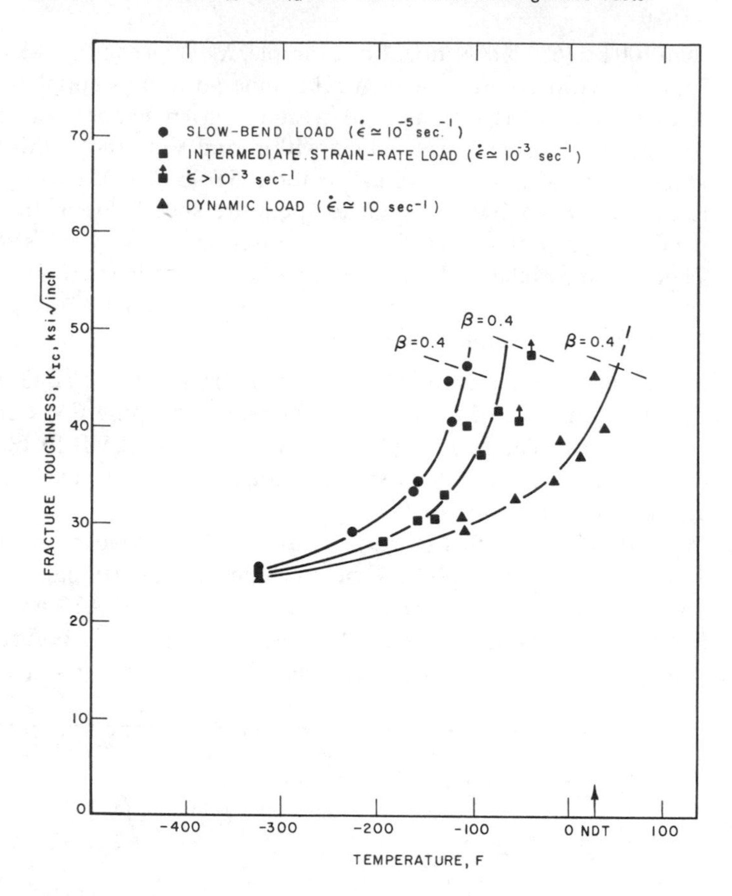

FIG. 6.11. Effect of temperature and strain rate on fracture toughness of A36 steel.

data. The reasonably good agreement between the experimentally determined behavior and the predicted behavior is an indication of the validity of Equation (6.2). The difference between the calculated CVN data and the experimentally determined CVN data at low temperatures (Figure 6.14) can be attributed primarily to deviations in the K_{Id} data, Figure 6.12, caused by the difficulties encountered in testing K_{Id} specimens and in interpreting the test results.

The K_{Ic} data obtained at a strain rate of 10^{-3} sec^{-1} (intermediate loading rate) can also be used in conjunction with Equation (6.2) to predict the CVN energy-absorption values measured by testing CVN specimens at the same strain rate. However, because of the double-shelf behavior (which is believed to be an artifact) of the CVN curves, agreement should not be expected between the predicted CVN values and the experimental values. On the other hand, the transition temperature defined by both sets of data should agree.

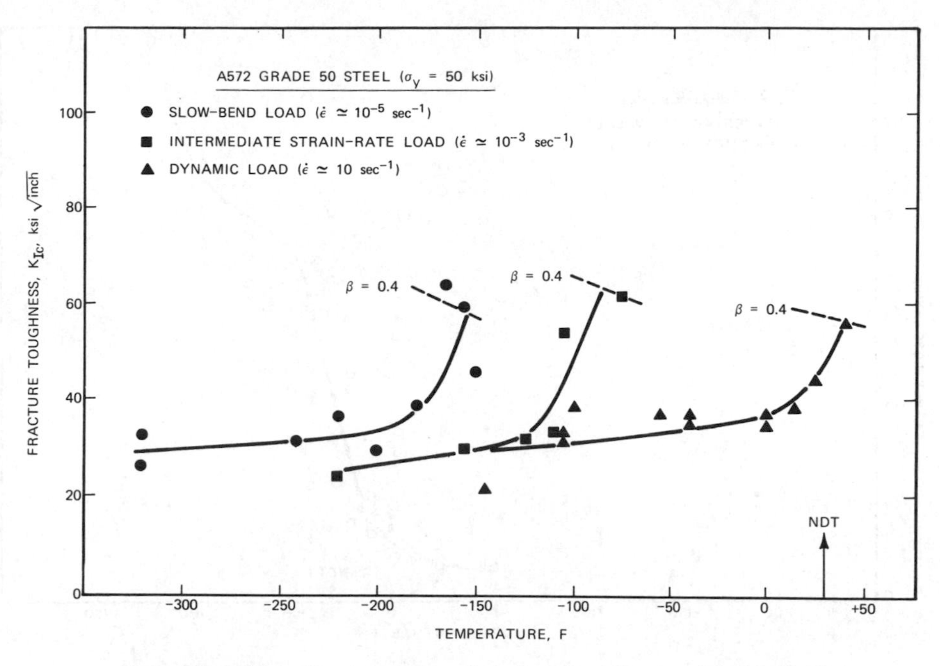

FIG. 6.12. Effect of temperature and strain rate on fracture toughness of A572 Grade 50 steel ($\sigma_{ys} = 50$ ksi).

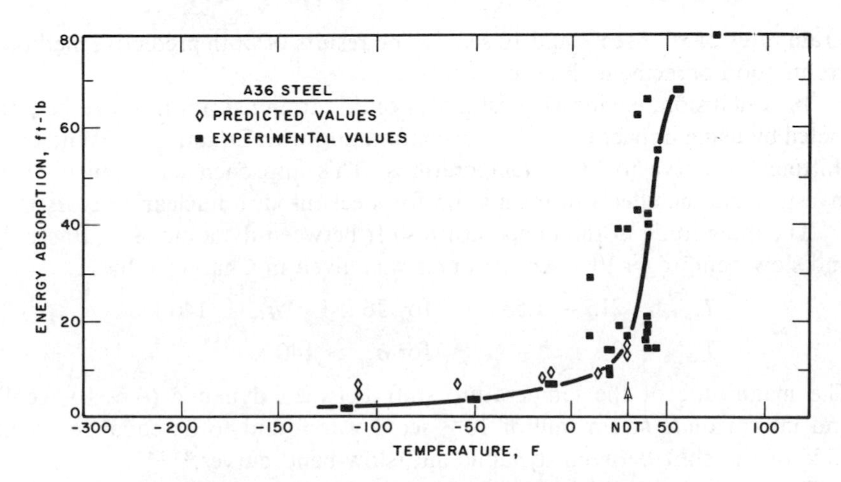

FIG. 6.13. Experimental data and predicted values of Charpy V-notch energy absorption for impact loading of standard CVN specimens.

Likewise, the CVN energy-absorption values at a strain rate of 10^{-3} sec^{-1} can be predicted by shifting the dynamic curve -120°F (-67°C).[23,24] This magnitude of the temperature shift for A36 and A572 Grade 50 steels has been shown to equal the shift between K_{Ic} tests or CVN tests conducted at

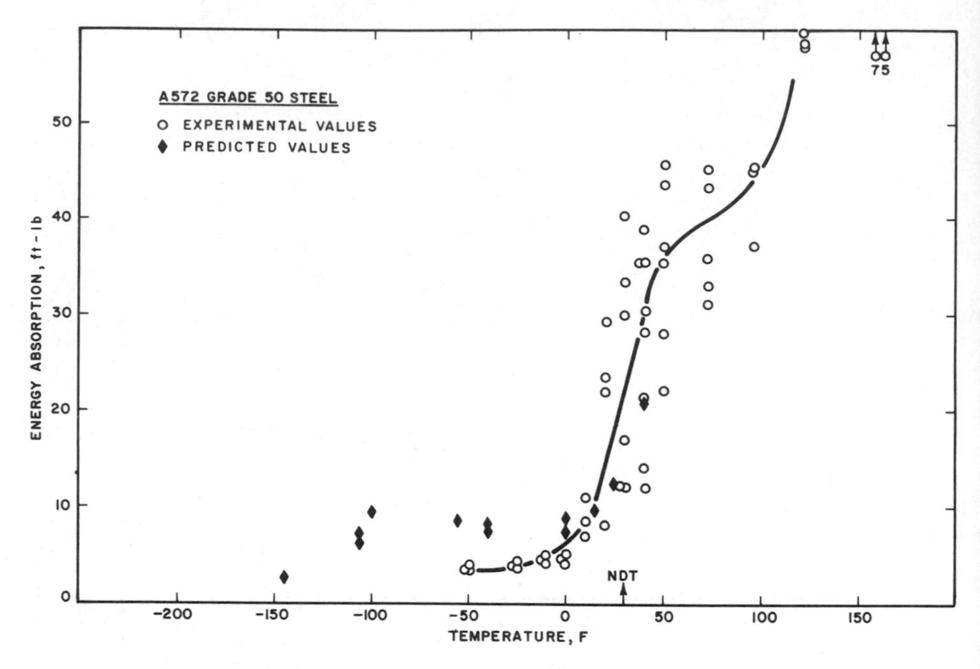

FIG. 6.14. Experimental data and predicted values of Charpy V-notch energy absorption for impact loading of standard CVN specimens.

strain rates of 10^{-3} sec^{-1} and 10 sec^{-1}. The results of both predictive methods are in good agreement (Figure 6.15).

In conclusion, engineering estimates of K_{Ic} at any strain rate can be predicted by using impact CVN data in conjunction with Equation (6.2) and then shifting the curve to lower temperatures. This approach has been used in investigating the effects of irradiation for steels used in nuclear reactors.[25]

The magnitude of the temperature shift between dynamic ($\dot{\epsilon} \approx 10$ sec^{-1}) and slow-bend ($\dot{\epsilon} \approx 10^{-5}$ sec^{-1}) curves was given in Chapter 4 by

$$T_{\text{shift}} = 215 - 1.5\sigma_{ys} \qquad \text{for } 36 \text{ ksi} < \sigma_{ys} < 140 \text{ ksi}$$

$$T_{\text{shift}} = 0 \qquad \text{for } \sigma_{ys} > 140 \text{ ksi}$$

The magnitude of the temperature shift between dynamic ($\dot{\epsilon} \approx 10$ sec^{-1}) and intermediate strain rate of 10^{-3} sec^{-1} was found to be equal to about 75% of the shift between dynamic and slow-bend curves.[23,24]

The general use of this procedure to estimate K_{Ic} values in the transition-temperature region from CVN impact results is summarized as follows:

1. Obtain standard CVN impact test results in the transition-temperature region.
2. Calculate K_{Id} values at each test temperature using

$$\frac{K_{Id}^2}{E} = 5(\text{CVN})$$

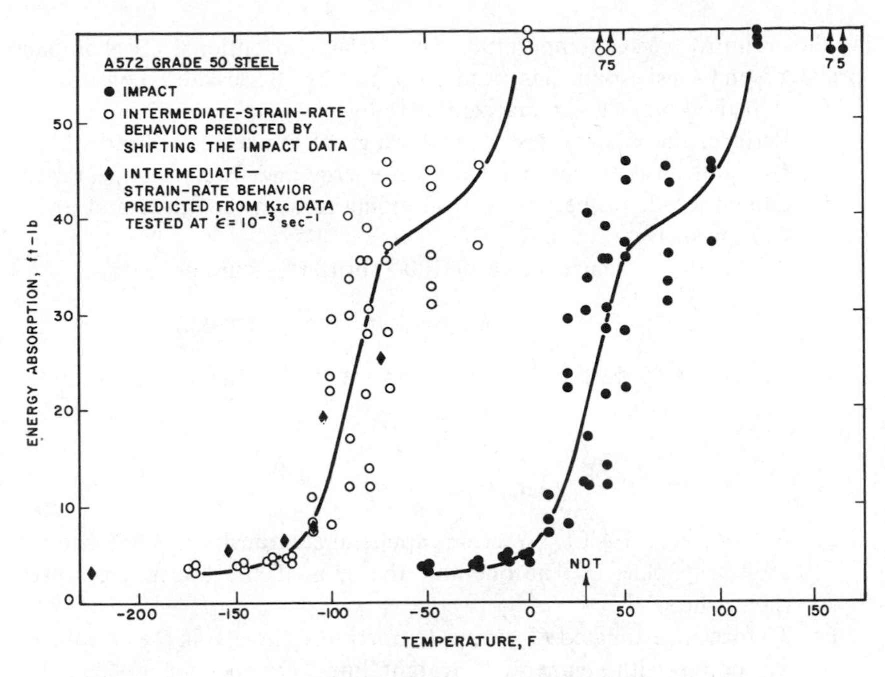

FIG. 6.15. Charpy V-notch energy-absorption behavior for impact loading and intermediate strain-rate loading of standard CVN specimens.

where the calculated K_{Id} values are at the same temperature as each CVN value.

3. Use the following equation for temperature shift to determine the temperature shift between K_{Id} and K_{Ic} values:

$$T_{\text{shift}} = 215 - 1.5\sigma_{ys} \qquad \text{for } 36 \text{ ksi} < \sigma_{ys} < 140 \text{ ksi}$$

4. Shift the K_{Id} values at each temperature by the temperature shift calculated in step 3 to obtain K_{Ic} (static) values as a function of temperature.

This procedure is limited to the lower end of the transition curve. As the upper shelf region is approached, where loading rate and notch acuity do not have so great an influence on the fracture-toughness behavior, the upper shelf correlation is recommended, even though it was developed primarily for steels with yield strengths greater than 80 ksi.

6.5. Approximation of Entire K_{Ic} Curve from CVN Impact Data

The correlations and procedures presented in the preceding section can be used to estimate the entire curves of K_{Ic}, K_{Id}, or K_{Ic} at an intermediate loading rate from CVN impact data. Another interesting method of predicting the entire K_{Ic} versus temperature curve from conventional CVN impact specimens and tensile data has been suggested by Begley and Logsdon.[26]

Their method may be summarized as follows:

1. Perform the Charpy test over a range of temperatures and obtain the full transition curves of impact energy and fracture appearance.
2. Obtain tensile properties corresponding to both the upper and lower Charpy shelves.
3. At the highest temperature of 100% brittle fracture appearance

$$K_{Ic} = .45\sigma_{ys}\sqrt{\text{in.}}$$

4. At the 100% ductile fracture temperature use the K_{Ic}-CVN upper shelf correlation:

$$\left(\frac{K_{Ic}}{\sigma_{ys}}\right)^2 = 5\left(\frac{\text{CVN}}{\sigma_{ys}} - 0.05\right)$$

5. At the 50% FATT (fracture appearance transition temperature) arbitrarily place K_{Ic} at one-half the sum of the upper and lower shelf values.
6. To form an estimated K_{Ic} versus temperature curve, join the calculated K_{Ic} points with segmented straight lines. On the upper shelf, K_{Ic} should be fairly constant. To estimate a full lower shelf extend a

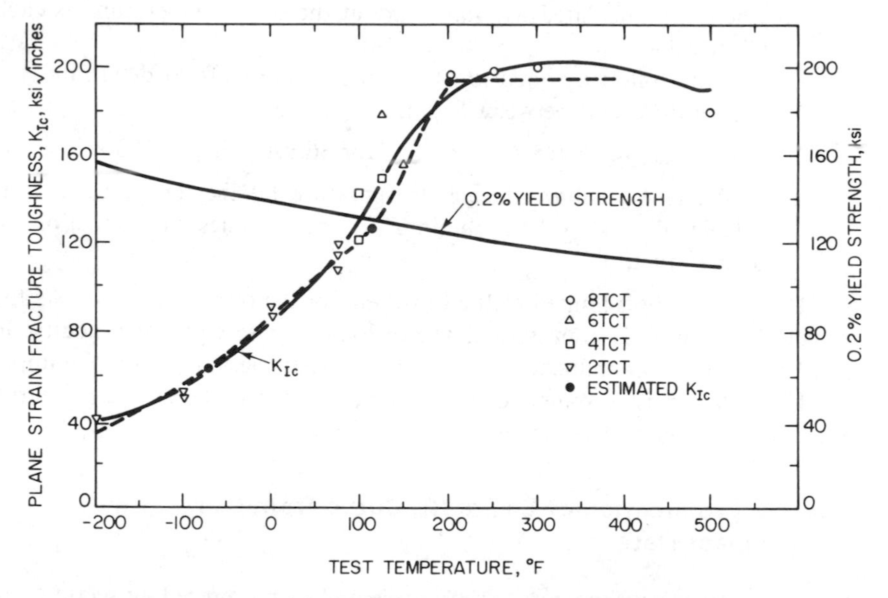

FIG. 6.16. Comparison of actual and estimated K_{Ic} values for a NiCrMoV steel (Reference 26).

line from the lower shelf value to about 25 ksi$\sqrt{\text{in.}}$, at -320°F. This is a reasonable value for steels. Experience with rotor steels place the best estimates of this point at 35 ksi$\sqrt{\text{in.}}$, 25 ksi$\sqrt{\text{in.}}$, and 25 ksi$\sqrt{\text{in.}}$ for NiCrMoV, NiMoV, and CrMoV alloys, respectively.

Begley and Logsdon state that "Agreement of estimated and measured K_{Ic} values is excellent. Charpy and tensile data defined K_{Ic} versus temperature curves for NiCrMoV, NiMoV, CrMoV steels and a 12 Cr stainless steel. The proposed method of estimating K_{Ic} over a range of temperatures from Charpy and tensile data is a useful engineering approach. This is especially true for instances where the size and amount of test material is limited."

Examples of their procedure for a NiCrMoV, a NiMoV, and a 12% Cr stainless steel are shown in Figures 6.16, 6.17, and 6.18, respectively. In these graphs, the dotted lines represent the toughness estimated from conventional Charpy and tensile tests. Estimated K_{Ic} values used to construct these lines are also shown. It is evident that the proposed method of correlation gives a very good representation of the entire K_{Ic} versus temperature curve, particularly when compared with the effort and expense of measuring K_{Ic} with 8-in.-thick specimens. This procedure should be restricted to steels having $\sigma_{ys} > 90$ ksi because the procedure does not account correctly for the effect of loading rate.

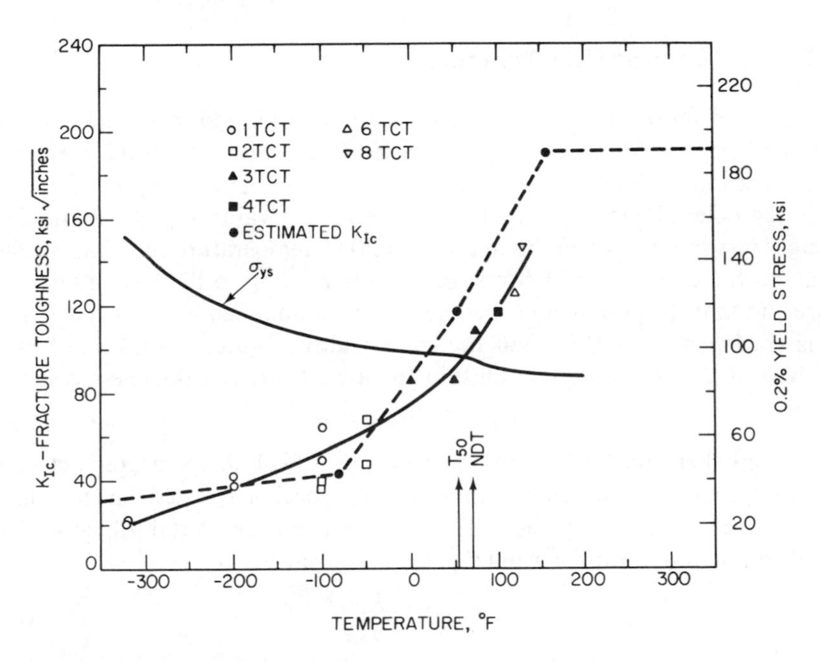

FIG. 6.17. Comparison of actual and estimated K_{Ic} values for NiMoV steel (Reference 26).

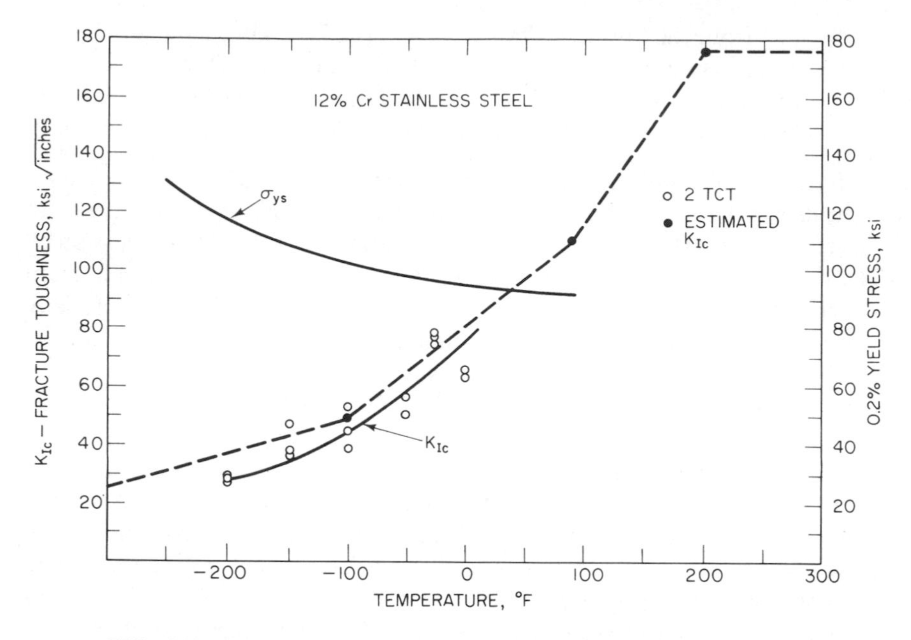

FIG. 6.18. Comparison of actual and estimated K_{Ic} values for a 12Cr stainless steel (Reference 26).

6.6. K_{Id} Value at NDT Temperature

In the drop weight NDT test, a specimen is subjected to a crack initiated from a brittle weld bead under impact-loading conditions. After examining the crack shape obtained in this type of test, Irwin[27] et al. proposed an analysis for determining a dynamic crack-toughness value, K_{Id}, from the drop weight test results. Assuming that at the NDT temperature the plate surface reached the dynamic yield stress, σ_{yd}, corresponding to the testing temperature and that the pop-in crack geometry was of an $a/2c$ ratio of 1 to 4 [where a is crack length and c is half the crack width (Figure 6.19)], Irwin[27] et al. arrived at the following relationship for a part-through-thickness crack:

$$K_{Id} = 0.78(\sqrt{\text{in.}})\sigma_{yd} \tag{6.3}$$

Shoemaker and Rolfe[28] observed that the NDT temperature is close to the temperature at which a 1-in.-thick K_{Ic} specimen tested under impact loading ceases to satisfy the ASTM requirements for valid K_{Ic} tests. The thickness requirement for valid K_{Ic} tests is given by

$$B \geq 2.5\left(\frac{K_{Ic}}{\sigma_{ys}}\right)^2 \tag{6.4}$$

where B = specimen thickness,
σ_{ys} = yield strength.

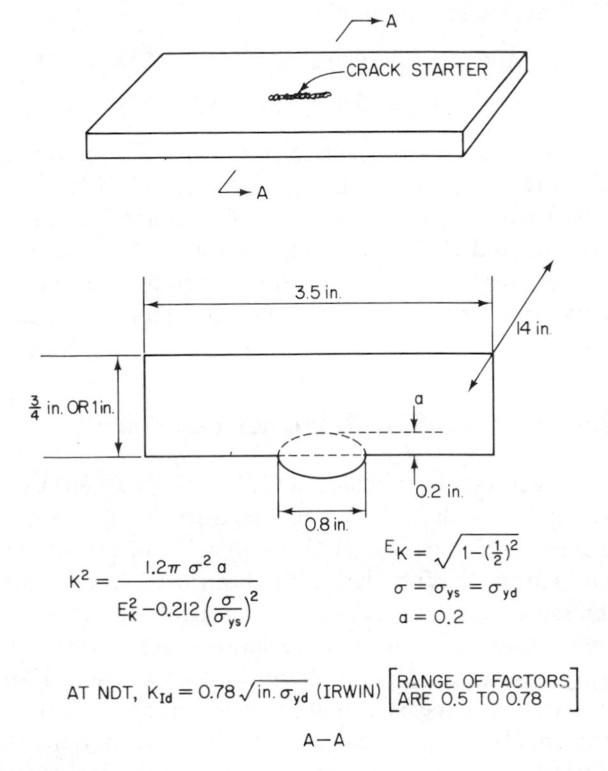

FIG. 6.19. Cross section of NDT test specimen showing initial crack and fracture-mechanics analysis.

This relationship can be used to represent this observation by Shoemaker and Rolfe[28] concerning the dynamic K_{Ic} value at NDT temperature. The resulting equation is

$$K_{Id} = 0.64(\sqrt{\text{in.}})\sigma_{yd} \tag{6.5}$$

where K_{Id} = critical plane-strain stress-intensity factor at NDT temperature and under dynamic loading ($\dot{\epsilon} \approx 10\ \text{sec}^{-1}$)

σ_{yd} = dynamic yield strength at NDT temperature.

Pellini has estimated that the factor relating K_{Id} and σ_{yd} should be 0.5. However, the differences in the various factors are slight, and the suggested relationship between K_{Id} and σ_{yd} at the NDT temperature is

$$K_{Id} = 0.6(\sqrt{\text{in.}})\sigma_{yd} \tag{6.6}$$

where K_{Id} = dynamic critical plane-strain stress-intensity factor at the NDT temperature, ksi$\sqrt{\text{in.}}$,

σ_{yd} = dynamic yield strength at the NDT temperature, ksi.

Using Equation (6.6), the calculated K_{Id} values at NDT for an A36 steel and an A572 Grade 50 steel would be

$$\text{A36:} \quad K_{Id} = 0.6\sigma_{yd} \simeq 0.6(40 + 25) \simeq 39 \text{ ksi}\sqrt{\text{in.}}$$

$$\text{A572 Grade 50:} \quad K_{Id} = 0.6\sigma_{yd} \simeq 0.6(55 + 25) \simeq 48 \text{ ksi}\sqrt{\text{in.}}$$

The values of the dynamic yield strength, σ_{yd}, are approximately equal to the static yield strength plus 25 ksi, i.e., $\sigma_{yd} \simeq \sigma_{ys} + 25$ ksi.

The K_{Id} test results presented in Figures 6.11 and 6.12 indicate measured values of about 40 and 50 ksi$\sqrt{\text{in.}}$ compared with the calculated values of 39 and 48 ksi$\sqrt{\text{in.}}$, respectively. Thus Equation (6.6) appears to give a realistic approximation to K_{Id} for low-strength structural steels at their NDT temperature.

6.7. K_{Id} from Precracked CVN Impact Test Results

Although impact tests on precracked CVN test specimens using uninstrumented impact machines have been conducted for many years,[29] the *instrumented* precracked CVN test is relatively new.[11] Its growth in popularity results primarily from the fact that it has been used to estimate K_{Id} values for nuclear pressure-vessel steels. The use of a lower-bound K_{Id} curve for the design of thick-section nuclear pressure vessels (see Chapter 15), the fact that conducting full-thickness K_{Id} tests (in some cases using 12-in.-thick test specimens) is extremely expensive, and the fact that space limitations restrict the engineer to small-size specimens in surveillance programs that monitor the effects of irradiation on these steels have led to growing use of the instrumented precracked CVN impact test for fracture control.

The primary advantage of the instrumented precracked Charpy test is that it has many of the features of the standard CVN test (small specimen, simple test, high strain rate, large sampling capability). It also has some of the advantages of the dynamic tear test (sharp energy transition and low initiation energy). As yet, it does not have the established correlation with service performance that the standard CVN impact test specimen has, particularly for the low-strength structural steels. By recording the loads during the precracked Charpy test, the total energy can be separated into initiation and propagation energies, and values of K_{Id} can be measured under appropriate conditions. Disadvantages of the instrumented precracked Charpy test are that it cannot measure full-thickness behavior and that control of the precracking procedure is difficult. Moreover, it does not appear to correlate with K_{Id} as well as the CVN impact test does.[30]

As part of the old Atomic Energy Commission (AEC) Heavy Section Steel Technology (HSST) program on nuclear pressure vessel steels, the static[31] and dynamic[32,33] fracture toughness of A533B steel ($\sigma_{ys} = 70$ ksi) has been measured as a function of temperature [using test specimens up to 12 in. thick (Figure 6.20)]. As shown in Figure 6.21, K_{Id} is less than K_{Ic} at any given

FIG. 6.20. Various size compact tension test specimens.

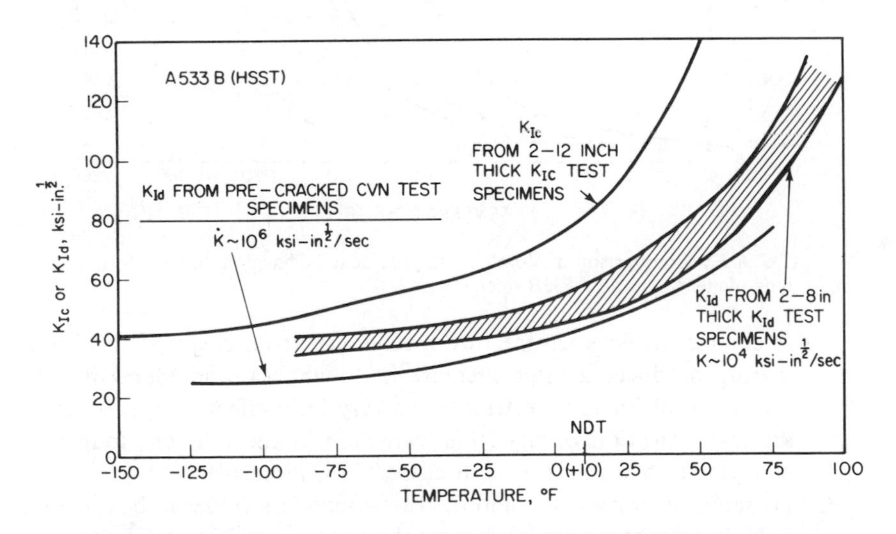

FIG. 6.21. Comparison of static, dynamic, and instrumented precracked CVN impact fracture toughness as a function of temperature for A533B steel.

temperature. At the NDT temperature of this A533B steel (+10°F), 6-in.-thick specimens were required for valid K_{Ic} tests, and 2-in.-thick specimens were required for valid K_{Id} tests ($\dot{K} \simeq 10^4$ ksi$\sqrt{\text{in.}}$/sec). Valid K_{Ic} values could not be obtained above 50°F even when using 12-in.-thick specimens, and valid K_{Id} values could not be obtained above 125°F for 8-in.-thick specimens.

There are several points to be made concerning the results presented in Figure 6.21:

1. K_{Ic} and K_{Id} show the same type of transition-temperature behavior as the CVN impact, dynamic tear, and precracked CVN curves for this steel (Figure 6.22).

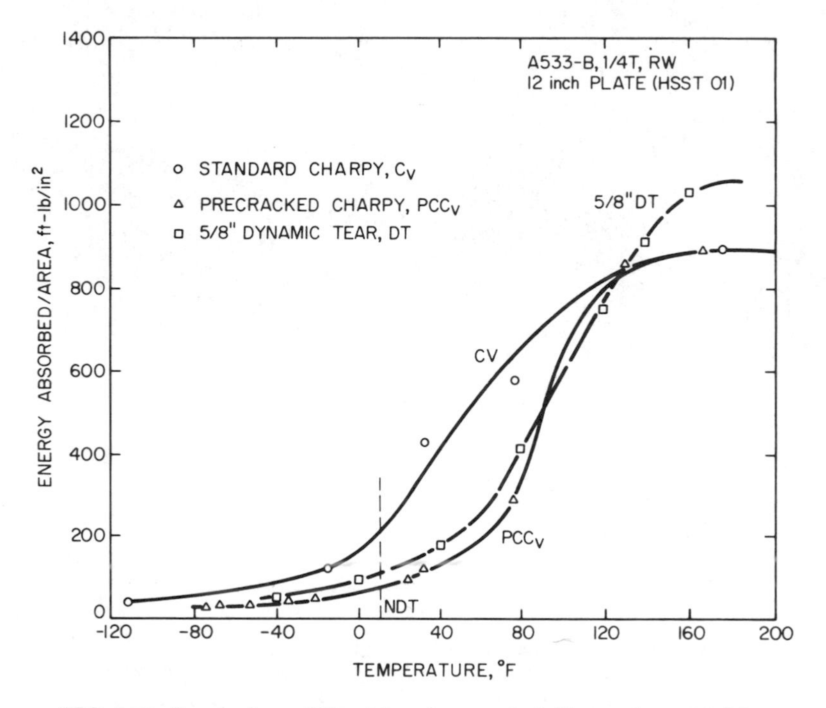

FIG. 6.22. Comparison of V-notch and precracked Charpy data with $\frac{5}{8}$-in. dynamic tear data on A533B steel.

2. A change in the microfracture mechanism from cleavage to fibrous tearing produces a large increase in toughness, and increasing the specimen thickness (constraint) has very little effect once this metallurgical transition occurs (this transition in the inherent toughness behavior of a material was discussed in Chapter 4).
3. Dynamic testing allows valid fracture-toughness values to be obtained at higher temperatures for a given thickness or allows use of a thinner specimen at a given temperature.

4. The K_{Ic} and K_{Id} test results presented in Figure 6.21 were very costly to obtain (in excess of $1 million).

The remaining curve in Figure 6.21 represents the dynamic fracture toughness of A533B steel determined from instrumented precracked Charpy impact tests.[34] The impact velocity of the Charpy hammer was approximately 200 in./sec, and this corresponds to $\dot{K} \simeq 10^6$ ksi$\sqrt{\text{in.}}$/sec. The higher strain rate of the Charpy impact test causes a greater increase in the yield strength of A533B steel and thus a slightly lower K_{Id}.

As discussed in Chapter 15, the Pressure Vessel Research Committee (PVRC) has proposed that a lower-bound fracture-toughness curve be used in the design of nuclear pressure vessels. Their recommended design curve (discussed in Chapter 15) essentially parallels the lower part of the scatter band for K_{Id} at $\dot{K} \simeq 10^4$ ksi$\sqrt{\text{in.}}$/sec. It is apparent from Figure 6.21 that a conservative lower-bound K_{Id} design curve can be obtained from instrumented precracked Charpy tests for temperatures up to 70°F. Most important of all, the K_{Id} curve from the instrumented precracked Charpy test can be determined very inexpensively compared to the presently accepted techniques of testing full-thickness specimens capable of satisfying the size requirements presented in Chapter 3.

The most prevalent use to date of the precracked Charpy specimen has been by Hartbower and associates.[35-38] Hartbower[35] found that slow-bend W/A values (W/A_{SB}) of a titanium (Ti-6A-4V) provided a good estimate of K_{Ic} through the relationship

$$K_{Ic} = 0.17(W/A)_{SB} + 16.2 \text{ psi}\sqrt{\text{in.}} \tag{6.7}$$

where (W/A_{SB}) is the precracked Charpy slow-bend value in in.-lb/in. and valid K_{Ic} values were obtained from center-notched tensile specimens.

Additional correlations between K_{Ic} and W/A from slow-bend and impact tests on precracked Charpy specimens of titanium and steel have been reported by Ronald et al.[39] Figure 6.23 shows a relation between K_{Ic} or K_{Id} and $(W/A)_{SB}$ or $(W/A)_{IMPACT}$ for titanium alloys. For a given loading rate, the relationship appears quite good. Figure 6.24 shows a correlation determined by Ronald et al.[39] between valid K_{Ic} values and slow-bend W/A measurements on precracked Charpy specimens of various high-strength titanium and steel alloys. The predicted line corresponds to the theoretical relationship expected between K_{Ic} and G_{Ic},

$$K_{Ic}^2 = \frac{EG_{Ic}}{1 - \nu^2} \tag{6.8}$$

where E is the elastic modulus and ν is Poisson's ratio. Ronald assumed that

$$G_{Ic} = \alpha(W/A)_{SB} \tag{6.9}$$

where the factor $\alpha = \frac{1}{2}$ was used to account for the fact that two fracture faces are formed even though the basic definition of G_{Ic} accounts for the

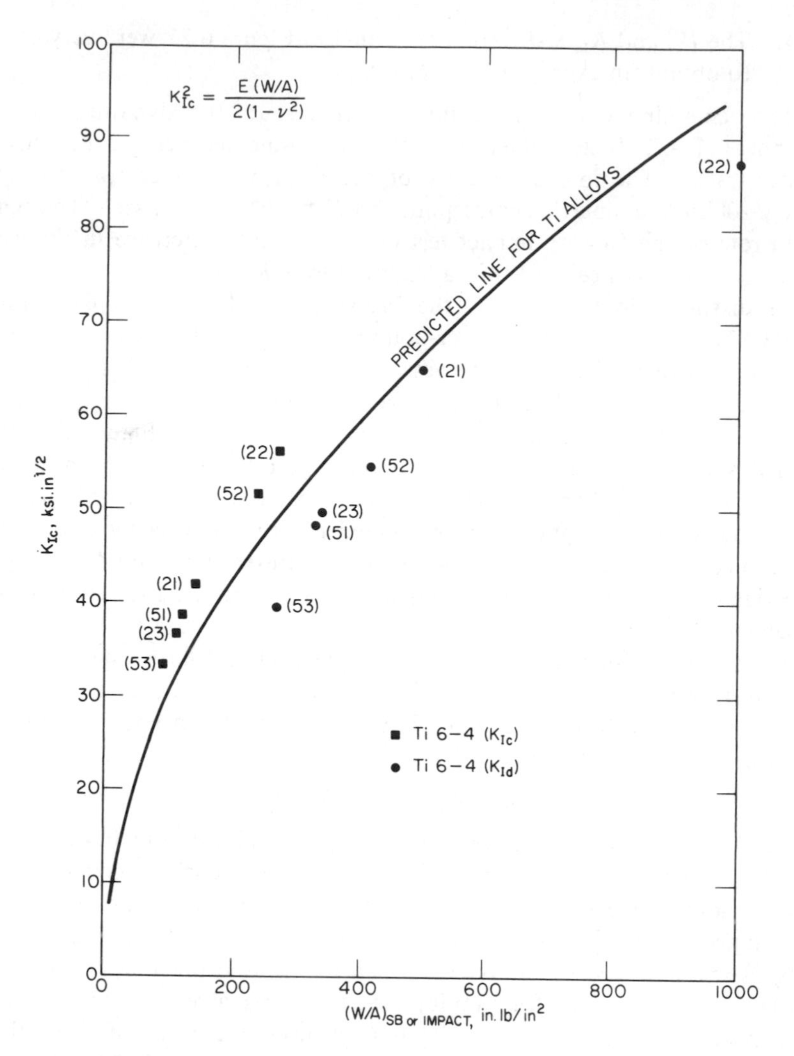

FIG. 6.23. K_{Ic} versus W/A for precracked specimens tested in slow bend and impact for titanium alloys.

two fracture surfaces. Combining Equations (6.8) and (6.9) results in the following relation between K_{Ic} and $(W/A)_{SB}$:

$$K_{Ic}^2 = \frac{E}{2(1-\nu^2)}(W/A_{SB}) \tag{6.10}$$

This relation between K_{Ic} and $(W/A)_{SB}$ is good up to K_{Ic}^2/E values of 300 psi·in. When shear-lip energies were subtracted from the W/A values, the corrected W/A values fell closer to the predicted line for the high-toughness materials.

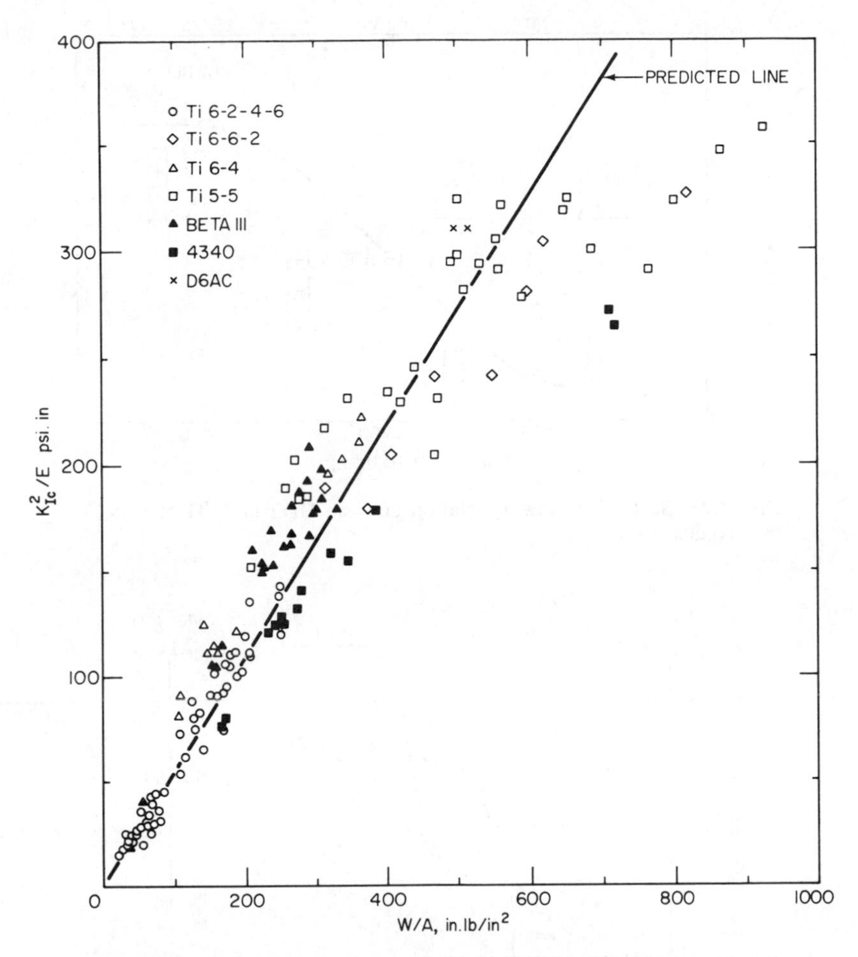

FIG. 6.24. K_{Ic}^2/E versus W/A for precracked specimens tested in slow bend.

6.8. NDT-DT-CVN-K_{Ic} Relations

As shown schematically in Figure 6.25, the beginning of the elastic-plastic transition as measured with dynamic tear (DT) test specimens occurs in the general vicinity of NDT. Actual test results substantiating this behavior for various structural steels are presented in Figures 6.26 through 6.28, as well as Figure 6.7 for a high-strength structural steel. Above NDT, the increase in toughness as measured by the DT specimen occurs rapidly as the constraint ahead of the crack tip is relaxed. Thus the NDT temperature appears to define the start of a rapid increase in crack toughness as measured with the DT specimen. However, as indicated by Equation (6.6) and the

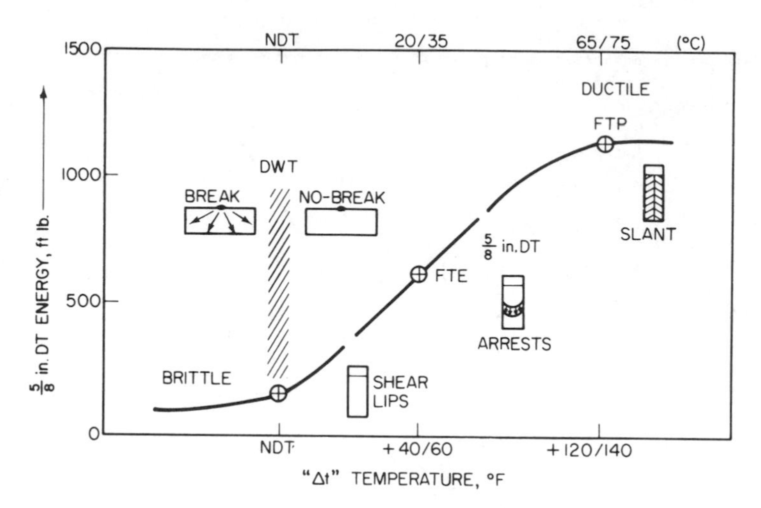

FIG. 6.25. Schematic showing relation between NDT and DT test specimen results.

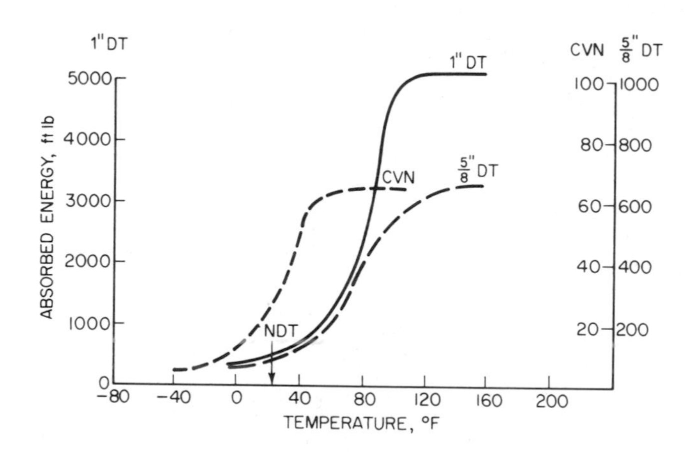

FIG. 6.26. Relation among NDT, CVN, and DT test results for ABS-C steel.

data in Figures 6.11 and 6.12, NDT is above the K_{Id} toughness transition indicating that the DT test is more severe than the K_{Id} test. The primary reason for this behavior is the exhaustion of ductility and residual tensile stresses at the tip of the DT notch caused by the pressed knife used to produce the notch.

At the upper shelf where plastic behavior occurs, the effect of notch acuity is less important, and thus a correlation between upper shelf CVN test results and DT test results might be expected. Such a correlation is shown in Figure 6.29. The scatter band is relatively wide because of the large number of mate-

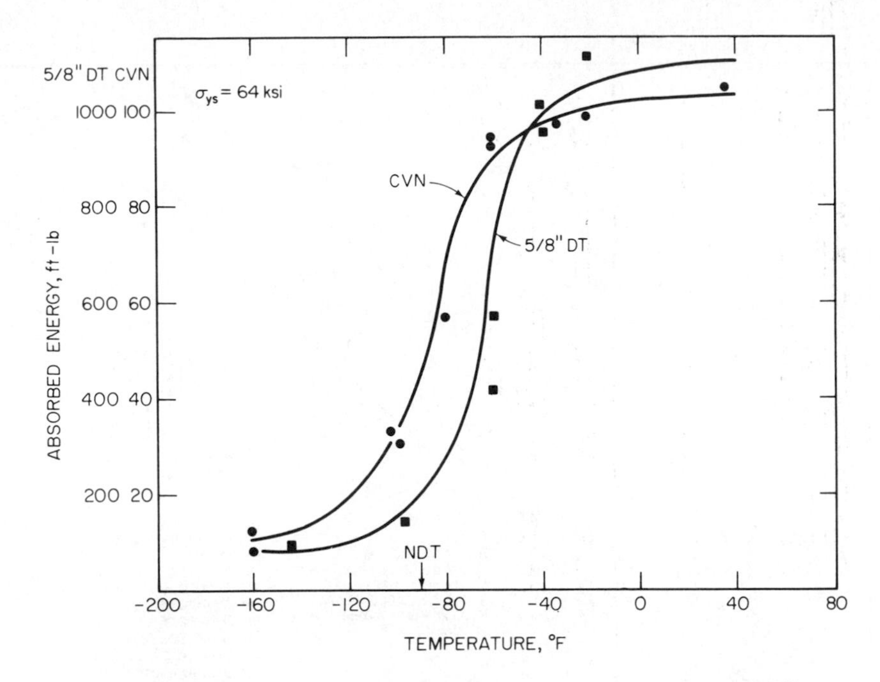

FIG. 6.27. DT and CVN test results for A537B steel; σ_{ys} = 64 ksi (441 MN/m²).

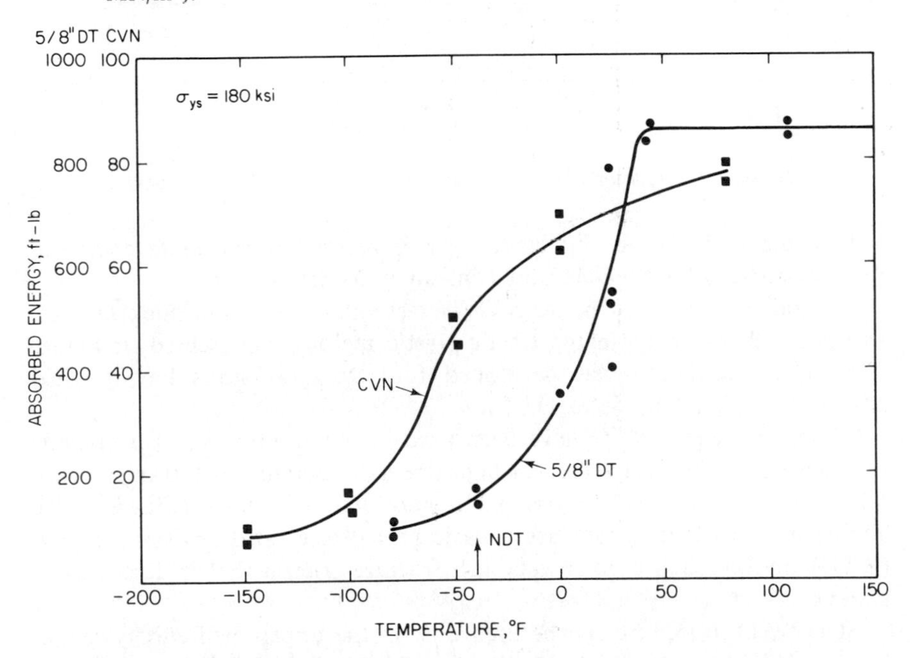

FIG. 6.28. DT and CVN test results for A517 steel; σ_{ys} = 108 ksi (745 MN/m²).

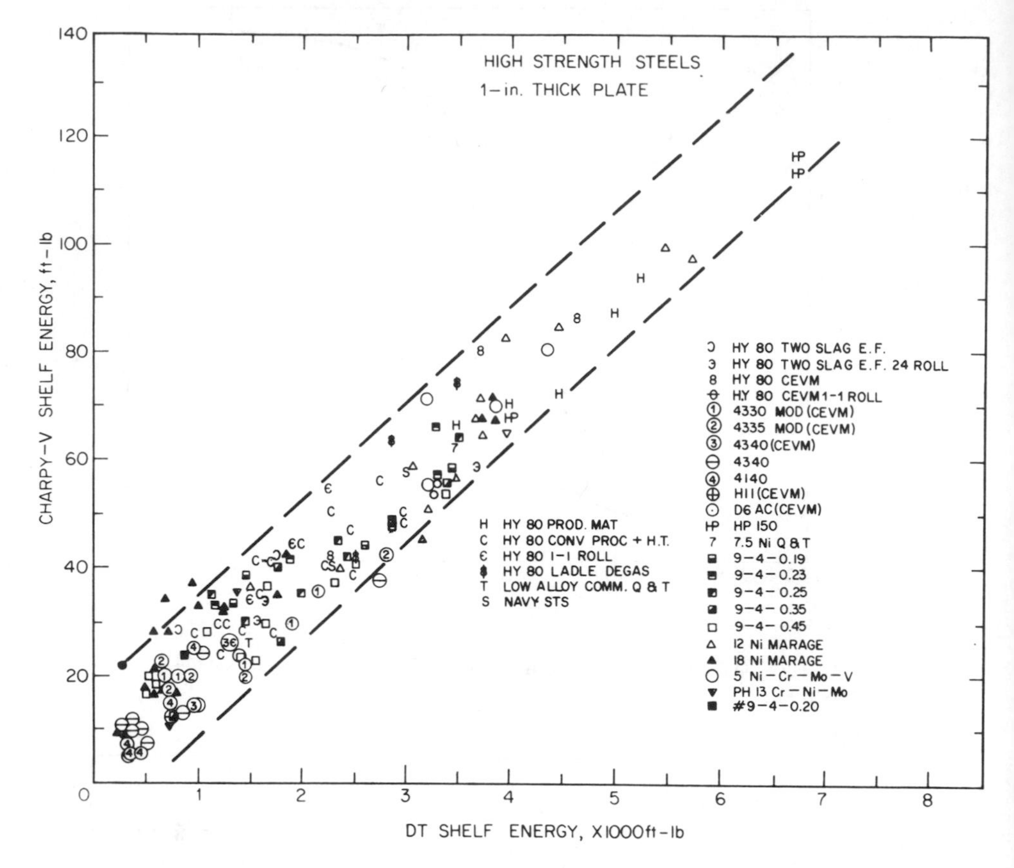

FIG. 6.29. Correlation of CVN impact shelf energy with DT shelf energy.

rials included; however, this correlation is helpful in providing a general relation between CVN values and 1-in.-thick DT test values.

A similar correlation between CVN test values and $\frac{5}{8}$-in.-thick DT test values at 32° or 75°F (in the elastic-plastic region) is presented in Figure 6.30. This correlation was developed for structural steels having yield strengths ranging from 55 to 111 ksi.

Empirical relations[40-42] have been used to obtain empirical correlations between K_{Ic} and DT test values for high-strength steels ($\sigma_{ys} = 150$ ksi) (Figure 6.31), aluminum alloys (Figure 6.32), and titanium alloys (Figure 6.33). These empirical correlations are a basic part of the ratio analysis diagram (RAD), an approach used to establish fracture criteria that is described in Chapter 13.

The RAD procedure can be used only in the upper shelf energy region and applies primarily to metals that are not rate-sensitive, i.e., materials

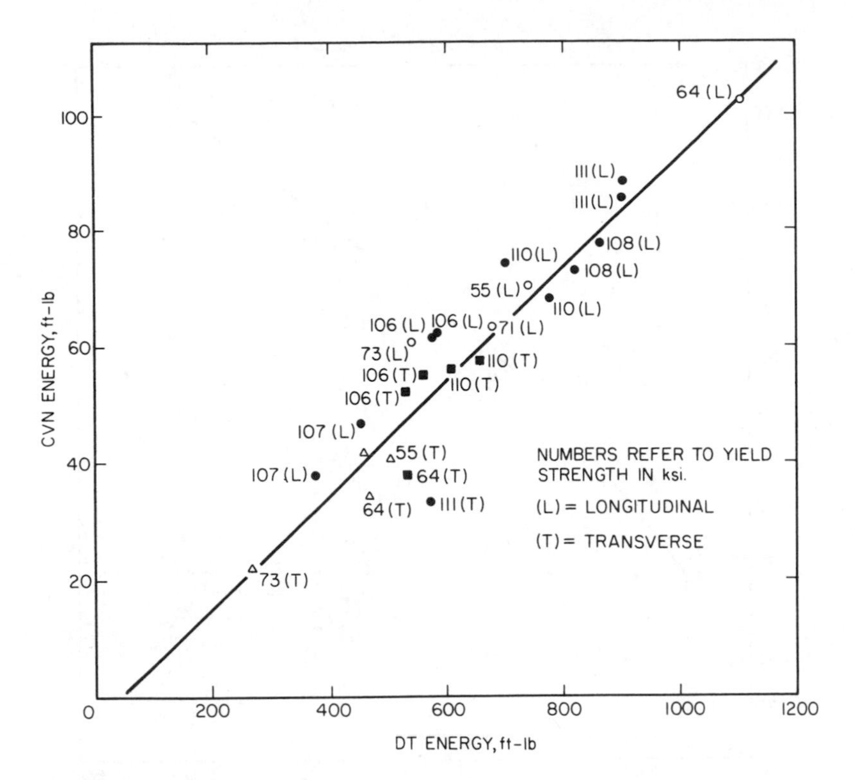

FIG. 6.30. Correlation between absorbed energy in $\frac{5}{8}$-in. DT and standard CVN test specimen at 32° or 75°F.

whose toughness does not change significantly with loading rate. Thus the correlations shown in Figures 6.31 through 6.33 appear reasonable because these materials are not rate-sensitive and the correlations are for shelf values, where notch acuity would not be very significant.

6.9. CVN Lateral Expansion

For many years, standards and code-writing bodies specified a constant value of minimum energy absorption as the criterion for toughness when judged by the Charpy impact test. The energy absorption value was selected arbitrarily without consideration of the strength level of the steel.

It is now widely recognized that energy absorption must be increased with strength in order to maintain a constant level of resistance to fracture toughness. Energy absorption is controlled by two factors: the strength of the steel, which regulates the force required to deform the Charpy specimen, and the ductility of the steel, which determines the distance through which the force acts during testing. Since loss of fracture toughness is due to loss

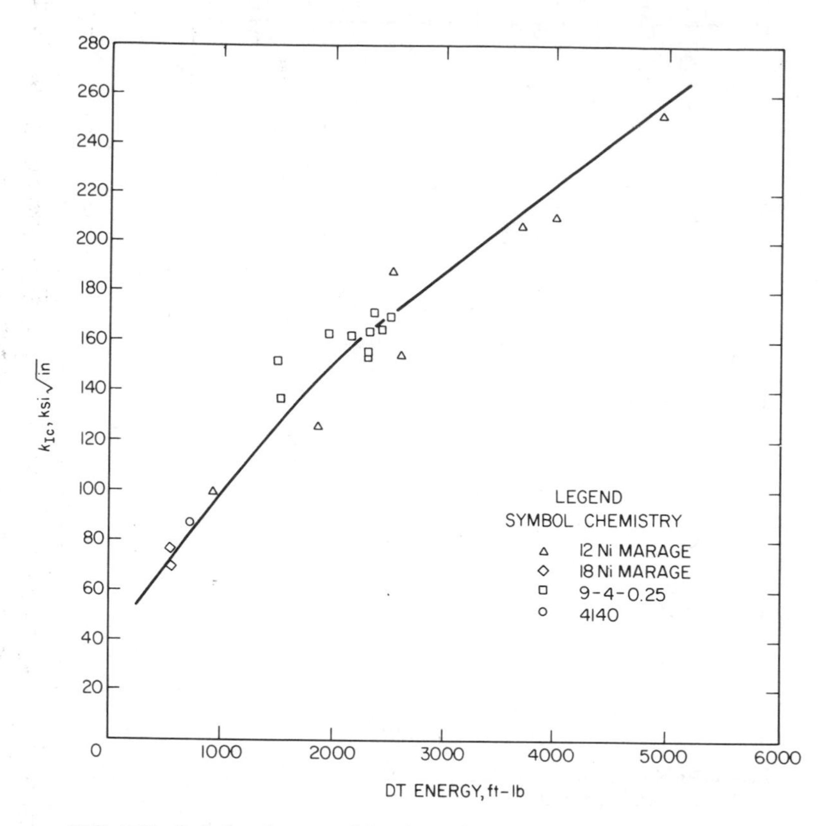

FIG. 6.31. Relation between 1-in. dynamic tear energy and K_{Ic} values for various high-strength steels.

in ductility rather than strength, a criterion which evaluates notch ductility rather than energy is a more significant and universal index.

An index of ductility can readily be obtained from Charpy specimens by measuring lateral expansion of the specimen at the compression side directly opposite the notch. The procedure is described in ASTM Specification A370-68. The ductility and energy relations in Charpy tests of carbon and alloy steels with yield strengths in the range 30–150 ksi were first described by Gross and Stout in 1958.[43] More extensive information was recently presented by Gross.[44,45] The results showed, for example, that 15 mils of lateral expansion corresponds to about 11 ft-lb for a 35-ksi yield strength (60-ksi tensile strength) steel, to about 15 ft-lb for a 55-ksi yield strength (85-ksi tensile strength) steel, and to about 22 ft-lb for a 120-ksi yield strength (133-ksi tensile strength) steel. These results and others[22,46] supported the adoption by the ASME Boiler and Pressure Vessel Committee of a lateral-expansion requirement for quenched and tempered steels.

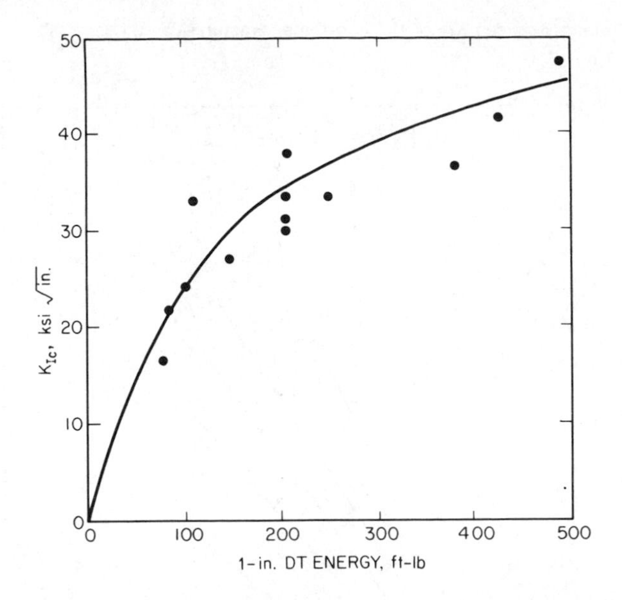

FIG. 6.32. Relation between K_{Ic} and DT energy for aluminum alloys.

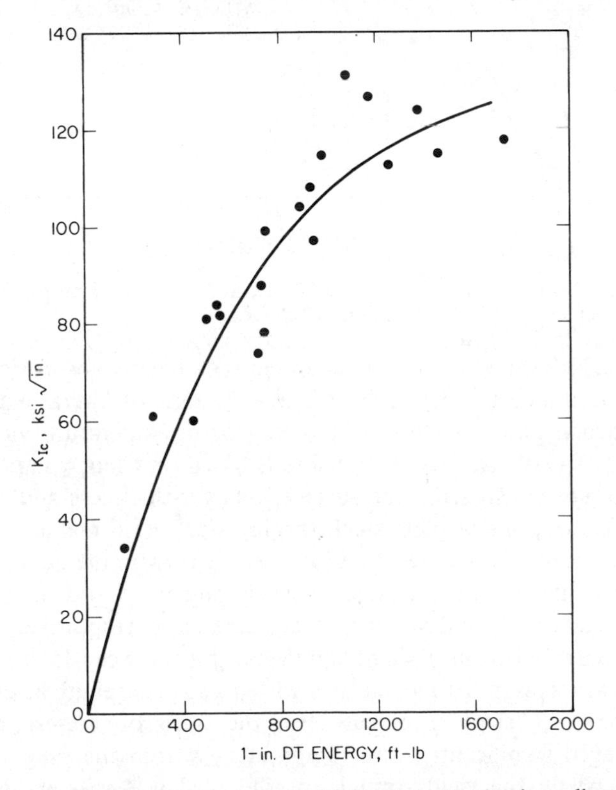

FIG. 6.33. Relation between K_{Ic} and DT energy for titanium alloys.

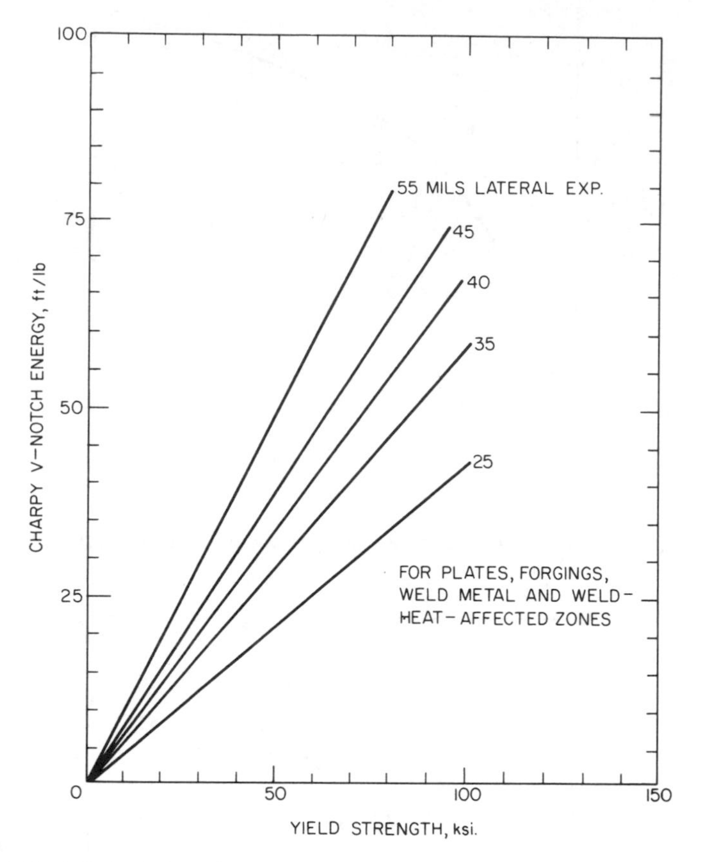

FIG. 6.34. Relationship at 25, 35, 40, 45, and 50 mils lateral expansion between energy and yield strength (Reference 47).

Figure 6.34[47] shows the relationship between energy absorption and yield strength determined at 25, 35, 40, 45, and 55 mils of lateral expansion for different carbon and low-alloy steels with room-temperature yield strength in the range 50–110 ksi. The relationship is based on Charpy impact and tensile data for specimens from the surface, quarter-thickness and center locations in thick sections of plate steel, forging steel, weld metal, and the weld heat-affected zone of plate steel and for specimens from the quarter-thickness location in 1- and 2-in.-thick plate steels. Figures 6.35 and 6.36[47] illustrate plots of the data at 35 and 40 mils lateral expansion, respectively. Each data point was obtained from a singular relationship between lateral expansion and energy absorption for a given steel with a known strength at the thickness location for the Charpy specimens. For the steels of interest, the Charpy values changed significantly with temperature within the range 50°–100°F, a range in which the yield strength of the steel was reasonably constant.

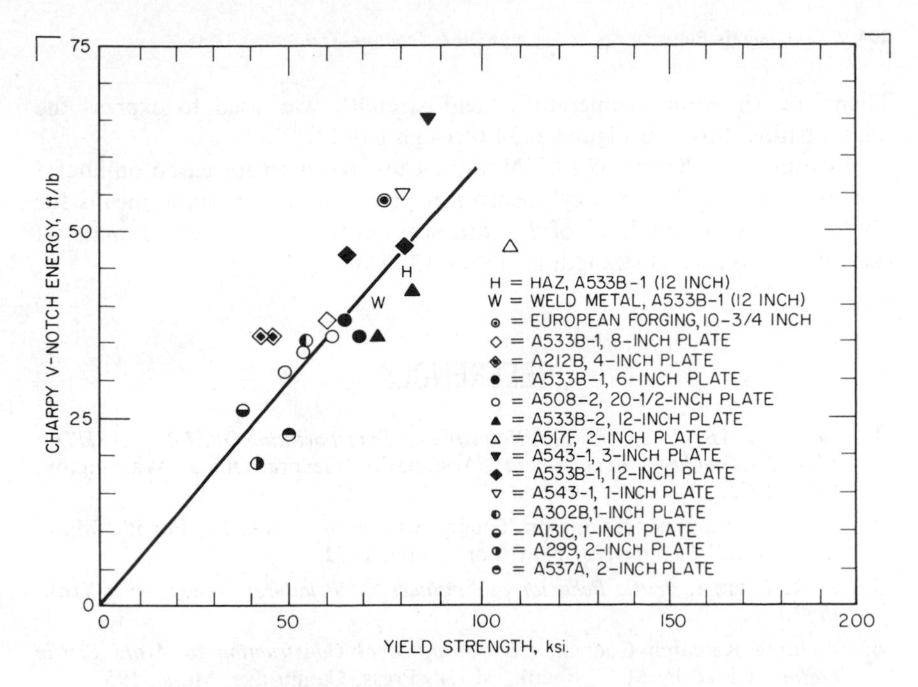

FIG. 6.35. Relationship at 35 mils lateral expansion between energy and yield strength (Reference 47).

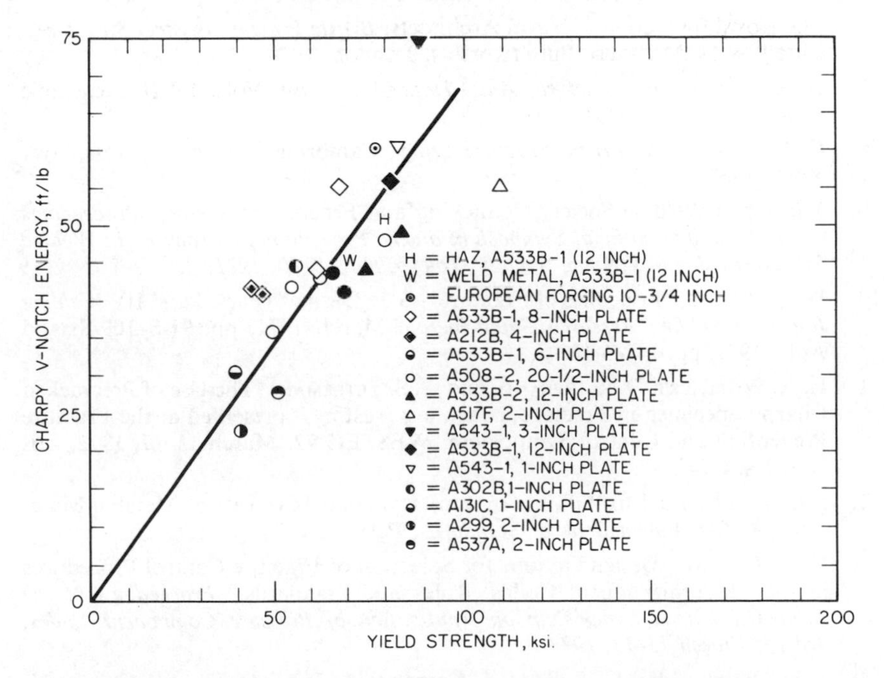

FIG. 6.36. Relationship at 40 mils lateral expansion between energy and yield strength (Reference 47).

Therefore, the room-temperature yield strength was used to express the relationships shown in Figures 6.34 through 6.36.

In summary, the use of a CVN impact test requirement based on lateral expansion rather than energy absorption provides an automatic means for obtaining a constant level of fracture toughness irrespective of material strength up to a yield strength of about 130 ksi.

REFERENCES

1. *American Association of State Highway and Transportation Officials (AASHTO) Material Toughness Requirements*, Association General Offices, Washington, D.C., 1973.
2. "PVRC Recommendations on Toughness Requirements for Ferritic Materials," *WRC Bulletin 175*, New York, Aug. 1972.
3. E. R. Parker, *Brittle Behavior of Engineering Structures*, Wiley, New York, 1957.
4. Welding Research Council, *Control of Steel Construction to Avoid Brittle Failure*, edited by M. E. Shank, M.I.T. Press, Cambridge, Mass., 1957.
5. W. J. Hall, H. Kihara, W. Soete, and A. A. Wells, *Brittle Fracture of Welded Plate*, Prentice-Hall, Englewood Cliffs, N.J., 1967.
6. The Royal Institution of Naval Architects, *Brittle Fracture in Steel Structures*, edited by G. M. Boyd, Butterworth's, London, 1970.
7. H. Libowitz, ed., *Fracture, An Advanced Treatise*, Vols. I–VII, Academic Press, New York, 1967.
8. C. F. Tipper, *The Brittle Fracture Story*, Cambridge University Press, New York, 1962.
9. The Japan Welding Society, "Cracking and Fracture in Welds," *Proceedings of the First International Symposium on the Prevention of Cracking in Welded Structures, Japan Welding Society, Tokyo, Nov. 8–10, 1971.*
10. W. S. Pellini, "Principles of Fracture—Safe Design" (Parts I and II), *Welding Journal (Welding Research Supplement)*, March 1971, pp. 91-S–109-S, and April 1971, pp. 147-S–162-S.
11. R. A. Wullaert, D. R. Ireland, and A. S. Tetelman, "The Use of Precracked Charpy Specimen in Fracture Toughness Testing," presented at the Fracture Prevention and Control Symposium, WESTEC 72, March 13–17, 1972, Los Angeles, Calif.
12. Proposed Method for 5/8-in. (16-mm) Dynamic Tear Test of Metallic Materials, *ASTM Annual Book of Standards*, Part 10, 1975.
13. W. S. Pellini, "Design Options for Selection of Fracture Control Procedures in the Modernization of Codes, Rules and Standards," *Proceedings of the Joint U.S.-Japan Symposium on Application of Pressure Component Codes, Tokyo, March 13–15, 1973.*
14. J. M. Barsom and S. T. Rolfe, "Correlations Between K_{Ic} and Charpy V-Notch Test Results in the Transition-Temperature Range," in "Impact Testing

of Metals," *ASTM STP 466*, American Society for Testing and Materials, Philadelphia, 1970, pp. 281–302.

15. D. P. CLAUSING, "Effect of Plane-Strain Sensitivity on the Charpy Toughness of Structural Steels," *International Journal of Fracture Mechanics, 6*, No. 1, March 1970.
16. J. H. HOLLOMAN, "The Notched-Bar Impact Test, *Transactions, American Institute of Mining, Metallurgical, and Petroleum Engineers, 158*, 1944, pp., 310–322.
17. J. H. GROSS, "Effect of Strength and Thickness on Notch Ductility," in "Impact Testing of Metals," *ASTM STP 466*, American Society for Testing and Materials, Philadelphia, 1970, p. 21.
18. J. A. BEGLEY and J. D. LANDES, "The J Integral as a Fracture Criterion," in "Stress Analysis and Growth of Cracks," *ASTM STP 514*, American Society for Testing and Materials, Philadelphia, 1972.
19. J. D. LANDES and J. A. BEGLEY, "The Effect of Specimen Geometry on J_{Ic}," in "Stress Analysis and Growth of Cracks," *ASTM STP 514*, American Society for Testing and Materials, Philadelphia, 1972.
20. S. T. ROLFE and S. R. NOVAK, "Slow-Bend K_{Ic} Testing of Medium-Strength High-Toughness Steels," in "Review of Developments in Plane Strain Fracture-Toughness Testing," *ASTM STP 463*, American Society for Testing and Materials, Philadelphia, 1970, pp. 124–159.
21. P. C. PARIS, discussion of paper by Begley and Landes: "The J Integral as a Fracture Criterion," *ASTM STP 514*, American Society for Testing and Materials, Philadelphia, 1972. (Ref. 18).
22. J. T. CORTEN and R. H. SAILORS, "Relationship Between Material Fracture Toughness Using Fracture Mechanics and Transition Temperature Tests," *T. & A. M. Report No. 346*, University of Illinois, Urbana, Aug. 1971. See also Paper No. 3, HSST Program Information Meeting, ORNL, March 25, 1971.
23. J. M. BARSOM, "The Development of AASHTO Fracture-Toughness Requirements for Bridge Steels," *American Iron and Steel Institute*, Washington, D. C., February 1975.
24. J. M. BARSOM, "Development of the AASHTO Fracture-Toughness Requirements for Bridge Steels," *Engineering Fracture Mechanics*, Vol. 7, No. 3, September 1975.
25. J. R. HAWTHORNE and T. R MAGER, "Relationship Between Charpy V and Fracture Mechanics K_{Ic} Assessments of A533-B Class 2 Pressure Vessel Steel," *ASTM STP 514*, American Society for Testing and Materials, Philadelphia, 1972.
26. J. A. BEGLEY and W. A. LOGSDON, "Correlation of Fracture Toughness and Charpy Properties for Rotor Steels," *Scientific Paper 71-1E7-MSLRF-P1*, Westinghouse Research Laboratories, Pittsburgh, July, 1971.
27. G. R. IRWIN, J. M. KRAFFT, P. C. PARIS, and A. A. WELLS, "Basic Aspects of Crack Growth and Fracture," *NRL Report 6598*, Washington, D.C., Nov. 21, 1967.
28. A. K. SHOEMAKER and S. T. ROLFE, "The Static and Dynamic Low-Temperature Crack-Toughness Performance of Seven Structural Steels," *Engineering Fracture Mechanics, 2*, No. 4, June 1971.

29. C. E. Hartbower, "Crack Initiation and Propagation in the V-Notch Charpy Impact Specimen," *Welding Journal*, *35*, No. 11, Nov. 1957, p. 494-s.
30. R. Roberts, G. R. Irwin, and Others, "Fracture Toughness of Bridge Steels. Phase II Report," Federal Highway Administration Report No. *FHWA-RD-74-59*, Washington, D.C., September, 1974.
31. W. O. Shabbits, W. H. Pryle, and E. T. Wessel, "Heavy Section Fracture Toughness Properties of A533 Grade B Class 1 Steel Plate and Submerged Arc Weldment," *WCAP-7414*, Westinghouse Electric Corp., Pittsburgh, Dec. 1969.
32. P. B. Crosley and E. J. Ripling, "Crack Arrest Fracture Toughness of A-533 Class 1 Pressure Vessel Steel," *HSSTP-TR-8*, Materials Research Laboratory, Glenwood, Ill., March 1970.
33. W. O. Shabbits, "Dynamic Fracture Toughness Properties of Heavy Section A533 Grade B Class 1 Steel Plate," *WCAP-6723*, Westinghouse Electric Corporation, Pittsburgh, Dec. 1970.
34. W. L. Server and A. S. Tetelman, "The Use of Precracked Charpy Specimens to Determine Dynamic Fracture Toughness," *Engineering Fracture Mechanics*, Vol. 4, No. 2, June 1972.
35. C. E. Hartbower, W. G. Reuter, and P. O. Crimmins, "Tensile Properties and Fracture Toughness of 6A1-4V Titanium," *AFML-TR-68-163*, Air Force Materials Laboratory, Dayton, Vol. 1, Sept. 1968; Vol. 2, March 1969.
36. C. E. Hartbower, "Crack Initiation and Propagation in the V-Notch Charpy Impact Specimen," *Welding Journal*, *36*, No. 11, Nov. 1957, p. 494-s.
37. G. E. Orner and C. E. Hartbower, "Sheet Fracture Toughness Evaluated by Charpy Impact and Slow Bend," *Welding Journal*, *40*, No. 9, Sept. 1961, p. 405-s.
38. C. E. Hartbower, "Materials Sensitive to Slow Rates of Straining," in "Impact Testing of Metals," *ASTM STP 466*, American Society for Testing and Materials, Philadelphia, 1970, pp. 113–147.
39. T. M. F. Ronald, J. A. Hall, and C. M. Pierce, "Some Observations Pertaining to Simple Fracture Toughness Screening Tests for Titanium," *AFML-TR-70-311*, Air Force Materials Laboratory, Dayton, March 1971.
40. E. A. Lange and J. J. Loss, "Dynamic Tear Energy—A Practical Performance Criterion for Fracture Resistance," in "Impact Testing of Metals," *ASTM STP 466*, American Society for Testing and Materials, Philadelphia, 1970, pp. 241–258.
41. R. W. Judy, Jr., R. J. Goode, and C. M. Freed, "Fracture Toughness Characterization Procedures and Interpretations to Fracture Safe Design for Structural Aluminum Alloys," *NRL Report 6879*, Naval Research Laboratory, Washington, Dec. 1968.
42. R. J. Goode, R. W. Judy, Jr., and R. W. Huber, "Procedures for Fracture Toughness Characterization and Interpretations to Failure-Safe Design for Structural Titanium Alloys," *NRL Report 6779*, Naval Research Laboratory, Washington, Dec. 1968.
43. J. H. Gross and R. D. Stout, "Ductility and Energy Relations in Charpy Tests of Structural Steels," *Welding Journal, Research Supplement*, *37*, No. 4, 1958, pp. 151-s–159-s.

44. J. H. GROSS, "The Effect of Strength and Toughness on Notch Ductility," *Welding Journal, Research Supplement*, *48*, No. 10, 1969, pp. 441-s–453-s.
45. J. H. GROSS, "Transition-Temperature Data for Five Structural Steels," *WRC Bulletin*, *155*, Oct. 1970.
46. A. K. SHOEMAKER, "Notch-Ductility Transition of Structural Steels of Various Yield Strengths," *ASME Paper No. 71-PVP-19*, presented at the First National Congress of Pressure Vessels and Piping, San Francisco, May 10–12, 1971.
47. PVRC Ad Hoc Task Group on Toughness Requirements, "PVRC Recommendations on Toughness Requirements for Ferritic Materials," *Welding Research Bulletin No. 175*, New York, Aug. 1972.

7

Fatigue-Crack Initiation

7.1. General

The fatigue life of structural components is determined by the sum of the elapsed cycles required to initiate a fatigue crack and to propagate the crack from subcritical dimensions to the critical size. Consequently, the fatigue life of structural components may be considered to be composed of three continuous stages: (1) fatigue-crack initiation, (2) fatigue-crack propagation, and (3) fracture. The fracture stage represents the terminal conditions (i.e., the particular combination of σ, a, and K_{Ic}) in the life of a structural component, as described in Chapter 5. The contribution of fatigue-crack initiation and fatigue-crack propagation to the fatigue life of structural components depends on the intended application. For example, structural components that contain stress concentrations or initial defects may be determined primarily by the fatigue-crack propagation characteristics of the components. On the other hand, the fatigue life of structural components intended for infinite-life application under constant-displacement fluctuation, i.e., cracks propagating in a decreasing stress field, may be governed by fatigue-crack initiation or fatigue-crack propagation or by both. Consequently, the useful life of cyclically loaded structural components can be determined only when the three stages in the life of the component are evaluated individually and the cyclic behavior in each stage is thoroughly understood.

Conventional procedures used to design structural components subjected to fluctuating loads provide a design fatigue curve which characterizes the basic unnotched fatigue properties of the material and a fatigue-strength-reduction factor. The fatigue-strength-reduction factor incorporates the effects of all the different parameters characteristic of the specific structural component that make it more susceptible to fatigue failure than the unnotched specimen, such as surface finish, geometry, defects, etc. The design fatigue curves are based on the prediction of cyclic life from data on nominal stress (or strain) versus elapsed cycles to failure (*S-N* curves), as determined from laboratory specimens. Such data are usually obtained by testing unnotched specimens and represent the number of cycles required to initiate a crack in the specimen plus the number of cycles required to propagate the crack from a subcritical size to a critical dimension. The dimensions of the critical crack required to cause failure depends on the magnitude of the applied stress and on the specimen size, as well as the particular testing conditions used.

Figure 7.1 is a schematic *S-N* curve divided into an initiation component and a propagation component. The number of cycles corresponding to the endurance limit represents initiation life primarily, whereas the number of cycles expended in crack initiation at a high value of applied alternating stress is negligible. As the magnitude of the applied alternating stress increases, the total fatigue life decreases and the percent of the fatigue life to crack initiation decreases. Consequently, *S-N*-type data do not provide complete information regarding safe-life predictions in structural components, particularly in components having surface irregularities different from those of the test specimens and in components containing crack-like imperfections, because the existence of surface irregularities and crack-like imperfections reduces and may eliminate the crack-initiation portion of the fatigue life of structural components.

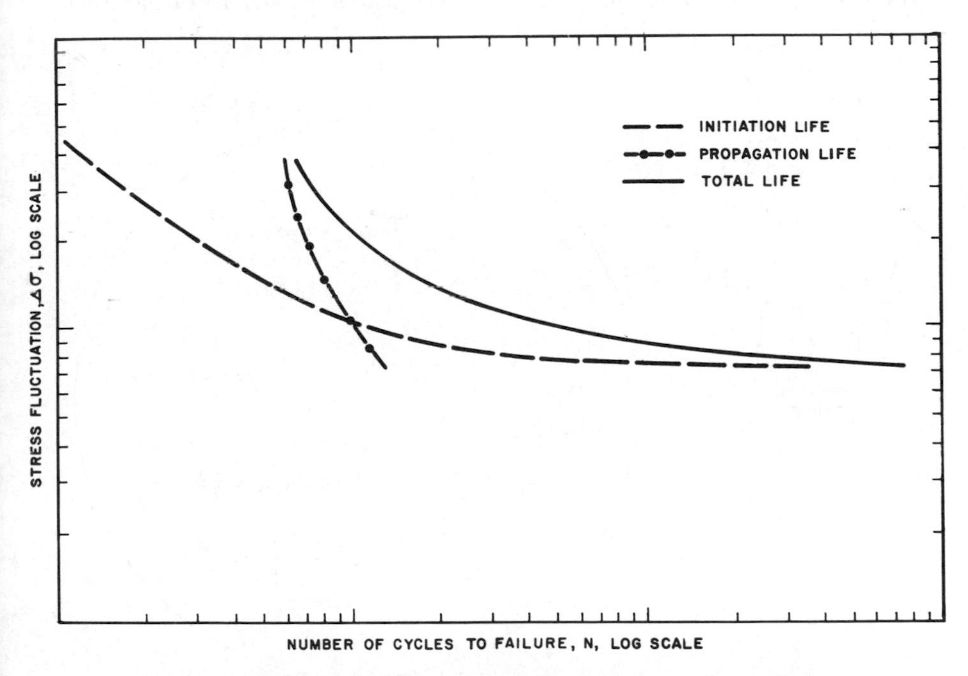

FIG. 7.1. Schematic *S-N* curve divided into initiation and propagation components.

Many attempts have been made to characterize the fatigue behavior of metals.[1-3] The results of some of these attempts have proved invaluable in the evaluation and prediction of the fatigue strength of structural components. However, these fatigue-strength-evaluation procedures are subject to limitations, caused primarily by the failure to adequately distinguish between fatigue-crack initiation and fatigue-crack propagation.

7.2. Fracture-Mechanics Methodology

The development of fracture-mechanics methodology offers considerable promise in improving our understanding of fatigue-crack initiation, fatigue-crack propagation, and unstable crack propagation and in solving the problem of designing to prevent failures caused by fatigue. As described in Chapter 2, linear-elastic fracture-mechanics technology is based on an analytical procedure that relates the stress-field magnitude and distribution in the vicinity of a crack tip to the nominal stress applied to the structure, to the size, shape, and orientation of the crack or crack-like imperfection, and to the material properties. Figure 7.2 presents the σ_x and σ_y components of the elastic-stress field in the vicinity of a crack tip in a body subjected to tensile stresses normal to the plane of the crack (Mode I deformation)[4]. These stress-field equations show that the distribution of the elastic-stress field in the vicinity of the crack

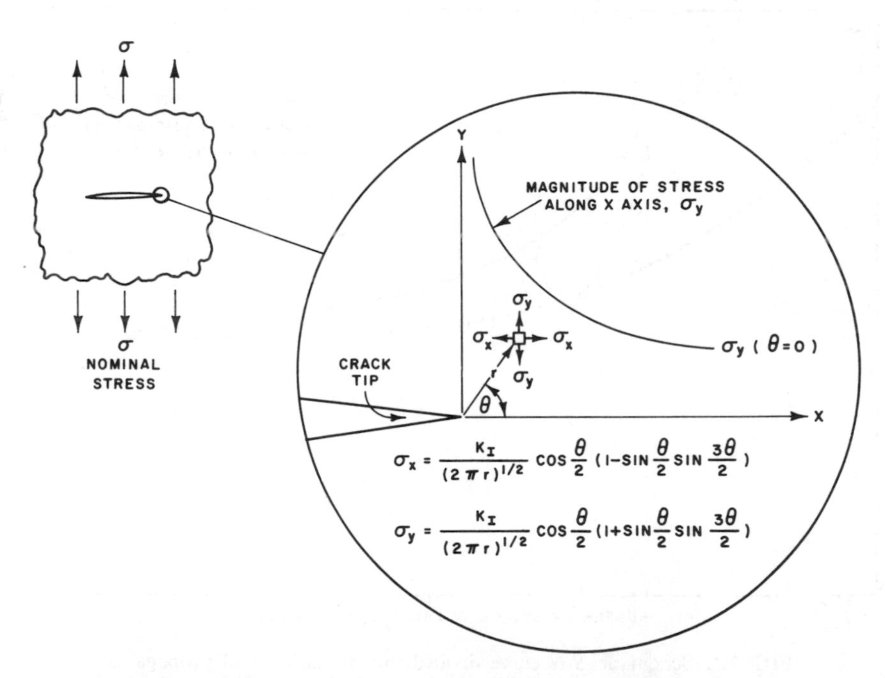

FIG. 7.2. Schematic illustration of the elastic-stress-field distribution near the tip of a fatigue crack (Mode I deformation).

tip is invariant in all structural components subjected to Mode I deformation and that the magnitude of the elastic-stress field can be described by a single parameter, K_I, designated the stress-intensity factor. Consequently, the applied stress, the crack shape, size, and orientation, and the structural configuration associated with structural components subjected to Mode I deformation affect the value of the stress-intensity factor but do not alter the stress-field distribution. Relationships between the stress-intensity factor and various body configurations, crack sizes, shapes, and orientations, and loading conditions have been established[5,6] and the more widely used ones were described in Chapter 2.

The critical-stress-intensity factor, K_c, K_{Ic}, or K_{Id}, represents the terminal conditions in the life of a structural component. The total useful life of the component is determined by the time necessary to initiate a crack and to propagate the crack from subcritical dimensions to the critical size, a_c. Crack initiation and subcritical crack propagation may be caused by cyclic stresses in the absence of an aggressive environment (fatigue), by an aggressive environment under sustained load (stress-corrosion cracking), or by the combined effects of cyclic stresses and an aggressive environment (corrosion fatigue). All these modes of crack initiation and subcritical crack propagation are localized phenomena that depend on the stress-field intensity at the tip of the notch or crack. Sufficient data are available to show that the rate of subcritical crack growth depends on the stress-intensity factor, K_I, which serves as a single-term parameter representative of the stress conditions in the vicinity of the crack tip.[7-12] The use of fracture-mechanics parameters to study the effect of stress concentration on fatigue-crack initiation is presented in this chapter.

7.3. Stress Field in the Vicinity of Stress Concentrations

The elastic-stress field in the vicinity of sharp elliptical or hyperbolic notches in a body subjected to tensile stresses normal to the plane of the notch is represented by the following equations:[13]

$$
\begin{aligned}
\sigma_x &= \frac{K_I}{(2\pi r)^{1/2}} \cos\frac{\theta}{2}\left[1 - \sin\frac{\theta}{2}\sin\frac{3\theta}{2}\right] - \frac{K_I}{(2\pi r)^{1/2}}\frac{\rho}{2r}\cos\frac{3\theta}{2} \\
\sigma_y &= \frac{K_I}{(2\pi r)^{1/2}} \cos\frac{\theta}{2}\left[1 + \sin\frac{\theta}{2}\sin\frac{3\theta}{2}\right] + \frac{K_I}{(2\pi r)^{1/2}}\frac{\rho}{2r}\cos\frac{3\theta}{2} \qquad (7.1) \\
\tau_{xy} &= \frac{K_I}{(2\pi r)^{1/2}} \sin\frac{\theta}{2}\cos\frac{\theta}{2}\cos\frac{3\theta}{2} - \frac{K_I}{(2\pi r)^{1/2}}\frac{\rho}{2r}\sin\frac{3\theta}{2}
\end{aligned}
$$

where the coordinates r, θ, and ρ are defined in Figure 7.3. The first term in Equations (7.1) defines the magnitude and distribution of the stress field in the vicinity of a fatigue crack. In these equations K_I is the crack-tip stress-intensity factor for a sharp crack of length equal to that of the sharp notch

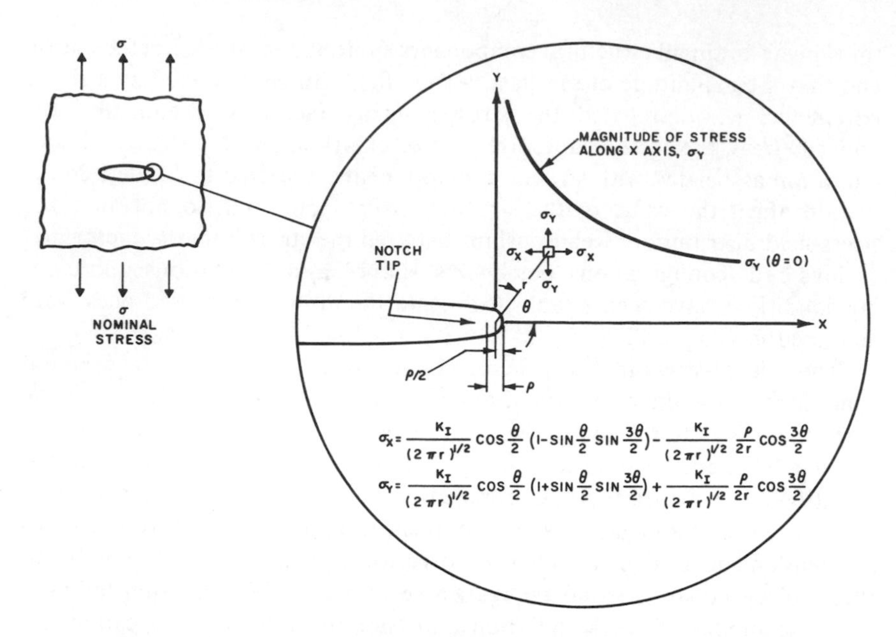

FIG. 7.3. Schematic illustration of the elastic-stress-field distribution near the tip of an elliptical notch (Mode I deformation).

and subjected to the same loading conditions as the notch. The second term in these equations represents the influence of a blunt-tip radius on this stress field. Equations (7.1) also show that on the crack center plane the stress singularity for narrow elliptical and hyperbolic notches is centered on a line located at $\rho/2$ behind the crack front (Figure 7.3).

Notches in structural components cause stress intensification in the vicinity of the notch tip. The material element at the tip of a notch in a cyclically loaded structural component is subjected to the maximum stress, σ_{max}, and to the maximum stress fluctuations, $\Delta\sigma_{max}$. Consequently, this material element is most susceptible to fatigue damage and is, in general, the origin of fatigue-crack initiation. For $r = \rho/2$, the maximum stress on this material element which can be derived from Equation (7.1) is

$$\sigma_{max} = \frac{2}{\sqrt{\pi}} \frac{K_I}{\sqrt{\rho}} \tag{7.2}$$

Although Equation (7.2) is considered exact only when ρ approaches zero, Wilson and Gabrielse[14] showed by using finite element analysis of blunt notches in compact-tension specimens that this relationship is accurate to within 10% for notch radii up to 0.180 in.

Thorough understanding of fatigue-crack initiation requires the development of accurate predictions of the localized stress and strain behavior in the vicinity of stress concentrations. Recent developments in elastic-plastic finite element stress analysis in the vicinity of notches should contribute significantly to the development of quantitative predictions of fatigue-crack-initiation behavior for structural components.[15,16]

7.4. Effect of Stress Concentration on Fatigue-Crack Initiation

The effect of a geometrical discontinuity in a loaded structural component is to intensify the magnitude of the nominal stress in the vicinity of the discontinuity. The localized stresses may cause the metal in that neighborhood to undergo plastic deformation. Because the nominal stresses in most structures are elastic, the zone of plastically deformed metal in the vicinity of stress concentrations is surrounded by an elastic-stress field. The deformations (strains) of the plastic zone are governed by the elastic displacements of the surrounding elastic-stress field. In other words, when the structure is stress-controlled, the localized plastic zones are strain-controlled. Consequently, to predict the effects of stress concentrations on the fatigue behavior of structures, the fatigue behavior of the localized plastic zones has been simulated by testing smooth specimens under strain-controlled conditions [Figure 7.4(a)]. A better simulation of the effects of stress concentrations on the fatigue behavior of structures is obtained by testing notched specimens under stress-controlled conditions [Figure 7.4(b)] because the applied stress can be more directly related to the structural loading.

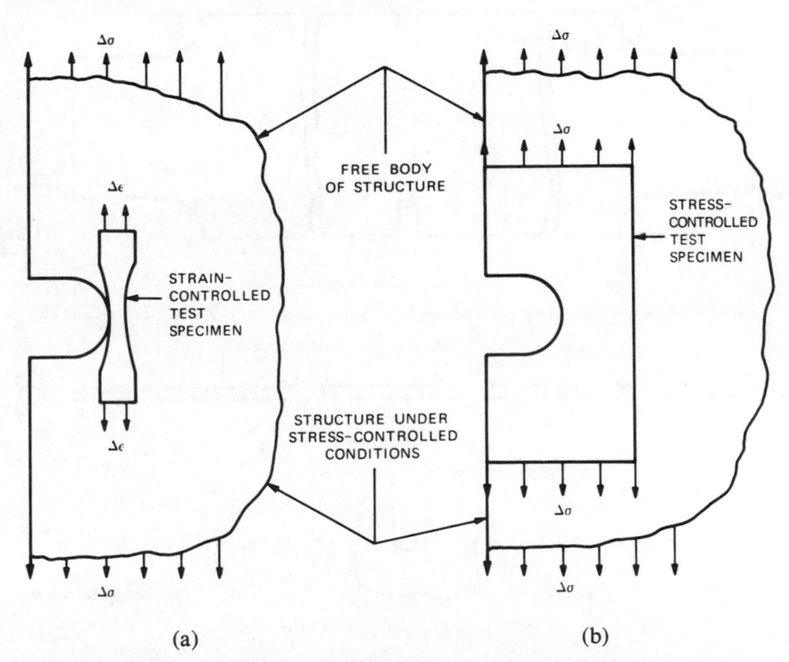

FIG. 7.4. Test specimen simulation of stress concentrations in structures. (a) Strain-controlled specimen. (b) Stress-controlled specimen.

The stress analyses presented in the preceding section facilitated the use of linear-elastic fracture-mechanics parameters in analyzing the fatigue-crack-initiation behavior of notched specimens. The applicability of these parameters to sharp notches has been demonstrated in limited work by Forman,[17] by Constable et al.,[18] and by Jack and Price.[19] Moreover, the applicability of these parameters to sharp and blunt notches has been verified by Barsom and McNicol[20] and by Clark.[21] The fatigue-initiation behavior of HY-130 steel specimens (Figure 7.5) tested by Barsom and McNicol[20] under axial zero-to-tension cyclic loads is presented in Figure 7.6 in terms of the number of cycles for fatigue-crack initiation, N_i, versus the nominal stress fluctuation, $\Delta\sigma$, and in Figure 7.7 in terms of N_i versus the stress-intensity-factor fluctuation divided by the square root of the notch radius, $\Delta K_I/\sqrt{\rho}$, for all the data presented in Figure 7.6. The stress-intensity factor, K_I, for notches was calculated as for a fatigue crack of length equal to the total notch depth. Figure 7.7 shows that a fatigue-crack-initiation threshold, $\Delta K_I/\sqrt{\rho})_{th}$, in HY-130 steel specimens subjected to zero-to-tension axial loads occurs at $K_I/\sqrt{\rho} \approx$ 85 ksi (586 MN/m²).

The maximum elastic stress at the root of the notch, σ_{max}, is equal to $2K_I/\sqrt{\pi\rho}$. Despite the fact that the term $\Delta K_I/\sqrt{\rho}$ loses its significance for an unnotched, polished specimen, let us assume that the fatigue-crack-initiation limit in such specimens occurs when $\Delta K_I/\sqrt{\rho} = 85$ ksi. Substituting this

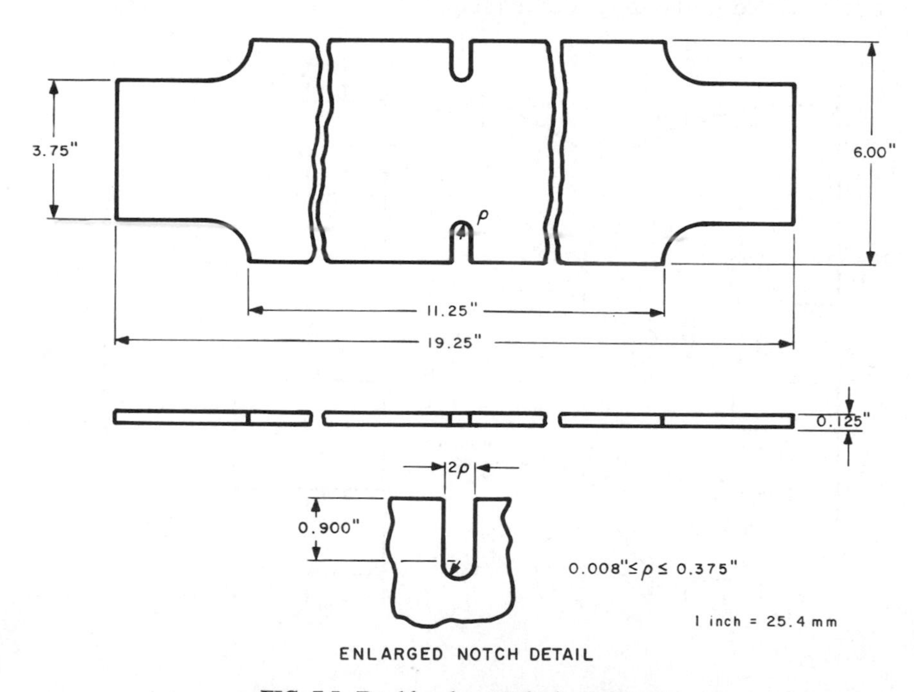

FIG. 7.5. Double-edge-notched specimens.

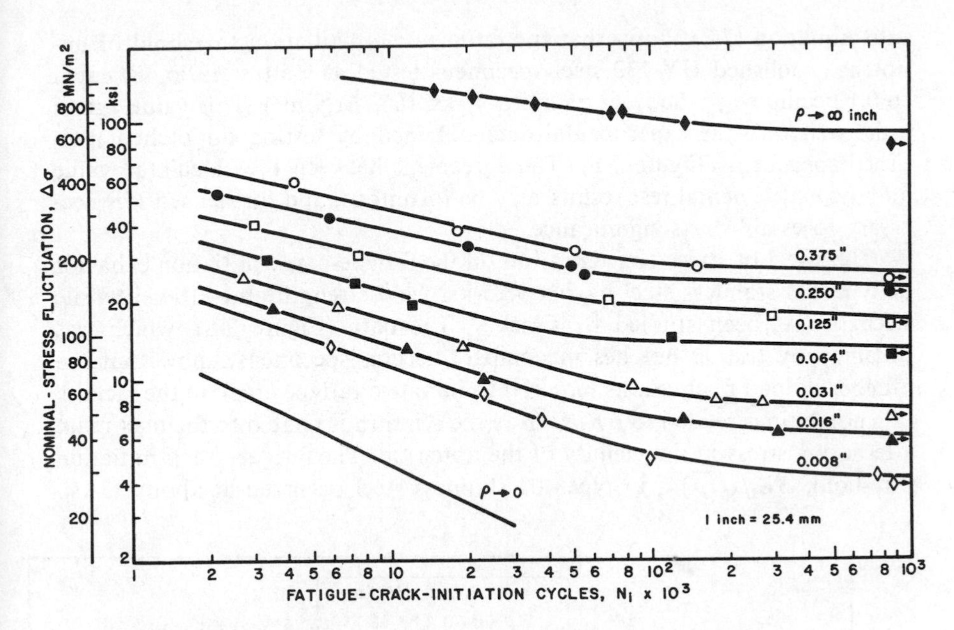

FIG. 7.6. Dependence of fatigue-crack initiation of HY-130 steel on nominal-stress fluctuations for various notch geometries.

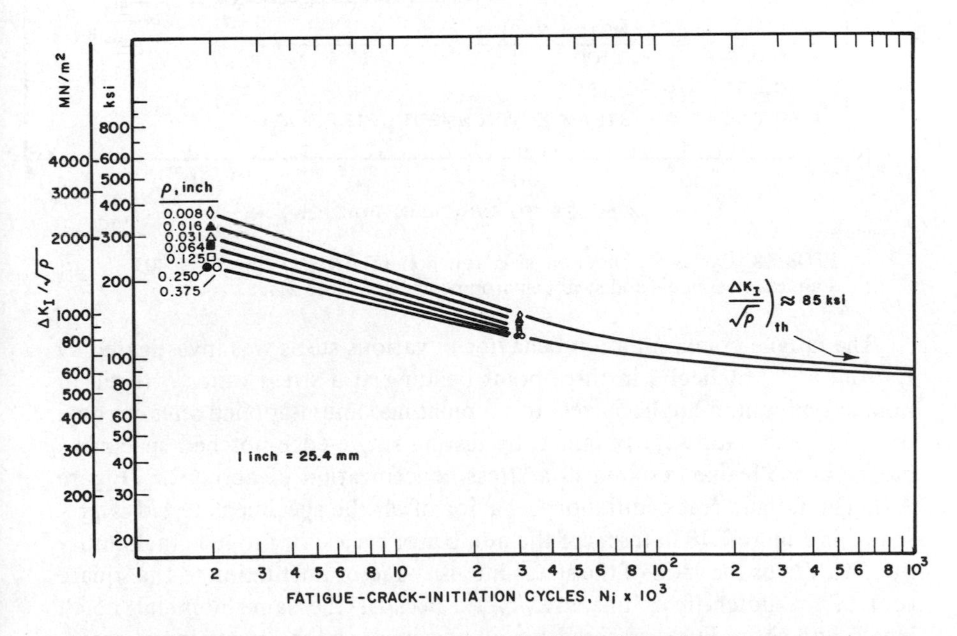

FIG. 7.7. Correlation of fatigue-crack-initiation life with the parameter $\Delta K_I/\sqrt{\rho}$ for HY-130 steel.

into Equation (7.2) shows that the fatigue-crack-initiation threshold of unnotched, polished HY-130 steel specimens tested at a stress ratio, R, equal to 0.1 occurs when $\Delta\sigma_{max}$ is equal to 96 ksi (662 MN/m^2). This value agrees quite well with the experimental data obtained by testing unnotched, polished specimens (Figure 7.6). The agreement between the calculated value and the experimental test results may be fortuitous, and further work is necessary to establish its significance.

The effect of stress concentration on the fatigue-crack-initiation behavior of type 403 stainless steel having 93.5-ksi yield strength and 110-ksi tensile strength has been studied by Clark.[21] The data (Figure 7.8), which was obtained by testing notches in compact-tension specimens, show that the number of load fluctuations required to initiate a fatigue crack in the vicinity of a notch tip is related to $\Delta K_I/\sqrt{\rho}$, which in turn is related to the maximum alternating stress in the vicinity of the notch tip. The fatigue-crack-initiation threshold, $(\Delta K_I/\sqrt{\rho})_{th}$, in type 403 stainless steel occurred at about 95 ksi.

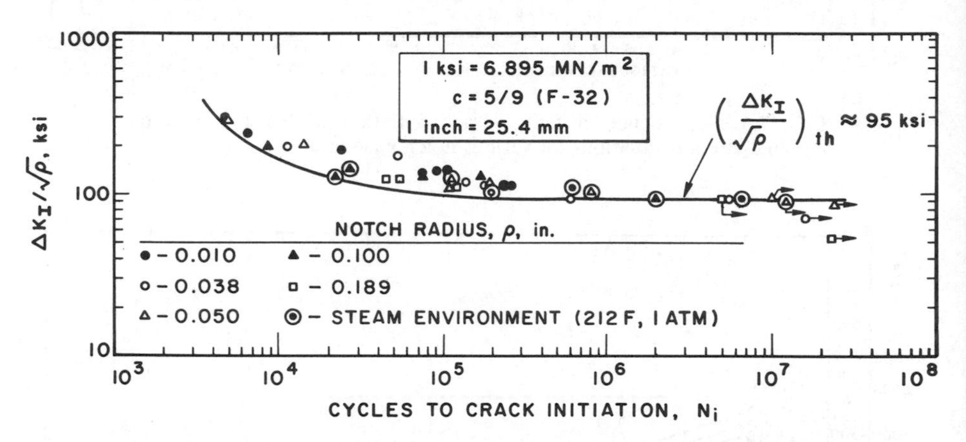

FIG. 7.8. Cycles to initiation as a function of $\Delta K/\sqrt{\rho}$ for type 403 stainless steel in air and steam environments (Reference 21).

The fatigue-crack-initiation behavior in various steels was investigated by Barsom and McNicol[22] in three-point bending at a stress ratio, R (ratio of nominal minimum-applied stress to nominal maximum-applied stress), equal to +0.1. The data were obtained by testing single-edge-notched specimens having a notch that resulted in a stress concentration of about 2.5 (Figure 7.9). The fatigue-crack-initiation behavior of all the specimens tested is presented in Figure 7.10 in terms of the number of cycles for fatigue-crack initiation, N_i, versus the ratio of the stress-intensity-factor fluctuation to the square root of the notch-tip radius, $\Delta K_I/\sqrt{\rho}$. Because the same nominal notch length and tip radius were used for all specimens of the steels investigated,

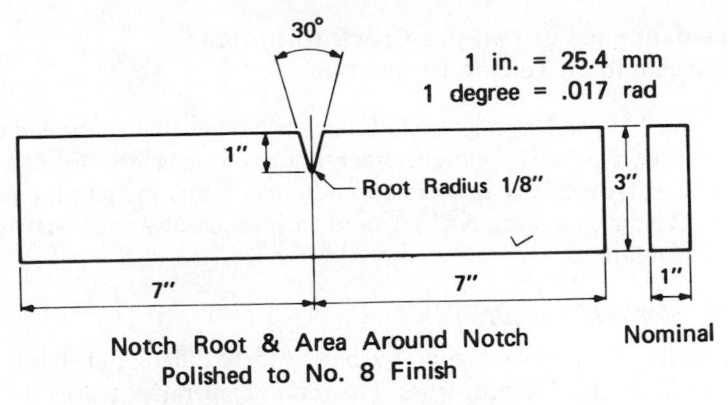

FIG. 7.9. Single-edge-notched bend test specimen.

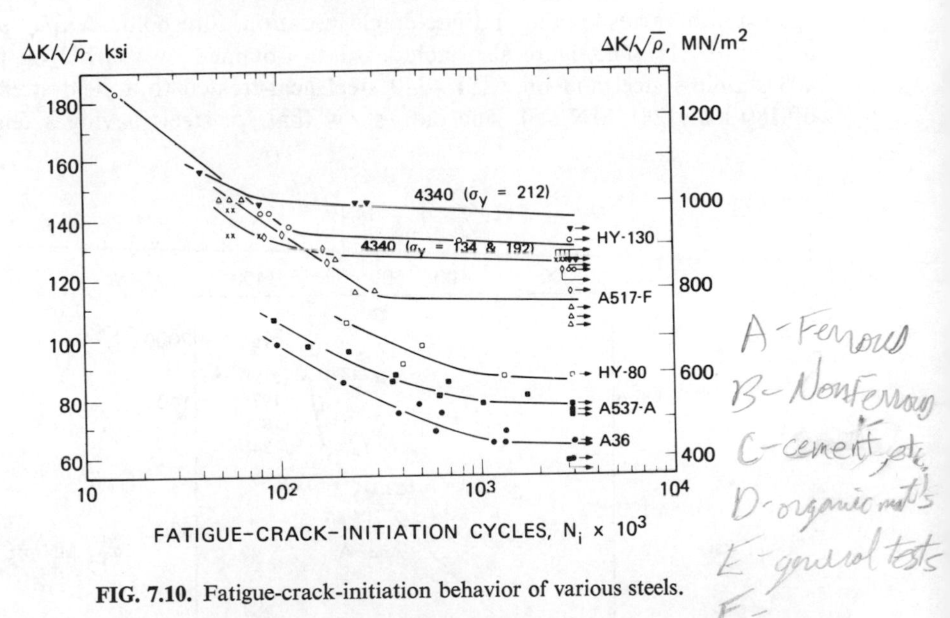

FIG. 7.10. Fatigue-crack-initiation behavior of various steels.

the differences in the fatigue-crack-initiation behavior shown in Figure 7.10 are related primarily to inherent differences in the fatigue-crack-initiation characteristics of the steels. The data show that fatigue cracks do not initiate in steel structural components when the body configuration, the notch geometry, and the nominal-stress fluctuations are such that the magnitude of the parameter $\Delta K_I/\sqrt{\rho}$ is less than a given value that is characteristic of the steel. In general, the value of this fatigue-crack-initiation threshold, $\Delta K_I/\sqrt{\rho})_{th}$, increased with increased yield strength or tensile strength of the steel.

7.5. Dependence of Fatigue-Crack-Initiation Threshold on Tensile Properties

As pointed out earlier, high-cycle fatigue behavior, that is, the endurance limit, represents primarily fatigue-crack-initiation behavior (Figure 7.1). Sufficient data are available to correlate endurance limits and tensile strengths of steels[3]. Relations between $\Delta K/\sqrt{\rho}$ and various tensile mechanical properties exist as follows.

7.5.1. Tensile Strength

The steels presented in Figure 7.9 have widely different chemical compositions and mechanical properties. The room-temperature tensile strengths ranged from 77 to 233 ksi (531 to 1606 MN/m²). These are plotted against the corresponding fatigue-crack-initiation threshold, $\Delta K_I/\sqrt{\rho})_{th}$, in Figure 7.11. The figure also includes data obtained by Clark[23] on type 403 stainless steel and on AISI 4340 steel heat-treated to a yield strength of 180 ksi (1241 MN/m²). The data show that for steels having a tensile

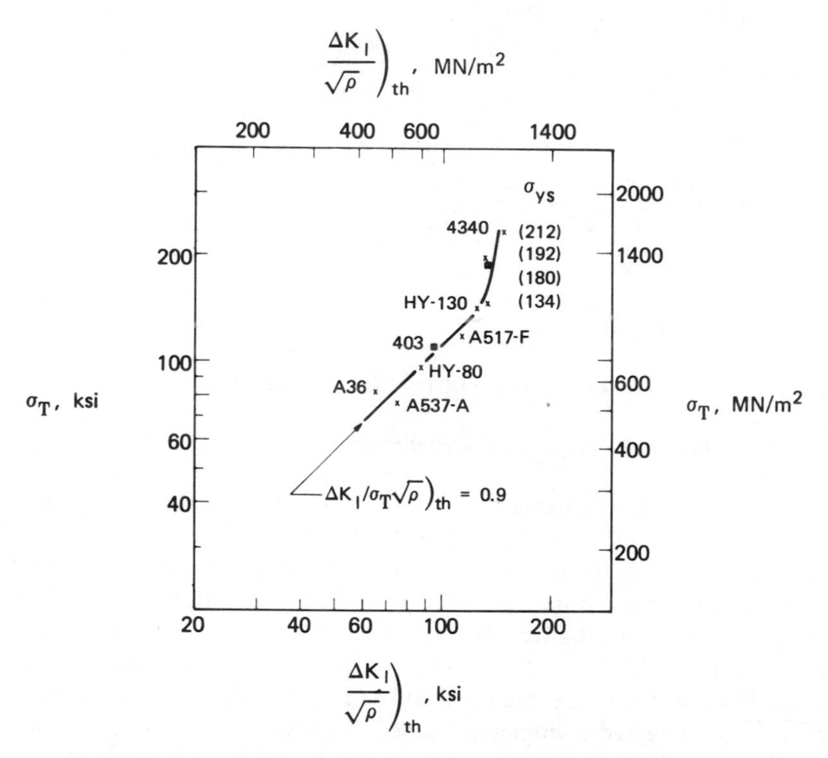

FIG. 7.11. Dependence of fatigue-crack-initiation threshold on tensile strength.

strength less than about 150 ksi (1034 MN/m^2) a given increase in the tensile strength corresponded, approximately, to an equal increase in the magnitude of $\Delta K_I/\sqrt{\rho})_{th}$. On the other hand, for steels having tensile strengths greater than about 150 ksi, an increase of about 100 ksi (689 MN/m^2) in the value of the tensile strength resulted in only a 20-ksi (138-MN/m^2) increase in the value of $\Delta K_I/\sqrt{\rho})_{th}$. This observation suggests that for high-strength ($\sigma_T >$ 150 ksi) steels, the fatigue-crack-initiation threshold is, essentially, independent of tensile strength.

The data presented in Figure 7.11 indicate that, under three-point bending at a stress ratio of 0.1, an approximate value of fatigue-crack-initiation threshold in steels having tensile strengths in the range of about 70 ksi (483 MN/m^2) to 150 ksi can be obtained by using

$$\left.\frac{\Delta K_I}{\sigma_T\sqrt{\rho}}\right)_{th} = 0.9 \tag{7.3}$$

However, for steels having tensile strengths greater than 150 ksi, the fatigue-crack-initiation-threshold values appear to be less than would be predicted by using Equation (7.3). Substituting Equation (7.2) into Equation (7.3) leads to the conclusion that the fatigue-crack-initiation threshold in steels of 150 ksi or less subjected to zero-to-tension bending stresses occurs when the maximum elastic stress calculated at the notch root is equal to the tensile strength of the steel.

Conventional fatigue-life data are usually obtained under fully reversed loading ($R = -1.0$) and show that, in general, the endurance limit occurs at an alternating stress, σ_{alt}, equal to about one half of the tensile strength of the steel. The alternating stress is equal to one half of the total stress range. Consequently, an endurance limit that occurs at an alternating stress of one half of the tensile strength corresponds to an applied maximum-stress range equal to the tensile strength.

The preceding observations indicate that the fatigue-crack-initiation threshold of various steels is governed, primarily, by the total maximum-stress range.

7.5.2. Yield Strength

Barsom and McNicol[20] applied the nondimensional parameter $K_I/\sigma_{ys}\sqrt{\rho}$, where σ_{ys} is the 0.2% offset yield strength, to normalize the fatigue-crack-initiation threshold behavior of various high-yield-strength martensitic steels. They suggested that the fatigue-crack-initiation threshold in various high-yield-strength martensitic steels is represented by

$$\frac{\Delta K_I}{\sigma_{ys}\sqrt{\rho}} = \text{constant} \tag{7.4}$$

Because the yield-to-tensile ratio in high-yield-strength steels is in the range $0.8 < \sigma_{ys}/\sigma_T < 1.0$, Equations (7.3) and (7.4) are very similar.

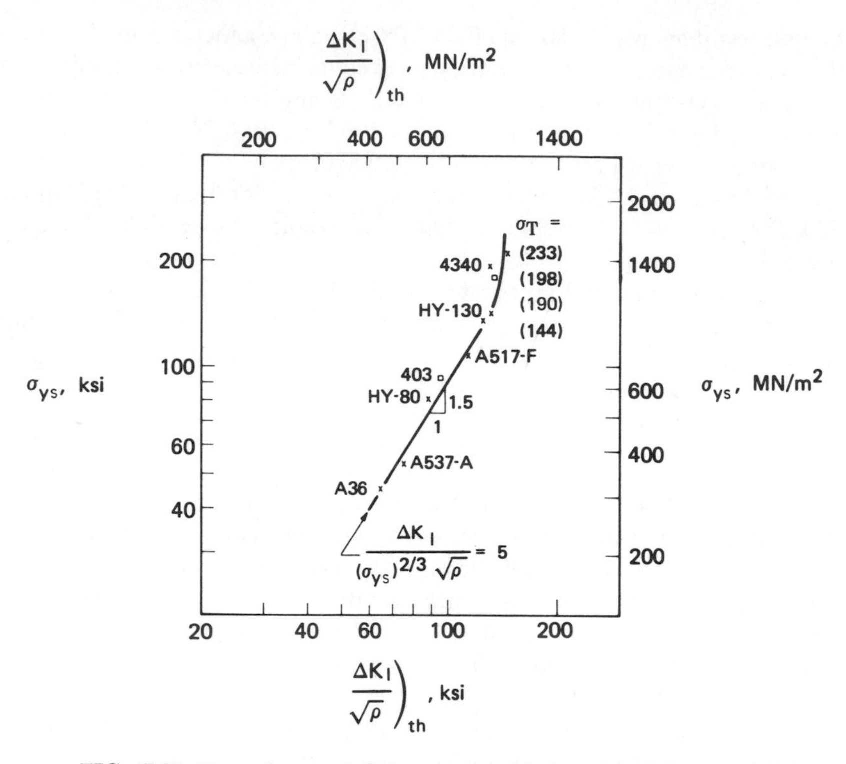

FIG. 7.12. Dependence of fatigue-crack-initiation threshold on yield strength.

The yield strength, σ_{ys}, and the corresponding fatigue-crack-initiation threshold, $\Delta K_I/\sqrt{\rho}$, of the steels shown in Figure 7.10 are presented in Figure 7.12. The figure also includes data obtained by Clark[23] on type 403 stainless steel and on AISI 4340 steel heat-treated to 180-ksi yield strength. The data indicate the existence of an excellent correlation of fatigue-crack-initiation threshold and yield strength in steels having yield strengths in the range of about 40–140 ksi (276–965 MN/m²). The correlation can be represented by

$$\frac{\Delta K_I}{(\sigma_{ys})^{2/3}\sqrt{\rho}} = 5 \tag{7.5}$$

where ΔK_I is in ksi$\sqrt{\text{in.}}$, σ_{ys} is in ksi, and ρ is in in. Tests conducted on steels of yield strengths greater than 140 ksi resulted in fatigue-crack-initiation threshold values less than would be predicted by using Equation (7.5).

7.5.3. Strain-Hardening Exponent

Clausing[24] investigated the tensile properties of various steels and showed the existence of a correlation of strain-hardening exponent with yield strength that was better than a correlation with tensile strength. The relationship between the fatigue-crack-initiation threshold and the strain-hardening exponent of the steels shown in Figure 7.10 was studied. Figure 7.13 shows the

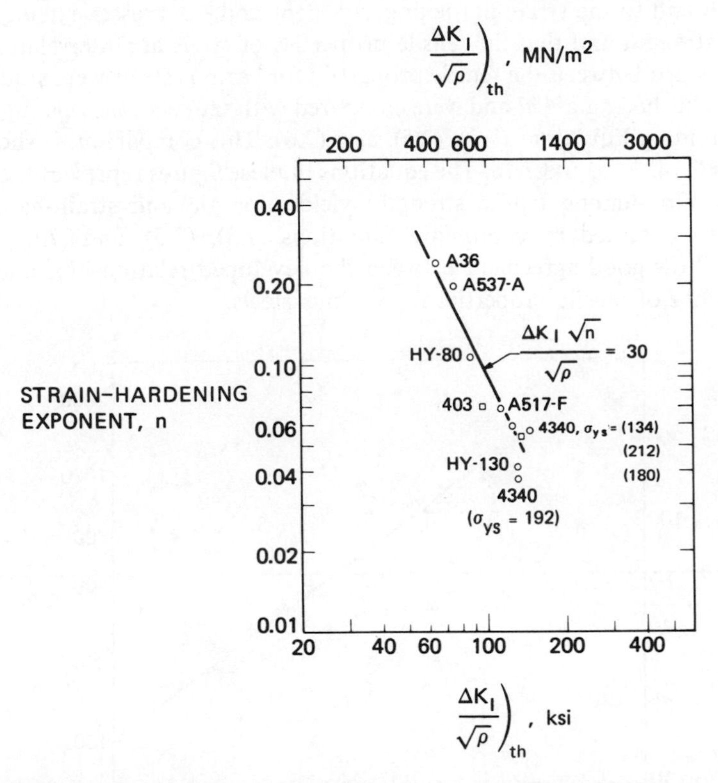

FIG. 7.13. Dependence of fatigue-crack-initiation threshold on strain-hardening exponent.

plot of the fatigue-crack-initiation threshold and the strain-hardening exponent. Included in this figure are data for type 403 and AISI 4340 steels tested by Clark.[23] The strain-hardening exponent of these two steels was approximated by using the uniform-elongation values. The data show that the fatigue-crack-initiation threshold of the various steels can be predicted by using the strain-hardening exponents and that the relationship between these parameters is given by

$$\frac{\Delta K_I \sqrt{n}}{\sqrt{\rho}} = 30 \tag{7.6}$$

Figure 7.13 shows that at low values of strain-hardening exponents, that is, for steels of yield strengths greater than about 140 ksi, the accuracy of Equation (7.6) decreases because of increased data scatter.

7.5.4. General Discussion

The correlations represented in the preceding sections show that the fatigue-crack-initiation threshold in various steels is related to the yield strength and to the strain-hardening exponent and, to a lesser extent, to the tensile strength and that the tensile properties of steels are interrelated. The relationships between the tensile properties for various steels were studied by using published data[24-27] and were compared with the relationships suggested by combining Equations (7.3), (7.5), and (7.6). This comparison is shown in Figures 7.14, 7.15, and 7.16. The equations in these figures represent the interrelationships among tensile strength, yield strength, and strain-hardening exponent obtained by combining Equations (7.3), (7.5), and (7.6). These figures show good agreement between the developed relationships and published data of tensile properties for various steels.

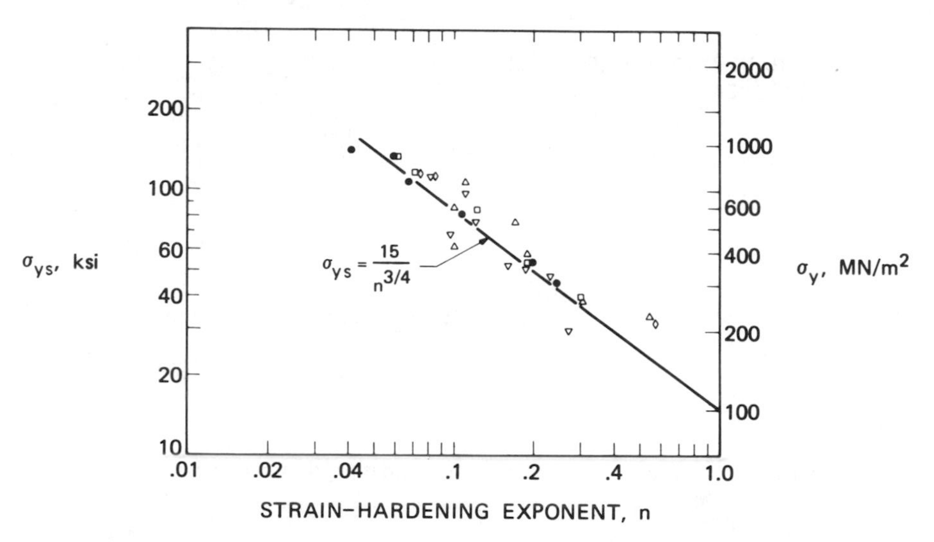

FIG. 7.14. Relationship between yield strength and strain-hardening exponent.

The interrelationships among tensile properties of various steels were developed from data of martensitic steels and ferrite-pearlite steels. The applicability of these equations to austenitic stainless steels has not been established. However, because of the metastable behavior of some austenitic stainless steels and because stable austenitic stainless steels do not usually conform to the same stress-strain relationship as martensitic and ferrite-pearlite steels, the relationships shown in Figures 7.14, 7.15, and 7.16 may not be useful to predict the fatigue-crack-initiation-threshold behavior in austenitic stainless steels. Further research is necessary to establish the fatigue-crack-initiation-threshold characteristics of austenitic stainless steels and of other metal alloys.

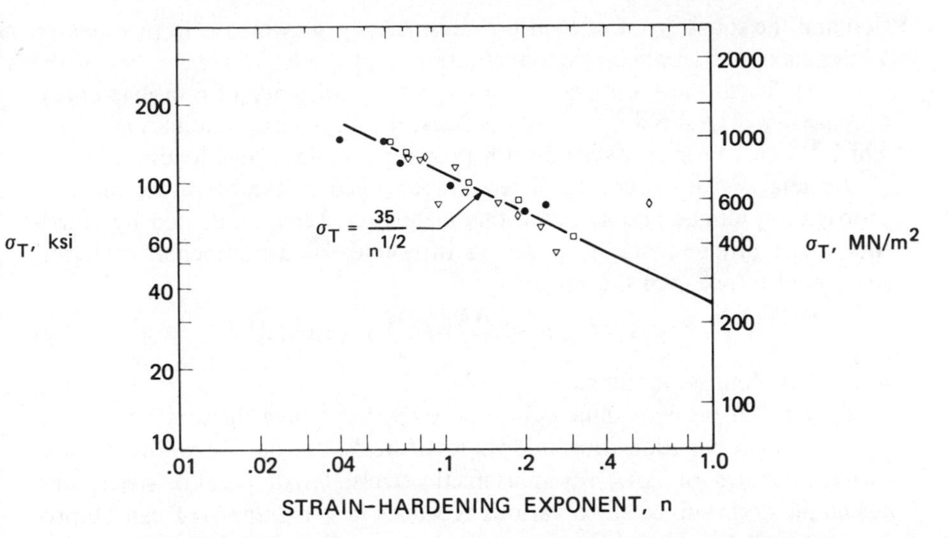

FIG. 7.15. Relationship between tensile strength and strain-hardening exponent.

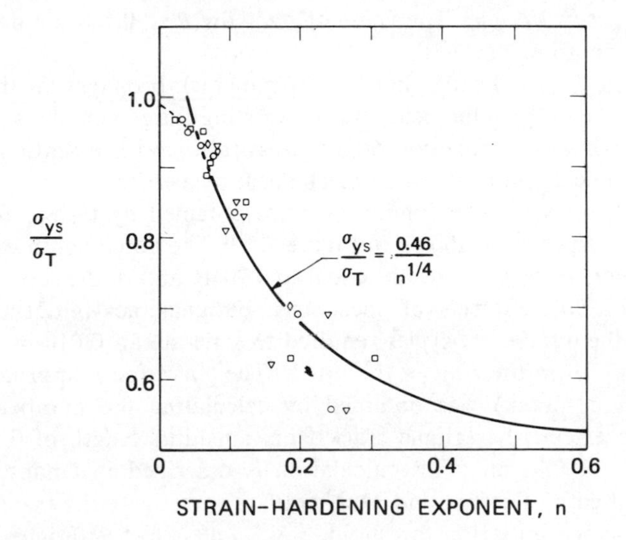

FIG. 7.16. Relationship between yield-to-tensile ratio and strain-hardening exponent.

7.6. Nonpropagating Fatigue Cracks

The preceding discussion indicates the existence of a fatigue-crack-initiation threshold in regions of stress concentrations. The magnitude of the fatigue-crack-initiation threshold for various martensitic and ferrite-pearlite steels appears to be related to the ratio of the stress-intensity-factor fluctuation and the square root of the tip radius, $\Delta K_I/\sqrt{\rho}$, which is in turn related to the maximum applied-stress fluctuation, and the tensile properties of the steels. Sufficient data are available to show the existence of a fatigue-crack-propagation threshold, ΔK_{th} (the stress-intensity-factor fluctuation below which existing fatigue cracks do not propagate under cyclic loading).[28-33]

An analysis of experimental results published in the literature on non-propagating fatigue cracks in various metals has been conducted by Harrison[31]. The fatigue-crack-propagation threshold for a number of materials was found to occur in the range

$$1.5 \times 10^{-4}\sqrt{\text{in.}} \leq \frac{\Delta K_{th}}{E} \leq 1.8 \times 10^{-4}\sqrt{\text{in.}} \qquad (7.7)$$

where E is Young's modulus.

Figure 7.17 presents data published by various investigators[31-33] on fatigue-crack-propagation-threshold values of steels. The data show that conservative estimates of ΔK_{th} for martensitic steels, ferrite-pearlite steels, and austenitic steels subjected to various R values larger than $+0.1$ can be predicted from[34]

$$\Delta K_{th} = 6.4(1 - 0.85R) \qquad (7.8)$$

where ΔK_{th} is in ksi$\sqrt{\text{in.}}$ The value of ΔK_{th} for $R < 0.1$ is a constant equal to 5.5 ksi$\sqrt{\text{in.}}$ (6 $MNm^{-3/2}$).

Equation (7.8) indicates that the fatigue-crack-propagation threshold of steels (i.e., the ΔK_{th} value below which existing fatigue cracks will not propagate) is primarily a function of the stress ratio and is essentially independent of chemical composition or mechanical properties.

The data presented in Figure 7.6 were obtained by testing the double-edge-notched specimens shown in Figure 7.5.[20] The notch depth was equal to $a_0 + \rho$, where a_0 was a constant equal to 0.90 in. and ρ, the root radius, was different for different sets of specimens. Fatigue-crack-initiation life was defined as the number of cycles required to generate an 0.010-in. (0.25-mm) fatigue crack from the root of the notch. The curve for ρ approaching zero (i.e., a fatigue crack) was obtained by calculating the number of cycles required to extend the fatigue crack from an initial length of 0.90 in. to a final length of 0.91 in. (This calculation is described in Chapter 8). This curve was used by Barsom and McNicol[20] to investigate the transition from the fatigue-crack-initiation threshold, $\Delta K_I/\sqrt{\rho})_{th}$, to the fatigue-crack-propagation threshold, ΔK_{th}, as ρ approaches zero.

The crack-opening displacement, δ, at the tip of a fatigue crack is given by (see Chapter 16)

$$\delta = \frac{K^2}{\sigma_{ys} E} \qquad (7.9)$$

where K = stress-intensity factor, psi$\sqrt{\text{in.}}$,
σ_{ys} = yield strength, psi,
E = Young's modulus, psi.

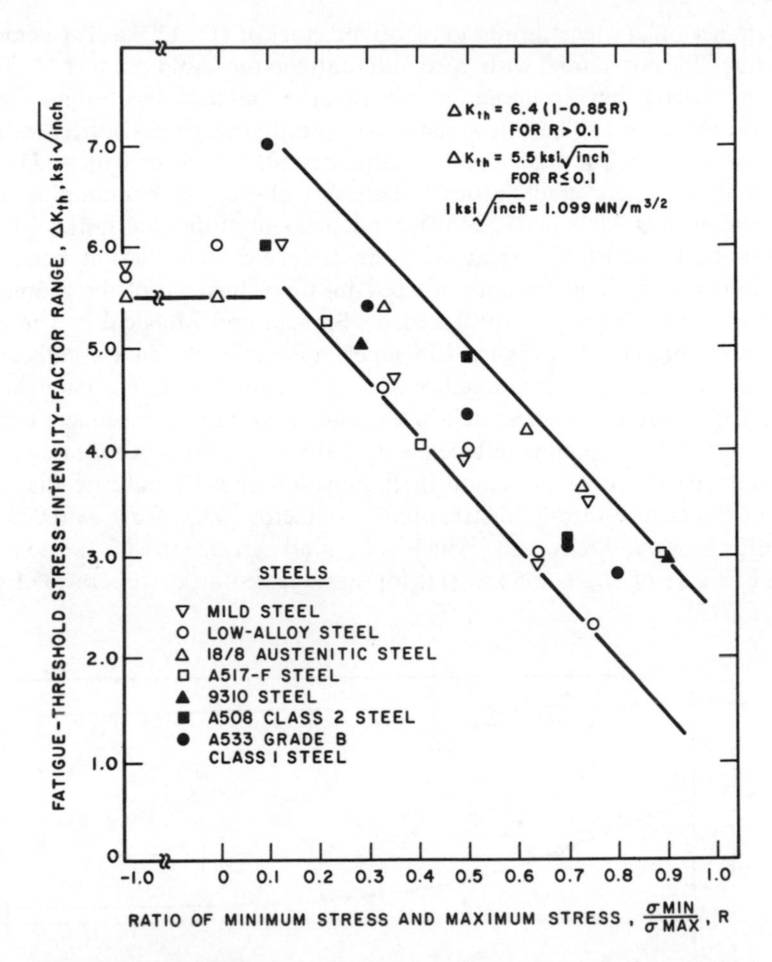

FIG. 7.17. Dependence of fatigue-threshold stress-intensity-factor range on stress ratio.

The value of δ at maximum load in a fatigue-cracked HY-130 steel specimen tested under 0–5-ksi$\sqrt{\text{in.}}$ fluctuation in the stress-intensity factor is equal to 0.6×10^{-5} in. (152 nm). The value of the tip radius is on the order of $\delta/2$, or $\rho = 0.3 \times 10^{-5}$ in. (76 nm). Substituting this value of ρ in the expression $\Delta K_I/\sqrt{\rho})_{th} = 85$ ksi for HY-130 steel (Figure 7.7) suggests that, under zero-to-tension axial loads, ΔK_{th} for HY-130 steel is equal to 0.15 ksi$\sqrt{\text{in.}}$ (0.16 MN/m$^{3/2}$), which appears to negate the existence of a fatigue-crack-propagation threshold in this steel.

Let us assume that a fatigue-crack-propagation threshold in HY-130 steel does occur under zero-to-tension loads at $\Delta K_{th} = 5$ ksi$\sqrt{\text{in.}}$. Then Equation (7.9) suggests that $\rho = 2.5 \times 10^{-3}$ in. at the fatigue-crack-propagation threshold. Judging by the value of ρ, the preceding observations appear to negate the existence of a fatigue threshold in fatigue-cracked HY-130 steel specimens and thus do not agree with available fatigue-threshold data.[28-33] This apparent discrepancy is based on the assumption that the fatigue-crack initiation behavior can be extrapolated to tip radii (i.e., stress concentration factors, k_t) well below the 0.008-in. radius tested. The discrepancy may be resolved if the fatigue-crack-initiation behavior of specimens containing root radii smaller than 0.008 in. ($k_t > 13$) is independent of the root radius (stress concentration factor). Consequently, the fatigue-crack-initiation behavior of specimens containing notches of 0.001-in. tip radius and of the geometry shown in Figure 7.5 was investigated by Barsom and McNicol.[20] The test results are represented in Figure 7.18 superimposed on the data obtained by testing specimens containing notches of 0.008-in. tip radius. The data show, conclusively, that the fatigue-crack-initiation behavior of the specimens tested by Barsom and McNicol is independent of the notch-tip root radius, ρ, for $\rho < 0.008$ in. The data presented in Figures 7.6 and 7.7 indicate that the value of the fatigue-threshold stress-intensity factor, ΔK_{th}, for $\rho = 0.008$ in. ($K_t \approx 13$) is about 7.6 ksi$\sqrt{\text{in.}}$, which is in good agreement with the conservative estimate of $K_{th} = 5.5$ ksi$\sqrt{\text{in.}}$ for steels tested at a stress ratio of 0.1 (Figure 7.17).

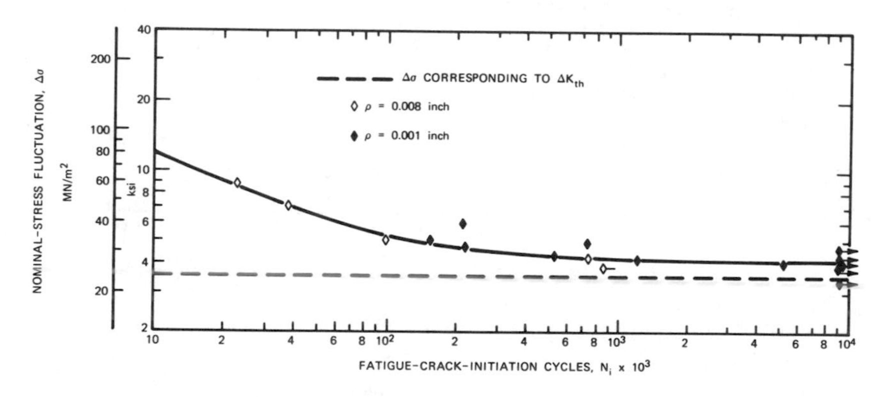

FIG. 7.18. Fatigue-crack initiation of HY-130 steel for very small root radii.

7.7. Finite-Initiation-Life Behavior

The fatigue-crack-initiation life of unnotched specimens is strongly dependent on the surface conditions of the specimen. Surface damage and surface irregularities can reduce the initiation life significantly because of the stress concentration. On the other hand, the fatigue-crack-initiation life of unnotched, polished specimens is related to the plastic deformation of the material. The plastic deformation causes the development of slip steps that become the nucleus of the fatigue-crack-initiation site. These observations are equally applicable to notched specimens. The fatigue-crack-initiation life of a notched specimen can be reduced significantly by surface irregularities in the vicinity of the notch tip. In the finite fatigue-crack-initiation-life region, the data presented in Figure 7.7 show that at a constant value of $\Delta K_I/\sqrt{\rho}$, N_i is primarily a function of $1/\rho^2$ (Figure 7.19). This functional relationship indicates that, at a constant value of $\Delta K_I/\sqrt{\rho}$, the number of elapsed cycles required to initiate a fatigue crack at the tip of a notch in a specimen of unit thickness is inversely related to the volume of an element at the notch tip rather than to the surface area of the notch root. In other words, the fatigue-crack-initiation life in the specimens tested by Barsom and McNicol[20] decreased as the volume of the plastically deformed material at the notch tip

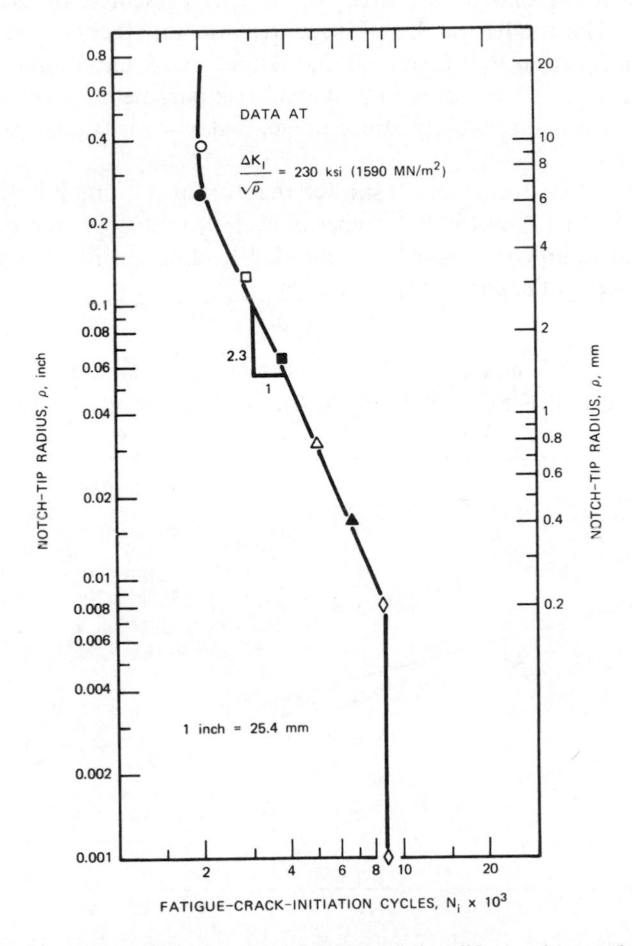

FIG. 7.19. Dependence of fatigue-crack-initiation life on the notch-tip radius in the region of finite-initiation life of HY-130 steel.

increased. Because of the high quality of the surface of the notches that were tested, the surface did not affect the number of elapsed cycles required to initiate a fatigue crack. The data presented in Figure 7.19 also suggest that the fatigue-crack-initiation life in the finite-life region is independent of notch-tip radius when ρ is large (that is, when the volume of plastically deformed metal is large) and when ρ is small. This behavior occurred in HY-130 steel when ρ was greater than 0.25 in. (6.4 mm) ($k_t < 3$) and when ρ was less than 0.008 in. (0.20 mm) ($k_t > 13$).

The fatigue-crack-initiation life of structural components is influenced by surface irregularities, by internal inclusions and other volumetric stress (strain) raisers, and by the volume of the plastically deformed metal. A mathematical model that depicts the dependence of fatigue-crack-initiation life on surface and volumetric characteristics was presented by Barsom and McNicol.[20] The model was based on a weakest-link theory and Weibull's distribution function.[35-38] It showed that fatigue-crack initiation in structural components can be represented by an areal-risk parameter, a volumetric-risk parameter, a zero-probability stress range, a flaw-density exponent, and a scale parameter.

Mowbray and McConnelee[15] showed that for fatigue-crack-initiation lives between 10^2 and 10^4 cycles the number of cycles to fatigue-crack initiation in notched and unnotched specimens subjected to different loading conditions can be related to the stress-strain parameter:

$$(E\sigma_{\max}\epsilon_a)^{1/2} \tag{7.10}$$

where E is Young's modulus and $\sigma_{\max}$ and ϵ_a are the maximum tensile stress and the maximum strain amplitude at the root of the notch. Figure 7.20[15]

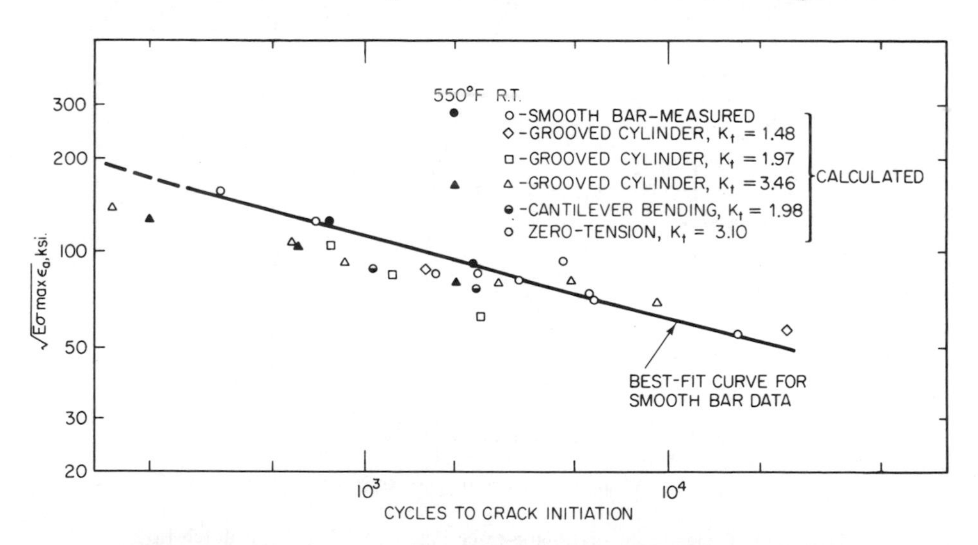

FIG. 7.20. Correlation of crack initiation life data for 2 1/4 Cr-1 Mo alloy steel.

shows the correlation of fatigue-crack initiation in $2\frac{1}{4}$Cr-1Mo steel ($\sigma_{ys} =$ 37 ksi) and the stress-strain parameter. Similar correlations were obtained for a low-carbon steel ($\sigma_{ys} = 29$ ksi) and for type 304 stainless steel ($\sigma_{ys} =$ 39 ksi).

The analysis and correlations discussed in this chapter can be used to quantitatively evaluate the fatigue-crack-initiation behavior of steels and can be incorporated in design considerations. However, considerable research is needed to determine the effects of various metallurgical and mechanical parameters on the fatigue-crack-initiation behavior of various metals.

REFERENCES

1. S. S. Manson, *Thermal Stress and Low-Cycle Fatigue*, McGraw-Hill, New York, 1966.
2. B. M. Wundt, "Effect of Notches on Low-Cycle Fatigue," *ASTM STP 490*, American Society for Testing and Materials, Philadelphia, May 1972.
3. E. G. Eeles and R. C. A. Thurston, "Fatigue Properties of Materials," *Ocean Engineering*, *1*, 1968.
4. P. C. Paris and G. C. Sih, "Stress Analysis of Cracks," in "Fracture-Toughness Testing and Its Applications," *ASTM STP 381*, American Society for Testing and Materials, Philadelphia, 1965.
5. H. Tada, P. C. Paris, and G. R. Irwin, *The Stress Analysis of Cracks Handbook*, Del Research Corporation, Hellertown, Pa., 1973.
6. G. C. Sih, *Handbook of Stress-Intensity Factors*, Institute of Fracture and Solid Mechanics, Lehigh University, Bethlehem, Pa., 1973.
7. P. C. Paris and F. Erdogan, "A Critical Analysis of Crack Propagation Laws," *Transactions of the ASME, Journal of Basic Engineering*, Series D, *85*, No. 3, 1963.
8. J. M. Barsom, "Fatigue-Crack Propagation in Steels of Various Yield Strengths," *Transactions of the ASME, Journal of Engineering for Industry*, Series B, *93*, No. 4, Nov. 1971.
9. J. M. Barsom, "Effect of Cyclic-Stress Form on Corrosion-Fatigue Crack Propagation Below K_{Iscc} in a High-Yield-Strength Steel," in *Corrosion Fatigue: Chemistry, Mechanics, and Microstructure*, International Corrosion Conference Series, Vol. NACE-2, National Association of Corrosion Engineers, Houston 1972.
10. W. G. Clark, Jr., and H. E. Trout, Jr., "Influence of Temperature and Section Size on Fatigue Crack Growth Behavior in Ni-Mo-V Alloy Steel," *Journal of Engineering Fracture Mechanics*, *2*, No. 2, Nov. 1970.
11. H. H. Johnson and P. C. Paris, "Sub-Critical Flaw Growth," *Journal of Engineering Fracture Mechanics*, *1*, No. 3, June 1968.
12. R. P. Wei, "Some Aspects of Environment-Enhanced Fatigue-Crack Growth," *Journal of Engineering Fracture Mechanics*, *1*, No. 4, April 1970.
13. M. Creager, "The Elastic Stress-Field Near the Tip of a Blunt Crack," Master of Science Thesis, Lehigh University, Bethlehem, Pa., 1966.

14. W. K. WILSON and S. E. GABRIELSE, "Elasticity Analysis of Blunt Notched Compact Tension Specimens," *Research Report 71-1E7-LOWFA-R1*, Westinghouse Research Laboratory, Pittsburgh, Feb. 5, 1971.
15. D. F. MOWBRAY and J. E. MCCONNELEE, "Applications of Finite Element Stress Analysis and Stress-Strain Properties in Determining Notch Fatigue Specimen Deformation and Life," in "Cyclic Stress-Strain Behavior—Analysis, Experimentation, and Failure Prediction," *ASTM STP 519*, American Society for Testing and Materials, Philadelphia, May 1973.
16. W. K. WILSON, "Elastic-Plastic Analysis of Blunt Notched CT Specimens and Applications," *Research Report 73-1E7-MAGRR-P1*, Westinghouse Research Laboratories, Pittsburgh Dec. 31, 1973.
17. R. G. FORMAN, "Study of Fatigue-Crack Initiation From Flaws Using Fracture Mechanics Theory," *Air Force Flight Dynamics Laboratory Technical Report AFFDL-TR-68-100*, Dayton, Sept. 1968.
18. I. CONSTABLE, L. E. CULBER, and J. G. WILLIAMS, "Notch-Root-Radii Effects in the Fatigue of Polymers," *International Journal of Fracture Mechanics, 6*, No. 3, Sept. 1970.
19. A. R. JACK and A. T. PRICE, "The Initiation of Fatigue Cracks from Notches in Mild-Steel Plates," *International Journal of Fracture Mechanics, 6*, No. 4, Dec. 1970.
20. J. M. BARSOM and R. C. MCNICOL, "Effect of Stress Concentration on Fatigue-Crack Initiation in HY-130 Steel," *ASTM STP 559*, American Society for Testing and Materials, Philadelphia, 1974.
21. W. G. CLARK, JR., "An Evaluation of the Fatigue Crack Initiation Properties of Type 403 Stainless Steel in Air and Steam Environments," *ASTM STP 559*, American Society for Testing and Materials, Philadelphia, 1974.
22. J. M. BARSOM and R. C. MCNICOL, unpublished data.
23. W. G. CLARK, JR., "How Fatigue Crack Initiation and Growth Properties Affect Material Selection and Design Criteria," *Metals Engineering Quarterly*, Aug. 1974.
24. D. P. CLAUSING, "Tensile Properties of Eight Constructional Steels Between 70 and −320°F," *Journal of Materials, 4*, No. 2, June 1969.
25. J. H. GROSS, "PVRC Interpretive Report of Pressure Vessel Research: Section 2—Materials Considerations," *Welding Research Council Bulletin No. 101*, New York, Nov. 1964.
26. B. F. LANGER, "Design-Stress Basis for Pressure Vessels," the William M. Murray Lecture, 1970, *Experimental Mechanics, 11*, No. 1, Jan. 1971.
27. C. P. ROYER, S. T. ROLFE, and J. T. EASLEY, "Effect of Strain Hardening on Bursting Behavior of Pressure Vessels," in *Second International Conference on Pressure Vessel Technology: Part II—Materials, Fabrication and Inspection*, the American Society of Mechanical Engineers, New York, 1973.
28. R. J. BUCCI, W. G. CLARK, JR., and P. C. PARIS, "Fatigue-Crack-Propagation Growth Rates Under a Wide Variation of ΔK for an ASTM A517 Grade F (T-1) Steel," *ASTM STP 513*, American Society for Testing and Materials, Philadelphia, 1972.
29. R. A. SCHMIDT, "A Threshold in Metal Fatigue," Master of Science Thesis, Lehigh University, Bethlehem, Pa., 1970.

30. P. C. Paris, "Testing for Very Slow Growth of Fatigue Cracks," *MTS Closed Loop Magazine*, *2*, No. 5, 1970.
31. J. D. Harrison, "An Analysis of Data on Non-Propagating Fatigue Cracks on a Fracture Mechanics Basis," *British Welding Journal*, *2*, No. 3, March 1970.
32. L. P. Pook, "Fatigue Crack Growth Data for Various Materials Deduced from the Fatigue Lives of Precracked Plates," *ASTM STP 513*, American Society for Testing and Materials, Philadelphia, 1972.
33. P. C. Paris, R. J. Bucci, E. T. Wessel, W. G. Clark, and T. R. Mager, "Extensive Study of Low Fatigue Crack Growth Rates in A533 and A508 Steels," *ASTM STP 513*, American Society for Testing and Materials, Philadelphia, 1972.
34. J. M. Barsom, "Fatigue Behavior of Pressure-Vessel Steels," *WRC Bulletin* 194, Welding Research Council, New York, May 1974.
35. W. Weibull, "A Statistical Theory of the Strength of Materials," *Ingenieur Vetenskaps Akademie Handlung*, *151*, 1939.
36. W. Weibull, "The Phenomenon of Rupture in Solids," *Ingenieur Vetenskaps Akademie Handlung*, *153*, 1939.
37. W. Weibull, "A Statistical Distribution Function of Wide Applicability," *Journal of Applied Mechanics*, *18*, 1951.
38. K. N. Smith, P. Watson, and T. H. Topper, "A Stress-Strain Function for the Fatigue of Metals," *Journal of Materials*, *5*, No. 4, Dec. 1970.

8

Fatigue-Crack Propagation Under Constant-Amplitude Load Fluctuation

8.1. General

The life of structural components that contain cracks or that develop cracks early in their life may be governed by the rate of subcritical crack propagation. Moreover, proof-testing or nondestructive testing procedures or both may provide information regarding the relative size and distribution of possible preexisting cracks prior to service. However, these inspection procedures are usually used to establish upper limits on undetectable defect size rather than actual crack size. These upper limits are determined by the maximum resolution of the inspection procedure. Thus, to establish the minimum fatigue life of structural components, it is reasonable to assume that the component contains the largest discontinuity that cannot be detected by the inspection method. The useful life of these structural components is determined by the fatigue-crack-growth behavior of the material. Therefore, to predict the minimum fatigue life of structural components and to establish safe inspection intervals, an understanding of the rate of fatigue-crack propagation is required. The most successful approach to the study of fatigue-crack propagation is based on fracture-mechanics concepts.

Most fatigue-crack-growth tests are conducted by subjecting a fatigue-cracked specimen to constant-amplitude cyclic-load fluctuations. Incremental increase of crack length is measured, and the corresponding number of elapsed load cycles is recorded. The data are presented on a plot of crack length, a, versus total number of elapsed load cycles, N (Figure 8.1). An increase in the magnitude of cyclic-load fluctuation results in a decrease of fatigue life of specimens having identical geometry (Figure 8.2). Furthermore, the fatigue life of specimens subjected to a fixed constant-amplitude cyclic-load fluctuation decreases as the length of the initial crack is increased (Figure 8.3). Consequently, under a given constant-amplitude stress fluctuation, most of the useful cyclic life is expended when the crack length is very small. Various a versus N curves can be generated by varying the magnitude

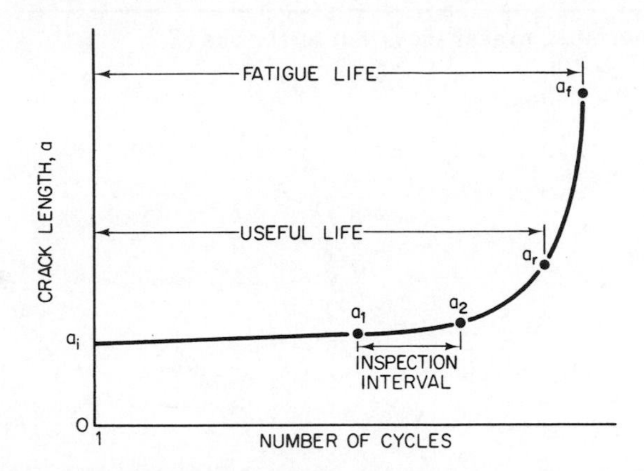

FIG. 8.1. Schematic representation of fatigue crack growth curve under constant amplitude loading.

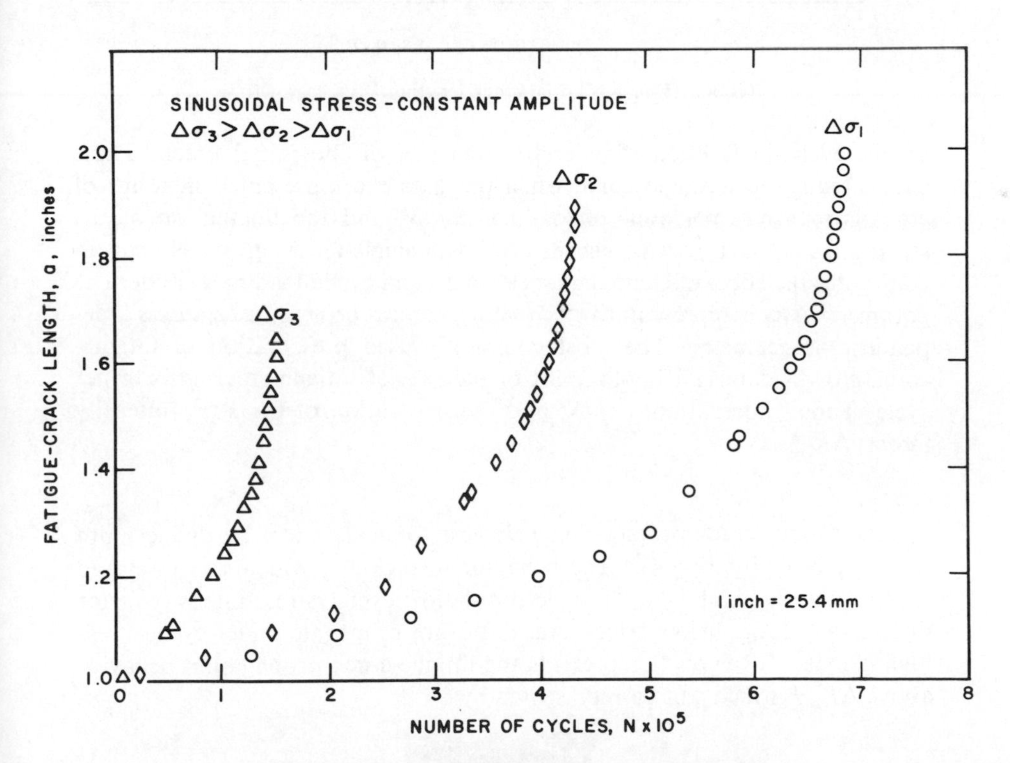

FIG. 8.2. Effect of cyclic-stress range on crack growth.

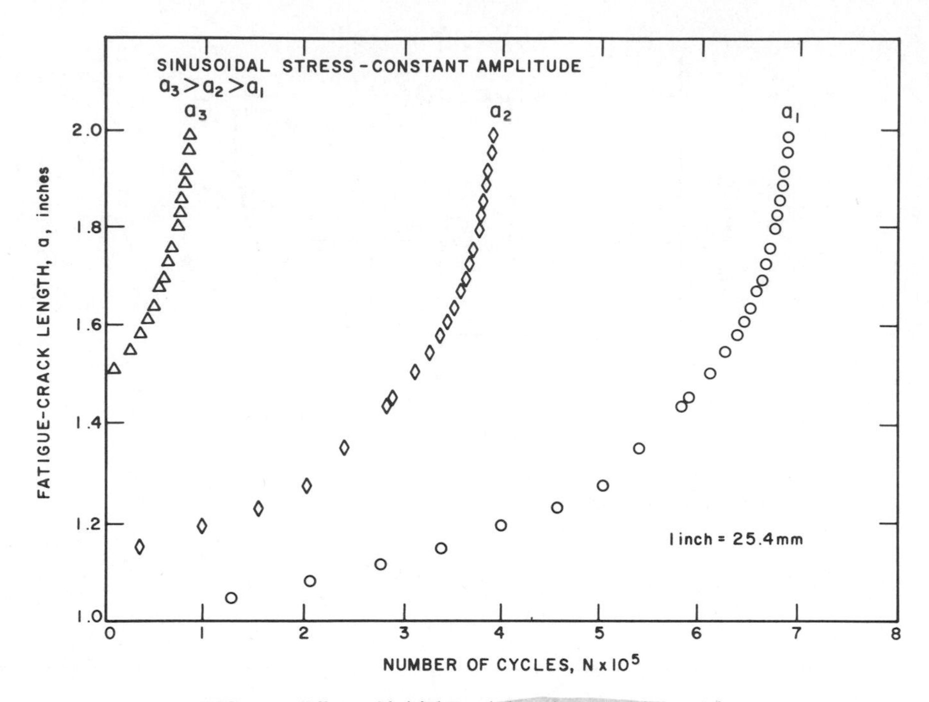

FIG. 8.3. Effect of initial crack length on crack growth.

of the cyclic-load fluctuation and/or the size of the initial crack. These curves reduce to a single curve when the data are represented in terms of crack-growth rate per cycle of loading, da/dN, and the fluctuation of the stress-intensity factor, ΔK_I, because ΔK_I is a single-term parameter that incorporates the effect of changing crack length and cyclic-load magnitude. The parameter ΔK_I is representative of the mechanical driving force and is independent of geometry. The most commonly used presentation of fatigue-crack-growth data is a log-log plot of the rate of fatigue-crack growth per cycle of load fluctuation, da/dN, and the fluctuation of the stress-intensity factor, ΔK_I.[1]

The fatigue-crack-propagation behavior for metals can be divided into three regions (Figure 8.4).[2] The behavior in region I, which was discussed in Chapter 7, exhibits a "fatigue-threshold" cyclic stress-intensity-factor fluctuation, ΔK_{th}, below which cracks do not propagate under cyclic-stress fluctuations.[3-8] Region II represents the fatigue-crack-propagation behavior above ΔK_{th},[9] which can be represented by

$$\frac{da}{dN} = A(\Delta K)^n \tag{8.1}$$

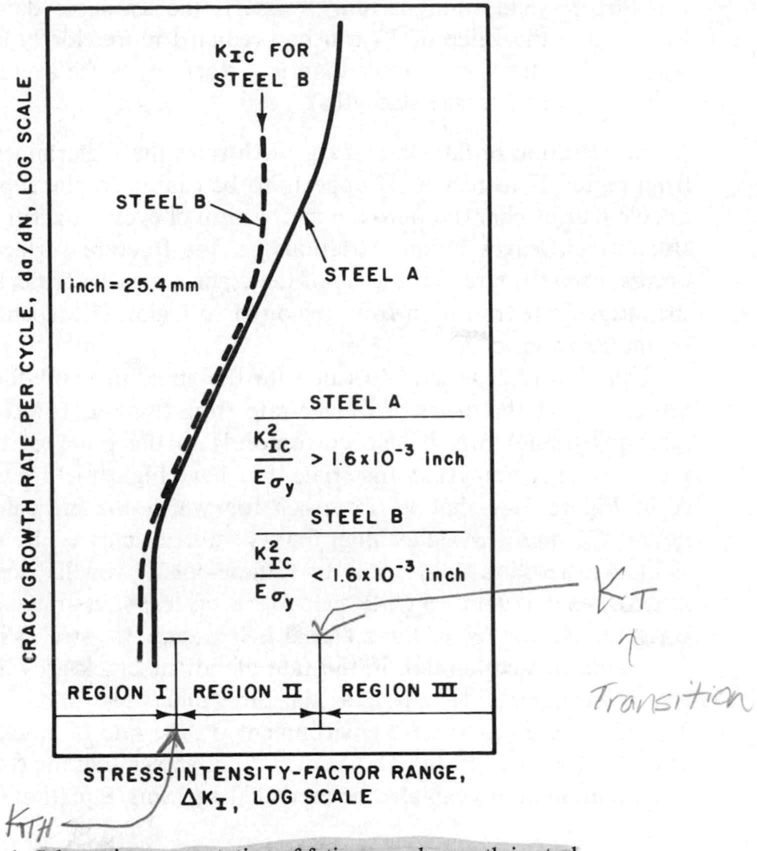

FIG. 8.4. Schematic representation of fatigue-crack growth in steel.

where a = crack length,

N = number of cycles,

ΔK = stress-intensity-factor fluctuation.

A and n are constants.

In region III the fatigue-crack growth per cycle is higher than predicted for region II. The data[9-13] show that the rate of fatigue-crack growth increases and that under zero-to-tension loading (that is, $\Delta K = K_{\max}$) this increase occurs at a constant value of crack-tip displacement, δ_T, and at a corresponding stress-intensity-factor value, K_T, given by

$$\delta_T = \frac{K_T^2}{E\sigma_{ys}} = 1.6 \times 10^{-3} \text{ in.} \quad (0.04 \text{ mm}) \tag{8.2}$$

where K_T = stress-intensity-factor-range value corresponding to onset of acceleration in fatigue-crack-growth rates,

E = Young's modulus,

σ_{ys} = yield strength (0.2% offset) (the available data indicate that the value of K_T can be predicted more closely by using a flow stress, σ_f, rather than σ_{ys}, where σ_f is the average of the yield and tensile strengths).

Acceleration of fatigue-crack-growth rates that determines the transition from region II to region III appears to be caused by the superposition of a ductile-tear mechanism onto the mechanism of cyclic subcritical crack extension, which leaves fatigue striations on the fracture surface. Ductile tear occurs when the strain at the tip of the crack reaches a critical value.[14] Thus, the fatigue-rate transition from region II to region III depends on K_{max} and on the stress ratio, R.

Equation (8.2) is used to calculate the stress-intensity-factor value corresponding to the onset of fatigue-rate transition, K_T (or ΔK_T for zero-to-tension loading), which also corresponds to the point of transition from region II to region III in materials that have high fracture toughness, steel A in Figure 8.4—that is, materials for which the critical-stress-intensity factor, K_{Ic} or K_c, is higher than the K_T value calculated by using Equation (8.2). Acceleration in the rate of fatigue-crack growth occurs at a stress-intensity-factor value slightly below the critical-stress-intensity factor, K_{Ic}, when the K_{Ic} (or K_c) of the material is less than K_T, steel B in Figure 8.4.[15] Furthermore, acceleration in the rate of fatigue-crack growth in an aggressive environment may occur at the threshold stress-intensity factor, K_{Iscc}. The effect of an aggressive environment on the rate of crack growth is discussed in Chapters 10 and 11. Crooker[12] has shown that the ΔK_T of aluminum and titanium alloys can also be predicted by using Equation (8.2).

8.2. Martensitic Steels

Extensive fatigue-crack-growth-rate data for various high-yield-strength (σ_{ys} greater than 80 ksi or 552 MN/m^2) martensitic steels show that the primary parameter affecting growth rate in region II is the range of fluctuation in the stress-intensity factor and that the mechanical and metallurgical properties of these steels have negligible effects on the fatigue-crack-growth rate in a room-temperature air environment.[13,16] The data for these steels fall within a single band, as shown in Figure 8.5, and the upper bound of the scatter of the fatigue-crack-propagation-rate data for martensitic steels in an air environment can be obtained from

$$\frac{da}{dN} = 0.66 \times 10^{-8}(\Delta K_I)^{2.25} \tag{8.3}$$

where a = in.,
K_I = ksi$\sqrt{\text{in.}}$

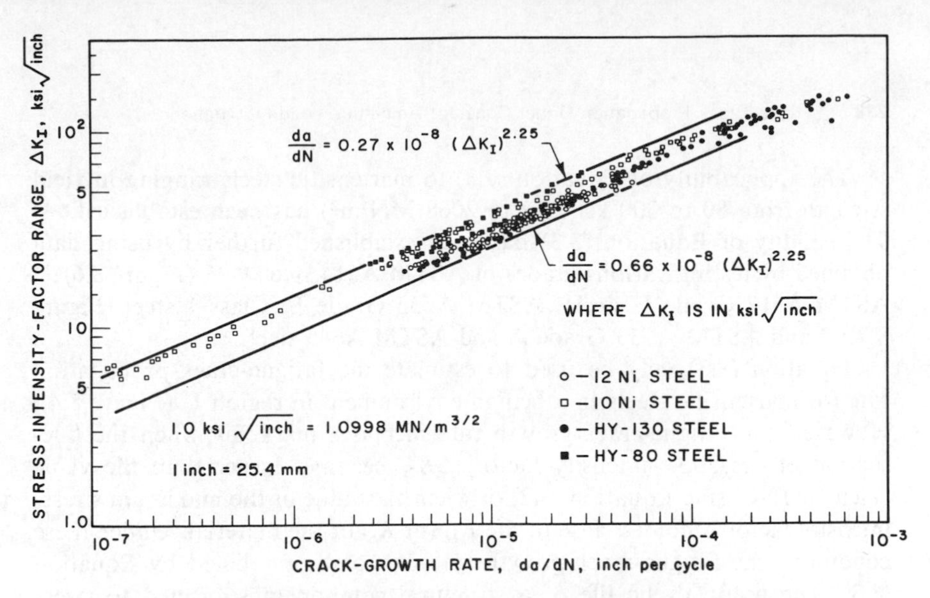

FIG. 8.5. Summary of fatigue-crack-propagation for martensitic steels.

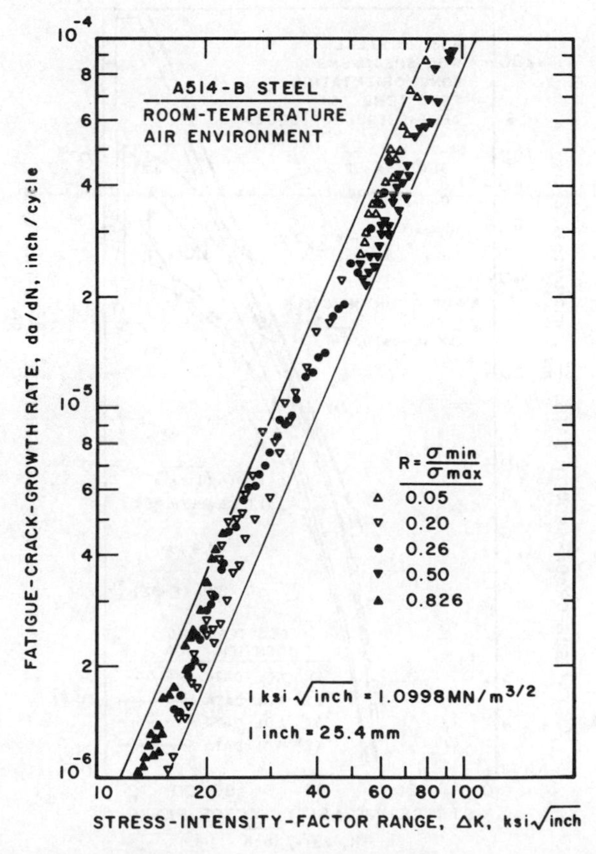

FIG. 8.6. Crack-growth rate as a function of stress-intensity range for A514-B steel.

The applicability of Equation (8.3) to martensitic steels ranging in yield strength from 80 to 300 ksi (552 to 2068 MN/m²) has been established.[9,10] The validity of Equation (8.3) has been established further by using data obtained by testing various grades of ASTM A514 steels[16,17] (Figure 8.6);[16] ASTM A517 Grade F steel;[3] ASTM A533 Grade B, Class 1 steel (Figure 8.7);[18] and ASTM A533 Grade A and ASTM A645 steels.[19]

Equation (8.3) may be used to estimate the fatigue-crack-propagation rate for martensitic steels in a benign environment in region II (Figure 8.4). However, the fatigue-crack-growth rate increases markedly when the fluctuation of the stress-intensity factor, ΔK_I, becomes larger than the value calculated by using Equation (8.2) or when the value of the maximum stress-intensity factor becomes close to the K_{Ic} (or K_c) of the material. Under these conditions the fatigue-crack-growth rate cannot be predicted by Equation (8.3). The useful cyclic life of a structural component subjected to stress

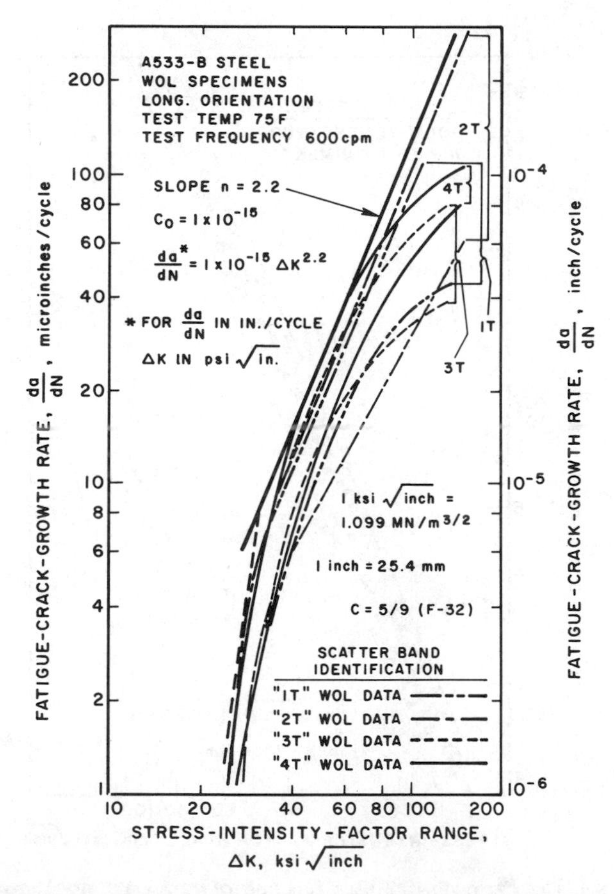

FIG. 8.7. Influence of specimen size on the da/dN vs ΔK relationship for A533 Gr. B, Cl. I steel (Reference 18).

fluctuations corresponding to region-III-type behavior is of practical interest only in relatively low-life structural applications. Further work is necessary to characterize the fatigue-crack-propagation behavior in region III.

8.3. Ferrite-Pearlite Steels

The fatigue-crack-growth-rate behavior in ferrite-pearlite steels[9] prior to the onset of fatigue-rate transition and above ΔK_{th} is presented in Figure 8.8. The data indicate that realistic estimates of the rate of fatigue-crack growth in these steels can be calculated from

$$\frac{da}{dN} = 3.6 \times 10^{-10}(\Delta K_I)^{3.0} \tag{8.4}$$

where a = in.,
$\Delta K_I = \text{ksi}\sqrt{\text{in.}}$

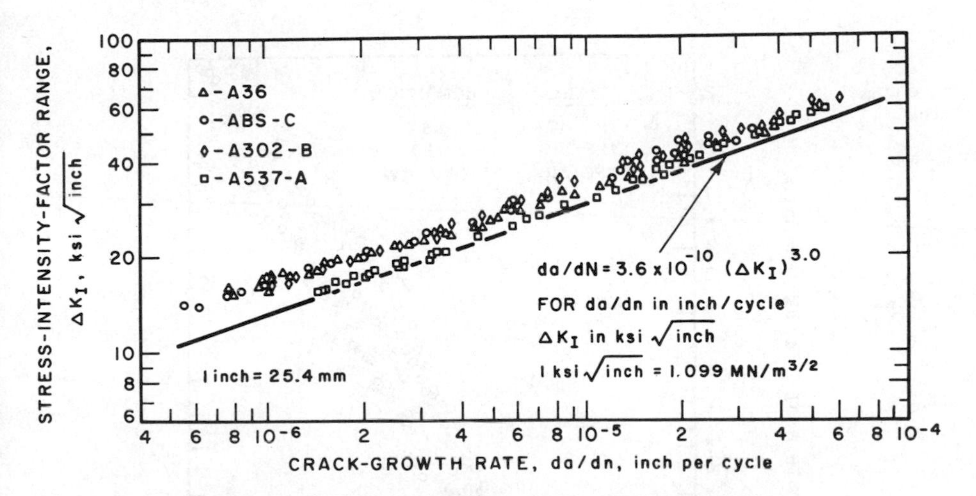

FIG. 8.8. Summary of fatigue-crack-growth data for ferrite-pearlite steels.

The data presented in Figures 8.5 and 8.8 and represented by Equations (8.3) and (8.4) show that the rate of fatigue-crack growth at a given ΔK_I value below the fatigue-rate transition is lower in ferrite-pearlite steels than in martensitic steels. The reason for the differences in the fatigue-crack-growth-rate behavior between martensitic steels and ferrite-pearlite steels is discussed later.

8.4. Austenitic Stainless Steels

Extensive fatigue-crack-growth-rate data for austenitic stainless steels in region II have been obtained. Data for solution-annealed and cold-worked type 316 stainless steel and for solution-annealed type 304 stainless steel were obtained by James[20] (Figure 8.9). Data on these types of steels were also obtained by Shahinian et al.[21] (Figure 8.10). Weber and Hertzberg[22] examined the effect of thermomechanical processing on fatigue-crack propagation in type 305 stainless steel. The steel was evaluated in the coarse-grained, cold-worked, and recrystallized conditions. A summary of their data is presented in Figure 8.11 and shows that crack propagation was essentially similar irrespective of thermomechanical processing. The data presented in Figures 8.9 through 8.11 indicate that conservative and realistic estimates of fatigue-crack-propagation rates for austenitic steels in a room-temperature air environment can be obtained from

$$\frac{da}{dN} = 3.0 \times 10^{-10}(\Delta K_{\mathrm{I}})^{3.25} \tag{8.5}$$

where a = in.,
$K_{\mathrm{I}} = \mathrm{ksi}\sqrt{\mathrm{in.}}$

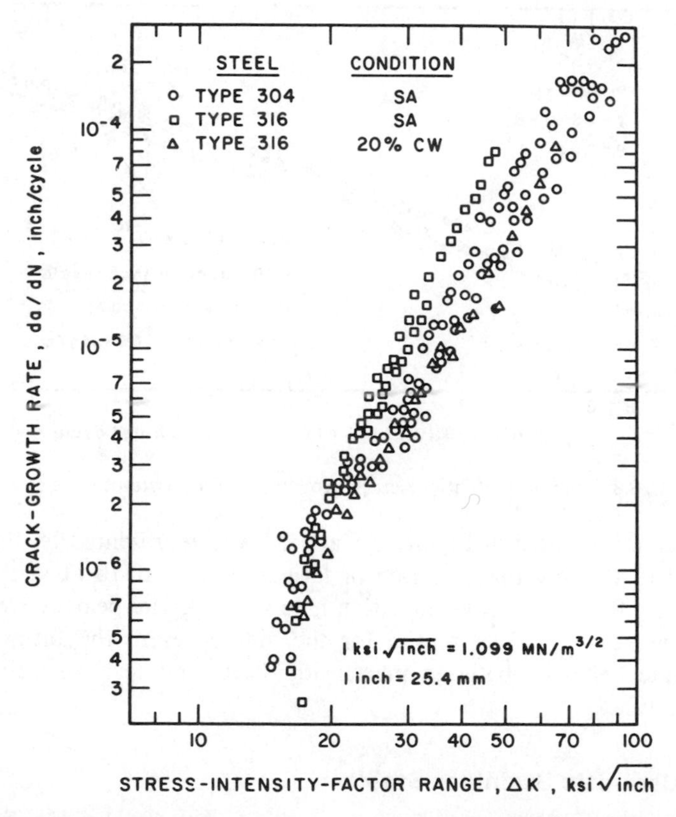

FIG. 8.9. Fatigue-crack-growth rate of solution-annealed type 304 and 316 stainless steels and 20 percent cold-worked type 316 stainless steel at 75°F.

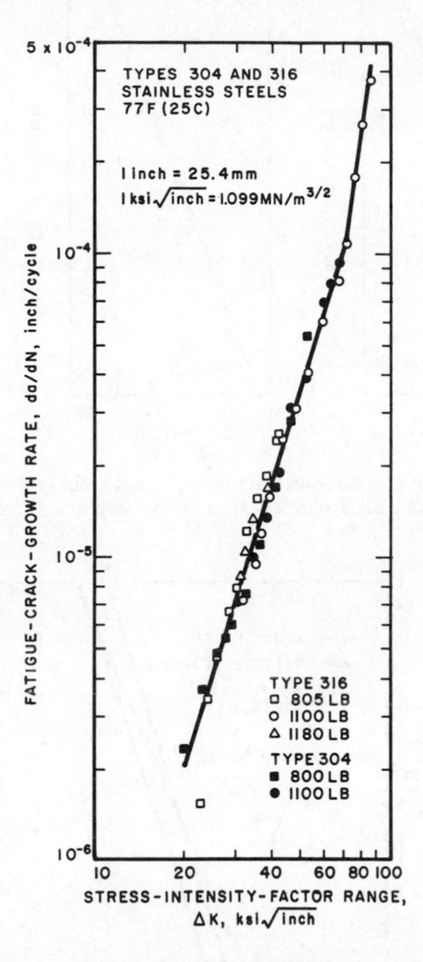

FIG. 8.10 Fatigue-crack-growth rates in type 304 and 316 stainless steels at room temperature as a function of range of stress-intensity factor.

8.5. General Discussion of Steels

Engineering estimates of fatigue-crack-growth rates in martensitic, ferrite-pearlite, and austenitic stainless steels can be obtained by using Equations (8.3), (8.4), and (8.5), respectively. These equations are plotted in Figure 8.12 to show the differences in the fatigue-crack-growth-rate behavior of these steels. The fatigue-crack-growth-rate behavior of a given steel may be changed from a behavior that is depicted by Equation (8.4) to that depicted by Equation (8.3). Such changes can be accomplished through thermomechanical processing.

The preceding discussion of fatigue-crack-growth rates in steels was

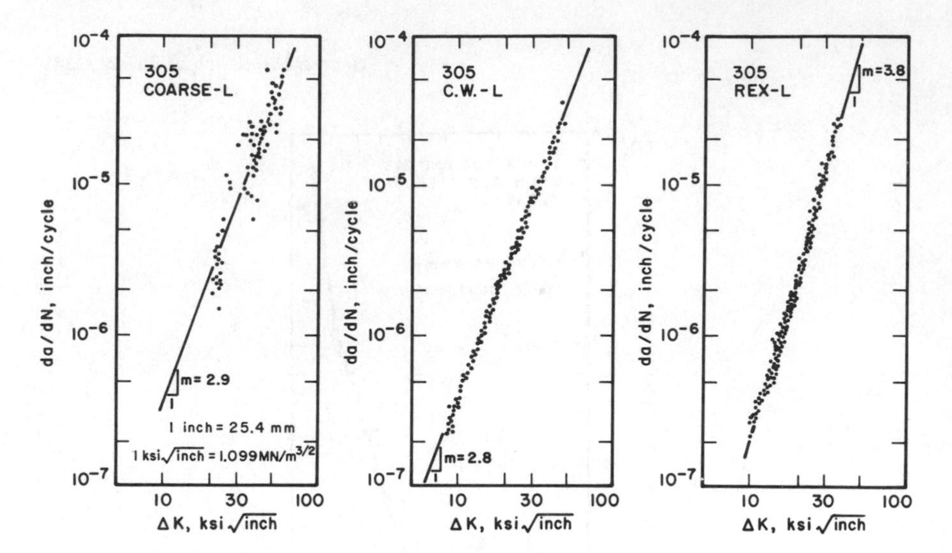

FIG. 8.11. Plot of stress-intensity range versus crack-propagation rate for "L" orientation of 305 stainless steel in coarse-grained, cold-worked, and recrystallized conditions.

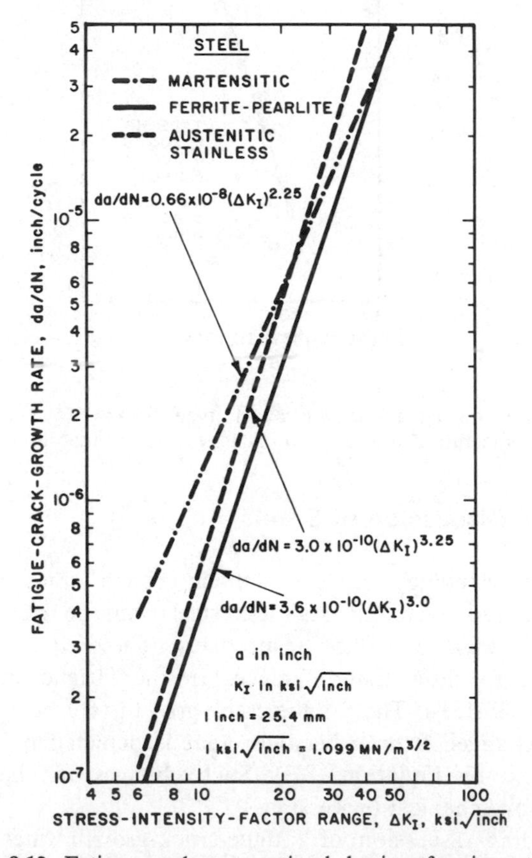

FIG. 8.12. Fatigue-crack-propagation behavior of various steels.

divided into three groups that reflect microstructural differences. These groups also reflect broad variations in mechanical properties. The ranges of the tensile strength, yield strength, and strain-hardening exponent of austenitic stainless steels, ferrite-pearlite steels, and martensitic steels for which Equations (8.3), (8.4), and (8.5) are valid are presented in Table 8.1. Considering the wide variation of the chemical compositions and tensile properties of ferrite-pearlite steels, martensitic steels, and austenitic stainless steels, and the relatively small variation in their fatigue-crack-growth rates (Figure 8.12), the use of a single equation to predict fatigue-crack-growth rates for these steels may be justified. Because of the difficulties encountered in the determination of the magnitude of stresses and stress fluctuations in complex structural details, the use of Equation (8.3) to calculate conservative estimates of the rate of fatigue-crack growth in various steels can be justified. Further justification for the use of Equation (8.3) to calculate conservative estimates of the rate of fatigue-crack growth in various steels was established by Clark.[23] A design example of the use of these equations is presented later in this chapter.

TABLE 8.1 Ranges in Mechanical Properties of Pressure-Vessel Steels

Steel	*Yield Strength,* σ_{ys} *(ksi)*	*Tensile Strength,* σ_T *(ksi)*	*Strain-Hardening Exponent, n*
Austenitic stainless	$30 < \sigma_{ys} < 50$	$75 < \sigma_T < 95$	$n > 0.30$
Ferrite-pearlite	$30 < \sigma_{ys} < 80$	$50 < \sigma_T < 110$	$0.15 < n < 0.30$
Martensitic	$\sigma_{ys} > 70$	$\sigma_T > 90$	$n < 0.15$

Conversion factor
1 ksi = 6.895 MN/m^2

8.6. Aluminum and Titanium Alloys

The preceding discussion shows that the region II fatigue-crack-growth-rate behavior of steels in a benign environment is essentially independent of mechanical and metallurgical properties of the material.

Figures 8.13 and 8.14[12] present the fatigue-crack-growth behavior of aluminum and titanium alloys, respectively. The appropriate crack-growth-rate expressions that define the upper scatter band behavior of these metal systems are also presented in these figures. The data indicate that, like steels, the fatigue-crack-growth rates for aluminum and titanium alloys fall within different but distinct scatter bands. However, the scatter bands for aluminum and titanium alloys are larger than for steels. Thus, it is apparent that for a given metal system subjected to a benign environment the fatigue-crack-

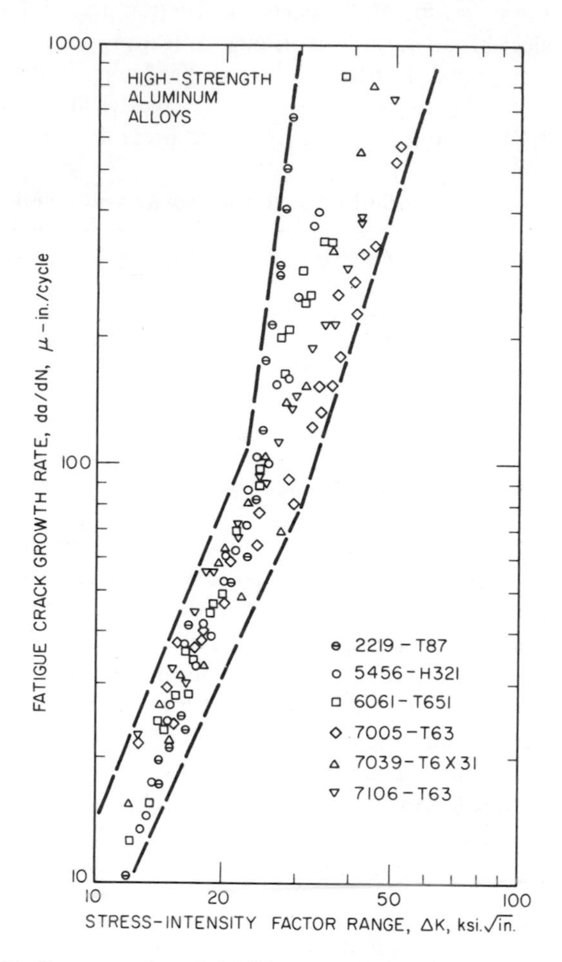

FIG. 8.13. Summary plot of *da*/*dN* versus Δ*K* for six aluminum alloys. The yield strengths of these alloys range from 34 to 55 ksi.

growth behavior in region II is not a pertinent factor in material selection considerations within a given metal system.

The fatigue-rate transition from region II to region III for aluminum and titanium alloys has been investigated by Clark and Wessel[24] and by Crooker.[12,25] The data presented in Figures 8.15[24] and 8.16[25] as well as other data obtained by Crooker[12] show conclusively that Equation (8.2) can be used to determine the stress-intensity-factor value corresponding to the transition from region II crack-growth-rate behavior to region III in aluminum and titanium alloys.

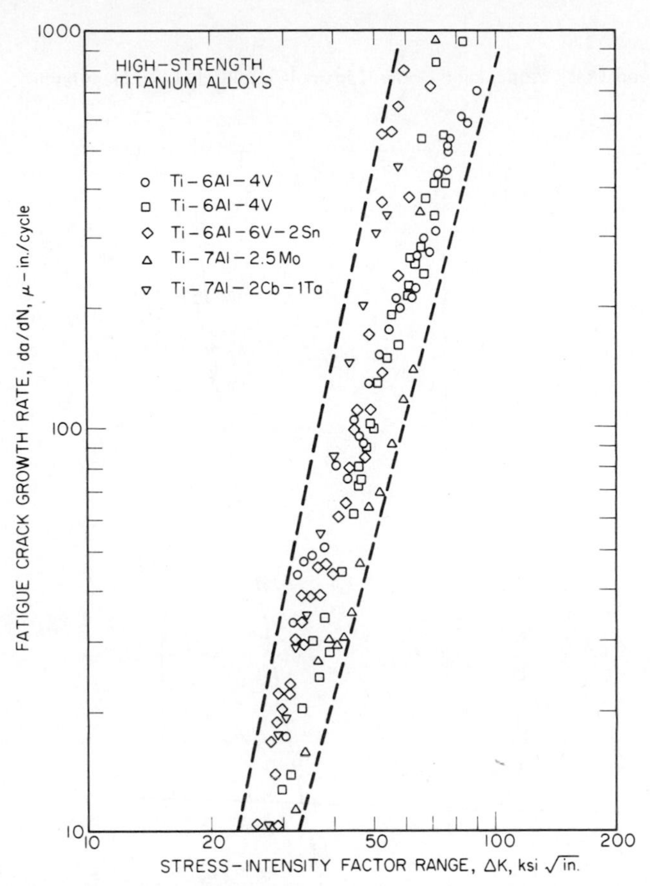

FIG. 8.14. Summary plot of da/dN versus ΔK data for five titanium alloys ranging in yield strength from 110 to 150 ksi.

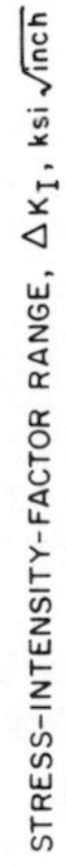

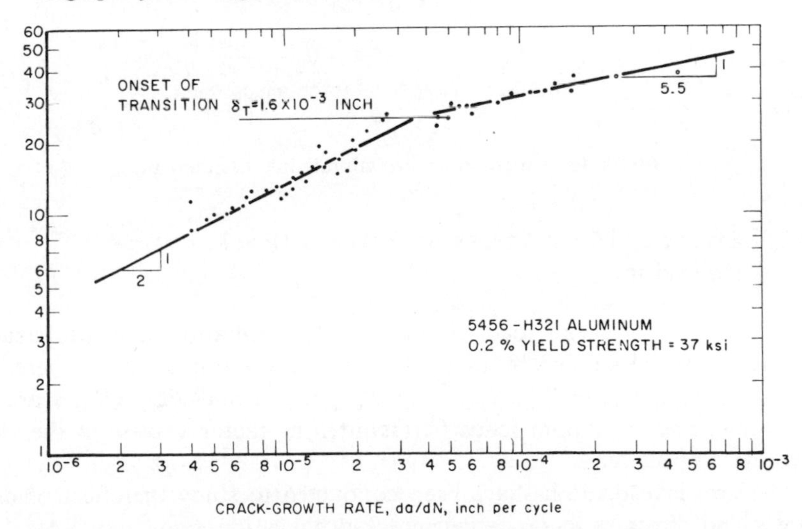

FIG. 8.15. Fatigue-crack-growth rate as a function of stress intensity for 5456-H321 aluminum.

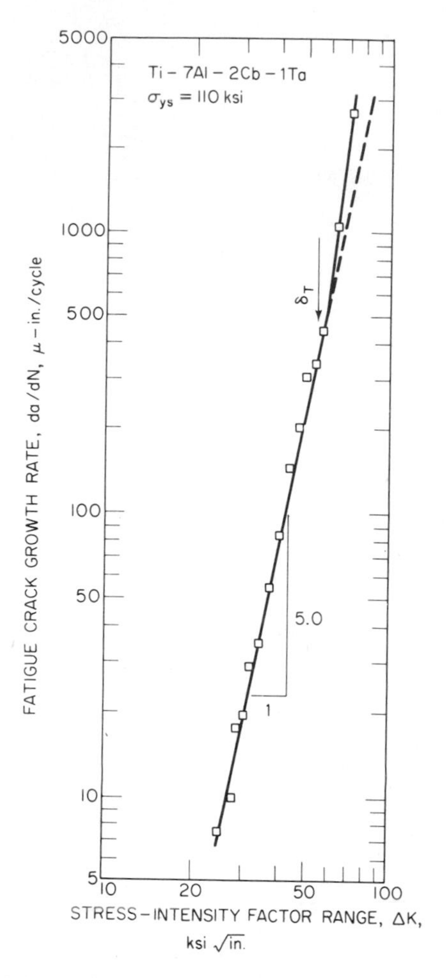

FIG. 8.16. Fatigue-crack-growth rate in a titanium alloy.

8.7. Effect of Mean Stress on Fatigue-Crack Behavior

The effect of mean load on fatigue-crack-initiation and propagation behavior can be studied by using the load ratio parameter, R, where R is equal to $P_{min}/P_{max} = K_{min}/K_{max}$, and P_{min} (K_{min}) and P_{max} (K_{max}) are the minimum and maximum loads (stress-intensity-factor values) in the cyclic process, respectively.

Several investigations have been conducted to study the effect of mean stress and stress ratio on fatigue-crack-propagation rate.[16,26-30] Available experimental data on ASTM A514 Grade B steel show no systematic change in fatigue-crack-growth rate with changes in R value from 0 to 0.82 (Figure 8.6).[16] The data also show that this change in R value has negligible effects on the rate of crack propagation in region II. On the other hand, Crooker[28] observed that the compression portion of fully reversed tension-compression cycling increased the growth rate in HY-80 steel by approximately 50% as compared with zero-tension cycling. Data on fatigue-crack growth in a 140-ksi (965-MN/m^2) yield strength martensitic steel[29] showed a systematic increase in growth rate with increase in R value from 0 to 0.75 and with decrease in R value from 0 to -2, but that maximum nominal stress levels of 0.39–0.94 of the yield strength had no apparent effect (Figure 8.17). The maximum increase in fatigue-crack-growth rate as a function of variation of R from -2 to 0.75 was less than a factor of 2.

Crooker et al.[28,29] studied the effect of stress ratio, R, in the range $-2 \leq R \leq 0.75$ by using part-through-crack specimens subjected to axial loads. Because of the difficulties encountered in measuring crack depth in these

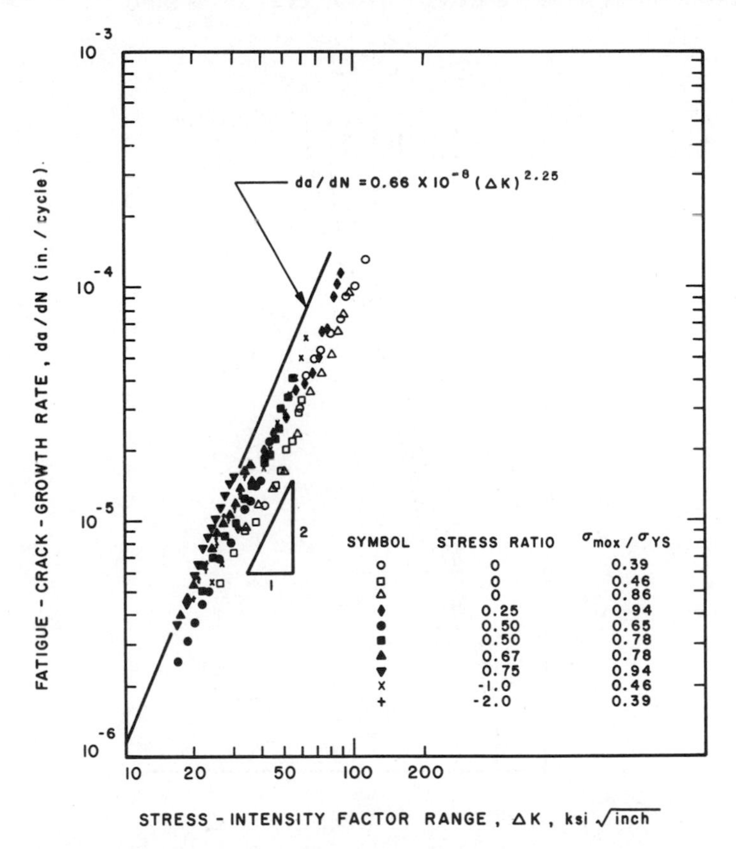

FIG. 8.17. Effect of stress ratio on the fatigue-crack-propagation rate in a 140 ksi yield strength martensitic steel.

specimens, the rate of fatigue-crack growth was calculated by measuring the rate of growth on the surface of the specimen and by assuming that the crack shape did not change (that is, a circular crack front remained circular) as the crack size increased. Despite reservations relating to the accuracy of this assumption and to the use of part-through-crack specimens for fatigue-crack-growth studies, analysis of the data presented in Figure 8.17 showed that the effect of positive stress ratio, $R \geq 0$, on the rate of fatigue-crack growth in the 140-ksi yield strength steel investigated can be predicted by using

$$\frac{da}{dN} = \frac{A(\Delta K)^{2.25}}{(1 - R)^{0.5}} \tag{8.6}$$

where A is a constant.

The upper bound of the scatter band of martensitic steels that is represented by Equation (8.3) is also presented in Figure 8.17. The data suggest that Equation (8.3) can be used to estimate fatigue-crack-growth rates in martensitic steels subjected to various values of stress ratio.

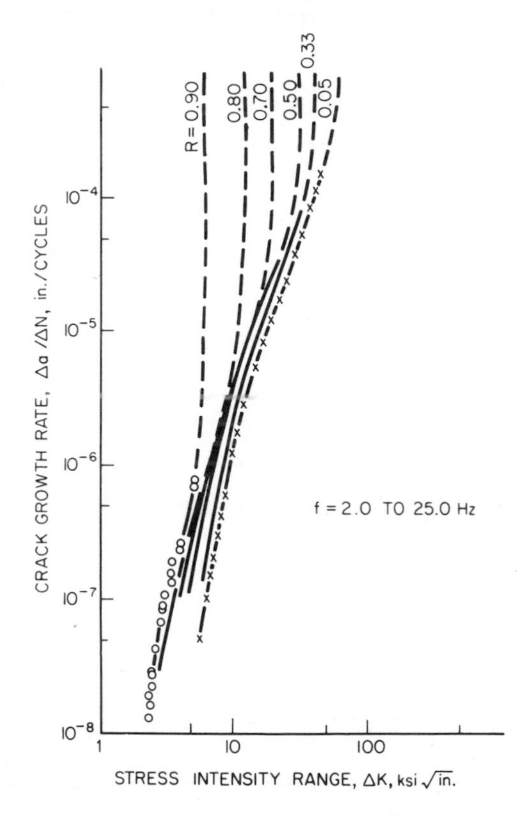

FIG. 8.18. Effect of stress ratio, R, on fatigue-crack-growth upper threshold.

The effect of stress ratio on the fatigue-crack stress-intensity-factor threshold can be predicted by using Equation (7.8). The effect of stress ratio on the fatigue behavior of unnotched specimens of pressure-vessel steels has been investigated by Dubuc et al.[31] under Pressure Vessel Research Committee sponsorship. The tests were conducted at ratios of mean strain to total strain ranging between approximately -1 and $+1$ and in the range of lives between 10^2 and 10^5 cycles. The test results were essentially independent of mean strain. A general discussion of the effect of stress ratio on the fatigue behavior of unnotched specimens of pressure-vessel steels has been presented by Langer.[32]

The preceding discussion indicates that stress ratio has a second-order effect on the rate of crack propagation in region II (Figure 8.4) and that Equations (8.3), (8.4), and (8.5) that represent the upper bound of data scatter can be used to estimate fatigue-crack-growth rates in steels subjected to various values of stress ratio.

The fatigue-crack initiation threshold, ΔK_{th}, has been shown in Chapter 7 to depend on the magnitude of the stress ratio, $R = K_{min}/K_{max}$. Moreover, because ΔK_T defines the region above which fatigue-crack propagation occurs by striation and ductile dimpling and because ductile dimpling depends on K_{max}, then ΔK_T must depend on K_{max} and on the stress ratio, R.

The general effect of stress ratio, R, on fatigue-crack-growth rates in regions I, II, and III are presented schematically in Figure 8.18.[33] Data in support of this generalized behavior has been published by Wei.[33]

8.8. Effects of Cyclic Frequency and Wave Form

The effect of loading rate on fracture toughness was presented in Chapter 4. The available data show that the rate of loading can affect the fracture toughness significantly and that the magnitude of this effect increases with decrease in yield strength. Because a change in cyclic frequency corresponds to a change in the rate of loading, the effect of cyclic frequency on fatigue-crack-growth rate should be established.

The fatigue-crack-growth rate for A36 steel ($\sigma_{ys} = 36$ ksi) under cyclic frequencies of 6–3,000 cycles/min (cpm) are presented in Figure 8.19. Similar data for 12Ni-5Cr-3Mo maraging steel ($\sigma_{ys} = 180$ ksi) were obtained under cyclic frequencies of 6–600 cpm.[2,14,34] The data show that the rates of fatigue-crack growth in a room-temperature benign environment were not affected by the frequency of the cyclic-stress fluctuations.

The discussions in the preceding sections show that the stress-intensity-factor fluctuation, ΔK_I, is the primary parameter that governs the rate of crack growth per cycle of loading. Because ΔK_I is related to the magnitude of the stress fluctuation and to the square root of crack length, this parameter does not account for possible differences in growth rates that may exist

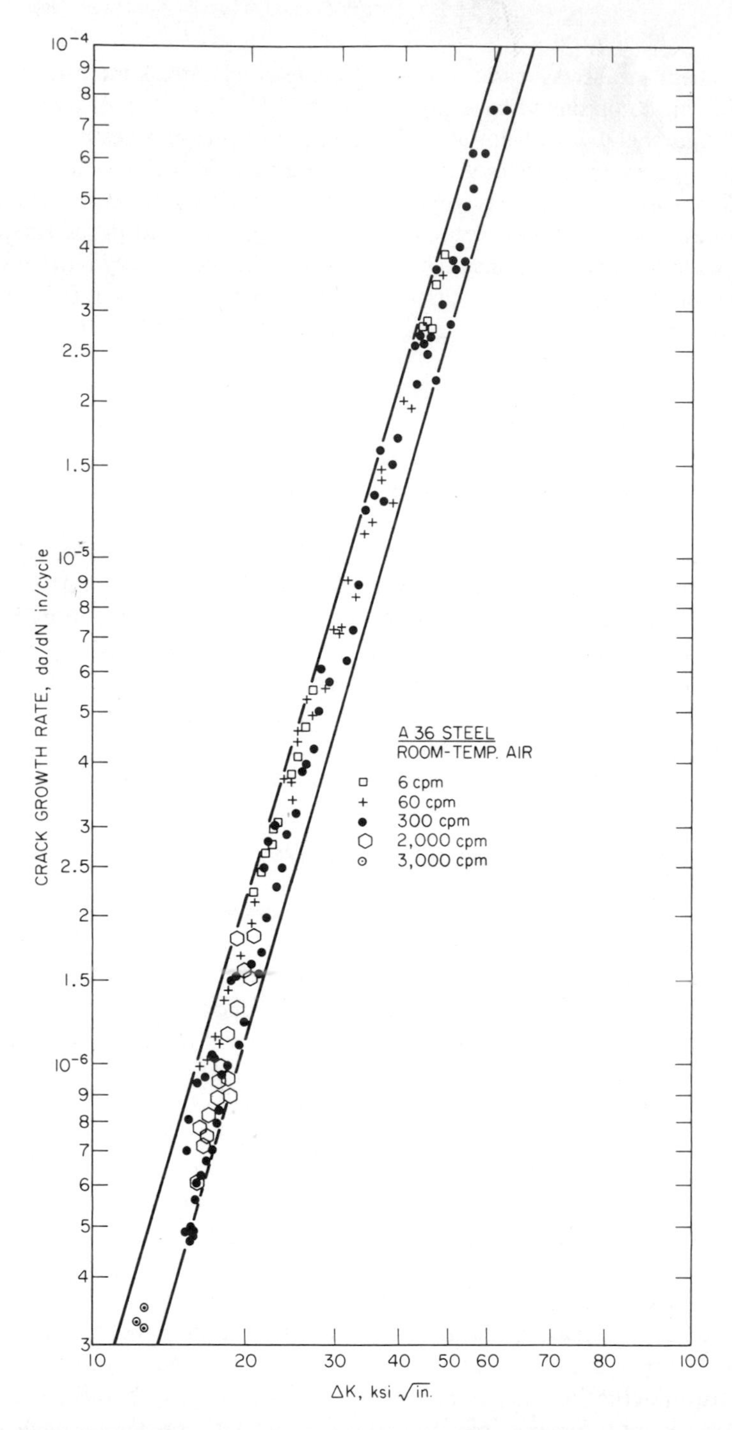

FIG. 8.19. Effect of cyclic frequency on crack growth rates in benign environment.

between various cyclic wave forms such as sine and square waves. Consequently, data on the fatigue-crack-growth rate of 12Ni-5Cr-3Mo maraging steel in a room-temperature air environment were obtained under sinusoidal, triangular, square, positive sawtooth (◿), and negative sawtooth (◺) cyclic-stress fluctuations. The combined data, presented in Figure 8.20,[2] show that the rates of fatigue-crack growth in a room-temperature air environment were not affected by the form of the cyclic-stress fluctuations.

The effects of cyclic frequency and wave form on corrosion-fatigue crack-growth rates are presented in Chapter 11.

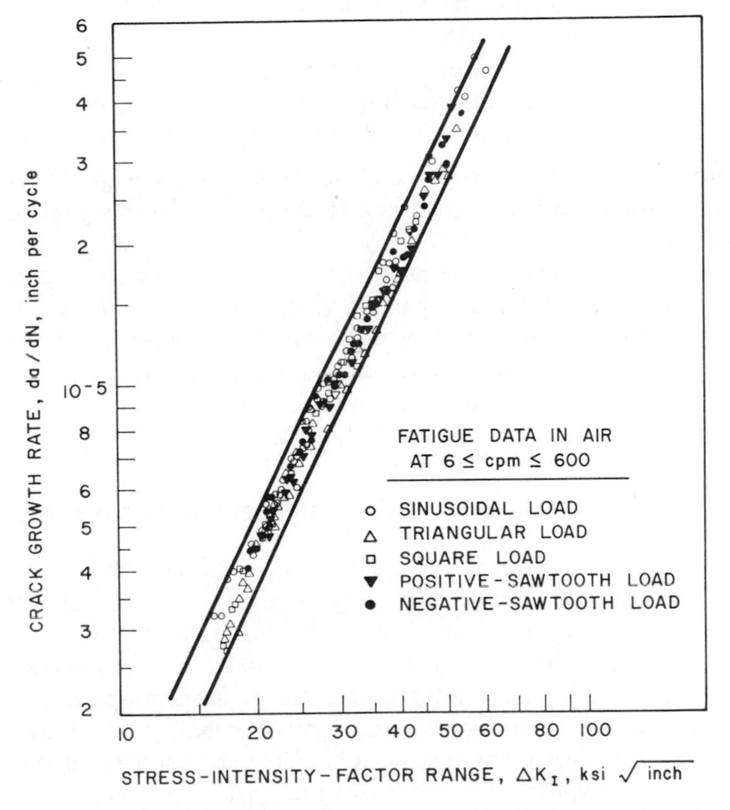

FIG. 8.20. Fatigue-crack-growth rates in 12Ni-5Cr-3Mo steel under various cyclic-stress fluctuations with different stress-time profiles.

8.9. Effects of Stress Concentration on Fatigue-Crack Growth

The effect of the stress-concentration factor on fatigue-crack initiation was discussed in Chapter 7. Unfortunately, the effect of the stress-concentration factor on the rate of crack propagation is yet to be investigated.

Because the fluctuation of the stress-intensity factor is the primary fatigue-crack-propagation force, it can be surmised that the rate of fatigue-crack propagation is governed by the local stress-intensity-factor fluctuation. That is, the fatigue-crack-propagation rate per cycle, da/dN, in the shadow of a notch must be governed by the relationship

$$\frac{da}{dN} = A(\Delta K_{eff})^n \qquad (8.7)$$

where A and n = constants,

$\Delta K_{eff} \propto k_t(a)\,\Delta\sigma\sqrt{a}$,

$\Delta\sigma$ = nominal stress fluctuation,

a = crack length,

$k_t(a)$ = stress-concentration factor; this factor is a function of crack length such that as the crack propagates outside the field of influence of the notch $k_t(a)$ approaches unity.

Equation (8.7) could be used to analyze the fatigue-crack-growth behavior of cracks emanating from holes, nozzle corners, or other regions of stress concentrations.

Novak and Barsom[35] summarized the state-of-the-art of all available theoretical K_I analyses for cracks in the vicinity of stress concentrations. The accuracy of some of these analyses was verified by their experimental results on the brittle-fracture behavior for cracks emanating from notches. These analyses can be used in conjunction with Equation (8.7) to predict the rate of growth of cracks in the shadow of stress concentrations.

8.10. Fatigue-Crack Propagation in Steel Weldments

Fatigue-crack-growth rates in weldments of various steels are available in the literature.[19,36-39] Fatigue-crack-growth rates in the weld metal and the heat-affected zone of submerged arc weldments of ASTM A533 Grade B, Class 1 steel have been obtained by Clark.[36] In general, fatigue cracks initiated in the heat-affected zone grew into the adjacent weld metal. Figures 8.21 and 8.22 present the fatigue-crack-growth behavior in the weld metal and in the heat-affected zone, respectively. The data show that the rate of fatigue-crack growth in the weld metal and in the heat-affected zone was equal to or less than that in the base metal and that the upper bound of the scatter band established by testing the base metal at room temperature may be used as a conservative estimate of fatigue properties of the weld metal and the heat-affected zone. This conclusion is confirmed by data obtained on HY-140 steel weldments,[37] on a 5% nickel steel (ASTM A645) weldment (Figure 8.23),[19] on type 308 weld metal (Figure 8.24),[38] and on type 316 weld metal.[38]

Extensive fatigue-crack-growth-rate data on various weldments has been published by Maddox.[39] Figure 8.25[39] presents data for structural C-Mn

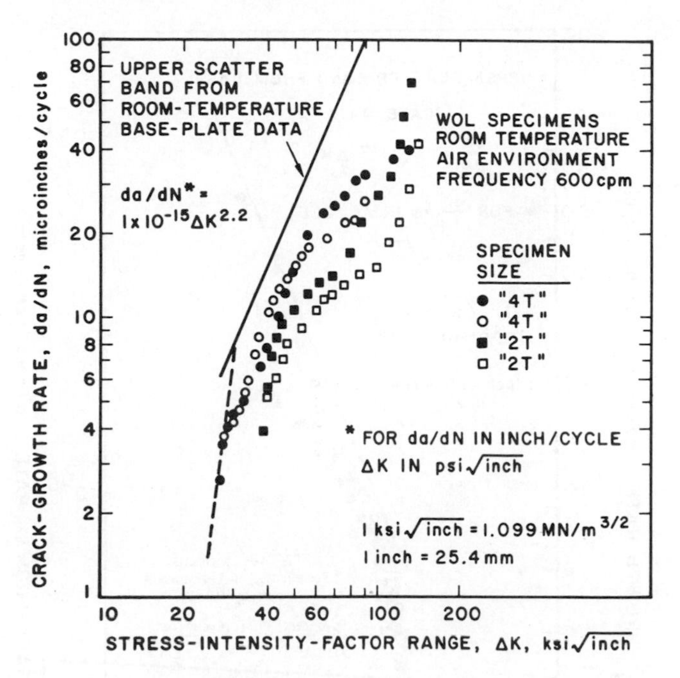

FIG. 8.21. Effect of specimen size on the crack-growth-rate behavior of A533 GR.B, CL. I steel weld metal.

steel weld metals, heat-affected zones, and base metals. A comparison of the data in Figure 8.25 and the equations in Figure 8.12 shows that the upper scatter band equations for base metals can be used to obtain conservative engineering estimates of the fatigue-crack-growth rates in base metals, weld metals, and heat-affected zones. Moreover, the data show that the rates of fatigue-crack growth in weld metals and in heat-affected zones are equal or less than that in the base metal.

As-welded structural components contain residual stresses often of yield stress magnitude.[40] Consequently, fatigue cracks in regions of tensile-residual stresses propagate under high stress ratios. Preceding discussions showed that high stress ratios affect the magnitude of ΔK_{th} and the magnitude of K_T but have negligible effect on the fatigue-crack-growth-rate behavior in region II. This observation is supported further by Maddox (Figure 8.26).[39]

8.11. Effects of Inhomogeneities on Fatigue-Crack Growth

As pointed out in a previous section, the fatigue-crack-growth rates prior to the fatigue-rate transition are slower in ferrite-pearlite steels than in martensitic steels. Moreover, unlike the fatigue-rate transitions in martensitic steels,

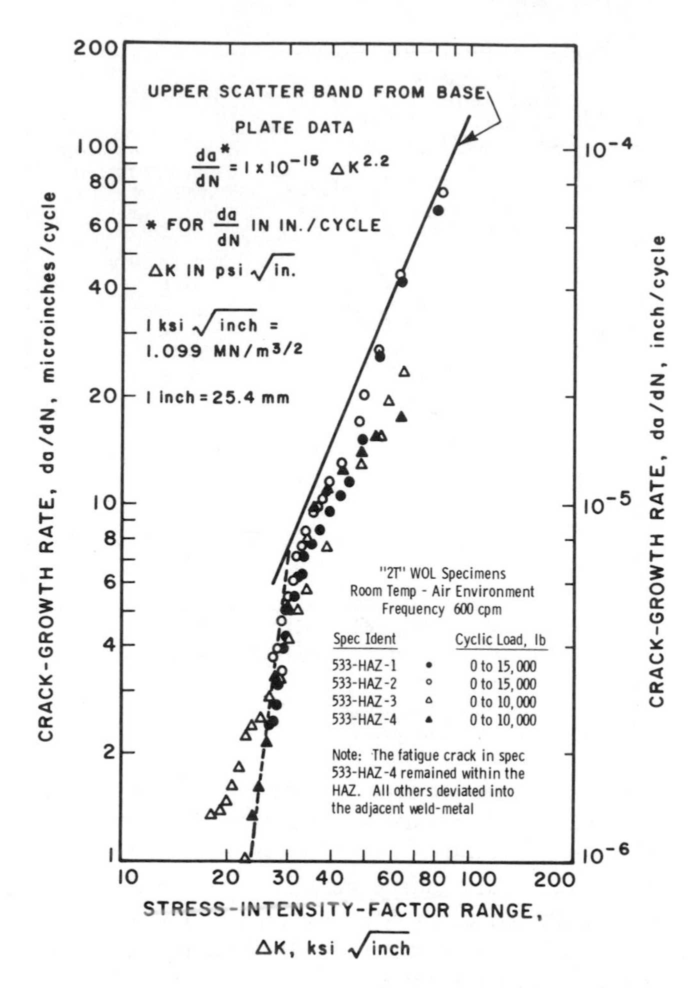

FIG. 8.22. Fatigue-crack growth-rate behavior observed in the vicinity of the heat-affected zone of an ASTM A533 Gr. B, CL. I weldment.

the fatigue-rate transitions in ferrite-pearlite steels occur at slightly higher ΔK_{I} values than predicted by Equation (8.2). The use of a flow stress, σ_f, rather than the yield strength, σ_{ys}, in Equation (8.2) is suggested to rectify this difference. Both the decrease in fatigue-crack-growth rates and the increase in the ΔK_{I} value at the onset of the fatigue-rate transition appear to be related to the composite character of the microstructure of ferrite-pearlite steel or to crack branching or to both.

8.11.1 Composite Behavior

A composite material is a material system made up of a mixture of two or more distinct constituents. Most composite constituents can be classified

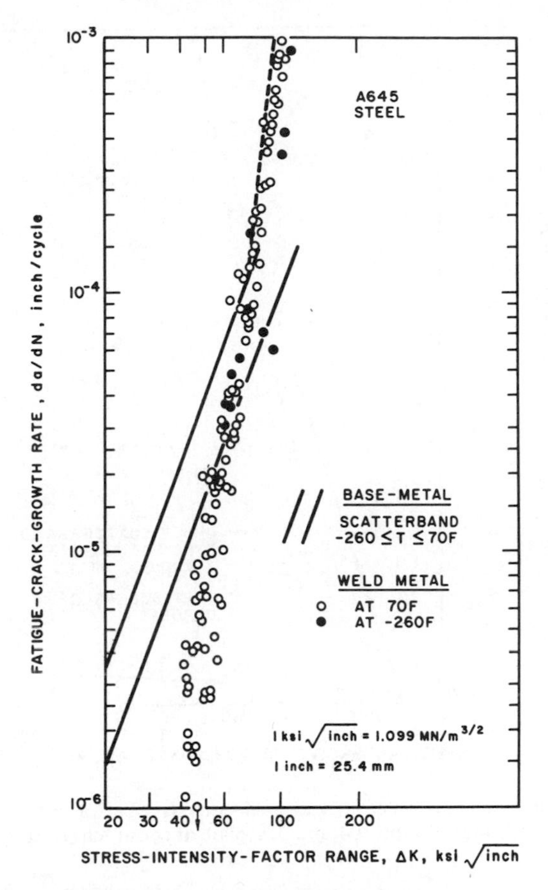

FIG. 8.23. Fatigue-crack-growth-rate data for A645 steel base metal and weld metal in room air at 70°F and −260°F.

as matrix formers or matrix modifiers. The matrix formers are those constituents which give the composite its bulk form and which hold the matrix modifiers in place. The matrix modifiers determine the character of the internal structure of the composite. In this context, ferrite-pearlite steels can be visualized as particulate composites in which the ferrite is the matrix former and the pearlite is the matrix modifier that is dispersed in the matrix as particles (that is, colonies or patches of irregular shapes).

Fatigue-crack-growth rates in particulate-composite materials might depend on the strength of the particles relative to the strength of the matrix, on the strength of the interface bond between the particles and the matrix, on the size and orientation of the particles, and on the distribution density of these particles. The last determines the mean-free path of the fatigue-crack front during its propagation. The larger the distribution density of the

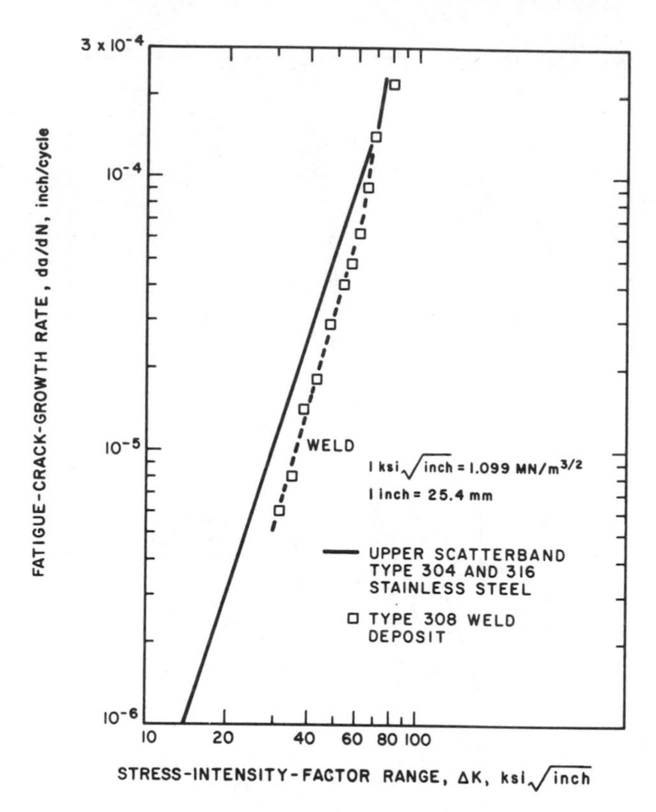

FIG. 8.24. Comparison of fatigue-crack-growth behavior of type 308 weld deposit with that of type 304 and 316 plate at room temperature.

particles, the smaller is the mean-free path, and therefore the larger is the effect of the particles on crack growth. The strengths of the particles and of the interface bond relative to the strength of the matrix determine whether the fatigue-crack-growth rates in the composite will be higher or lower than the rates in the absence of these matrix modifiers.

In ferrite-pearlite-steel composites the interface bond between the constituents is quite strong. Furthermore, because of the iron carbides in the pearlite, the strength of these matrix modifiers is higher than the strength of the matrix, and the ductility of the matrix is greater than the ductility of the matrix modifier. These properties of the constituents and of the interface tend to improve the strength and toughness of the matrix because the ductile ferrite can distribute the stresses to the stronger pearlite colonies.

The effect of pearlite colonies on the rate of fatigue-crack growth in ferrite-pearlite steels was investigated by using scanning electron micrographs of the fracture profile.[9] The micrographs showed that the plastic deformation under cyclic loading is much more extensive in the ferrite matrix than in

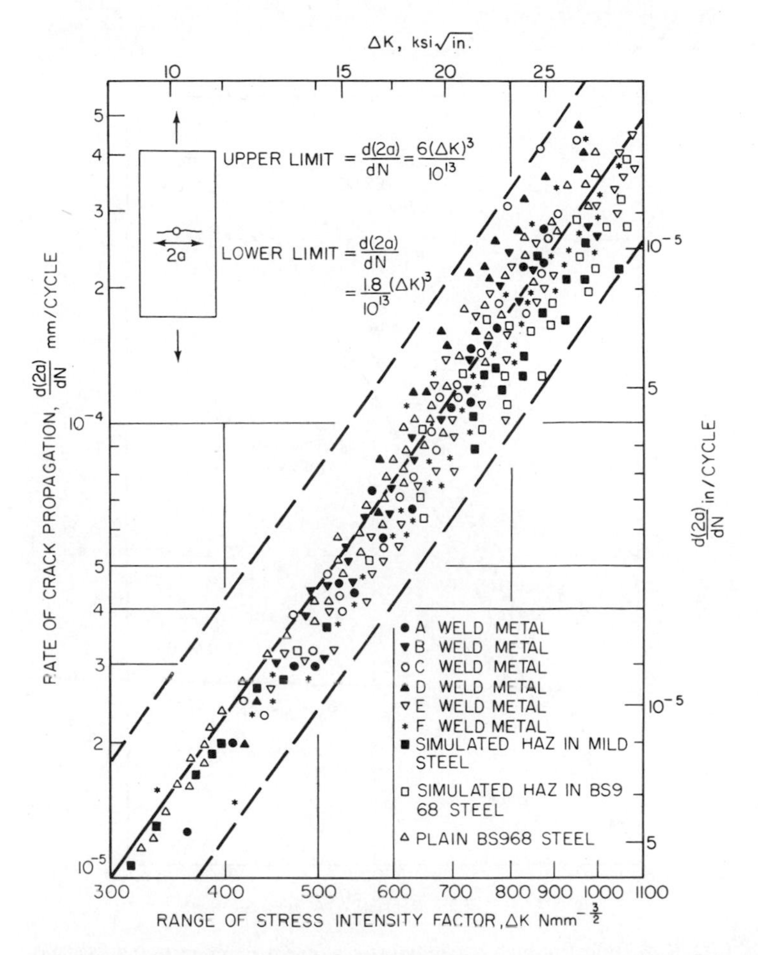

FIG. 8.25. Plane-strain fatigue-crack-propagation data for C-Mn steel weld metals, heat-affected zones, and base metals.

the pearlite colonies. The micrographs also showed secondary fatigue cracks preferentially seeking to propagate around a pearlite colony rather than through it. These observations suggest that the path of least resistance to fatigue-crack growth is through the ferrite matrix and that the pearlite colonies tend to retard crack growth.

Because the properties of the constituents in a composite may affect the rate of fatigue-crack growth, acceleration in the fatigue-crack-growth rates caused by undesirable properties of the matrix modifiers or of the interface or of both would be expected in some composite materials. This behavior has been observed[41] in two 7178 aluminum alloys that differed markedly only in impurity contents.

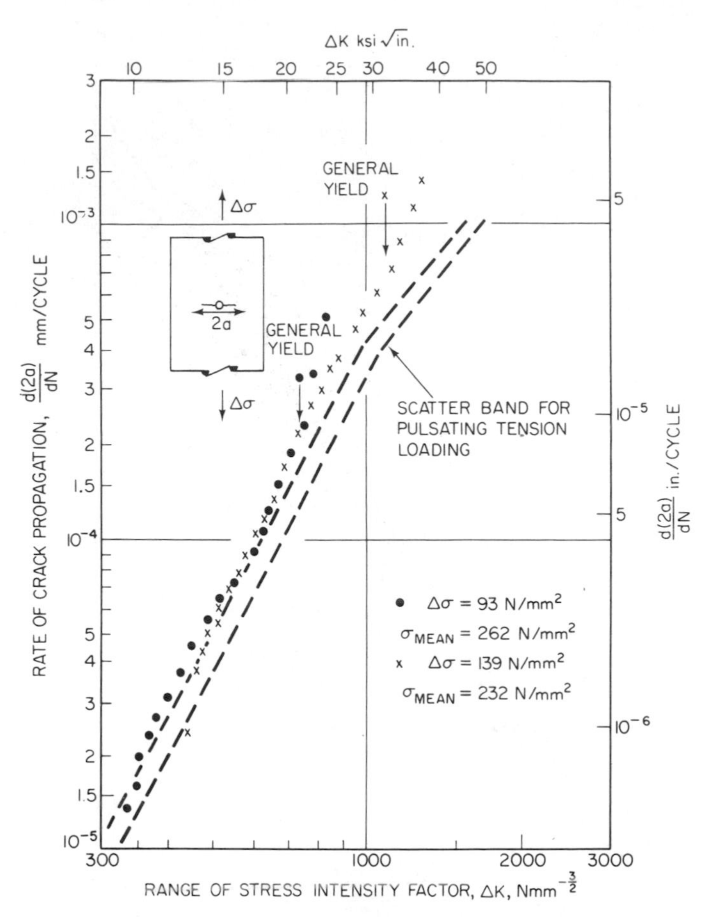

FIG. 8.26. Effect of high maximum stress (corresponding to 0.8 × yield) on fatigue-crack propagation in C-Mn steel to BS 4360 Grade 50B.

The preferential propagation of fatigue cracks around a pearlite colony rather than through it causes crack branching. Crack branching was observed in ferrite-pearlite steels but not in martensitic steels.[9] Crack branching in ferrite-pearlite steels is also caused by cracks initiating ahead of the fatigue-crack tip in the large plastically deformed zone whose large size is determined by the low yield strength of the ferrite matrix. These secondary cracks tend to share the crack-opening displacement at the crack tip with the main crack and thereby cause a reduction in the stress-intensity factor. Consequently, the actual fatigue-crack-growth rate is less for a branched crack than the rate estimated for a single crack front. This observation appears to explain the differences in fatigue-crack-growth rates between martensitic and ferrite-pearlite steels [Equations (8.3) and (8.4)]. Similarly, at the same crack-opening displacement, the strain at the tip of a branched crack is less than that at the tip of a single crack. Therefore, the value of the crack-opening displacement at the onset of fatigue-rate transition [Equation (8.2)] for a branched crack would be expected to be slightly higher than 1.6×10^{-3} in., which was shown to be applicable to martensitic steels where severe crack branching was not observed. Examination of the data[9,13] indicated that onset of fatigue-rate transition in martensitic steels and in ferrite-pearlite steels can be predicted more accurately by using the flow stress, σ_f, in Equation (8.2) rather than the yield strength, σ_{ys}.

8.12. Significance of Fatigue-Rate Transition

The fatigue-rate transition from region II behavior to region III behavior can be predicted by using Equation (8.2). The stress-intensity-factor value, K_T, that defines this region of transition under zero-to-tension loads can be calculated by rewriting Equation (8.2) in the form

$$K_T = 0.04\sqrt{E\sigma_{ys}} \tag{8.8}$$

where K_T is in ksi$\sqrt{\text{in.}}$ and E and σ_{ys} are in ksi.

Acceleration in the rate of fatigue-crack growth occurs at a stress-intensity-factor value slightly below the critical stress-intensity factor, K_{Ic} (or K_c), when the K_{Ic} of the material is less than K_T (steel B in Figure 8.4). Equation (8.8) is used to calculate the stress-intensity-factor value corresponding to onset of fatigue-rate transition, K_T, which also corresponds to the point of transition from region II to region III in materials that have high fracture toughness (steel A in Figure 8.4)—that is, materials for which the critical-stress-intensity factor, K_{Ic} (or K_c), is higher than the K_T value calculated by using Equation (8.8). This equation shows that acceleration in fatigue-crack-growth rate in materials having high fracture toughness, K_{Ic} (or K_c) $> K_T$, is governed by Young's modulus, E, and by the yield strength, σ_{ys} (or the flow stress, σ_f) of the material and is independent of the fracture toughness. Consequently, a significant increase in the fracture toughness of a metal above K_T may have a negligible effect on the fatigue life of a structural component. Moreover, extrapolation of the behavior in region II to K_{Ic} (or K_c) values that are higher than K_T could seriously overestimate the useful fatigue life of a structural component.[42] Accurate predictions of finite fatigue lives of structural components subjected to high stress fluctuations require an exact characterization of the fatigue-crack-growth behavior above and below the fatigue-rate transition region and cannot be based on extrapolations.

8.13. Design Example

For most structural materials, the tolerable flaw sizes are much larger than any initial undetected flaws. However, for structures subjected to fatigue loading (or stress-corrosion cracking), these initial cracks can grow throughout the life of the structure. Thus, an overall approach to preventing fracture or fatigue failures in large welded structures assumes that a small flaw of certain geometry exists after fabrication and that this flaw can either cause brittle fracture or grow by fatigue to the critical size. To ensure that the structure does not fail by fracture, the calculated critical crack size, a_{cr}, at design load must be sufficiently large, and the number of cycles of loading required to grow a small crack to a critical crack must be greater than the design life of the structure.

Thus, although *S-N* curves have been widely used to analyze the fatigue behavior of steels and weldments, closer inspection of the overall fatigue process in complex welded structures indicates that a more rational analysis of fatigue behavior is possible by using concepts of fracture mechanics. Specifically, small (possibly large) fabrication flaws are invariably present in welded structures, even though the structure has been inspected. Accordingly, a conservative approach to designing to prevent fatigue failure would be to *assume the presence of an initial flaw* and analyze the fatigue-crack-growth behavior of the structural member. The size of the initial flaw is obviously dependent on the detail geometry, quality of fabrication and inspection.

A schematic diagram showing the general relation between fatigue-crack initiation and propagation is shown in Figure 8.27. The question of when does a crack "initiate" to become a "propagating" crack is somewhat

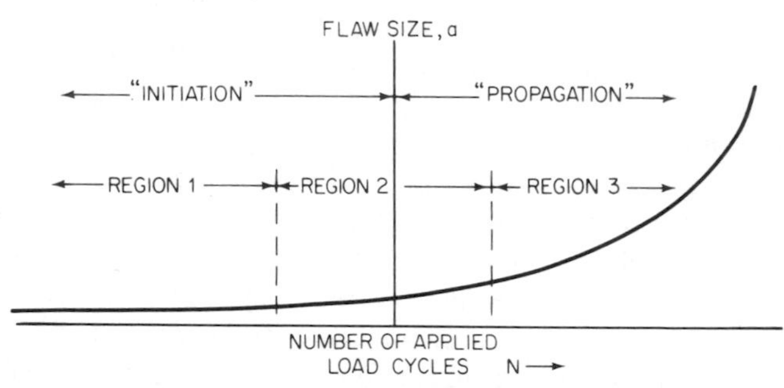

FIG. 8.27. Schematic showing relation between "initiation" life and "propagation" life.

philosophical and depends on the level of observation of a crack, i.e., crystal imperfection, dislocation, microcrack, lack of penetration, etc. One approach to fatigue would be to assume an initial flaw size on the basis of the quality of inspection used and then to calculate the number of cycles it would take for this crack to grow to a size critical for brittle fracture.

The procedure to analyze the crack-growth behavior in steels and weld metals using fracture-mechanics concepts is as follows:

1. On the basis of quality of inspection estimate the maximum initial flaw size, a_0, present in the structure and the associated K_I relation (Chapter 2) for the member being analyzed.
2. Knowing K_c or K_{Ic} and the nominal maximum design stress, calculate the critical flaw size, a_{cr}, that would cause failure by brittle fracture.
3. Obtain an expression relating the fatigue-crack-growth rate of the steel or weld metal being analyzed. The following conservative estimates of the fatigue-crack growth per cycle of loading, da/dN, have been determined for martensitic steels (for example, A514/517) as well as ferrite-pearlite steels (for example, A36) in a room-temperature air environment and were discussed previously:

$$\textit{martensitic steels:} \quad \frac{da}{dN} = 0.66 \times 10^{-8}(\Delta K_I)^{2.25} \tag{8.3}$$

$$\textit{ferrite-pearlite steels:} \quad \frac{da}{dN} = 3.6 \times 10^{-10}(\Delta K_I)^3 \tag{8.4}$$

where da/dN = fatigue-crack growth per cycle of loading, in./cycle,
ΔK_I = stress-intensity-factor range, ksi$\sqrt{\text{in.}}$ (MN/m$^{3/2}$).

4. Determine ΔK_I using the appropriate expression for K_I, the estimated initial flaw size, a_0, and the range of live-load stress (cyclic stress range).
5. Integrate the crack-growth-rate expression between the limits of a_0 (at the initial K_I) and a_{cr} (at K_{Ic}) to obtain the life of the structure prior to failure.

A numerical example of this procedure is as follows:

1. Assume the following conditions:
   a. A514 steel, $\sigma_{ys} = 100$ ksi (689 MN/m^2).
   b. $K_{Ic} = 150$ ksi$\sqrt{\text{in.}}$ (165 MN/mn$^{3/2}$).
   c. $a_0 = 0.3$ in. (7.6 mm), edge crack in tension (Chapter 2).
   d. $\sigma_{max} = 45$ ksi (310 MN/m^2);
      $\sigma_{min} = 25$ ksi (172 MN/m^2);
      $\Delta\sigma = 20$ ksi (138 MN/m^2) (live-load stress range).
   e. $K_I = 1.12\sqrt{\pi}\,\sigma\sqrt{a}$ edge crack in tension (Chapter 2).

2. Calculate a_{cr} at $\sigma = 45$ ksi (310 MN/m²):

$$a_{cr} = \left(\frac{K_{Ic}}{1.12\sqrt{\pi}\,\sigma_{max}}\right)^2 = \left(\frac{150}{1.12(1.77)(45)}\right)^2$$
$$= 2.8 \text{ in.} \quad (71.1 \text{ mm})$$

3. Assume an increment of crack growth, Δa. In this case assume that $\Delta a = 0.1$ in. (2.5 mm). If smaller increments of crack growth were assumed, the accuracy would be increased slightly.
4. Determine expression for ΔK_I, where a_{avg} represents the average crack size between the two crack increments a_i and a_j:

$$\Delta K_I = 1.12\sqrt{\pi}\,\Delta\sigma\sqrt{a_{avg}}$$
$$= 1.98(20)\sqrt{a_{avg}}$$

5. Using the appropriate expression for crack-growth rate,

$$da/dN = 0.66 \times 10^{-8}(\Delta K_I)^{2.25}$$

solve for ΔN for each increment of crack growth, replacing da/dN by $\Delta a/\Delta N$:

$$\Delta N = \frac{\Delta a}{0.66 \times 10^{-8}[1.98(20)\sqrt{a_{avg}}]^{2.25}}$$
$$= 12{,}500 \text{ cycles}$$

6. Repeat for $a = 0.4$–0.5 in. (10.2 to 12.7 mm), etc., by numerical integration, as shown in Table 8.2. The flaw size–life results for this example are presented in Figure 8.28. If only the desired total life is required, the expression for ΔN can be integrated directly. In this example, direct integration yielded a life of 87,600 cycles, while the numerical technique gave a life of 86,700 cycles.

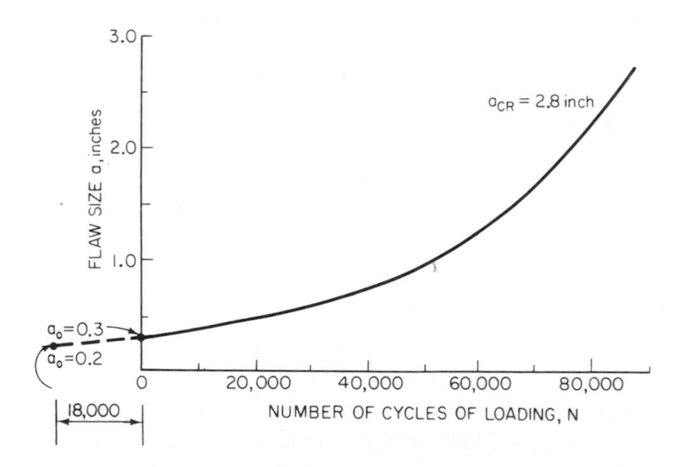

FIG. 8.28. Fatigue-crack-growth curve.

TABLE 8.2 Fatigue-Crack-Growth Calculations

$$\Delta N = \frac{\Delta a}{0.66 \times 10^{-8}[1.98(\Delta\sigma)\sqrt{a_{avg}}]^{2.25}}$$

where $\Delta a = 0.10$ in. (2.54 mm)
$\Delta\sigma = 20$ ksi (138 MN/m²)

a_0 (*in.*)	a_f (*in.*)	a_{avg} (*in.*)	ΔK (*ksi*$\sqrt{in.}$)	ΔN (*cycles*)	$\sum N$ (*cycles*)
0.3	0.4	0.35	23.5	12,500	12,500
0.4	0.5	0.45	26.7	9,750	22,250
0.5	0.6	0.55	29.4	7,550	29,800
0.6	0.7	0.65	32.2	6,150	35,950
0.7	0.8	0.75	34.6	5,200	41,150
0.8	0.9	0.85	36.6	4,600	45,750
0.9	1.0	0.95	38.8	4,100	49,850
1.0	1.1	1.05	40.5	3,700	53,550
1.1	1.2	1.15	42.5	3,300	56,850
1.2	1.3	1.25	44.5	2,950	59,800
1.3	1.4	1.35	46.1	2,700	62,500
1.4	1.5	1.45	47.7	2,550	65,050
1.5	1.6	1.55	49.3	2,350	67,400
1.6	1.7	1.65	51.0	2,200	69,600
1.7	1.8	1.75	52.5	2,050	71,650
1.8	1.9	1.85	54.0	1,900	73,550
1.9	2.0	1.95	55.6	1,800	75,350
2.0	2.1	2.05	56.8	1,700	77,050
2.1	2.2	2.15	58.5	1,600	78,650
2.2	2.3	2.25	59.6	1,500	80,150
2.3	2.4	2.35	60.8	1,450	81,600
2.4	2.5	2.45	62.5	1,400	83,000
2.5	2.6	2.55	63.5	1,350	84,350
2.6	2.7	2.65	64.8	1,200	85,550
2.7	2.8	2.75	66.0	1,150	86,700

1 in. = 25.4 mm
1 ksi$\sqrt{\text{in.}}$ = 1.1 MN/m$^{3/2}$

Note that the total life to propagate a crack from 0.3 to 2.8 in. (7.6 to 71.1 mm) in this example is 86,700 cycles. If the required life is 100,000 cycles, then this design would be inadequate and one or more of the following changes should be made:

1. Increase the critical crack size at failure [$a_{cr} = 2.8$ in. (71.1 mm)] by using a material with a higher K_{Ic} value.
2. Lower the design stress, σ_{max}, to increase the critical crack size at failure.
3. Lower the stress range ($\Delta\sigma$) to decrease the *rate* of crack growth, thereby increasing the number of cycles required for the crack to grow

to the critical size. Note that because the rate of crack growth is a power function of $\Delta\sigma$, or actually ΔK, lowering the stress range slightly has a significant effect on the life.

4. Improve the fabrication quality and inspection capability so that the initial flaw size (a_0) is reduced. It is clear from Table 8.2 and Figure 8.28 that most of the life is taken up in the early stages of crack propagation. In fact, to double the initial crack size during the early stages of propagation requires almost half the total number of cycles. Therefore, any decrease in initial flaw size has a very significant effect on the fatigue life of a structural member.

In this example, if a_0 were only 0.2 in. (5.1 mm), the design would be satisfactory. That is, the number of cycles to grow a crack 0.2–0.3 in. (5.1–7.6 mm) is about 18,000 cycles, as indicated in Figure 8.28, which [added to the 86,700 cycles required to grow the crack from 0.3 to 2.8 in. (7.6 to 71.1 mm)] would make the total life equal to 104,700 cycles. It should be noted that for steels with high toughness levels the state-of-stress ahead of large cracks may be plane stress, and thus larger cracks could be tolerated than are calculated on the basis of plane-strain behavior. However, because the crack-growth rate is increasing rapidly for large cracks as illustrated in Figure 8.28, the life may not be increased significantly. Moreover, because the crack-growth rate is increasing rapidly for large cracks, a significant increase in the fracture toughness of the material (that is, in the size of the tolerable crack at failure) may result in a negligible increase in the fatigue life of the structural member.

REFERENCES

1. P. C. Paris and F. Erdogan, "A Critical Analysis of Crack Propagation Laws," *Transactions of the ASME, Journal of Basic Engineering*, Series D, *85*, No. 3, 1963.
2. J. M. Barsom, "Effect of Cyclic-Stress Form on Corrosion-Fatigue Crack Propagation Below K_{Iscc} in a High-Yield-Strength Steel," in *Corrosion Fatigue: Chemistry, Mechanics, and Micro-Structure*, International Corrosion Conference Series, Vol. NACE-2, National Association of Corrosion Engineers, Houston,1972.
3. R. J. Bucci, W. G. Clark, Jr., and P. C. Paris, "Fatigue-Crack-Propagation Growth Rates Under a Wide Variation of ΔK for an ASTM A517 Grade F (T-1) Steel," *ASTM STP 513*, American Society for Testing and Materials, Philadelphia,1972.
4. R. A. Schmidt, "A Threshold in Metal Fatigue," Master of Science Thesis, Lehigh University, Bethlehem, Pa., 1970.
5. P. C. Paris, "Testing for Very Slow Growth of Fatigue Cracks," *MTS Closed Loop Magazine*, *2*, No. 5, 1970.

6. J. D. Harrison, "An Analysis of Data on Non-Propagating Fatigue Cracks on a Fracture Mechanics Basis," *British Welding Journal*, *2*, No. 3, March 1970.
7. L. P. Pook, "Fatigue Crack Growth Data for Various Materials Deduced From the Fatigue Lives of Precracked Plates," *ASTM STP 513*, American Society for Testing and Materials, Philadelphia,1972.
8. P. C. Paris, R. J. Bucci, E. T. Wessel, W. G. Clark, and T. R. Mager, "Extensive Study of Low Fatigue Crack Growth Rates in A533 and A508 Steels," *ASTM STP 513*, American Society for Testing and Materials, Philadelphia,1972.
9. J. M. Barsom, "Fatigue-Crack Propagation in Steels of Various Yield Strengths," *Transactions of the ASME, Journal of Engineering for Industry*, Series B, *93*, No. 4, Nov. 1971.
10. J. M. Barsom, E. J. Imhof, Jr., and S. T. Rolfe, "Fatigue-Crack Propagation in High-Yield-Strength Steels," *Engineering Fracture Mechanics*, *2*, No. 4, June 1971.
11. J. M. Barsom, "The Dependence of Fatigue Crack Propagation on Strain Energy Release Rate and Crack Opening Displacement," *ASTM STP 486*, American Society for Testing and Materials, Philadelphia,1971.
12. T. W. Crooker, "Crack Propagation in Aluminum Alloys Under High-Amplitude Cyclic Load," *Naval Research Laboratory Report 7286*, Washington, D.C.,July 12, 1971.
13. J. M. Barsom, "Fatigue Behavior of Pressure-Vessel Steels," *WRC Bulletin 194*, Welding Research Council, New York,May 1974.
14. J. M. Barsom, "Investigation of Subcritical Crack Propagation," Doctor of Philosophy Dissertation, University of Pittsburgh, Pittsburgh, 1969.
15. E. J. Imhof and J. M. Barsom, "Fatigue and Corrosion-Fatigue Crack Growth of 4340 Steel at Various Yield Strengths," *ASTM STP 536*, American Society for Testing and Materials, Philadelphia,1973.
16. J. M. Barsom, "Fatigue-Crack Growth Under Variable-Amplitude Loading in ASTM A514 Grade B Steel," *ASTM STP 536*, American Society for Testing and Materials, Philadelphia,1973.
17. M. Parry, H. Nordberg, and R. W. Hertzberg, "Fatigue Crack Propagation in A514 Base Plate and Welded Joints," *Welding Journal*, *51*, No. 10, Oct. 1972.
18. W. G. Clark, Jr., "Effect of Temperature and Section Size on Fatigue Crack Growth in A533 Grade B, Class 1 Pressure Vessel Steel," *Journal of Materials*, *6*, No. 1, March 1971.
19. R. J. Bucci, B. N. Greene, and P. C. Paris, "Fatigue Crack Propagation and Fracture Toughness of 5Ni and 9Ni Steel at Cryogenic Temperatures," *ASTM STP* 536, American Society for Testing and Materials, Philadelphia,1973.
20. L. A. James, "The Effect of Elevated Temperature upon the Fatigue-Crack Propagation Behavior of Two Austenitic Stainless Steels," in *Mechanical Behavior of Materials*, Vol. III, The Society of Materials Science, Tokyo, Japan, 1972.
21. P. Shahinian, H. E. Watson, and H. H. Smith, "Fatigue Crack Growth in Selected Alloys for Reactor Applications," *Journal of Materials*, *7*, No. 4, 1972.

22. J. H. Weber and R. W. Hertzberg, "Effect of Thermomechanical Processing on Fatigue Crack Propagation," *Metallurgical Transactions*, *4*, Feb. 1973.

23. W. G. Clark, Jr., "How Fatigue Crack Initiation and Growth Properties Affect Material Selection and Design Criteria," *Metals Engineering Quarterly*, Aug. 1974.

24. W. G. Clark, Jr., and E. T. Wessel, "Interpretation of the Fracture Behavior of 5456-H321 Aluminum With WOL Toughness Specimens," *Scientific Paper 67-1D6-BTLFR-P4*, Westinghouse Research Laboratory, Pittsburgh, Sept. 1967.

25. T. W. Crooker, "Factors Determining the Performance of High Strength Structural Metals (Slope Transition Behavior of Fatigue Crack Growth Rate Curves)," *NRL Report of Progress*, Naval Research Laboratory, Washington, Dec. 1970.

26. K. Walker, "The Effect of Stress Ratio During Crack Propagation and Fatigue for 2024-T3 and 7075-T6 Aluminum," *ASTM STP 462*, American Society for Testing and Materials, Philadelphia,1970.

27. W. Elber, "The Significance of Fatigue Crack Closure," *ASTM STP 486*, American Society for Testing and Materials, Philadelphia,1971.

28. T. W. Crooker, "Effect of Tension-Compression Cycling on Fatigue Crack Growth in High-Strength Alloys," *Naval Research Laboratory Report 7220*, Naval Research Laboratory, Washington,Jan. 1971.

29. T. W. Crooker and D. J. Krause, "The Influence of Stress Ratio and Stress Level on Fatigue Crack Growth Rates in 140-ksi YS Steel," *Report of NRL Progress*, Naval Research Laboratory, Washington,Dec. 1972.

30. D. J. Krause and T. W. Crooker, "Effect of Constant-Amplitude Loading Parameters on Low-Cycle Fatigue-Crack Growth in a 140- to 150-ksi Yield Strength Steel," *Report of NRL Progress*, Naval Research Laboratory, Washington,March 1973.

31. J. Dubuc, J. R. Vanasse, A. Biron, and A. Bazergui, "Evaluation of Pressure Vessel Design Criteria for Effect of Mean Stress in Low-Cycle Fatigue," in *First International Conference on Pressure Vessel Technology: Part II—Materials and Fabrication*, American Society of Mechanical Engineers, New York, 1969.

32. B. F. Langer, "Design of Vessels Involving Fatigue," in *Pressure Vessel Engineering Technology*, edited by R. W. Nichols, Applied Science Publishers, Ltd., London, 1971, Chapter 2.

33. R. P. Wei, "Fracture Mechanics Approach to Fatigue Analysis in Design," *ASME Paper No. 73-DE-22*, New York,April 1973.

34. J. M. Barsom, "Corrosion Fatigue Crack Propagation Below K_{Iscc}," *Journal of Engineering Fracture Mechanics*, *3*, No. 1, July 1971.

35. S. R. Novak and J. M. Barsom, "AISI Project 168—Toughness Criteria for Structural Steels: Brittle Fracture (K_{Ic}) Behavior for Cracks Emanating from Notches," Presented at Ninth Annual Fracture Mechanics Symposium, Pittsburgh, 1975. (To be published.)

36. W. G. Clark, Jr., "Fatigue Crack Growth Characteristics of Heavy Section ASTM A533 Grade B Class 1 Steel Weldments," *ASME Paper No. 70-PVP-24*, American Society of Mechanical Engineers, New York,1970.

37. W. G. CLARK, JR., and D. S. KIM, "Effect of Synthetic Seawater on the Crack Growth Properties of HY-140 Steel Weldments," *Engineering Fracture Mechanics*, *4*, 1972.

38. P. SHAHINIAN, H. H. SMITH, and J. R. HAWTHORNE, "Fatigue Crack Propagation in Stainless Steel Weldments at High Temperature," *Welding Journal*, *51*, No. 11, 1972.

39. S. J. MADDOX, "Assessing the Significance of Flaws in Welds Subject to Fatigue," *Welding Journal*, *53*, No. 9, Sept. 1974.

40. T. R. GURNEY, *Fatigue of Welded Structures*, Cambridge University Press, New York, 1968.

41. R. M. N. Pelloux, "Fractographic Analysis of the Influence of Constituent Particles on Fatigue-Crack Propagation in Aluminum Alloys," *ASM Transactions Quarterly*, *57*, No. 2, June 1964.

42. C. M. CARMAN and J. M. KATLIN, "Low Cycle Fatigue Crack Propagation of High-Strength Steels," *ASME Paper No*. 66-MET-3, New York,April 1966.

9

Fatigue-Crack Propagation Under Variable-Amplitude Load Fluctuation

9.1. Introduction

Many engineering structures are subjected to complex fluctuating-load environments. Research conducted since the early 1960s has shown that the variation in the magnitude of the load cycles may affect the fatigue life of components significantly. A thorough understanding of the effect of various variables on the fatigue-crack initiation and propagation under variable-amplitude load fluctuations is essential to the development of accurate prediction methods of fatigue lives for engineering structures. Unfortunately, the effects of variable-amplitude loading on fatigue life are presently not well established. The results of the traditional *S-N* approach for variable-amplitude loading, like constant-amplitude loading, combine fatigue-crack initiation and propagation in different proportions.[1-5] The fracture-mechanics investigations of fatigue lives under variable-amplitude loading have concentrated exclusively on the fatigue-crack-propagation behavior of metals.[6-11] Some significant results of these investigations are presented in this chapter.

9.2. Stress Histories

The development of accurate fatigue-life prediction procedures for variable-amplitude cyclic-load fluctuations requires an understanding and characterizing of stress environments.

The simplest cyclic-stress history is the constant-amplitude sinusoidal-stress fluctuation shown in Figure 9.1. Such a stress history can be represented by a steady mean stress, σ_{mean}, and a fluctuating stress range, $\Delta\sigma$. The stress range is the algebraic difference between the maximum stress, σ_{max}, and the minimum stress, σ_{min}, in the cycle

$$\Delta\sigma = \sigma_{max} - \sigma_{min} \tag{9.1}$$

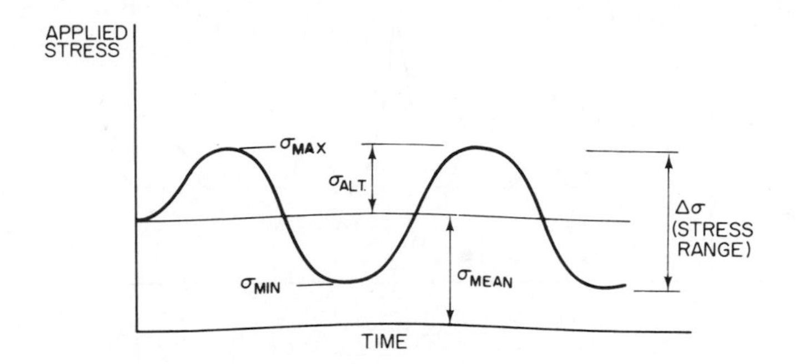

FIG. 9.1. Constant-amplitude sinusoidal stress loading.

The mean stress is the algebraic mean of σ_{max} and σ_{min} in the cycle

$$\sigma_{mean} = \frac{\sigma_{max} + \sigma_{min}}{2} \tag{9.2}$$

The alternating stress or stress amplitude is half of the stress range in the cycle

$$\sigma_{alt} = \frac{\Delta\sigma}{2} = \frac{\sigma_{max} - \sigma_{min}}{2} \tag{9.3}$$

A constant-amplitude sinusoidal-stress history can be represented by an analytical function and can be described by various parameters. Unfortunately, many structural components are subjected to random-stress histories that cannot be represented by an analytical function and that lack a describable pattern, as shown in Figure 9.2.

Between the extremes of a simple, constant-amplitude sinusoidal-stress history and a complex, random-stress history, there is a multitude of stress patterns of varying degrees of complexity which can be described by analytic

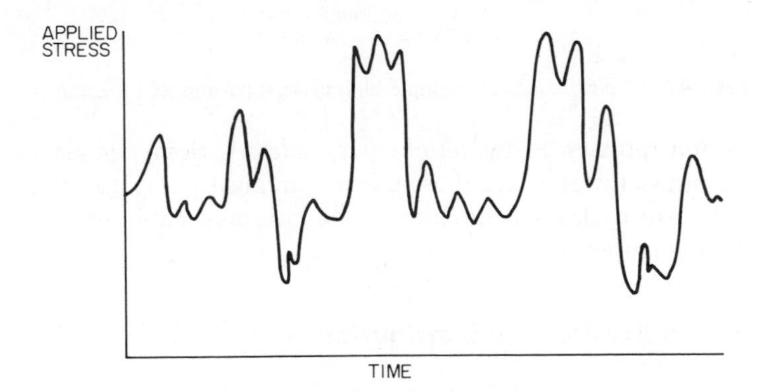

FIG. 9.2. Random-stress loading.

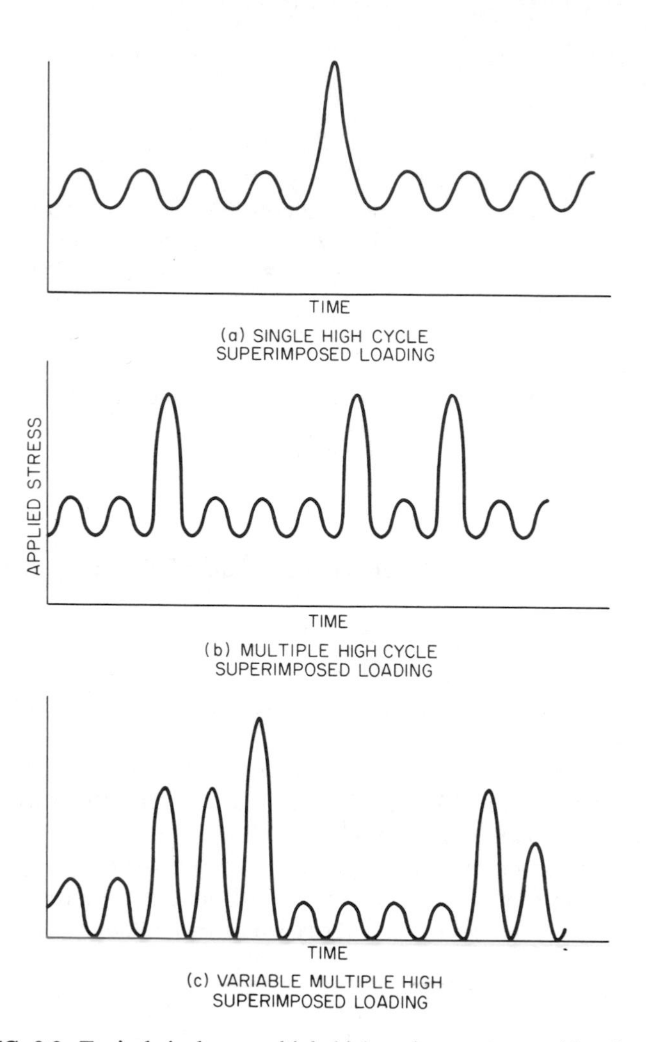

FIG. 9.3. Typical single or multiple high cycle superimposed loading.

functions and represented by various parameters.[12] Some simple variable-amplitude stress histories are those corresponding to a single or multiple high-tensile-load cycles superimposed upon constant-amplitude cyclic-load fluctuations (Figure 9.3).

9.3. Probability-Density Distribution

Many engineering structures such as bridges, ships, and others are subjected to variable-amplitude random-sequence load fluctuations. The probability of occurrence of the same sequence of stress fluctuations for a given detail in such structures obtained during a given time interval is very small. Consequently, the magnitude of stress fluctuations must be characterized to study the fatigue behavior of components subjected to variable-amplitude random-sequence stress fluctuations. The magnitude of the stress fluctuations should be characterized and described by analytic functions. The use of probability-density curves to characterize variable-amplitude cyclic-stress fluctuations appear to be very useful.[10,13-15]

Stress history, or stress spectrum, for a particular location in a structure subjected to variable-amplitude stress fluctuation can be defined in terms of the frequency of occurrence of maximum (peak) stresses. Usually, frequency-of-occurrence data are presented as a histogram, or bar graph (Figure 9.4),

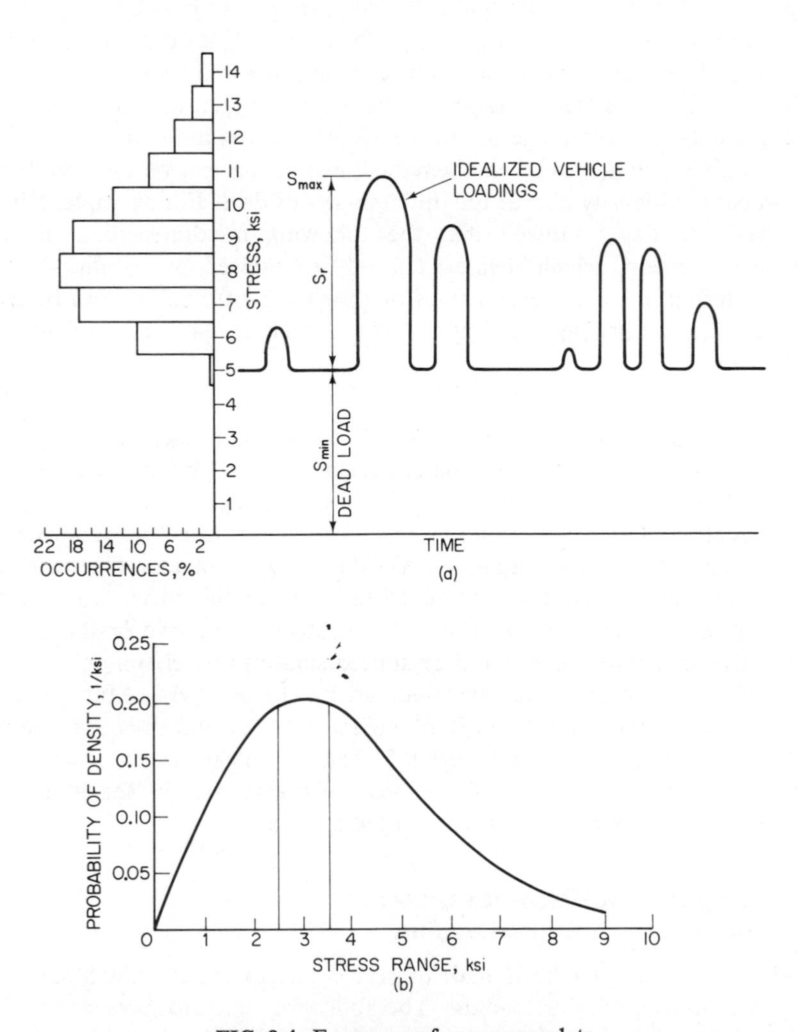

FIG. 9.4. Frequency-of-occurrence data.

in which the height of the bar represents the percentage of recorded maximum stresses that fall within a certain stress interval represented by the width of the bar. For example, 20.2% of the maximum stresses in Figure 9.4(a) fall within the interval between 7.5 and 8.5 ksi (52 to 59 MN/m^2). The frequency of occurrence of stress ranges can be represented by similar plots with the vertical scale changed according to the relationship between σ_{max}, σ_{min}, and stress range, σ_r or $\Delta\sigma$. Since stress range is the most important stress parameter controlling the fatigue life of structural components, stress range is used to define the major stress cycles in the following discussion.

The frequency-of-occurrence data can be presented in a more general form by dividing the percentage of occurrence for each interval, i.e., the height of each bar, in Figure 9.4(a) by the interval width to obtain a probability-density curve such as shown in Figure 9.4(b). Thus, data from sources that use different stress-range intervals can be compared by using the probability-density curve. The area under the curve between any two values of $\Delta\sigma$ represents the percentage of occurrence within that interval.

A single nondimensional mathematical expression can be used to define the probability-density curves for different sets of data. For example, Klippstein and Schilling[14] showed that the following nondimensional mathematical expression, which defines a family of skewed probability-density curves referred to as Rayleigh curves or distribution functions, can be used to accurately fit a probability-density curve to each available set of field data for bridges:

$$p' = 1.011x'e^{-1/2(x')^2} \tag{9.4}$$

where $x' = (\sigma_r - \sigma_{r\min})/\sigma_{rd}$, σ_r(i.e. $\Delta\sigma$) is the stress range, and $\sigma_{r\min}$ (i.e. $\Delta\sigma_{\min}$) and σ_{rd} (i.e. $\Delta\sigma_d$) are constant parameters that define any particular probability-density curve from the family of curves represented by Equation (9.4). Equation (9.4) is plotted in Figure 9.5(a). As illustrated in Figure 9.5(b), a particular curve from the family is defined by two parameters: (1) the modal stress range, σ_{rm}, which corresponds to the peak of the curve; and (2) the parameter σ_{rd}, which is a measure of the width of the curve or the dispersion of the data. The curve could be shifted sideways by changing σ_{rm}, and the width of the curve could be modified by changing σ_{rd}. Mathematical expressions for the modal, median, mean, and root-mean-square values of the spectrum are given in Figure 9.5. The root-mean-square (rms) value is defined as the square root of the mean of the squares of the individual values of x' or σ_r. Stress (σ) is represented as S in Fig. 9.5.

9.4. Fatigue-Crack Growth Under Variable-Amplitude Loading

Many attempts have been made to predict fatigue-crack-growth behavior under variable-amplitude loading. The following sections present the behavior under simple and complex variable-amplitude loading. The simple

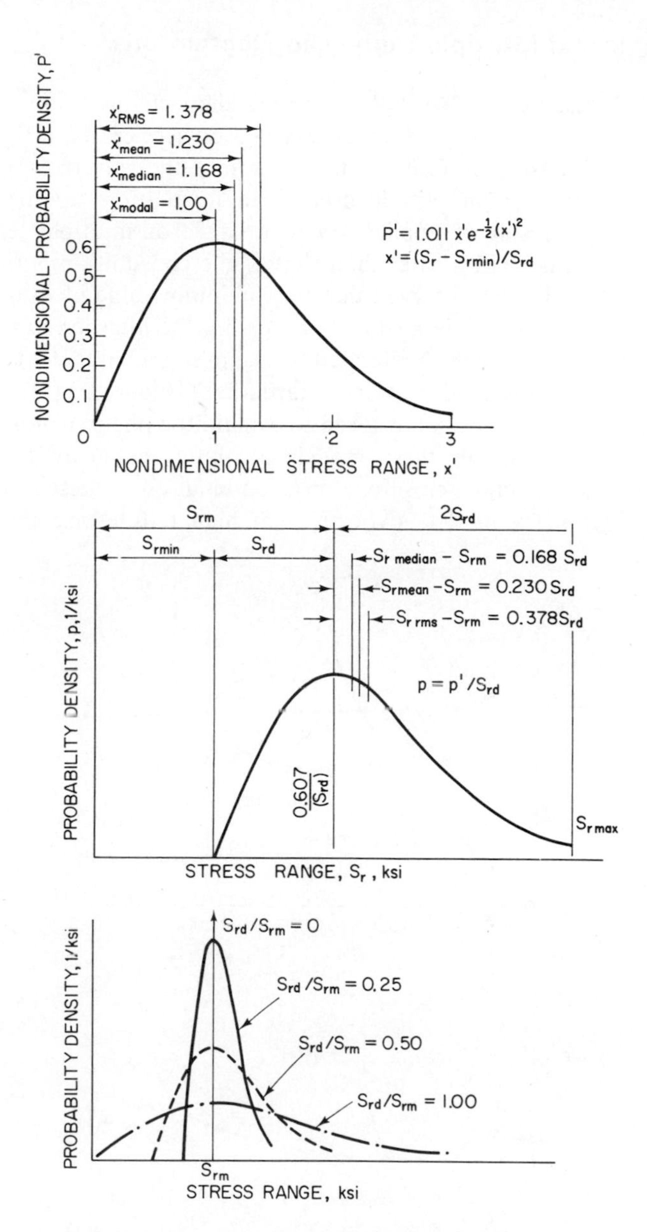

FIG. 9.5. Characteristics of Rayleigh probability curves.

loadings correspond to a single or multiple high tensile-load fluctuations superimposed upon constant-amplitude cyclic-load fluctuations (Figure 9.3). Complex variable-amplitude loadings correspond to multivalue cyclic-load fluctuations.

9.5. Single and Multiple High Load Fluctuations

Several investigators[6-9,11,16-18] observed that changes in cyclic-load magnitude (Figure 9.3) may result in retarded or accelerated fatigue-crack-growth rate. Extensive published data show that the rate of fatigue-crack growth under constant-amplitude cyclic-load fluctuation can be retarded significantly as a result of application of a single or multiple tensile-load cycle having a peak load greater than that of the constant-amplitude cycles (Figure 9.6). Von Euw[19] observed that the minimum value of fatigue-crack-growth rate did not occur immediately following the high tensile-load cycle but that the rate of growth decelerated to a minimum value. This deceleration region has been termed "delayed retardation" [Figure 9.6(b)].

Several models have been advanced to explain the phenomenon of crack-growth delay. In general, these models attribute the delay behavior to residual stresses,[20,21] crack closure,[22] or a combination of these mechanisms. A crack-tip-geometry model advocates that high tensile-load cycles cause

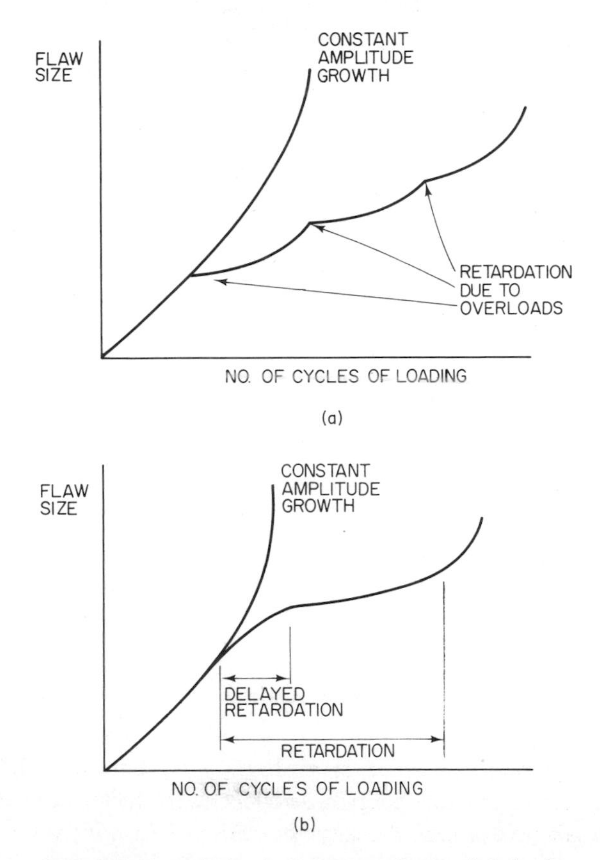

FIG. 9.6. Retardation in fatigue-crack-growth behavior.

crack-tip blunting, which in turn causes retardation in fatigue-crack growth at the lower cyclic-load fluctuations until the crack is resharpened. The residual-stress model suggests that the application of a high tensile-load cycle forms residual compressive stresses in the vicinity of the crack tip that reduce the rate of fatigue-crack growth. Finally, the crack-closure model postulates that the delay in fatigue-crack growth is caused by the formation of a zone of residual tensile deformation left in the wake of a propagating crack that causes the crack to remain closed during a portion of the applied tensile-load cycle. Consequently, fatigue-crack-growth delay occurs because only the portion of the tensile-load cycles that is above the crack-opening level is effective in extending the crack. These models are useful to predict trends in fatigue-crack-growth-rate behavior caused by single or multiple high tensile-load cycles but are of little value to predict fatigue lives under these conditions. Fatigue-crack-growth delay has been shown to be strongly dependent on all the loading variables, such as the stress-intensity-factor fluctuation, the ΔK_I of the high tensile-load cycle, the ΔK_I of the constant-amplitude cycles (Figure 9.7[11]), the stress ratios of these ΔK_I values, and the number of constant-amplitude cycles between the high tensile-load cycles.[8,11,23,24] Extensive research is necessary to further our understanding of the significance of these variables in order to develop equations that can be used to predict accurately the fatigue life of components subjected to single or multiple high tensile-load cycles.

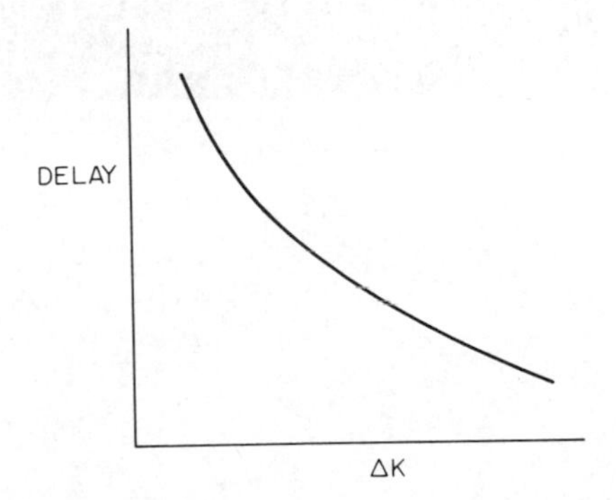

FIG. 9.7. Schematic showing effect of ΔK on fatigue-crack-growth delay.

9.6. Variable-Amplitude Load Fluctuations

Extensive investigations are currently being conducted to develop methods to predict fatigue lives under variable-amplitude load fluctuations. Some of these investigations resulted in models that can be used to predict trends in fatigue-crack-propagation rates but have varying degrees of success in predicting fatigue lives under variable-amplitude loads.[10,11,20,21] Presently, the models presented by Barsom[10,13,25] and by Wei and Shih[11] appear to be promising. The model advocated by Wei and Shih is a superposition model where the delay cycles caused by a change in stress magnitude are superimposed on the cycles obtained under constant-amplitude loading assuming no load interactions. The delay cycles in this model must be estimated from experimental data.

The model advanced by Barsom[10] relates fatigue-crack-growth rate per cycle to an effective stress-intensity factor that is characteristic of the probability-density curve. The development of this model, which was designated the rms (root-mean-square) model, and the supporting experimental data are presented in the following sections.

9.6.1. The Root-Mean-Square (rms) Model

Incremental increase of crack length and the corresponding number of elapsed load cycles can be measured under variable-amplitude random-

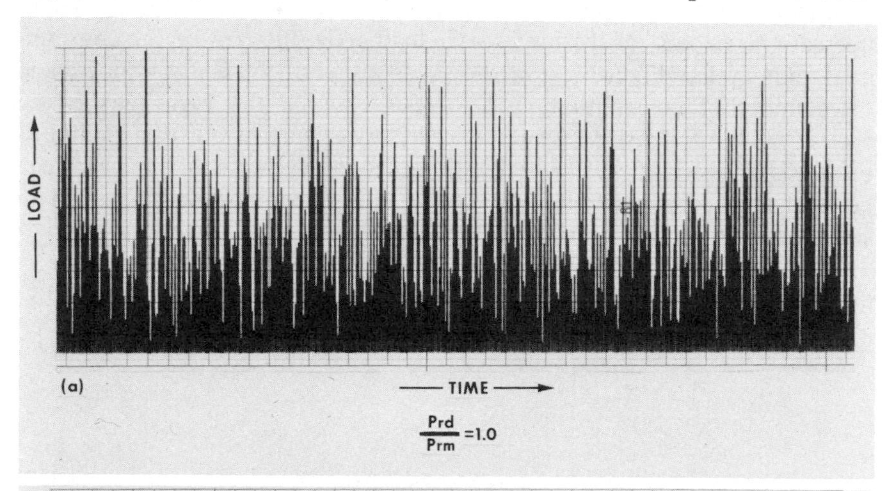

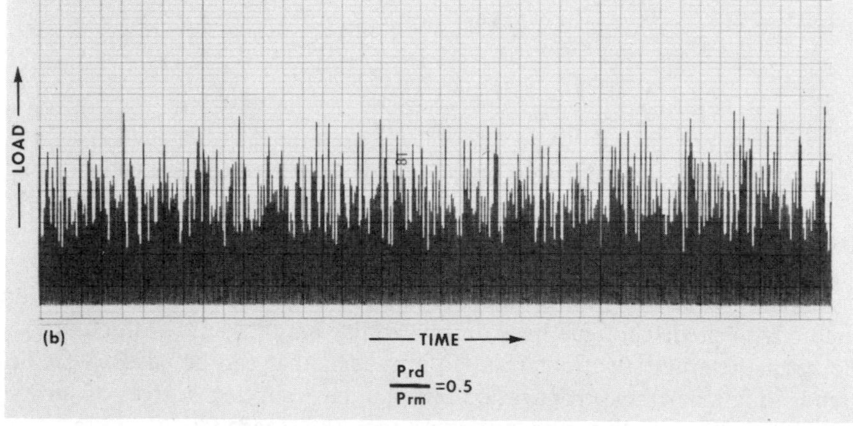

FIG. 9.8. Two variable-amplitude random-sequence load fluctuations investigated.

sequence load spectra. However, unlike constant-amplitiude cyclic-load data, the magnitude of ΔK_I changes for each cycle. Reduction of data in terms of fracture-mechanics concepts requires the establishment of a correlation parameter that incorporates the effects of crack length, cyclic-load amplitude, and cyclic-load sequence.

Barsom[10] attempted to determine the magnitude of constant-amplitude cyclic-load fluctuation that results in the same a versus N curve obtained under variable-amplitude cyclic-load fluctuation when both spectra are applied to identical specimens (including initial crack length). In other words, one of the objectives of his investigation was to find a single stress-intensity parameter, such as mean, modal, or root mean square, that can be used to define the crack-growth rate under both constant- and variable-amplitude loadings. The selected parameter *must* characterize the probability-density curve.

Variable-amplitude random-sequence load spectra having Rayleigh probability-density curves of P_{rd}/P_{rm} (or σ_{rd}/σ_{rm}) values equal to 0.5 and 1.0 were investigated as part of Project 12–12 of the National Cooperative Highway Research Program (NCHRP) of the National Academy of Sciences.[10,13,15] A typical portion of the 500-cycle loading block for each is shown in Figure 9.8. Data for crack length versus the corresponding number of elapsed load cycles obtained by subjecting identical specimens to these random-sequence load spectra are presented in Figure 9.9. Figure 9.9 also includes data

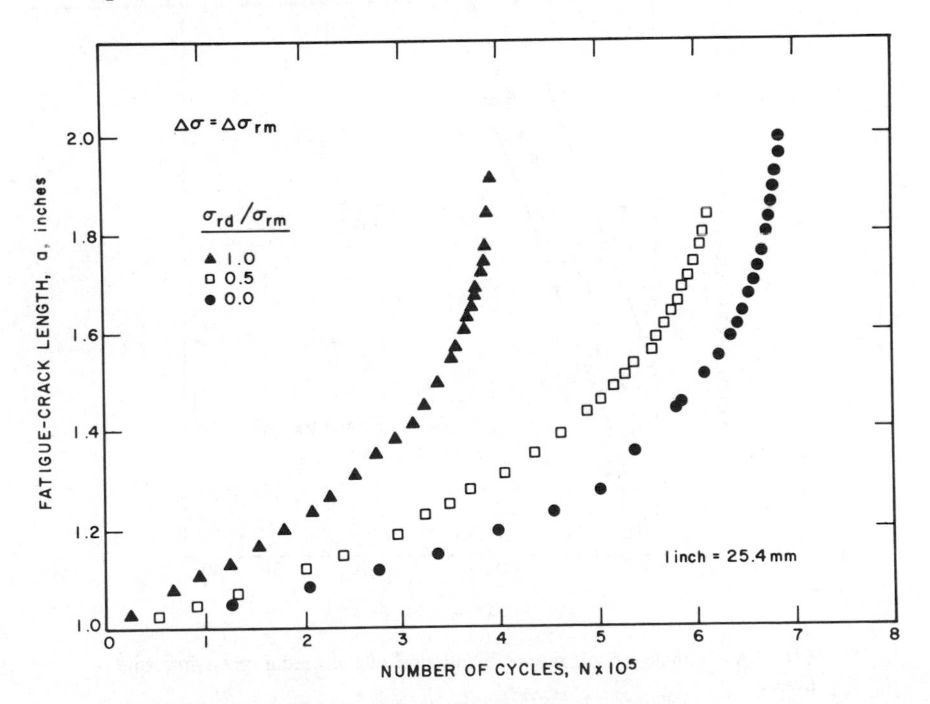

FIG. 9.9. Crack growth under spectra loads.

obtained under constant-amplitude cyclic-load fluctuation ($P_{rd}/P_{rm} = 0$). The load range, P_r, for every cycle in the constant-amplitude tests was equal to P_{rm}. The data show that the fatigue life under constant-amplitude cyclic-load fluctuation is longer than the life obtained under random-sequence load spectra having the same value of P_{rm}. The data are re-presented in Figure 9.10 in terms of crack-growth rate and the modal stress-intensity factor, K_{rm}.

A better correlation between data obtained under constant-amplitude and variable-amplitude random-sequence load spectra was obtained on the basis of the root mean square of the load distribution, where the root mean square is the square root of the mean of the squares of the individual load cycles in a spectrum. The combined crack-growth-rate data are presented in Figure 9.11 as a function of ΔK_{rms}. The data show that, within the limits of the experimental work, the average fatigue-crack-growth rates per cycle,

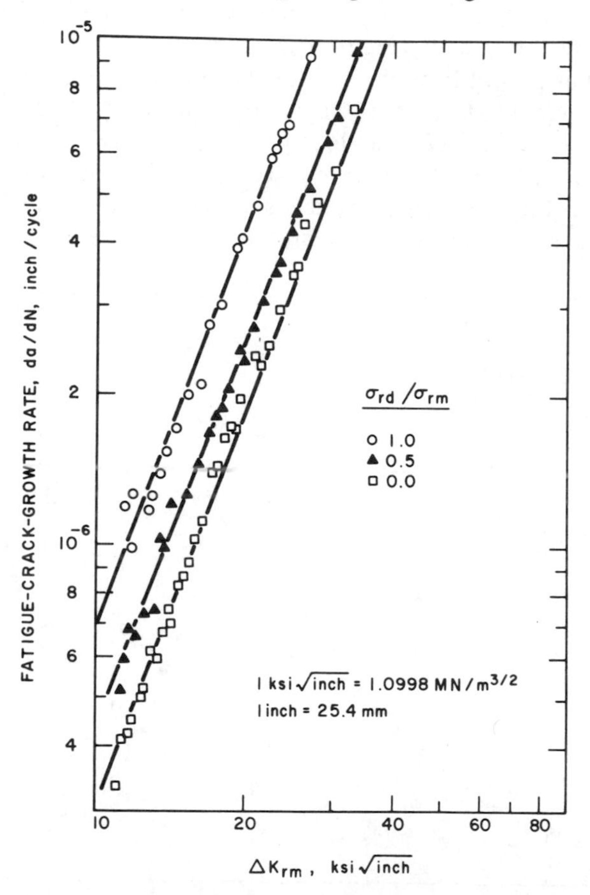

FIG. 9.10. Crack-growth rate as a function of the modal stress-intensity factor.

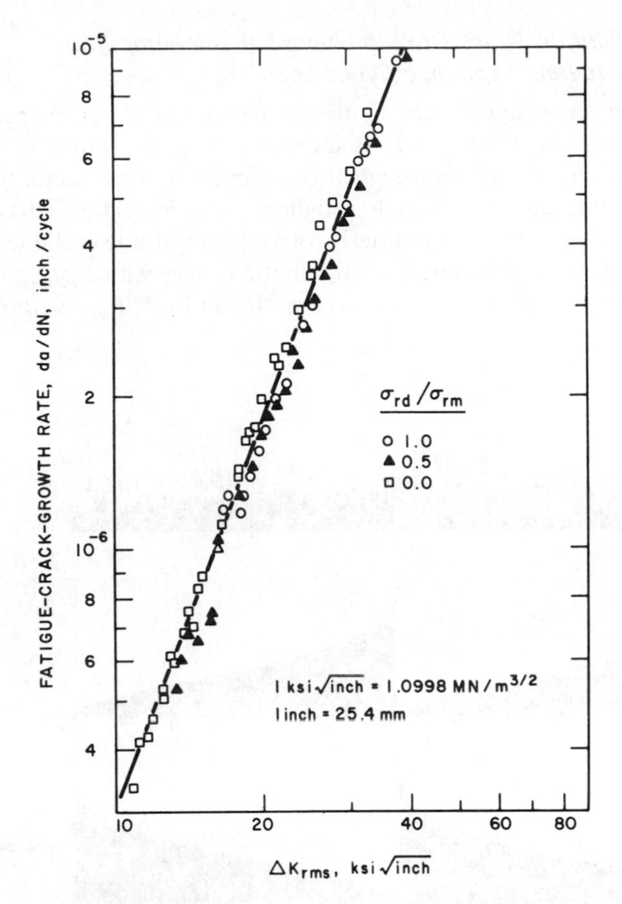

FIG. 9.11. Crack-growth rate as a function of the root-mean-square stress-intensity factor.

da/dN, under variable-amplitude random-sequence stress spectra can be represented by

$$\frac{da}{dN} = A(\Delta K_{\text{rms}})^n \tag{9.5}$$

where A and n are constants and

$$\Delta K_{\text{rms}} = \sqrt{\frac{\sum_{i=1}^{k} \Delta K_i^2}{n}}$$

The root-mean-square value of the stress-intensity factor under constant-amplitude cyclic-load fluctuation is equal to the stress-intensity-factor fluctuation. Consequently, the average fatigue-crack-growth rate can be predicted from constant-amplitude data by using Equation (9.5).

9.6.2 Fatigue-Crack Growth Under Variable-Amplitude Ordered-Sequence Cyclic Load

The root-mean-square stress-intensity factor, ΔK_{rms}, is characteristic of the load-distribution curve and is independent of the order of the cyclic-load fluctuations. To determine whether the order of load fluctuations affects the average rate of crack growth, fatigue tests were performed on identical specimens under random and ordered variable-amplitude cyclic-load fluctuations that represent the same load-distribution curve with $P_{rd}/P_{rm} = 1.0$.[10,13] The tests were conducted at a constant minimum load, $P_{\min}$, with $P_{\min}/P_{rm} = 0.25$.

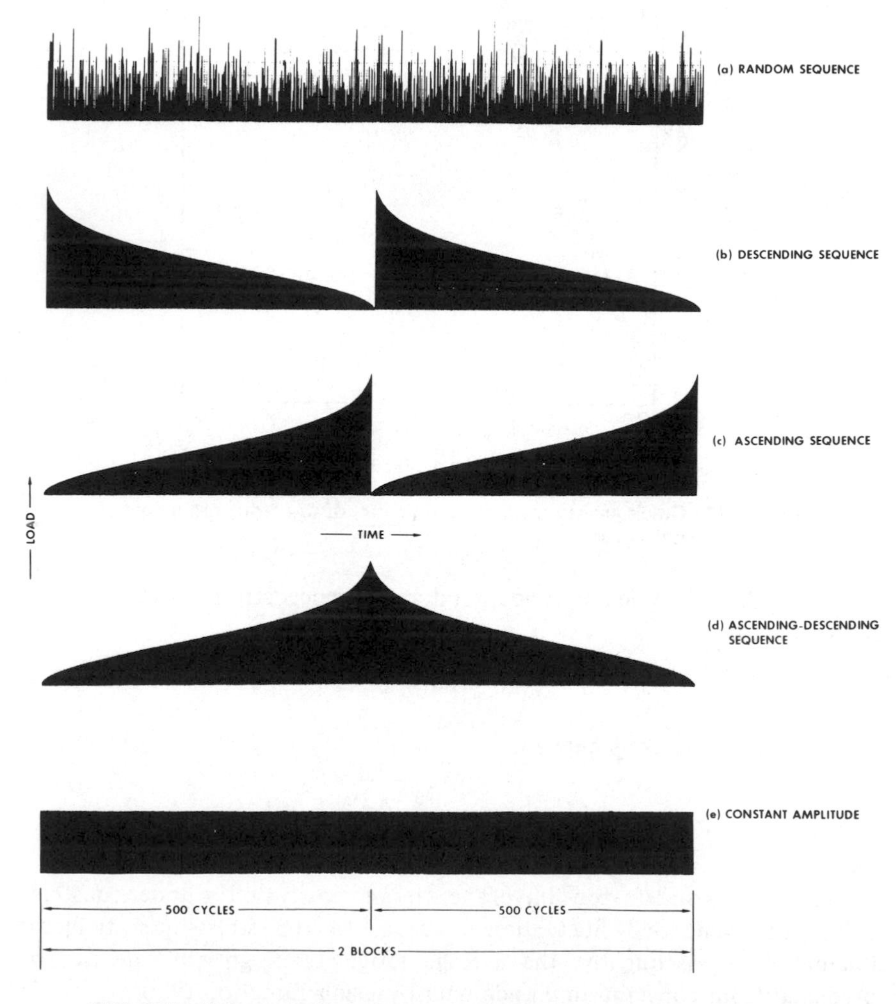

FIG. 9.12. Various random-sequence and ordered-sequence load fluctuations studied to establish the ΔK_{rms} analysis.

Fatigue-crack-growth-rate tests were conducted by using the variable-amplitude random-sequence load fluctuations shown in Figure 9.12(a). In other tests these same load fluctuations were arranged in ascending magnitudes [Figure 9.12(b)], descending magnitudes [Figure 9.12(c)], and combined ascending-descending magnitudes [Figure 9.12(d)]. The fatigue-crack-growth data obtained under these various conditions and under a constant-amplitude cyclic-load fluctuation of $P_r = \Delta P_{\text{rms}}$ [Figure 9.12(e)] are presented in Figure 9.13. The data show that the average rate of fatigue-crack growth is represented accurately by Equation (9.5) regardless of the order of occurrence of the cyclic-load fluctuations.

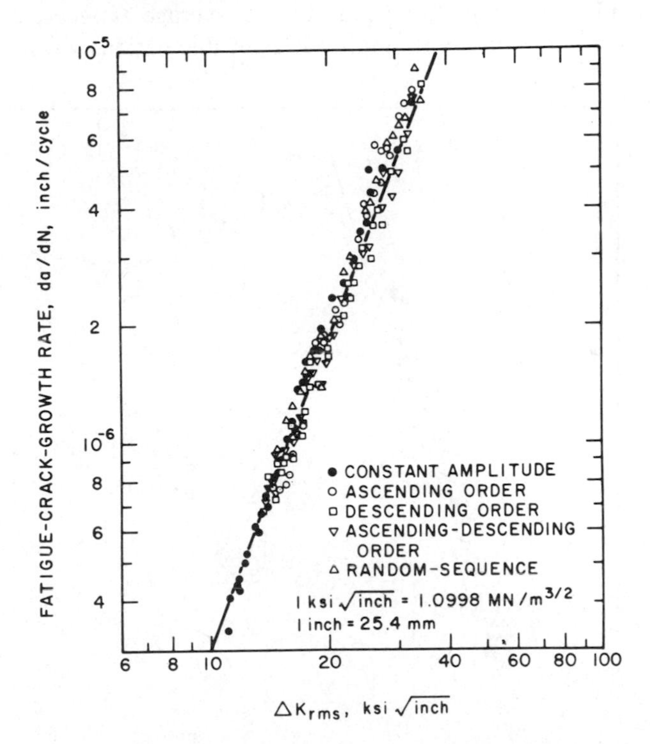

FIG. 9.13. Summary of crack-growth-rate data under random-sequence and ordered-sequence load fluctuations.

9.7. Fatigue-Crack Growth in Various Steels

The preceding results were obtained by testing A514 Grade B steel under variable-amplitude random- and ordered-sequence cyclic-load fluctuations. Because several investigators[6-9,11,16-18] have noted that changes in cyclic-load magnitude can lead to accelerated or retarded rates of fatigue-crack growth in various metals, the applicability of the RMS model to steels of various yield strengths was investigated under NCHRP Project 12-14.[25]

Fatigue-crack-growth rates under constant-amplitude and variable-amplitude random-sequence load fluctuations were investigated for A36, A588 Grade A, A588 Grade B, A514 Grade E, and A514 Grade F steels. All loadings followed a Rayleigh probability-density curve, with the ratio of the load-range deviation to the model (peak) load (P_{rd}/P_{rm}) equal to either 0 or 1.0. The data presented in Figures 9.14 through 9.18 show that, within the limits of the experimental work, the average fatigue-crack-growth rates per cycle, da/dN, in various steels subjected to variable-amplitude load spectra can be represented by Equation (9.5). Moreover, the average fatigue-crack-growth rates for the various steels studied under variable-amplitude random-sequence load fluctuations are equal to the average fatigue-crack-growth rates obtained under constant-amplitude load fluctuations when the stress-

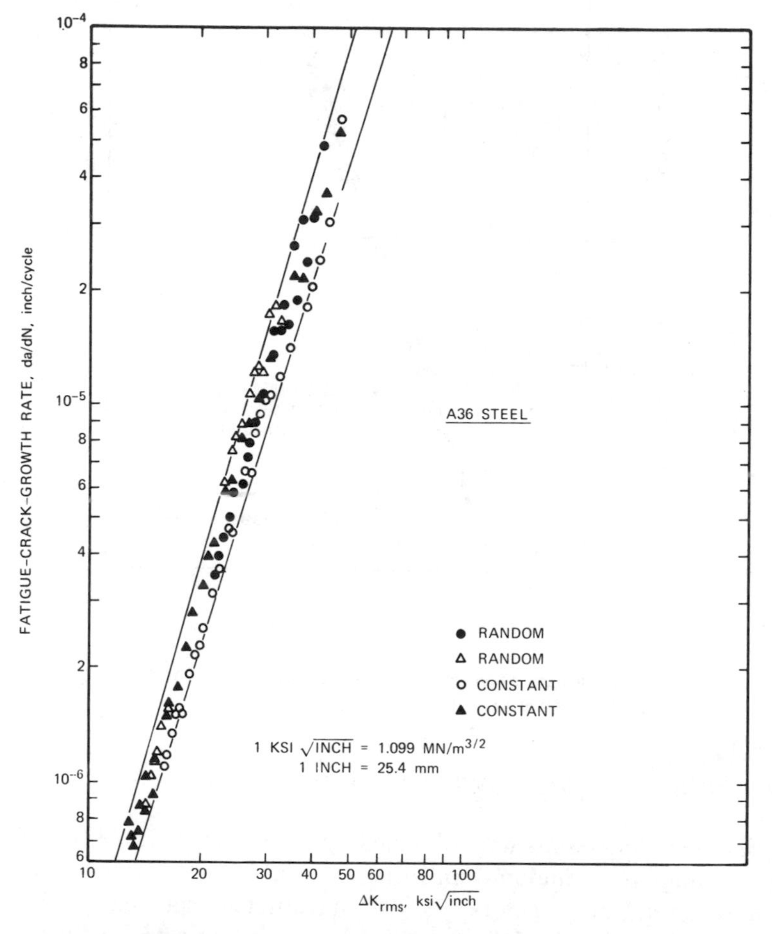

FIG. 9.14. Crack-growth rate as a function of the root-mean-square stress-intensity factor for A36 steel.

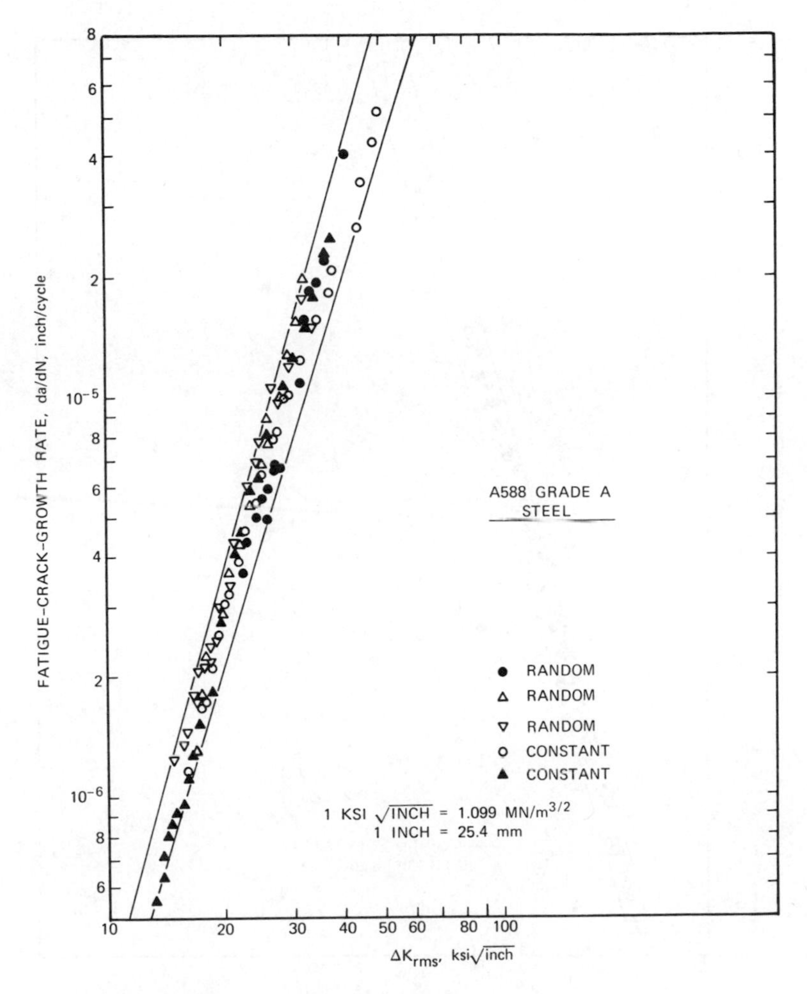

FIG. 9.15. Crack-growth rate as a function of the root-mean-square stress-intensity factor for A588 grade A steel.

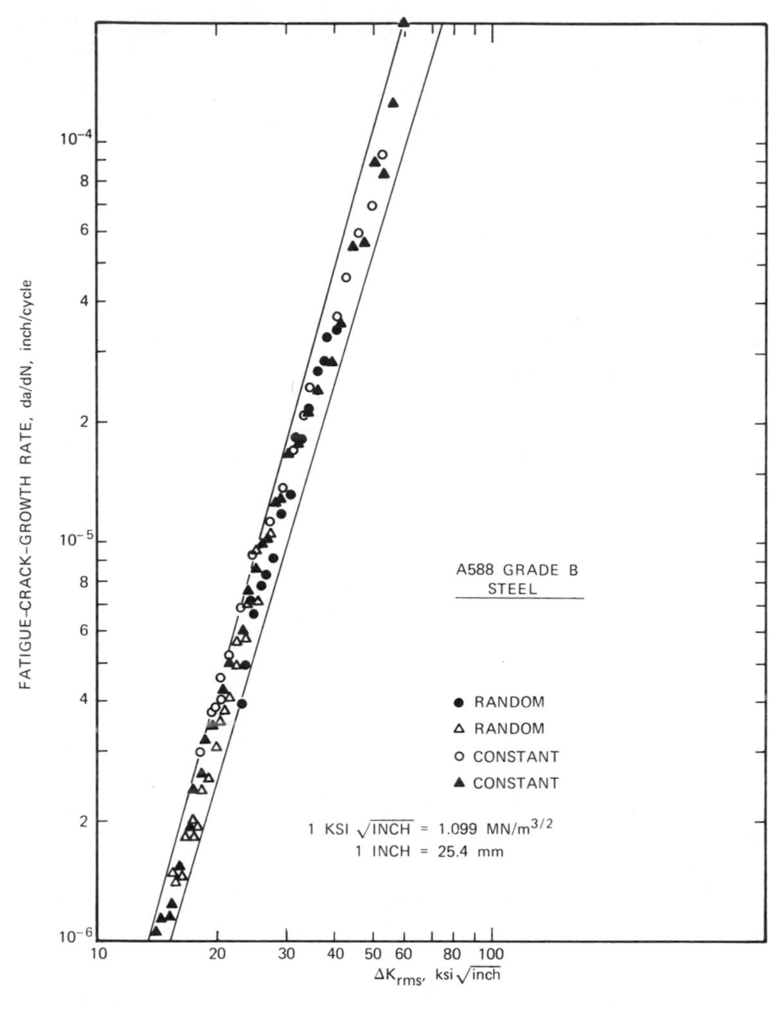

FIG. 9.16. Crack-growth rate as a function of the root-mean-square stress-intensity factor for A588 grade B steel.

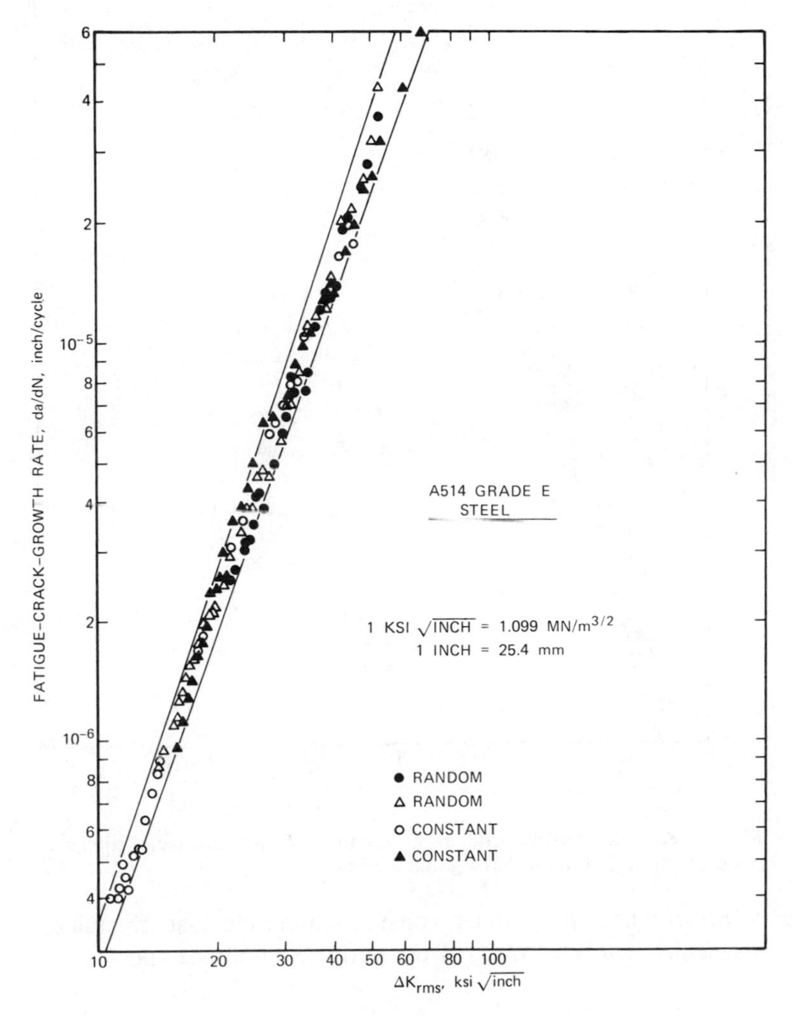

FIG. 9.17. Crack-growth rate as a function of the root-mean-square stress-intensity factor for A514 grade E steel.

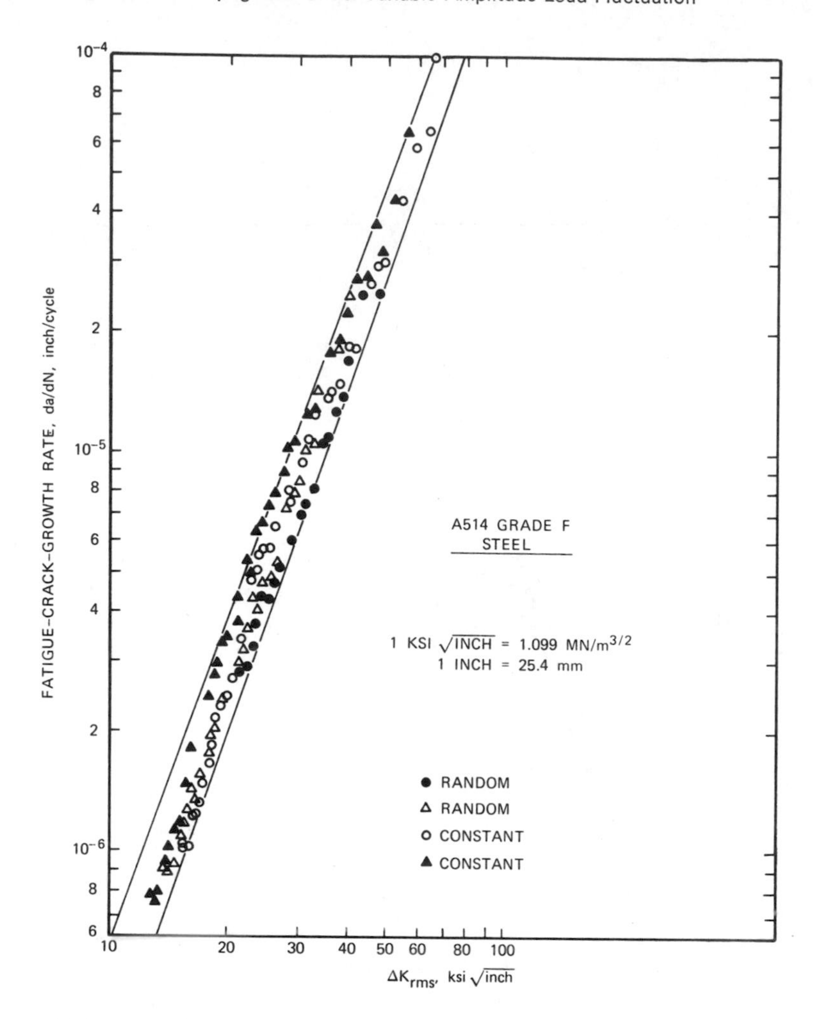

FIG. 9.18. Crack-growth rate as a function of the root-mean-square stress-intensity factor for A514 grade F steel.

intensity-factor range, ΔK, under constant-amplitude load fluctuations is equal in magnitude to the ΔK_{rms} of the variable-amplitude spectra.

9.8. Fatigue-Crack Growth Under Various Unimodal Distribution Curves

The applicability of the RMS model for correlating crack-growth rates under variable-amplitude random-sequence load fluctuations that follow unimodal distribution curves different from the Rayleigh type have been studied. Fatigue-crack-growth rates under constant-amplitude load fluctuations and under four unimodal variable-amplitude random-sequence load fluctuations were investigated by Barsom.[26] A block of variable-amplitude random-sequence load fluctuations of each unimodal distribution curve and of constant-amplitude load fluctuations is presented in Figure 9.19. Each block was applied repeatedly to a single specimen until the test was terminated. The number of cycles in the blocks corresponding to the four distribution curves investigated varied between 302 and 500 cycles. Schematics of

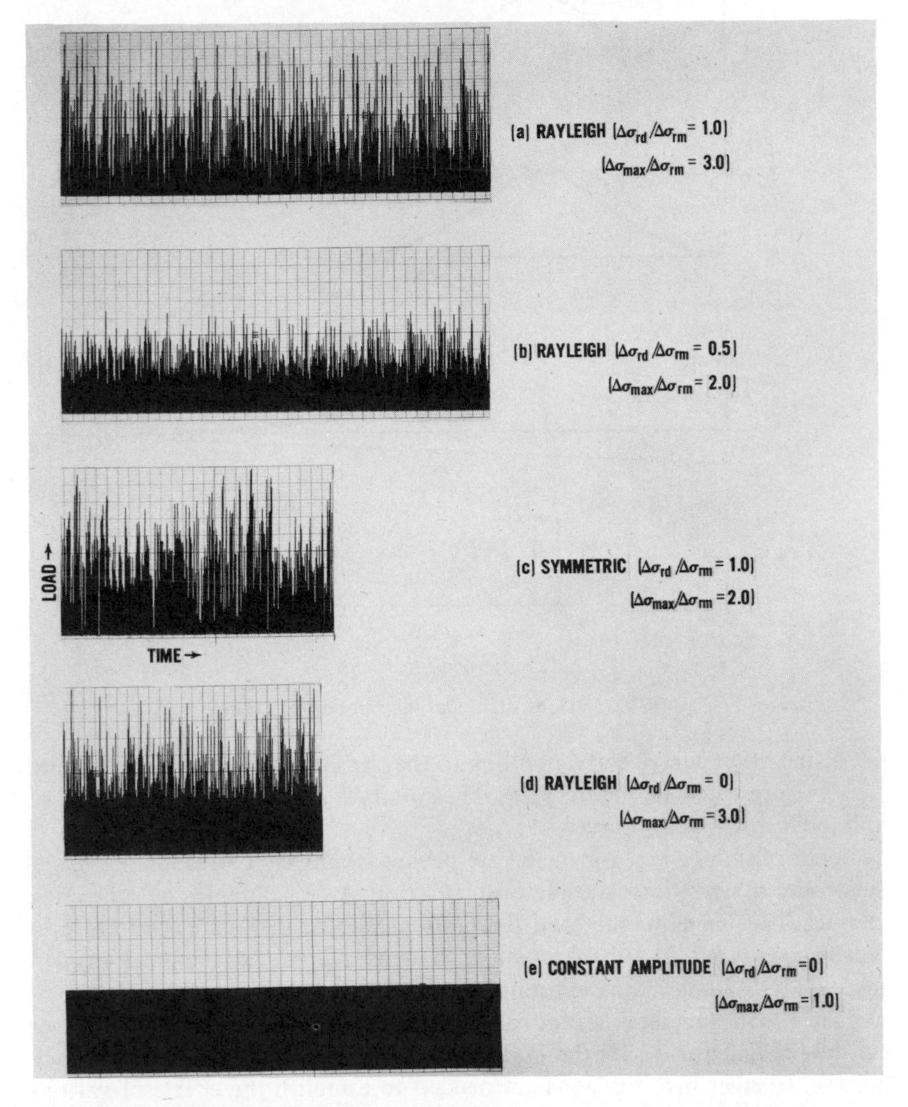

FIG. 9.19. Single blocks of load fluctuations for various distribution functions.

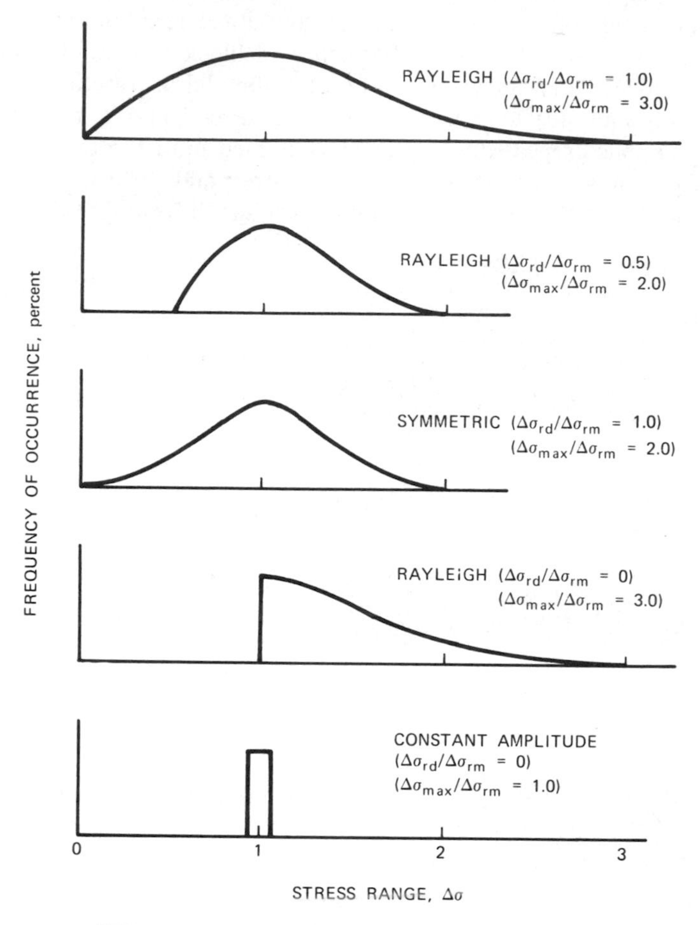

FIG. 9.20. Various unimodal distribution functions.

the distribution curves corresponding to the various blocks shown in Figure 9.19 are presented in Figure 9.20. These curves cover a wide variation of unimodal distribution curves.

Data of fatigue-crack-growth rate per cycle and the corresponding root-mean-square stress-intensity-factor fluctuation, ΔK_{rms}, obtained by subjecting identical specimens to the load fluctuations shown in Figure 9.19 are presented in Figure 9.21. The data show that the average fatigue-crack-growth rate, da/dN, under constant-amplitude and variable-amplitude random-sequence load fluctuations that follow various unimodal distribution curves can be predicted by using the RMS model [Equation (9.5)] of fatigue-crack growth. Further investigations are needed to establish the effects of various parameters, such as variable minimum load, on the rate of fatigue-crack growth under variable-amplitude load fluctuations and the necessary modifi-

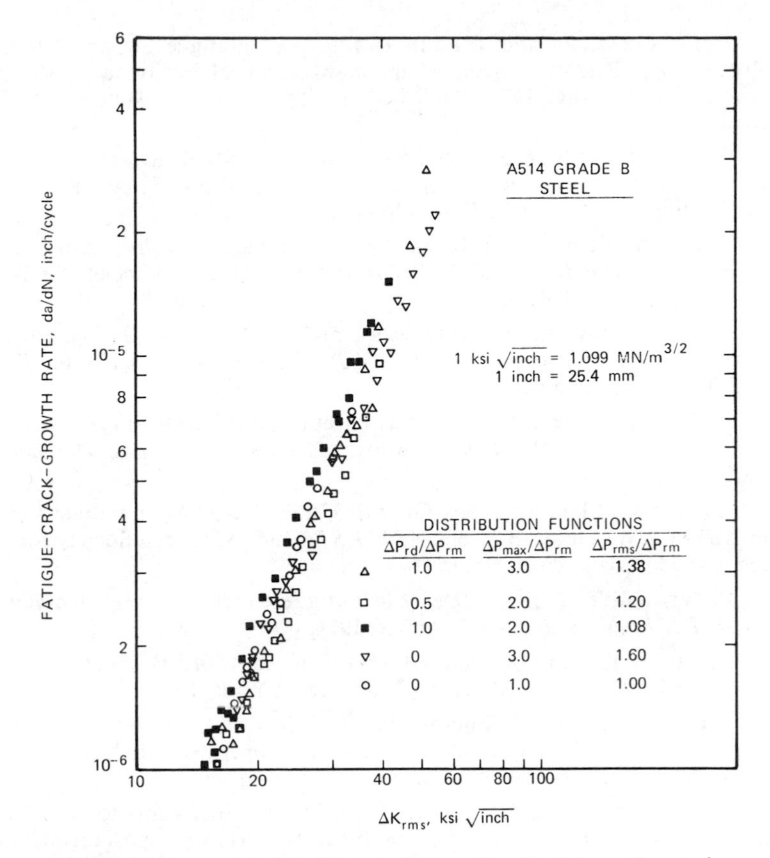

FIG. 9.21. Summary of fatigue-crack growth-rate data under various unimodal distribution functions.

cations to the RMS model to account for these effects or the development of a better model.

REFERENCES

1. M. A. MINER, "Cumulative Damage in Fatigue," *Journal of Applied Mechanics*, Sept. 1945.
2. A. F. MADAYAG, ED., *Metal Fatigue: Theory and Design*, Wiley, New York, 1969.
3. T. H. TOPPER, B. I. SANDOR, and JoDEAN MORROW, "Cumulative Fatigue Damage Under Cyclic Strain Control," *Journal of Materials, JMLSA, 4*, No. 1, March 1969.
4. "Effects of Environment and Complex Load History on Fatigue Life," *ASTM STP 462*, American Society for Testing and Materials, Philadelphia 1970.

5. HIROSHI NAKAMURA and TAKESHI HORIKAWA, "Fatigue Strength of Steel," Proceedings: The 1974 Symposium on Mechanical Behavior of Materials, Kyoto, Japan, Aug. 1974. Published by the Society of Materials Science, Japan.
6. J. SCHIJVE, "Significance of Fatigue Cracks in Micro-Range and Macro-Range," in "Fatigue Crack Propagation," *ASTM STP 415*, American Society for Testing and Materials, Philadelphia 1967.
7. J. C. MCMILLAN and R. M. N. PELLOUX, "Fatigue Crack Propagation Under Program and Random Loads," *ASTM STP 415*, American Society for Testing and Materials, Philadelphia 1967.
8. E. F. J. VONEUW, R. W. HERTZBERG, and R. ROBERTS, "Delay Effects in Fatigue Crack Propagation," *ASTM STP 513*, American Society for Testing and Materials, Philadelphia 1972.
9. R. E. JONES, "Fatigue Crack Growth Retardation After Single-Cycle Peak Overload in Ti-6A1-4V Titanium Alloy," *Engineering Fracture Mechanics, 5*, 1973.
10. J. M. BARSOM, "Fatigue-Crack Growth Under Variable-Amplitude Loading in ASTM A514 Grade B Steel," *ASTM STP 536*, American Society for Testing and Materials, Philadelphia 1973.
11. R. P. WEI and T. T. SHIH, "Delay in Fatigue Crack Growth," *International Journal of Fracture, 10*, No. 1, March 1974.
12. H. L. LEVE, "Cumulative Damage Theories," in *Metal Fatigue: Theory and Design*, edited by A. F. Madayag, Wiley, New York, 1969.
13. "NCHRP Project 12-12: Interim Report," *U.S. Steel Research Laboratory Report 76.019–001*, Oct. 1, 1972; National Cooperative Highway Research Program-National Academy of Sciences, Washington, D.C.
14. K. H. KLIPPSTEIN and C. G. SCHILLING, "Stress Spectrums for Short-Span Steel Bridges," submitted for presentation at the ASTM Symposium on Fatigue Crack Growth under Spectrum Loads, Montreal, June 1975.
15. C. G. SCHILLING, K. H. KLIPPSTEIN, J. M. BARSOM, and G. T. BLAKE, "Fatigue of Welded Steel Bridge Members Under Variable-Amplitude Loadings," *Research Results Digest*, Digest 60, National Cooperative Highway Research Program, April 1974.
16. H. F. HARDRATH and A. T. MCEVILY, "Engineering Aspects of Fatigue-Crack Propagation," in *Proceedings of the Crack Propagation Symposium*, Vol. 1, Cranfield, England, Oct. 1961.
17. J. SCHIJVE, F. A. JACOBS, and P. J. TROMP, "Crack Propagation in Clad 2024-T3A1 Under Flight Simulation Loading. Effect of Truncating High Gust Loads," *NLR TR-69050-U*, National Lucht-En Ruimtevaart-laboratorium (National Aeorspace Laboratory NLR—The Netherlands), June 1969.
18. C. M. HUDSON and H. F. HARDRATH, "Effects of Changing Stress Amplitude on the Rate of Fatigue-Crack Propagation of Two Aluminum Alloys," *NASA Technical Note D-960*, NASA, Cleveland Sept. 1961.
19. E. F. J. VONEUW, "Effect of Single Peak Overloading on Fatigue Crack Propagation," Master's dissertation, Lehigh University, Bethlehem, Pa 1968.
20. O. E. WHEELER, "Spectrum Loading and Crack Growth," *General Dynamics Report FZM-5602*, Fort Worth June 30, 1970.

21. J. Willenborg, R. M. Engle, and H. A. Wood, "A Crack Growth Retardation Model Using an Effective Stress Concept," *Technical Memorandum 71-1-FBR*, Air Force Flight Dynamics Laboratory, Jan. 1971.
22. W. Elber, "The Significance of Fatigue Crack Closure," *ASTM STP 486*, American Society for Testing and Materials, Philadelphia 1971.
23. F. H. Gardner and R. I. Stephens, "Subcritical Crack Growth Under Single and Multiple Periodic Overloads in Cold-Rolled Steel," *ASTM STP 559*, American Society for Testing and Materials, Philadelphia 1974.
24. V. W. Trebules, Jr., R. Roberts, and R. W. Hertzberg, "Effect of Multiple Overloads on Fatigue Crack Propagation in 2024-T3 Aluminum Alloy," *ASTM STP* 536, American Society for Testing and Materials, Philadelphia 1973.
25. J. M. Barsom and S. R. Novak, "NCHRP Project 12–14: Subcritical Crack Growth in Steel Bridge Members," U.S. Steel Research, Final Report for Highway Research Board, National Cooperative Highway Research Program, Washington, D.C., Sept. 1974.
26. J. M. Barsom, unpublished data.

10

Stress-Corrosion Cracking

10.1. Introduction

Delayed failure of structural components subjected to an aggressive environment may occur under statically applied stresses well below the yield strength of the material. Failure of structural components under these conditions is caused by stress-corrosion cracking. The traditional approach to studying the stress-corrosion susceptibility of a material in a given environment is based on the time required to cause failure of smooth or mildly notched specimens subjected to different stress levels. This time-to-failure approach, like the traditional *S-N* approach to fatigue, combines the time required to initiate a crack and the time required to propagate the crack to critical dimensions. The need to separate stress-corrosion cracking into initiation and propagation stages was emphasized by experimental results for titanium alloys[1]. These results showed that some materials that appear to be immune to stress-corrosion in the traditional smooth-specimen tests may be highly susceptible to stress-corrosion cracking when tested under the same conditions using precracked specimens. The behavior of such materials was attributed to their immunity to pitting (crack initiation) and to their high intrinsic susceptibility to stress-corrosion cracking (crack propagation). The following discussion presents the use of fracture-mechanics concepts to study the stress-corrosion cracking of environment-material systems by using precracked specimens.

10.2. Fracture-Mechanics Approach

The application of linear-elastic fracture-mechanics concepts to study stress-corrosion cracking has met with considerable success. Because environmentally enhanced crack growth and stress-corrosion attack would be expected to occur in the highly stressed region at the crack tip, it is logical to use the stress-intensity factor, K_I, to characterize the mechanical component of the driving force in stress-corrosion cracking. Sufficient data have been published to support this observation.[1-7]

The use of the stress-intensity factor, K_I, to study stress-corrosion cracking is based on assumptions and is subject to limitations similar to those encountered in the study of fracture toughness. The primary assumption that must be satisfied when K_I is used to study the stress-corrosion-cracking behavior of materials is the existence of a plane-strain state of stress at the crack tip. This assumption requires small plastic deformation at the crack tip relative to the geometry of the test specimen and leads to size limitations on the geometry of the test specimen. Because these limitations must be established experimentally, and because there does not exist at present a standard test method for stress-corrosion cracking, the limitations established for plane-strain fracture-toughness, K_{Ic}, tests (Chapter 3) are usually applied to stress-corrosion-cracking tests.

10.3. Experimental Procedures

Experimental procedures for stress-corrosion-cracking tests of precracked specimens may be divided into two general categories. They are time-to-failure tests and crack-growth-rate tests. The time-to-failure tests are similar to the conventional stress-corrosion tests for smooth or notched specimens[8-12]. This type of test using precracked specimens has been widely used since the early work of Brown and Beachem[1]. The crack-growth-rate tests are more complex and require more sophisticated instruments than the time-to-failure tests. However, data obtained by using crack-growth-rate tests should provide information necessary to enhance the understanding of the kinetics of stress-corrosion cracking and to verify the threshold behavior K_{Iscc}, as described below.

Various precracked specimens and methods of loading can be used to study the stress-corrosion-cracking behavior of materials in both time-to-failure and crack-growth-rate tests. However, the two most widely used combinations are the cantilever-beam specimen under constant load and the wedge-opening-loading (WOL) specimen under constant displacement conditions (modified WOL specimen) that was developed by Novak and Rolfe[13].

10.3.1. Cantilever-Beam Test Specimen

Various investigators[1-7,12-18] have used fracture-mechanics concepts to study the effects of environment on precracked specimens. However, the fracture mechanics approach to environmental testing did not become widely used until Brown[1,19,20] introduced the K_{Iscc} threshold concept by using precracked cantilever-beam specimens. The K_{Iscc} value for a particular material and environment is the plane-strain stress-intensity-factor threshold below which subcritical cracks will not propagate. Since that time, the cantilever-beam test specimen has been used widely to study the stress-corrosion-cracking characteristics of steels,[3,7,14,15,17,19,21] titanium alloys,[18,22-24] and aluminums alloys.[12,25,26]

Figure 10.1 presents a geometry of a cantilever-beam specimen that has been used to study the stress-corrosion-cracking behavior of materials.[3,17,27,28] The specimen is usually face-notched 5–10% of the thickness, and the notch is extended by fatigue-cracking the specimens at low stress-intensity-factor levels. Then the specimens are tested in a stand similar to that shown in Figure 10.2. Usually, two specimens are monotonically loaded to failure in air to establish the critical stress-intensity factor for failure in the absence of environmental effects (K_{Ic} if ASTM requirements[29] are satisfied and K_{Ix} if they are not satisfied). Subsequently, specimens are immersed in the environment and dead-weight-loaded to successively lower initial stress-intensity-factor, K_{Ii}, levels. If the material is susceptible to the test environment, the fatigue crack will propagate. As the crack length increases under constant load, the stress-intensity factor at the crack tip increases to the K_{Ic} (or K_{Ix}) level, and the specimen fractures. The lower the value of the initial K_I, the longer the time to failure. Specimens that do not fail after a long period of test time, usually 1,000 hr for steels, should be fractured and inspected for

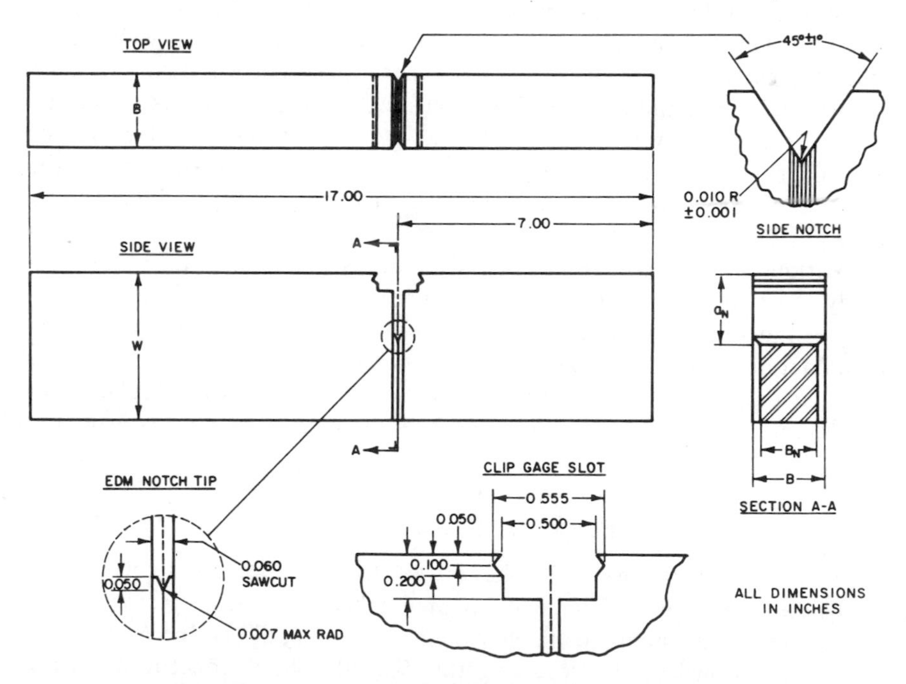

FIG. 10.1. Fatigue-cracked cantilever-beam test specimen.

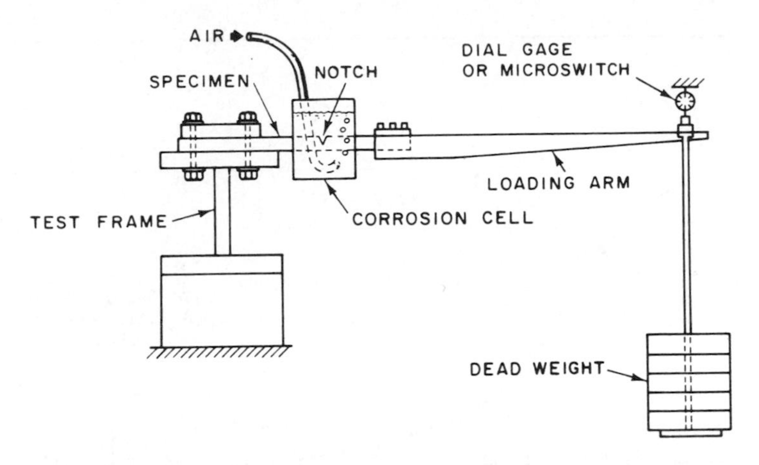

FIG. 10.2. Schematic drawing of fatigue-cracked cantilever-beam test specimen and fixtures.

possible crack extension. The highest plane-strain K_{Ii} level at which crack extension does not occur after a long test time corresponds to the stress-corrosion-cracking threshold, K_{Iscc}. Thus, K_{Iscc} for a given material and environment corresponds to the plane-strain stress intensity factor below which statically loaded structural components arc expected to have infinite life when subjected to the particular test environment.

Stress-intensity-factor values for cantilever-beam specimens (Figure 10.1) can be calculated by using the following equation developed by Bueckner[30] for an edge crack in a strip subjected to in-plane bending and modified to account for reduction of thickness due to face notches:

$$K_I = \frac{6M}{(B \cdot B_N)^{1/2}(W - a)^{3/2}} F\left(\frac{a}{W}\right) \tag{10.1}$$

where K_I = stress-intensity factor,
M = bending moment,
B = gross specimen width,
B_N = net specimen width,
W = specimen depth,
a = total crack length, and

$F\left(\frac{a}{W}\right)$ =	for $\frac{a}{W}$ =
0.36	0.05
0.49	0.10
0.60	0.20
0.66	0.30
0.69	0.40
0.72	0.50
0.73	0.60 and larger.

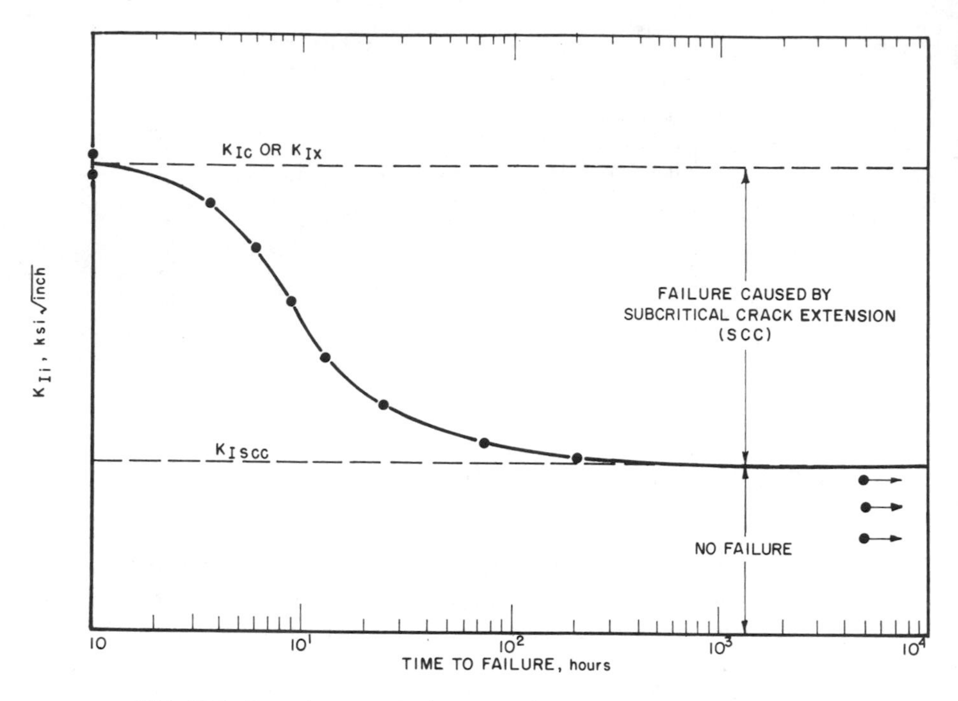

FIG. 10.3. Procedure to obtain K_{Iscc} with precracked cantilever-beam specimens.

Figure 10.3 is a schematic representation of test results obtained by using cantilever-beam test specimens. Approximately ten precracked cantilever specimens are needed to establish K_{Iscc} for a particular material and environment.

10.3.2. Wedge-Opening-Loading (WOL) Test Specimen

Extensive analytical and experimental investigations have been conducted to study the behavior of WOL specimens.[31-33] These specimens have been used to study the fracture toughness,[34] fatigue-crack initiation[35] and propagation,[27] stress-corrosion-cracking,[13] and corrosion-fatigue-crack-growth[27,36] behavior of various materials. The geometry of 1-in.-thick (1T) WOL specimens is shown in Figure 10.4. The stress-intensity factor, K_I, at the crack tip is calculated from[25]

$$K_I = \frac{C_3 P}{B\sqrt{a}} \tag{10.2}$$

where P = applied load,
B = specimen thickness,
a = crack length measured from the loading plane,
C_3 = a function of the dimensionless crack length, a/W (Figure 10.5),

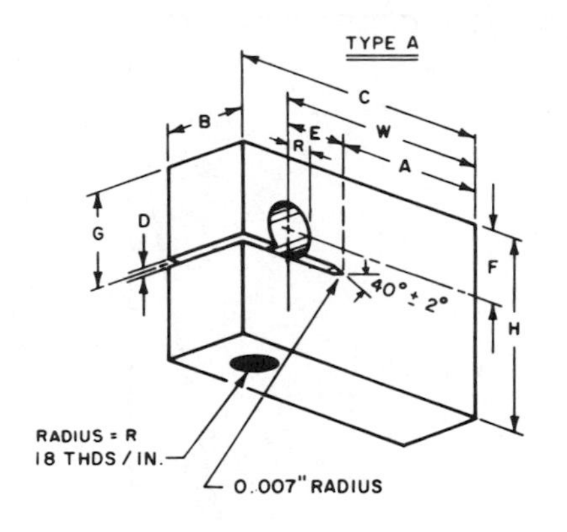

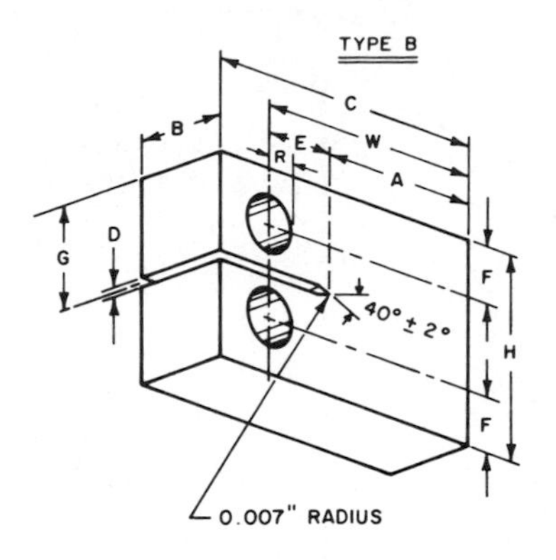

SPEC	B	W	C	A	E	H	G	D	R	F
IT-A	1.000	2.550	3.200	1.783	0.767	2.480	1.240	0.094	0.350	1.000
IT-B	1.000	2.550	3.200	1.783	0.767	2.480	1.240	0.094	0.250	0.650

1 inch = 25.4 mm
1 degree = 0.017 rad

FIG. 10.4. Two types of 1T WOL specimens.

where W is the specimen length measured from the loading plane. Expressed in a polynomial form, C_3 for the WOL specimen geometry presented in Figure 10.4 can be represented as

$$30.96\left(\frac{a}{W}\right) - 195.8\left(\frac{a}{W}\right)^2 + 730.6\left(\frac{a}{W}\right)^3 - 1186.3\left(\frac{a}{W}\right)^4 + 754.6\left(\frac{a}{W}\right)^5$$

In the range $0.25 \leq a/W \leq 0.75$, the polynomial is accurate to within 0.5% of the experimental compliance.

The WOL specimen was modified by the use of a bolt and loading tup (Figure 10.6) so that it can be self-stressed without using a tensile machine.[13] The crack opening is fixed by the bolt, and the loading is by constant displacement rather than by constant load as in the cantilever-beam specimen. Because a constant crack-opening displacement is maintained throughout the test, the force, P, decreases as the crack length increases (Figure 10.6). In cantilever-beam testing, the K_I value increases as the crack length increases under constant load, which leads to fracture for each specimen. In contrast, for the modified WOL specimen, the K_I value decreases as the crack length increases under a decreasing load. The decrease in load more than compensates for the increase in crack length and leads to crack arrest at K_{Iscc}. A comparison of these two types of behavior is shown schematically in Figure

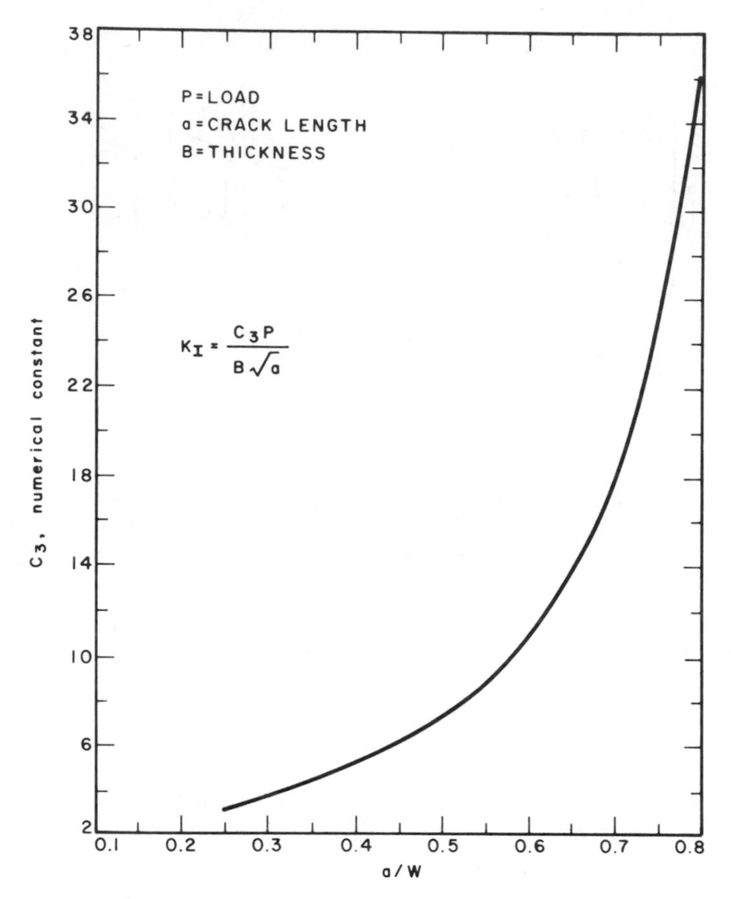

FIG. 10.5. Relation between constant C_3 used in calculating K_I and a/W ratio.

10.7. Thus, only a single WOL specimen is required to establish the K_{Iscc} level because K_I approaches K_{Iscc} in the limit. However, duplicate specimens are usually tested to demonstrate reproducibility. Because the bolt-loaded WOL specimen is self-stressed and portable, it can be used to study the stress-corrosion-cracking behavior of materials under actual operating conditions in field environments.

10.4. K_{Iscc}—A Material Property

Brown and Beachem[37] investigated the K_{Iscc} for environment-material systems by using various specimen geometries. Their results (Figure 10.8) show that indentical K_{Iscc} values were obtained for a given environment-material system by using center-cracked specimens, surface-cracked speci-

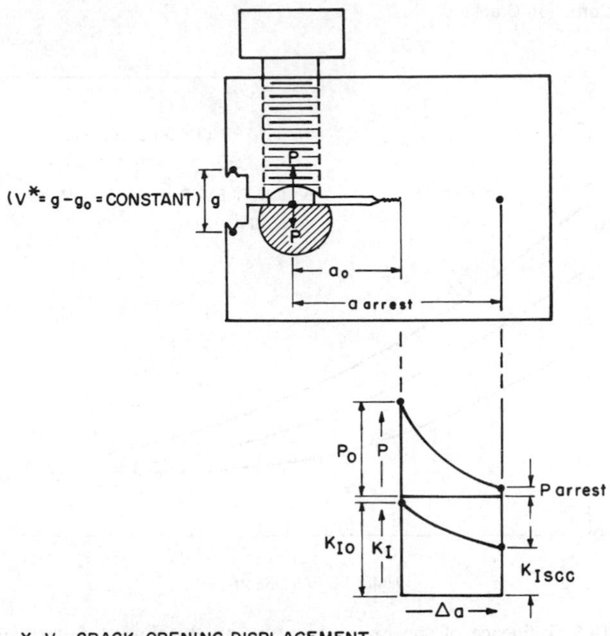

FIG. 10.6. Schematic showing basic principle of modified WOL specimen.

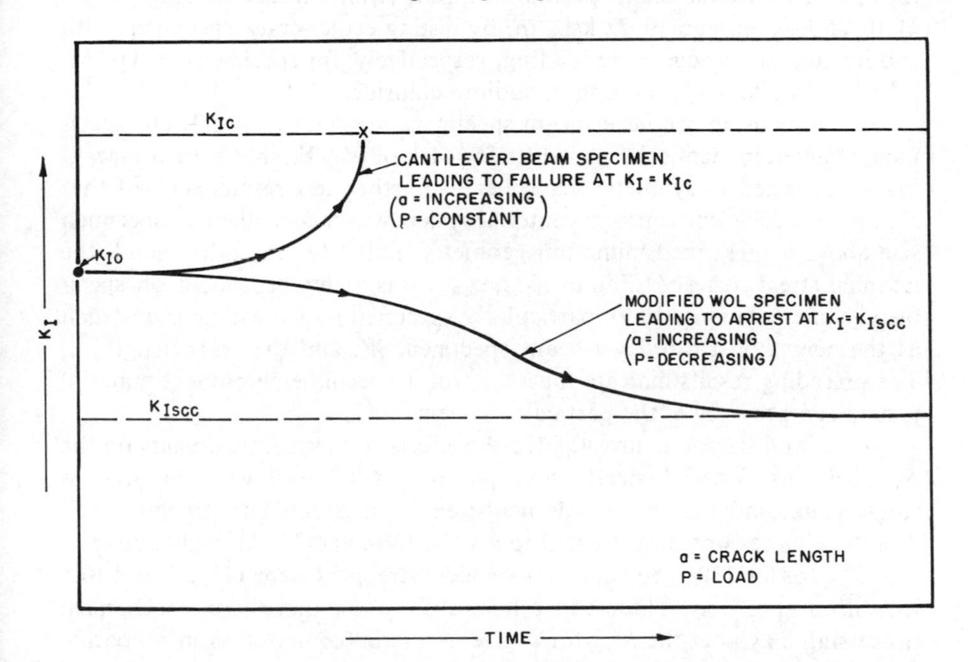

FIG. 10.7. Difference in behavior of modified WOL and cantilever-beam specimens.

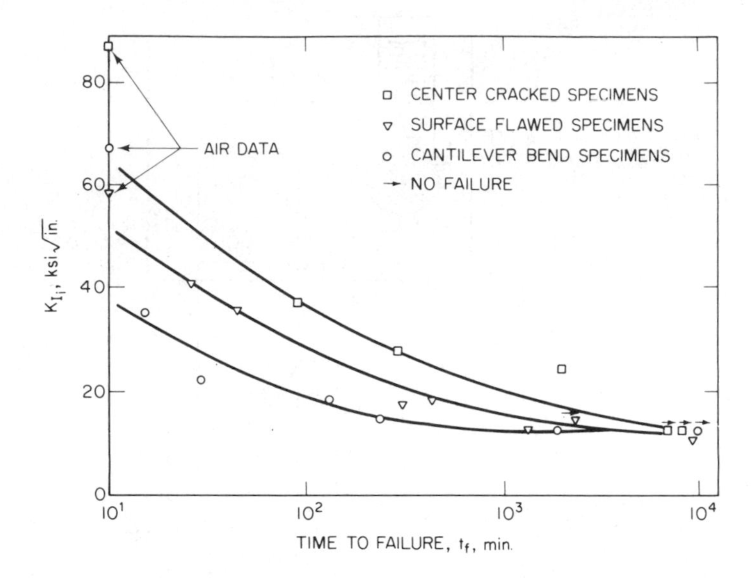

FIG. 10.8. Influence of specimen geometry on the time to failure (AISI 4340 steel).

mens, and cantilever-beam specimens. Smith et al.[22] measured K_{Iscc} values of 10–25 ksi$\sqrt{in.}$ and 10–22 ksi$\sqrt{in.}$ by testing center-crack specimens with end loading and wedge-force loading, respectively, for specimens of Ti-8Al-1Mo-1V alloy in 3.5% solution of sodium chloride.

K_{Iscc} tests using cantilever-beam specimens and bolt-loaded WOL specimens resulted in identical K_{Iscc} values for each of two 12Ni-5Cr-3Mo maraging steels tested in synthetic sea water.[13] Further test results showed that K_{Iscc} for a specific environment-material system was independent of specimen size above a prescribed minimum geometry limit.[3] On the other hand, the nominal stress corresponding to K_{Iscc}, σ_{Nscc}, was highly dependent on specimen geometry (Figure 10.9), particularly specimen in-plane dimensions such as the height of a cantilever-beam specimen, W, and the crack length, a. The preceding results indicate that K_{Iscc} for a specific environment-material system is a property of the particular system.

Imhof and Barsom[27] investigated the effects of thermal treatments on the K_{Iscc} behavior for 4340 steel. Three pieces of 4340 steel were cut from a single plate, and each piece was heat-treated to a different strength level. The three pieces were heat-treated to a 130-, 180-, and 220-ksi yield strength. The K_{Iscc} for the 130-, 180-, and 220-ksi yield strengths were 111, 26, and 10.5 ksi$\sqrt{in.}$, respectively. These and other results show that thermomechanical processing may alter the K_{Iscc} for a given material composition in a specific environment. Available data also show that K_{Iscc} for a given material composition, thermomechanical processing, and environment may be different for different orientations of test specimens. An example of this behavior is

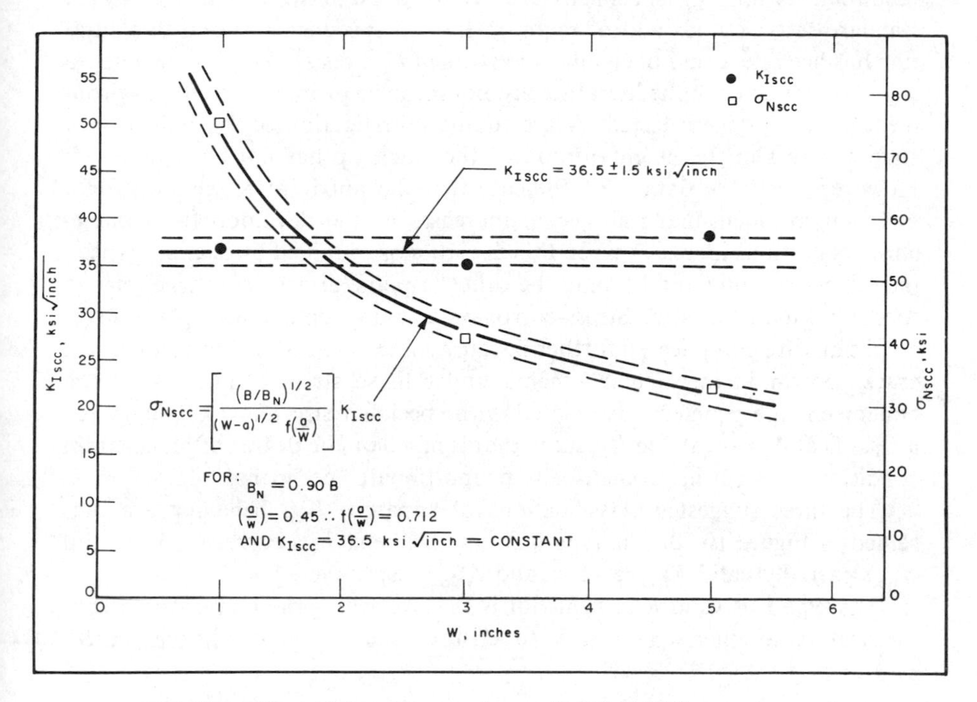

FIG. 10.9. Effect of W on K_{Iscc} and σ_{Nscc} for 18Ni (250) maraging steel.

observed in aluminum alloys where the susceptibility to stress-corrosion cracking in the short-transverse direction (crack plane parallel to plate surfaces) is greater than in the other directions. Consequently, although K_{Iscc} is a unique property of the tested environment-material system, extreme care should be exercised to ensure the use of the correct K_{Iscc} value for a specific application.

10.5. Validity of K_{Iscc} Data

Standard test methods for stress-corrosion cracking are yet to be established. Discussions of the important parameters that must be considered to ensure valid stress-corrosion cracking have been presented in various publications.[2,3,38] The two most important parameters are specimen size and duration of the test.

10.5.1. Specimen Size

K_{Iscc} is the threshold stress-intensity factor for stress-corrosion cracking under plane-strain conditions. To ensure the existence of plane-strain conditions at the tip of the crack, K_{Iscc} specimens, like K_{Ic} specimens, must satisfy minimum size requirements. However, unlike K_{Ic} specimens, the size requirements for K_{Iscc} specimens are yet to be established. The absence of standard test methods and of requirements on specimen size and test duration has been the cause of significant errors in K_{Iscc} tests. Many of the alleged K_{Iscc} data reported in the literature are not intrinsic properties of the environment-material system tested. A systematic investigation of the influence of state of stress in the neighborhood of the crack tip has not been reported. However, available data[2,20,39] indicate that the apparent K_{Iscc} value for a given environment-material system increases as the deviation from plane-strain conditions increases when this deviation is obtained by changing specimen thickness only and keeping the other in-plane dimensions large relative to the plastic-zone size. Stress-corrosion-cracking data under plane-stress conditions are complicated further because some materials exhibit subcritical crack growth in inert environments under these stress conditions.[2,40] To circumvent the problems associated with specimen size, Novak[17] suggested a classification to evaluate K_{Iscc} data that is based on the degree of plane-strain conditions existent in geometrically proportionate specimens.

The three suggested classifications of apparent K_{Iscc} behavior are presented in Figure 10.10. The type I, II, and III behaviors correspond to valid K_{Iscc}, partially valid K_{Iscc}, and invalid K_{Iscc}, respectively.

The type I or valid K_{Iscc} behavior is that for which the conservative geometrical requirements necessary to ensure plane-strain conditions in K_{Ic}

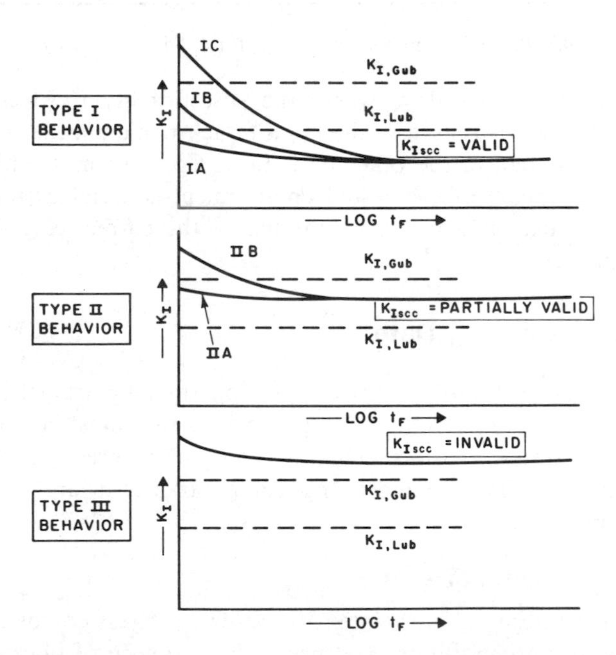

FIG. 10.10. Three basic types of apparent K_{Iscc} behavior and classification.

fracture-toughness test specimens are satisfied. Thus, the boundary between type I and type II behaviors, designated $K_{I,Lub}$ in Figure 10.10, corresponds to the highest K_I value that simultaneously satisfies the four equations

$$a_{min} = B_{min} = (W - a)_{min} = \frac{W_{min}}{2} = 2.5\left(\frac{K_I}{\sigma_{ys}}\right)^2 \tag{10.3}$$

where a_{min} = minimum crack length,
B_{min} = minimum specimen thickness,
W_{min} = minimum specimen width,
σ_{ys} = material yield strength.

The boundary between type II and type III behaviors, designated $K_{I,Gub}$ in Figure 10.10, was selected to correspond to the highest K_I value that represents the initial occurrence of a plastic hinge for an elastic-perfectly plastic material subjected to bending and that simultaneously satisfies the four equations

$$a_{min} = B_{min} = (W - a)_{min} = \frac{W_{min}}{2} = 1.0\left(\frac{K_I}{\sigma_{ys}}\right)^2 \tag{10.4}$$

These classifications were proposed to discern the effect of specimen size on K_{Iscc} behavior and may be used to evaluate published data.

10.5.2. Test Duration

Test duration is the second primary parameter that must be understood to ensure correct test results. The schematic representation for obtaining K_{Iscc} by using precracked cantilever-beam specimens (Figure 10.3) suggests that the true K_{Iscc} level was established with test duration greater than 1,000 hr. Test durations less than 200 hr (Figure 10.3) would result in apparent K_{Iscc} values that are greater than the true K_{Iscc} value obtained after 1,000-hr test duration. The influence of test duration on the apparent K_{Iscc} value obtained by using cantilever-beam test specimens of a 180-ksi yield strength, high-alloy steel in room-temperature synthetic seawater is shown in Table 10.1. The data show that an increase of test duration from 100 hr to 10,000 hr decreased the apparent K_{Iscc} value from 170 ksi$\sqrt{\text{in.}}$ to 25 ksi$\sqrt{\text{in.}}$ Proper test durations depend on specimen configuration, specimen size, and nature

TABLE 10.1 Influence of Cutoff Time on Apparent K_{Iscc}; Constant-Load Cantilever Bend Specimens (Increasing K_I) (Reference 38)

Elapsed Time (hr)	*Apparent K_{Iscc} (ksi$\sqrt{\text{in.}}$)*
100	170
1,000	115
10,000	25

of loading, as well as the environment-material system. Test durations for bolt-loaded WOL specimens are longer than for cantilever-beam specimens.[13] In general, test durations for titanium, steel, and aluminum alloys are on the order of 100, 1,000, and 10,000 hr, respectively. The differences in test duration for different metal alloys are related partly to the incubation-time behavior in stress-corrosion cracking for the particular environment-material system. The incubation-time behavior represents the test time prior to crack extension during which a fatigue crack under sustained load in an aggressive environment appears to be dormant. The existence of incubation periods for precracked specimens has been demonstrated by various investigators.[35-38] Benjamin and Steigerwald[41] demonstrated the dependence of incubation time on prior loading history. Novak demonstrated the dependence of incubation time on the magnitude of the stress-intensity factor (Table 10.2).[38] It is apparent that as the applied stress-intensity factor, K_I, approaches K_I at fracture, the incubation time must approach zero, and as the applied K_I approaches K_{Iscc}, the incubation time approaches infinity. Consequently, specimens subjected to K_I values between K_{Ic} (or K_I) and K_{Iscc} can exhibit initiation times that may be of 1,000-hr duration or greater.

TABLE 10.2 Influence of K_I on Incubation Time; Constant-Displacement WOL Specimens (Decreasing K_I)

	Extent of Crack Growth (in.)					
K_I (*ksi*$\sqrt{in.}$)	*200 hr*	*700 hr*	*1,400 hr*	*2,200 hr*	*3,500 hr*	*5,000 hr*
180	ND*	0.35	0.76	1.00	1.12	—
150	ND	ND	ND	0.28	0.52	0.61
120	ND	ND	ND	ND	0.03	0.045
90	ND	ND	ND	ND	ND	0.045

*ND: no detectable growth.

10.6. General Observations

In general, the higher the yield strength for a given material, the lower the K_{Iscc} value in a given environment.[24,27] The K_{Iscc} for a single plate of 4340 steel tested in 3.5% solution of sodium chloride decreased from 111 to 10.5 ksi$\sqrt{in.}$ as the yield strength was increased from 130 to 220 ksi. In general, the K_{Iscc} in room-temperature sodium chloride solutions for steels having a yield strength greater than about 200 ksi is less than 20 ksi$\sqrt{in.}$ Similar generalizations cannot be made for steels of lower yield strengths. Moreover, stress-corrosion-cracking data for steels having yield strengths less than 130 ksi are very sparse. The results of 5,000-hr (30-week) stress-corrosion tests for five steels having yield strengths less than 130 ksi obtained by testing precracked cantilever-beam specimens in room-temperature 3% sodium chloride solution are presented in Figures 10.11 through 10.15.[42] The results presented in these figures show apparent K_{Iscc} values that ranged from 80 to 106 ksi$\sqrt{\text{in.}}$ (88 to 117 $MNm^{-3/2}$). The apparent K_{Iscc} values measured for most of these steels corresponded to conditions involving substantial crack-tip plasticity. Consequently, linear-elastic fracture-mechanics concepts cannot be used for quantitative analysis of the respective stress-corrosion-cracking behavior.[17] Furthermore, the apparent K_{Iscc} values are suppressed

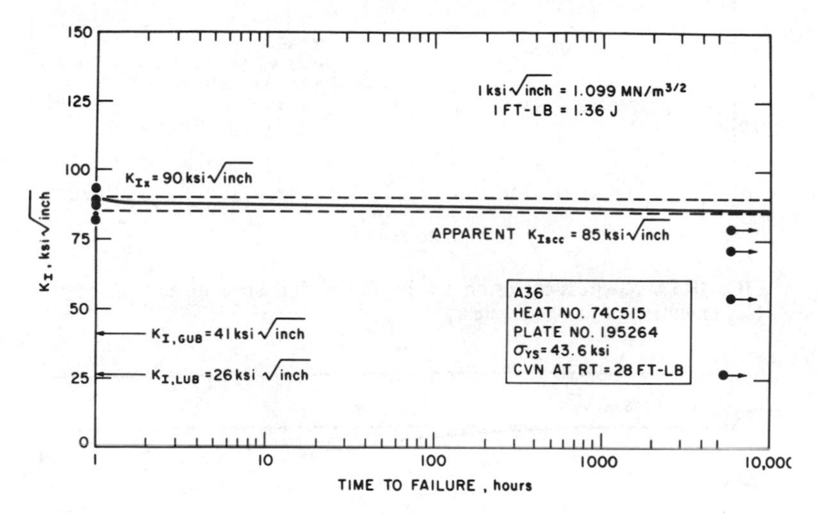

FIG. 10.11. K_I-stress-corrosion results for A36 steel in aerated 3% NaCl solution of distilled water.

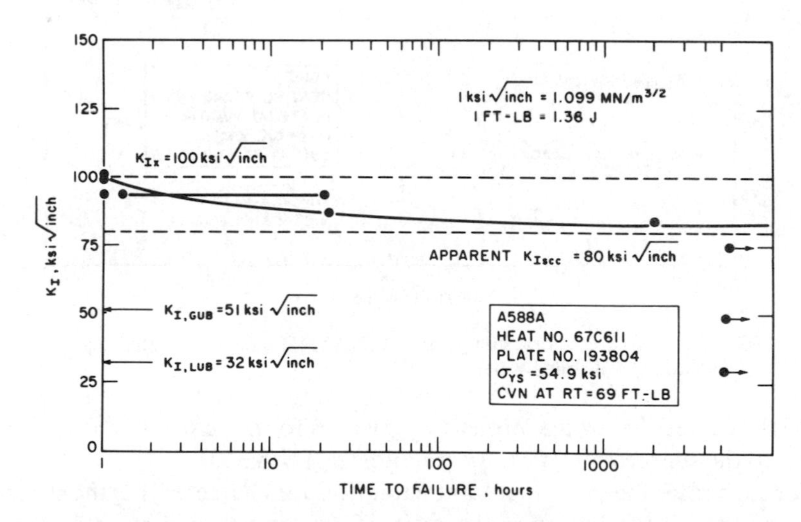

FIG. 10.12. K_I-stress-corrosion results for A588A steel in aerated 3% NaCl solution of distilled water.

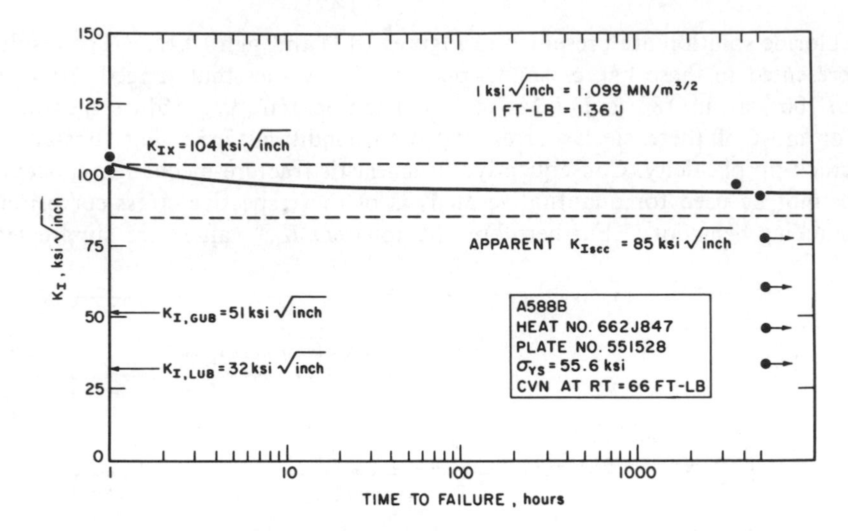

FIG. 10.13. K_I-stress-corrosion results for A558B steel in aerated 3% NaCl solution of distilled water.

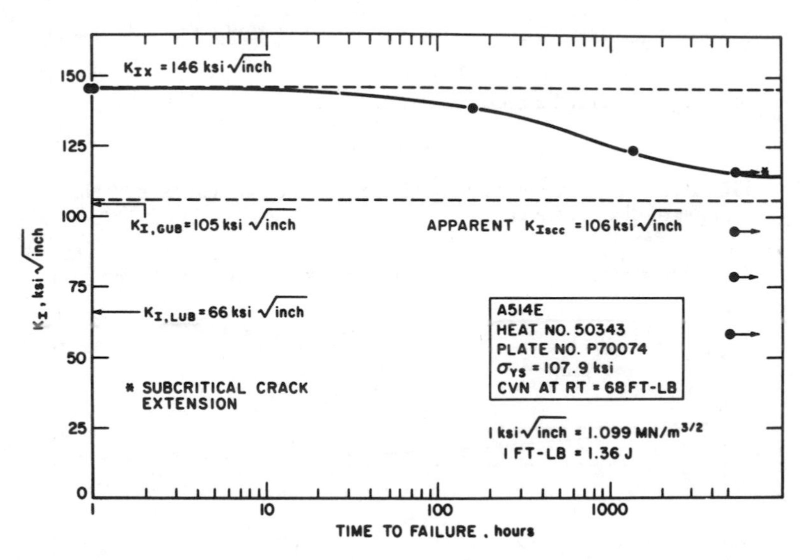

FIG. 10.14. K_I-stress-corrosion results for A514E steel in aerated 3% NaCl solution of distilled water.

to various degrees below the intrinsic K_{Iscc} values for these steels in a manner similar to the suppression effect for fracture (K_{Ic}) behavior.[17]

For such cases the most important parameter for characterizing the stress-corrosion-cracking behavior is the ratio of the apparent threshold, K_{Iscc},

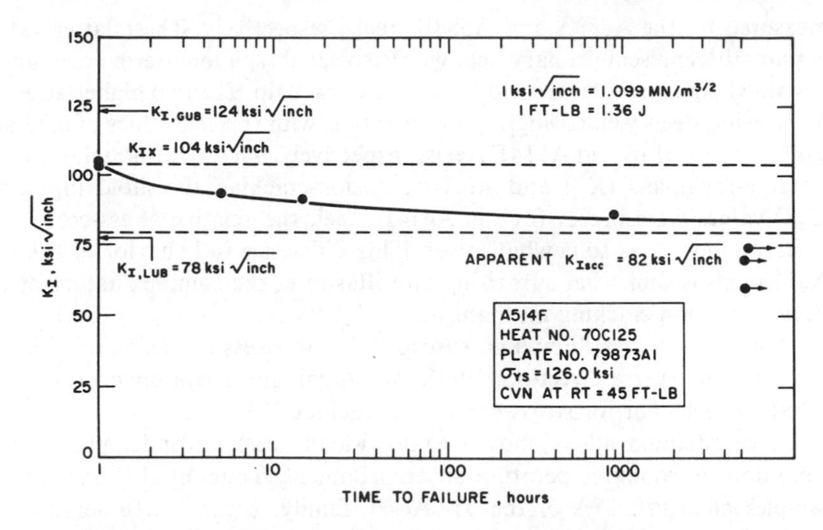

FIG. 10.15. K_I-stress-corrosion results for A514E steel in aerated 3% NaCl solution of distilled water.

to the value at fracture, K_{Ix}, obtained with an identical size specimen (apparent K_{Iscc}/K_{Ix}).[17] This ratio, known as the relative index of stress-corrosion-cracking susceptibility, can be used to ascertain quickly the degradation in fracture behavior as a result of the test solution and the inherent stress-corrosion-cracking characteristics of each steel.

General experience with many steels and weldments, some of which are used successfully in long-time environmental service applications, dictates the following broad-based interpretation for assessing the relative stress-corrosion-cracking index: (1) values in the range 0.95–1.00 represent material behaviors that are immune to stress-corrosion cracking, (2) values in the range 0.80–0.95 represent material behaviors that are either moderately susceptible to stress-corrosion cracking if the K_{Iscc} value is valid or primarily the result of long-time creep that occurs at the crack tip (under the presence of intensely high stress levels) if the apparent K_{Iscc} value measured is not valid and high levels of crack-tip plasticity are involved, and (3) values less than 0.80 are generally the result of true susceptibility to stress-corrosion cracking for steels regardless of whether the apparent K_{Iscc} value is valid or not.

The results presented in Figures 10.11 through 10.15 show that the five steels yielded relative stress-corrosion-cracking indices of 0.73 or higher for the 5,000-hr tests in continuously aerated 3% sodium chloride solution. In particular, a value of 0.95, representing essentially immune behavior, was found to occur for the A36 steel. Nearly identical values of 0.80 and 0.82 were measured for the A588A and A588B steels, respectively. These latter values apparently represent primary behavior associated with long-term creep under sustained high-stress-level conditions at the crack tip. The two higher-strength martensitic steels yielded slightly lower ratios, with specific values of 0.73 and 0.79 for the A514E and A514F steels, respectively. That is, although both the fracture-toughness (K_{Ix}) and stress-corrosion-cracking threshold (apparent K_{Iscc}) values were higher for the A514E steel, the relative stress-corrosion-cracking index was somewhat lower. This difference in behavior of the two A514 steels is somewhat surprising and illustrates the complex nature of the stress-corrosion-cracking mechanism.

Aluminum alloys show high susceptibility to stress corrosion cracking in the short-transverse direction.[11,12] Some aluminum alloys do not exhibit a threshold behavior for stress-corrosion cracking.[25,26]

Some titanium alloys show sustained-load cracking under plane-strain conditions in room-temperature air environment. Yoder et al.[43] tested plate samples of eight alloys of the Ti-6Al-4V family. Figure 10.16 summarizes results for one of these alloys tested by using cantilever-bend specimens and part-through crack tension specimens. They concluded that sustained-load cracking in room-temperature air environment is widespread and serious in these alloys and that the resulting degradations ranged from 11 to 35%. The magnitude of degradation did not appear to correlate with interstitial contents, processing variables, strength level, or toughness level, but it was orientation-dependent.

Novak[17] conducted a systematic study to determine the effect of prior plastic strain on the mechanical and environmental properties of four steels ranging in yield strength from 40 to 200 ksi (550 to 1400 N/mm²). Each steel was evaluated first in the unstrained condition and then after 1 and either 3 or 5% plastic strain. The results showed that the value of the stress-intensity factor at fracture decreased with increased magnitude of prior plastic strain. However, the corresponding change in the apparent K_{Iscc} value did not follow any consistent pattern of behavior.

Extensive investigations have been conducted to evaluate the K_{Iscc} behavior of various metals in seawater.[2-7,17,18,24-27] The test environments used were synthetic seawater, 3 or 3.5% solution of sodium chloride in distilled water, or natural seawater. In general, the results suggest that natural seawater may be more severe than either the synthetic seawater or the solutions of sodium chloride. Tests in natural seawater environments can result in K_{Iscc} values that are 20–30% lower than values obtained in synthetic seawater or solutions of sodium chloride.

A systematic investigation of the influence of test temperature on the K_{Iscc} behavior of environment-material systems is yet to be conducted.

K_{Iscc} data for various environment-material systems have been gathered and published.[44-48]

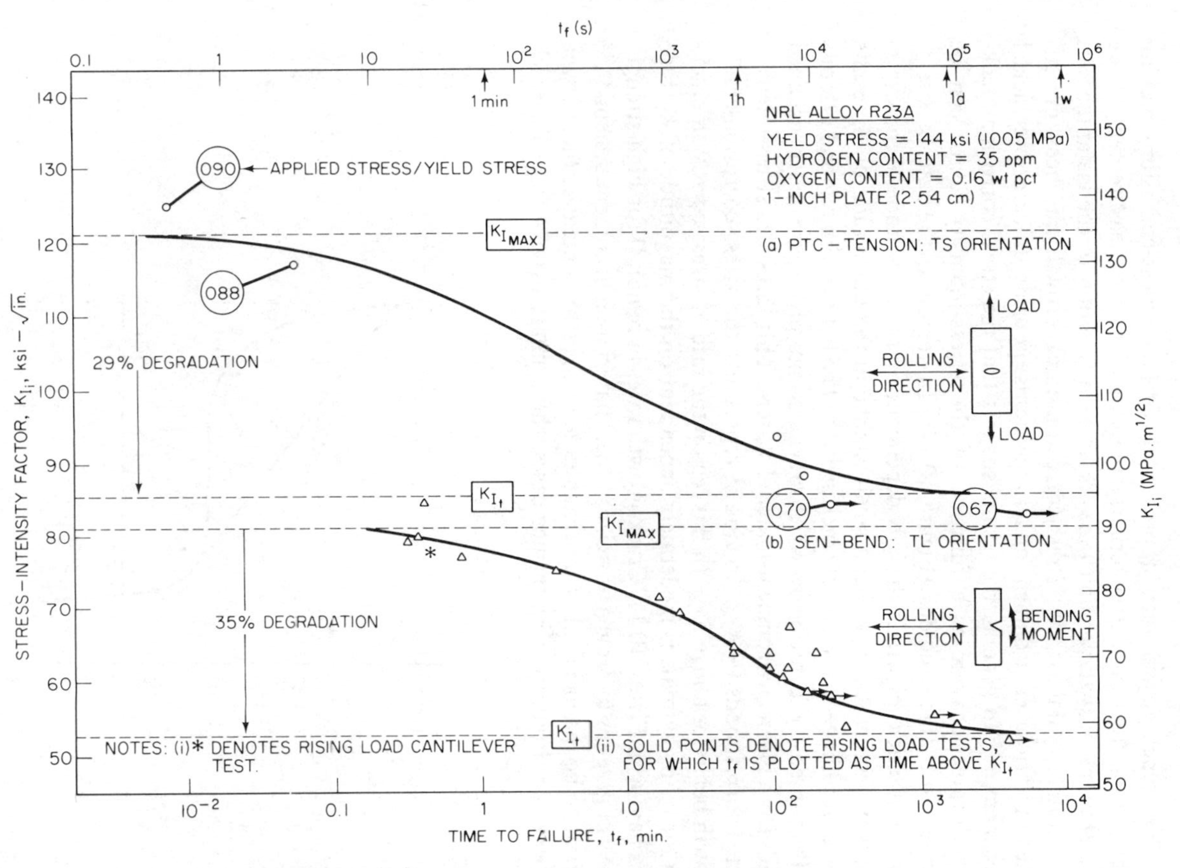

FIG. 10.16. Sustained-load cracking of a commercial grade Ti-6A1-4V material air environment.

10.7. Crack-Growth-Rate Tests

The crack-growth-rate approach to study stress-corrosion-cracking behavior of environment-material systems involves the measurement of the rate of crack growth per unit time, da/dt, as a function of the instantaneous stress-intensity factor, K_I. Stress-corrosion-crack growth has been investigated in various environment-material systems by using different precracked specimens.[2,38] In general, the results suggest that the stress-corrosion-crack-growth-rate behavior as a function of the stress-intensity factor can be divided into three regions (Figure 10.17). In region I, the rate of stress-corrosion-crack growth is strongly dependent on the magnitude of the stress-intensity factor, K_I, such that a small change in the magnitude of K_I results in a large change in the rate of crack growth. The behavior in region I exhibits a stress-intensity factor below which cracks do not propagate under sustained loads for a given environment-material system. This threshold stress-intensity factor corresponds to K_{Iscc}. Region II represents the stress-corrosion-crack-growth behavior above K_{Iscc}. In this region the rate of stress-corrosion cracking for many systems is moderately dependent on the magnitude of K_I, type A behavior in Figure 10.17. Crack-growth rates in region II for high-strength steels in gaseous hydrogen as well as other material-environment systems[45-48] appear to be independent of the magnitude of the stress-intensity factor, type B behavior (Figure 10.17). In such cases, the primary driving force for crack

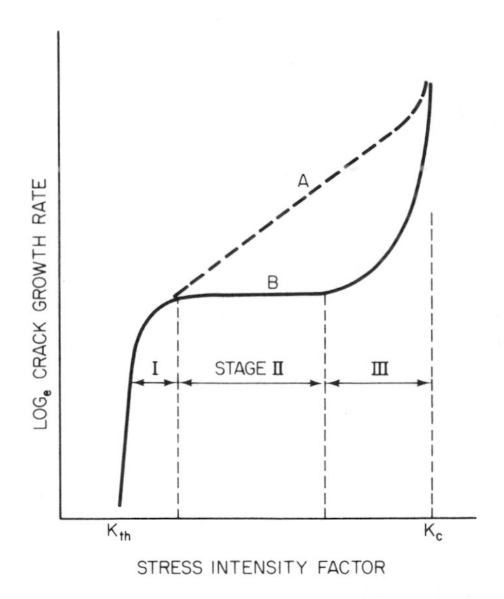

FIG. 10.17. Schematic illustration of the functional relationship between stress-intensity factor (K) and subcritical crack growth rate (da/dt).

growth is not mechanical (K_I) in nature but is related to other processes occurring at the crack tip such as chemical, electrochemical, mass-transport, diffusion, and adsorption processes. The crack growth rate in region III increases rapidly with K_I as the value of K_I approaches K_{Ic} (or K_c) for the material.

The characteristic crack-growth behavior for a given environment-material system is determined by the mechanical properties, chemical properties, or both, for the system. The crack-growth approach to study stress-corrosion cracking is of great value in determining the mechanism of crack extension. It can be used to analyze the safety and reliability of structures subjected to stress-intensity-factor values that are above K_{Iscc} and for environment-material systems that do not exhibit a threshold behavior. The bolt-loaded WOL specimen can be used best to determine the stress-corrosion-crack-growth-rate behavior and the K_{Iscc}. The rate of growth at K_{Iscc} should be equal to or less than 10^{-5} in./hr.

10.8. Effect of Composition and Applied Potential

The primary factors that influence K_{Iscc} (aside from strength level, σ_{ys}) are currently unknown. Less is known about the corresponding factors that influence the kinetics of crack propagation (from a fatigue crack) due to stress-corrosion-cracking at K_I levels above K_{Iscc}. For example, the very existence of incubation times (t_{inc}) that range to 1,000 hr or more for precracked steel specimens has only recently been documented.[38] Whereas t_{inc} is known to depend strongly on initial stress-intensity level (K_{Ii}), the crack-propagation rate (da/dt) appears to be relatively insensitive to stress-intensity level (K_{Ii}) for many materials, until K_{Iscc} is approached.[38] However, both t_{inc} and da/dt can vary significantly from one material to another, even at the same strength level (σ_{ys}) for tests in the same environment. As an example, the kinetics of stress-corrosion cracking for precracked-specimen tests of 18Ni 250-grade maraging steel tested in seawater-type solutions are substantially slower than those of lean-alloy high-strength steels at the same strength level (D6AC, H-11, AISI 4340). Although a total test time between 10 and 100 hr is generally sufficient to establish K_{Iscc} for such lean-alloy steels,[27,38] corresponding test times of 2,000 hr or more are required for 18Ni 250-grade maraging steel.[3,5] Reasons for such differences are currently unknown.

Investigations to study the effect of steel cleanliness and purity on fracture toughness and stress-corrosion cracking have been limited to materials having yield strengths greater than about 180 ksi. As tensile strength increases from 250 to 310 ksi, the improvements in fracture toughness that can be realized by producing ultra-high-purity steel decrease with strength until at a tensile strength of 310 ksi, steels with ultra-high purity have no higher fracture toughness than steels of normal commercial purity.[49] Studies on stress-corrosion cracking in 18Ni maraging steel indicate that there may be a similar relation between strength and resistance to stress-corrosion cracking as determined by K_{Iscc} measurements.[3-5] For example, at 180-ksi yield strength, increases in the purity of 18Ni maraging steel resulted in increases in K_{Iscc} from 108 ksi$\sqrt{\text{in.}}$ to 150 ksi$\sqrt{\text{in.}}$, while at a yield strength level of 250 ksi, a reduction in C, Mn, P, S, and Si contents of an 18Ni maraging steel resulted in an increase of fracture toughness from a K_{Ic} value of 67 ksi$\sqrt{\text{in.}}$ to 105 ksi$\sqrt{\text{in.}}$, but the resistance to stress-corrosion cracking remained essentially unaffected—exhibiting a K_{Iscc} of 36 ksi$\sqrt{\text{in.}}$ in the lower-toughness steel and a value of 25 ksi$\sqrt{\text{in.}}$ in the higher-toughness steel. Thus, in maraging steels the improvement in K_{Iscc} that can be realized by improved steel purity appears to be insignificant for the very high strength level steels, e.g. 250 ksi.

If the above relations between strength, toughness, and resistance to stress-corrosion were found to hold regardless of steel type, composition, or degree of purity achieved, the use of high-strength steels in aircraft applications could be limited by the relatively low resistance to stress-corrosion cracking even in the useful tensile-strength range of 240–270 ksi, where improvements in steel melting and processing have been found to produce marked improvements in fracture toughness. Insufficient work has been conducted at very high-purity levels and on steels other than 18Ni maraging steels to determine to what extent the behavior observed to date in the maraging steels is found in other types of steel.

Steels with tensile strengths greater than 240 ksi can have K_{Ic} values greater than 70 ksi$\sqrt{\text{in.}}$ but rarely have a K_{Iscc} greater than 20 ksi$\sqrt{\text{in.}}$.[27,50,51] Proctor and Paxton[52] obtained K_{Iscc} values between 8 and 13 ksi$\sqrt{\text{in.}}$ for 300-grade 18Ni maraging steels of varying purity. Similar data was obtained by Carter,[51] who reported that values of K_{Ic} as high as 71 ksi$\sqrt{\text{in.}}$ for 300-grade 18Ni maraging steels were obtained by controlling residual elements but the K_{Iscc} values remained very low at about 7 ksi$\sqrt{\text{in.}}$ Low-alloy steels such as D6AC, 4340, and 300M of commercial purity also have K_{Iscc} values between 10 and 20 ksi$\sqrt{\text{in.}}$

Dautovich and Floreen[53] studied the stress-corrosion-cracking behavior of maraging steels and suggested that increases in K_{Iscc} values can be obtained at carbon and sulfur levels below 20 ppm. There is also limited evidence that control of residual elements can lead to improved toughness and stress-corrosion behavior of 9-4-45 steel and 10Ni-Cr-Mo-Co steel.[50,54]

Stress-corrosion-cracking tests conducted on steels having yield strengths less than or equal to 180 ksi indicate that the stress-corrosion-cracking kinetics (t_{inc} and da/dt) are more rapid and K_{Iscc} values less with cathodic potentials compared with identical evaluation under open-circuit conditions.[55,56] In particular, a 180-ksi-strength 17-4 pH steel evaluated in salt water has been observed[56] under cathodic potentials to exhibit a dramatic decrease in K_{Iscc} to a value close to half of that observed under open-circuit conditions. The extent of reduction in K_{Iscc} behavior due to applied electrical potential for higher-strength steels (240–270 ksi) has not yet been investigated under any conditions and may be substantially greater.

Despite the significant progress that has been achieved in stress-corrosion-cracking over the past several years, it can be said that, in general, "there presently is no reliable fundamental theory of stress-corrosion cracking in any alloy-environment system which can be used to predict the performance of equipment even in environments where conditions are readily defined".[57] Thus test results, preferably using the methods described in this chapter, must still be used to design to prevent failures by stress-corrosion.

REFERENCES

1. B. F. Brown and C. D. Beachem, "A Study of the Stress Factor in Corrosion Cracking by Use of the Precracked Cantilever-Beam Specimen," *Corrosion Science*, *5*, 1965.
2. H. H. Johnson and P. C. Paris, "Subcritical Flaw Growth," *Engineering Fracture Mechanics*, *1*, No. 1, 1968.
3. S. R. Novak and S. T. Rolfe, "Comparison of Fracture Mechanics and Nominal Stress Analysis in Stress Corrosion Cracking," *Corrosion*, *26*, No. 4, April 1970.
4. S. R. Novak and S. T. Rolfe, "K_{Ic} Stress-Corrosion Tests of 12Ni-5Cr-3Mo and 18Ni-8Co-3Mo Maraging Steels and Weldments," *U.S. Steel Applied Research Laboratory Report No. 39.018-002(34)* (*S-23309-2*), Defense Documentation Center, Arlington, Va., Jan. 1, 1966.
5. S. R. Novak, "Comprehensive Investigation of the K_{Iscc} Behavior of Candidate HY-180/210 Steel Weldments," *U. S. Steel Applied Research Laboratory Report No.89.021-024(1)* (*B-63105*) Defense Documentation Center, Arlington, Va., Dec. 31, 1970.
6. S. R. Novak and S. T. Rolfe, "Fatigue-Cracked Cantilever Beam Stress-Corrosion Tests of HY-80 and 5Ni-Cr-Mo-V Steels," *AD482783L*, Defense Documentation Center, Arlington, Va., Jan. 1, 1966.
7. S. R. Novak and S. T. Rolfe, "K_{Ic} Stress-Corrosion Tests of 12Ni-5Cr-Mo and 18Ni-8Co-3Mo Maraging Steels and Weldments", *AD482761L*, Defense Documentation Center, Arlington, Va., Jan. 1, 1966.
8. H. H. Uhlig, *The Corrosion Handbook*, Wiley, New York, 1963.
9. H. L. Logan, *The Stress Corrosion of Metals*, Wiley, New York, 1966.
10. A. W. Loginow, "Stress Corrosion Testing of Alloys," *Materials Protection*, *5*, No. 5, May 1966.
11. D. O. Sprowls and R. H. Brown, "What Every Engineer Should Know About Stress Corrosion of Aluminum," *Metals Progress*, *81*, Nos. 4 and 5, 1962.

12. D. O. Sprowls, M. B. Shumaker, J. D. Walsh and J. W. Coursen, "Evaluation of Stress-Corrosion Cracking Using Fracture Mechanics Techniques," *Contract NAS 8-21487*, *Final Report*, George C. Marshall Space Flight Center, Huntsville, Alabama, May 31, 1973.

13. S. R. Novak and S. T. Rolfe, "Modified WOL Specimen for K_{Iscc} Environmental Testing," *Journal of Materials*, *JMLSA*, ASTM, Philadelphia, *4*, No. 3, Sept. 1969.

14. E. A. Steigerwald, "Delayed Failure of High-Strength Steel in Liquid Environments," in *Proceedings of the American Society for Testing Materials*, Philadelphia, Vol. 60, 1960.

15. H. H. Johnson and A. M. Willner, "Moisture and Stable Crack Growth in a High-Strength Steel," *Applied Materials Research*, Jan. 1965.

16. C. F. Tiffany and J. N. Masters, "Applied Fracture Mechanics," *Fracture Toughness Testing and Its Applications*, *ASTM STP 381*, American Society for Testing and Materials, Philadelphia, April 1965.

17. S. R. Novak, "Effect of Prior Uniform Plastic Strain on the K_{Iscc} of High-Strength Steels in Sea Water," *Engineering Fracture Mechanics*, *5*, No. 3, 1973.

18. R. W. Judy, Jr., and R. J. Goode, "Stress-Corrosion Cracking Characteristics of Alloys of Titanium in Salt Water," *NRL Report 6564*, Naval Research Laboratory, Washington, D.C., July 21, 1967.

19. B. F. Brown, "Stress-Corrosion Cracking and Corrosion Fatigue of High-Strength Steels," in "Problems in the Load-Carrying Application of High-Strength Steels," *DMIC Report 210*, Defense Metals Information Center, Battelle Memorial Institute, Columbus, Ohio, Oct. 26–28, 1964.

20. B. F. Brown, "A New Stress-Corrosion Cracking Test for High-Strength Alloys," *Materials Research and Standards*, *6*, No. 3, March, 1966.

21. H. P. Leckie, "Effect of Environment on Stress Induced Failure of High-Strength Maraging Steels," *Fundamental Aspects of Stress Corrosion Cracking*, *NACE-1*, National Association of Corrosion Engineers, Houston, 1969.

22. H. R. Smith, D. E. Piper, and F. K. Downey, "A Study of Stress-Corrosion Cracking by Wedge-Force Loading," *Engineering Fracture Mechanics*, *1*, No. 1, 1968.

23. R. W. Huber, R. J. Goode, and R. W. Judy, Jr., "Fracture Toughness and Stress-Corrosion Cracking of Some Titanium Alloy Weldments," *Welding Journal*, *46*, No. 10. Oct. 1967.

24. M. H. Peterson, B. F. Brown, R. L. Newbegin, and R. E. Groover, "Stress Corrosion Cracking of High Strength Steels and Titanium Alloys in Chloride Solutions at Ambient Temperature," *Corrosion*, *23*, 1967.

25 M. O. Speidel, "Stress Corrosion Cracking of Aluminum Alloys," *Metallurgical Transactions*, *6A*, No. 4, April 1975.

26. M. O. Speidel and M. V. Hyatt, "Stress Corrosion Cracking of High-Strength Aluminum Alloys," *Advances in Corrosion Science and Technology*, Vol. II, edited by M. G. Fontana and R. W. Staehle, Plenum, New York, 1972.

27. E. J. Imhof and J. M. Barsom, "Fatigue and Corrosion-Fatigue Crack Growth of 4340 Steel at Various Yield Strengths," *ASTM STP 536*, American Society for Testing and Materials, Philadelphia, 1973.

28. J. M. Barsom, "Corrosion-Fatigue Crack Propagation Below K_{Iscc}" *Engineering Fracture Mechanics, 3*, No. 1, July 1971.

29. "Standard Method of Test for Plane-Strain Fracture Toughness of Metallic Materials," *ASTM E399, Annual Book of ASTM Standards*, American Society for Testing and Materials, Philadelphia, 1974.

30. P. C. Paris and G. C. Sih, "Stress Analysis of Cracks," *ASTM STP 381*, American Society for Testing and Materials, Philadelphia, 1965.

31. W. K. Wilson, "Review of Analysis and Development of WOL Specimen," *67-7D7-BTLPV-R1*, Westinghouse Research Laboratories, Pittsburgh, March 1967.

32. W. K. Wilson, "Analytical Determination of Stress Intensity Factors for the Manjoine Brittle Fracture Test Specimen," *WERL-0029-3*, Westinghouse Research Laboratories, Pittsburgh, Aug. 1965.

33. M. M. Leven, "Stress Distribution in the M4 Biaxial Fracture Specimen," *65-1D7-STRSS-S1*, Westinghouse Research Laboratories, Pittsburgh, March 1965.

34. E. T. Wessel, "State of the Art of the WOL Specimen for K_{Ic} Fracture Toughness Testing," *Engineering Fracture Mechanics, 1*, No. 1, June 1968.

35. W. G. Clark, Jr., "Evaluation of the Fatigue Crack Initiation Properties of Type 403 Stainless Steel in Air and Steam Environments," *ASTM STP 559*, American Society for Testing and Materials, Philadelphia, 1974.

36. J. M. Barsom, "Effect of Cyclic-Stress Form on Corrosion-Fatigue Crack Propagation Below K_{Iscc} in a High-Yield-Strength Steel," in *Corrosion Fatigue: Chemistry, Mechanics and Microstructure, NACE-2*, National Association of Corrosion Engineers, Houston, 1972.

37. B. F. Brown and C. D. Beachem, *Specimens for Evaluating the Susceptibility of High Strength Steels to Stress Corrosion Cracking*, Internal Report, U.S. Naval Research Laboratory, Washington, D.C., 1966.

38. R. P. Wei, S. R. Novak, and D. P. Williams, "Some Important Considerations in the Development of Stress Corrosion Cracking Test Methods," *Materials Research and Standards, MTRSA, 12*, No. 9, 1972.

39. D. E. Piper, S. H. Smith, and R. V. Carter, "Corrosion Fatigue and Stress-Corrosion Cracking in Aqueous Environments," *A.S.M. National Metal Congress*, Oct. 1966.

40. G. G. Hancock and H. H. Johnson, "Subcritical Crack Growth in AM350 Steel," *Materials Research and Standards, 6*, 1966.

41. W. D. Benjamin and E. A. Steigerwald, "An Incubation Time for the Initiation of Stress-Corrosion Cracking in Precracked 4340 Steel," *Transactions of the American Society for Metals, 60*, No. 3, September, 1967.

42. J. M. Barsom and S. R. Novak, "Subcritical Crack Growth in Steel Bridge Members," *Final Report, NCHRP Project 12–14, National Cooperative Highway Research Program*, Washington, D.C., Sept. 1974.

43. G. R. Yoder, C. A. Griffis, and T. W. Crooker, "Sustained-Load Cracking of Titanium—A Survey of 6Al-4V Alloys," *NRL Report 7596*, Naval Research Laboratory, Washington, D.C., 1973.

44. "Stress Corrosion Testing," *ASTM STP 425*, American Society for Testing and Materials, Philadelphia, Dec. 1967.

45. R. W. Staehle, A. J. Forty, and D. Van Rooyen, eds., *Fundamental Aspects of Stress-Corrosion Cracking*, *NACE-1*, National Association of Corrosion Engineers, Houston, 1969.

46. B. F. Brown, ed., *Stress-Corrosion Cracking in High-Strength Steels and in Titanium and Aluminum Alloys*, Naval Research Laboratory, Washington, D.C., 1972.

47. A. Agrawal, B. F. Brown, J. Kruger, and R. W. Staehle, eds., *U.R. Evans Conference on Localized Corrosion*, *NACE-3*, National Association of Corrosion Engineers, Houston, 1971.

48. M. O. Speidel, M. J. Blackburn, T. R. Beck, and J. A. Feeney, "Corrosion Fatigue and Stress Corrosion Crack Growth in High Strength Aluminum Alloys, Magnesium Alloys, and Titanium Alloys Exposed to Aqueous Solutions," *Corrosion Fatigue: Chemistry, Mechanics and Microstructure*, *NACE-2*, National Association of Corrosion Engineers, Houston, 1972.

49. L. F. Porter, "A Discussion of the Paper 'The Role of Inclusions on Mechanical Properties in High-Strength Steels,' " *Journal of Vacuum Science and Technology*, *9*, No. 6, Nov.–Dec. 1972.

50. R. T. Ault, C. M. Waid, and R. B. Bertole, "Development of an Improved Ultra-High Strength Steel for Forged Aircraft Component," *AFML-TR-71-27*, Air Force, Dayton, Feb. 1971.

51. C. S. Carter, "Evaluation of a High Purity 18 percent Ni (300) Maraging Steel Forging," *AFML-TR-70-139*, Air Force, Dayton, June 1970.

52. R. P. M. Proctor and H. W. Paxton, "The Effect of Trace Impurities on the Stress Corrosion Cracking Susceptibility and Fracture Toughness of 18Ni Maraging Steel," *Corrosion Science*, *11*, 1971.

53. D. P. Dautovich and S. Floreen, "The Stress Corrosion and Hydrogen Embrittlement Behavior of Maraging Steels," presented at NACE Conference, UNIEUX-FIRMINY, France, 1973 (to be published in NACE-4, National Association for Corrosion Engineers), Houston.

54. B. Mravic and J. H. Smith, "Development of Improved High-Strength Steels for Aircraft Structural Components," *AFML-TR-71-213*, Air Force, Dayton, Oct. 1971.

55. H. P. Leckie and A. W. Loginow, "Stress Corrosion Behavior of High Strength Steels," *Corrosion 24*, No. 9, Sept. 1968, pp. 291–297.

56. R. W. Judy, Jr., C. T. Fujii, and R. J. Goode, "Properties of 17-4 pH Steel," *Naval Research Laboratory* (*NRL*) *Report 7639*, Washington, D.C., Dec. 18, 1973.

57. R. W. Staehle, "Evaluation of Current State of Stress Corrosion Cracking," *Fundamental Aspects of Stress Corrosion Cracking*, *NACE-1*, National Association of Corrosion Engineers, Houston, 1969.

11

Corrosion Fatigue

11.1. Introduction

The fatigue-crack-initiation and -propagation behaviors in benign environments were discussed in Chapters 7, 8, and 9. Because structures operate in various environments, the effect of environments on the behavior of structures must be established. The effect of environments on the fracture behavior of statically loaded structural components that contain fatigue cracks was presented in Chapter 10. However, since most structural components are subjected to fluctuating loads, the corrosion-fatigue behavior of metals in various environments is of primary importance.

Several investigators have studied the corrosion fatigue behavior of various environment-material systems.[1-15] The results of these investigations have helped greatly in the selection of proper materials for a given application. Despite the significant progress that has been achieved to establish the effects of various mechanical parameters on the corrosion-fatigue behavior of environment-material systems, little has been achieved to establish mechanisms of corrosion fatigue in these systems. The available information shows the high complexity of the corrosion-fatigue behavior and suggests that a significant understanding of this behavior can be achieved only by a synthesis of contributions from various fields.

11.2. General Behavior

Corrosion-fatigue behavior of a given environment-material system refers to the characteristics of the material under fluctuating loads in the presence of the particular environment. Different environments have different effects on the cyclic behavior of a given material. Similarly, the corrosion-fatigue behavior of different materials is different in the same environment.

The corrosion-fatigue behavior of metals subjected to load fluctuation in the presence of an environment to which the metal is immune is identical to the fatigue behavior of the metal in the absence of that environment. Consequently, the corrosion-fatigue behavior of an environment-material system can be studied by establishing the deviation of the corrosion-fatigue behavior for the environment-material system from the fatigue behavior of the material in a benign environment.

Stress-corrosion crack growth in a statically loaded structure is caused by interactions of chemical and mechanical processes at the crack tip. The highest plane-strain stress-intensity-factor value at which subcritical crack growth does not occur in a material loaded statically in an aggressive environment is designated K_{Iscc}. Consequently, to establish the effect of an environment on the fatigue-crack-growth behavior of a material, the fatigue-

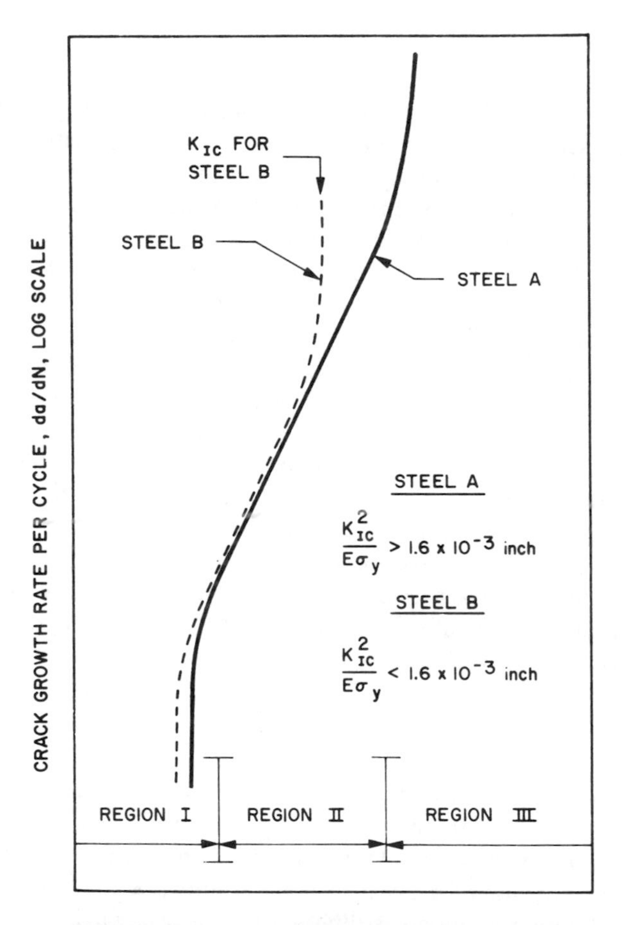

FIG. 11.1. Schematic representation of fatigue crack growth in steel.

crack-growth behavior for the material in a benign environment and the K_{Iscc} for the environment-material system should be established first as references.

The generalized fatigue-crack-growth behavior in a benign environment (Figure 11.1), presented in Chapter 8, is a special case of the corrosion-fatigue behavior for metals. It represents the "corrosion-fatigue" behavior of metals subjected to load fluctuations in the presence of any environment that does not affect the fatigue-crack-growth behavior for the metal. Thus, the corrosion-fatigue behavior for a given environment-material system could be investigated by establishing the base-line fatigue behavior and then by determining the effect of the environment on the fatigue behavior regions I, II, and III (Figure 11.1). However, because K_{Iscc} for an environment-material system defines the plane-strain K_{I} value above which stress-corrosion crack growth can occur under static loads, the corrosion-fatigue behavior for the environment-material system could be altered when the maximum value of K_{I}, $K_{\text{I max}}$, in a given load cycle becomes greater than K_{Iscc}. Consequently, the corrosion-fatigue behavior should be divided into below-K_{Iscc} and above-K_{Iscc} behaviors.

11.3. Corrosion-Fatigue Behavior Below K_{Iscc}

The first systematic investigation into the effect of environment and loading variables on the rate of fatigue-crack growth below K_{Iscc} was conducted on 12Ni-5Cr-3Mo maraging steel (yield strength = 180 ksi) in 3% solution of sodium chloride.[4,16,17] The data showed that environmental acceleration of fatigue-crack growth does occur below K_{Iscc} (Figure 11.2) and that the magnitude of this acceleration is dependent on the frequency of the cyclic-stress-intensity fluctuations. The test results also showed that the fatigue-crack-growth rates in 12Ni-5Cr-3Mo maraging steel tested in a room-temperature air environment and in a room-temperature 3% solution of sodium chloride can be represented by

$$\frac{da}{dN} = D(t)(\Delta K)^2 \tag{11.1}$$

where $D(t)$ depends on the environment-material system and on the sinusoidal cyclic stress-intensity frequencies. That is, the magnitude of the environmental accelerations of fatigue-crack growth can be increased or decreased substantially by changing the environment, the material, and the frequency of loading. In air, $D(t)$ was a constant, independent of frequency. In the sodium chloride solution at high frequencies (cpm > 600), $D(t)$ had essentially the same values as in air, Figure 11.3; thus the environment had negligible effects on the fatigue-crack-growth rate. In the sodium chloride solution at 6 cpm, however, $D(t)$ was three times higher than the value in air, which indicates that the fatigue-crack-growth rate was increased significantly by

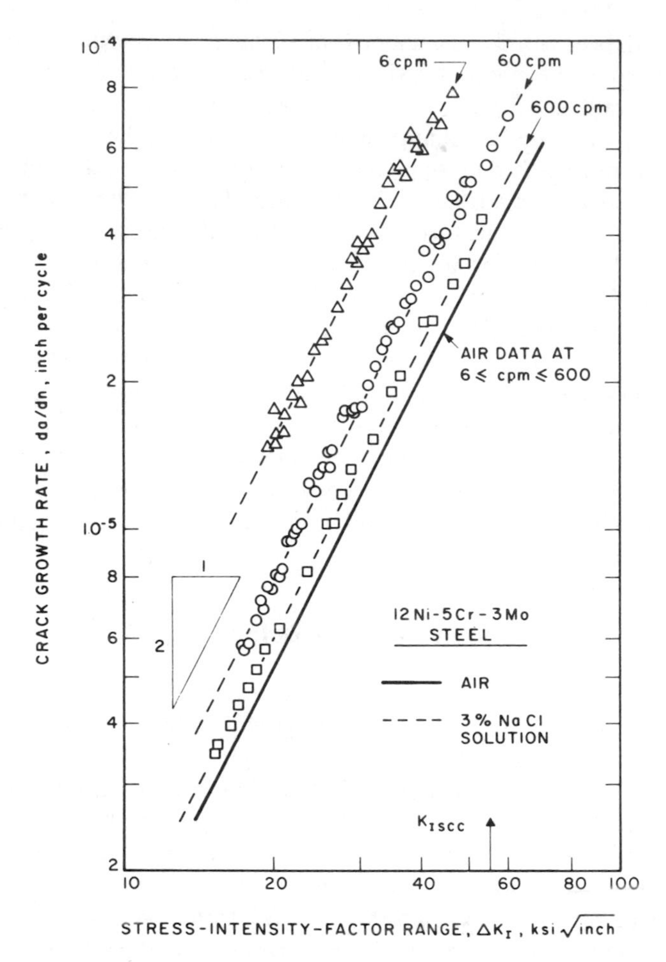

FIG. 11.2. Corrosion-fatigue-crack growth data as a function of test frequency.

the environment. The data in Figure 11.3 may be used to predict the value of $D(t)$ for any sinusoidal frequency equal to or greater than about 6 cpm in the environment-material system investigated. Because these results were obtained below K_{Iscc}, Barsom[16,17] concluded that the corrosion-fatigue-crack-growth rate for 12Ni-5Cr-3Mo maraging steel in a 3% solution of sodium chloride increases to a maximum value and then decreases as the sinusoidal cyclic-stress frequency decreases from 600 cpm to frequencies below 6 cpm. Similar behavior has been established for HY-80 steel[10] and for Ti-8Al-1V-1Mo alloy.[13] The dependence of corrosion-fatigue-crack-growth rate below K_{Iscc} on cyclic frequency has been established for steels,[4,5,10,11,16,17] aluminum alloys,[7,8,14] and titanium alloys.[8,13] However,

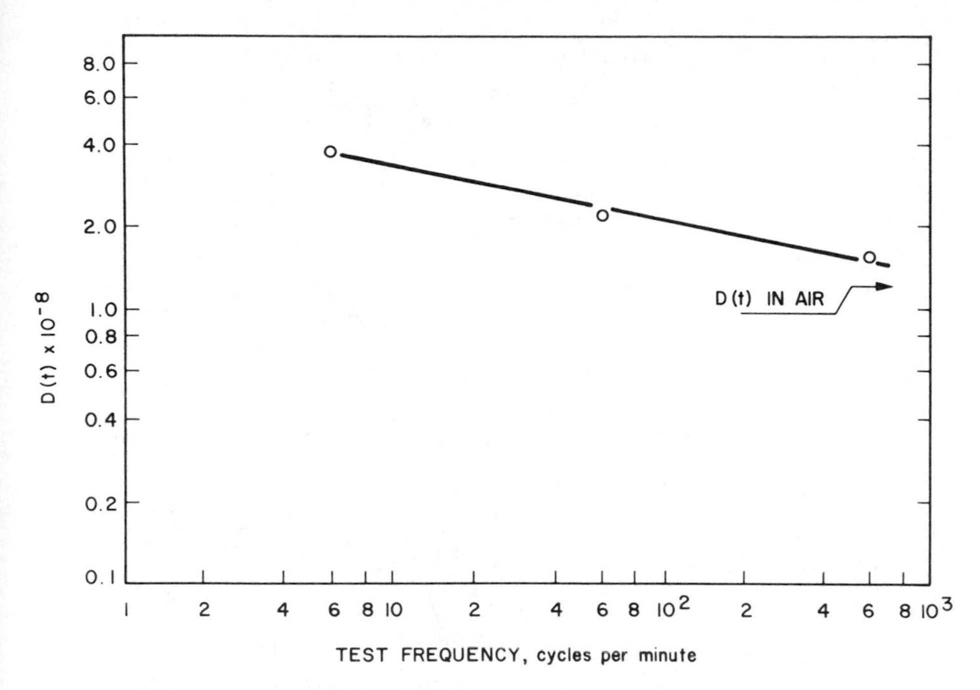

FIG. 11.3. Correlation between time-dependent function $D(t)$ and test frequency for 12Ni-5Cr-3Mo steel tested in sodium chloride solution.

unlike the data presented in Figure 11.2, curves obtained for different cyclic frequencies in various environment-metal systems do not appear to be parallel to each other (Figure 11.4).[18]

The magnitude of the effect of cyclic frequency on the rate of corrosion-fatigue-crack growth depends strongly on the environment-material system.[19,20] The data presented in Figure 11.5[19] show that the 10Ni-Cr-Mo-Co steels tested were highly resistant to the 3% solution of sodium chloride and that of the four steels tested, the 12Ni-5Cr-3Mo steel was the least resistant to the 3% solution of sodium chloride. Similarly, fatigue-crack-growth rates below K_{Iscc} were accelerated by a factor of 2 when 4340 steel of 130-ksi yield strength was tested at 6 cpm in a sodium-chloride solution (Figure 11.6).[20] Under identical test conditions, the corrosion-fatigue-crack-growth rates in the same 4340 steel heat-treated to 180-ksi yield strength were five to six times higher than the fatigue-crack-growth rates in room-temperature air environments (Figure 11.6).

The effect of various additions to aqueous solutions on the stress-corrosion cracking and on the corrosion-fatigue-crack-growth rates at a given cyclic frequency have been investigated for high-strength aluminum alloys (Figure 11.7), mangesium alloys (Figure 11.8), and titanium alloys.[8] A sum-

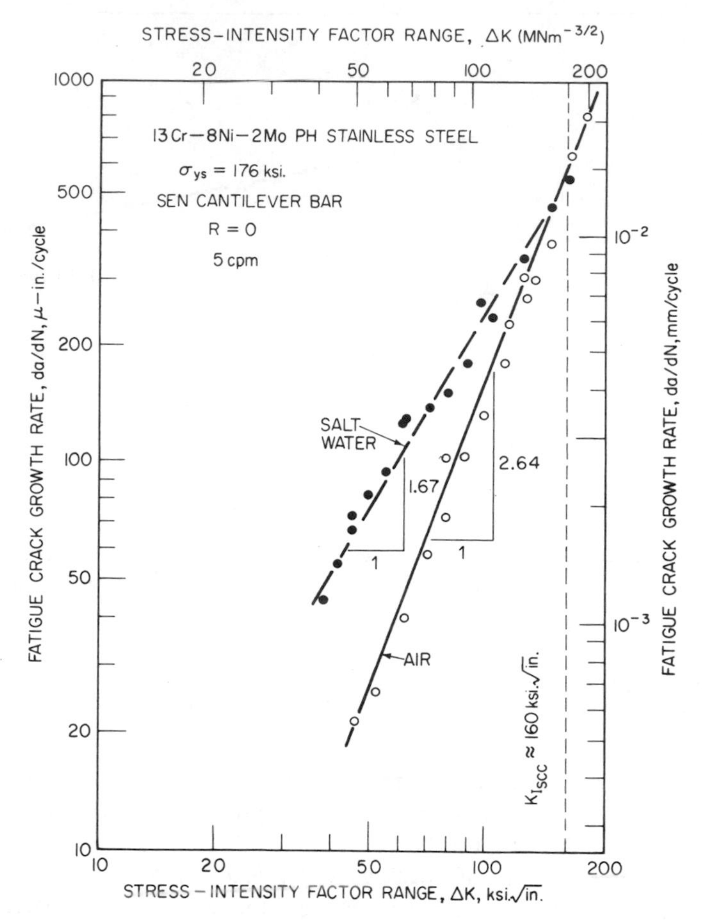

FIG. 11.4. Air and salt water fatigue crack growth rate behavior of 13Cr-8 Ni-2 Mo PH stainless steel.

mary of the effects of the additions is presented in Table 11.1.[8] The data suggest that the additions to aqueous solutions that accelerate stress-corrosion-crack growth also accelerate the corrosion-fatigue-crack growth and those additions that do not affect stress-corrosion cracking have no effect on corrosion-fatigue-crack growth.

Extensive corrosion-fatigue data have been obtained for various aluminum alloys in water and water-vapor environments. The results indicate that these environments have a significant effect on the fatigue-crack-growth rate for aluminum alloys.[7,14,21-25] The effect of water and water vapor on the rate of fatigue-crack growth for a 7075 aluminum alloy is shown in Figure 11.9.

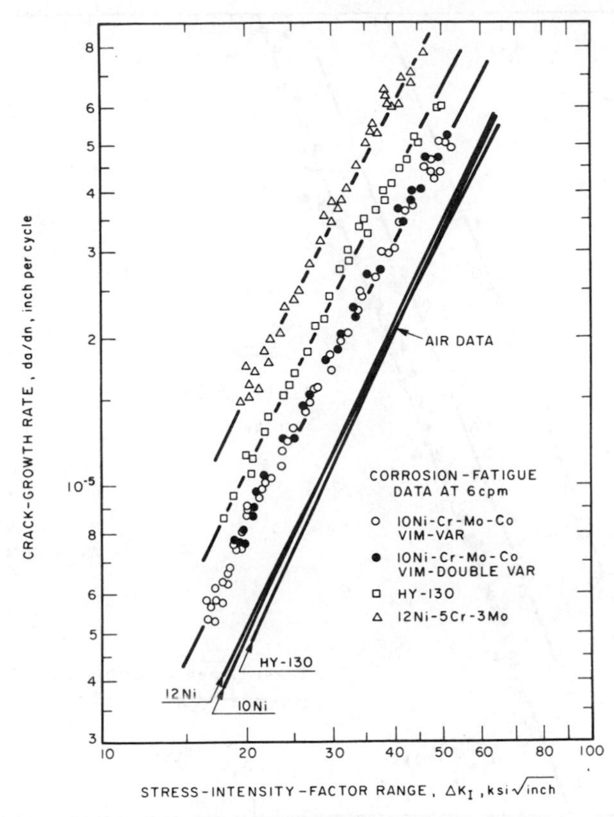

FIG. 11.5. Fatigue-crack-growth rates in air and in 3% solution of sodium chloride below K_{Iscc} for various high-yield-strength steel.

TABLE 11.1. Effect of Various Additions to Aqueous Environments on Acceleration of Subcritical Crack Growth in High-Strength Light Metals

	Stress Corrosion		*"True" Corrosion Fatigue (Region II)*	
Alloys	*Additions Which* Can *Accelerate Crack Growth*	*Additions Which* Do Not *Accelerate Crack Growth*	*Additions Which* Can *Accelerate Crack Growth*	*Additions Which* Do Not *Accelerate Crack Growth*
Aluminum base (7079-T651)	Cl^-, Br^-, I^-	$SO_4^=$	Cl^-, Br^-, I^-	$SO_4^=$
Titanium base (Ti-6Al-4V)	Cl^-, Br^-, I^-	$SO_4^=$	Cl^-, Br^-, I^-	$SO_4^=$
Magnesium base (ZK60A-T5)	Cl^-, Br^-, I^- $SO_4^=$		Cl^-, Br^-, I^- $SO_4^=$	

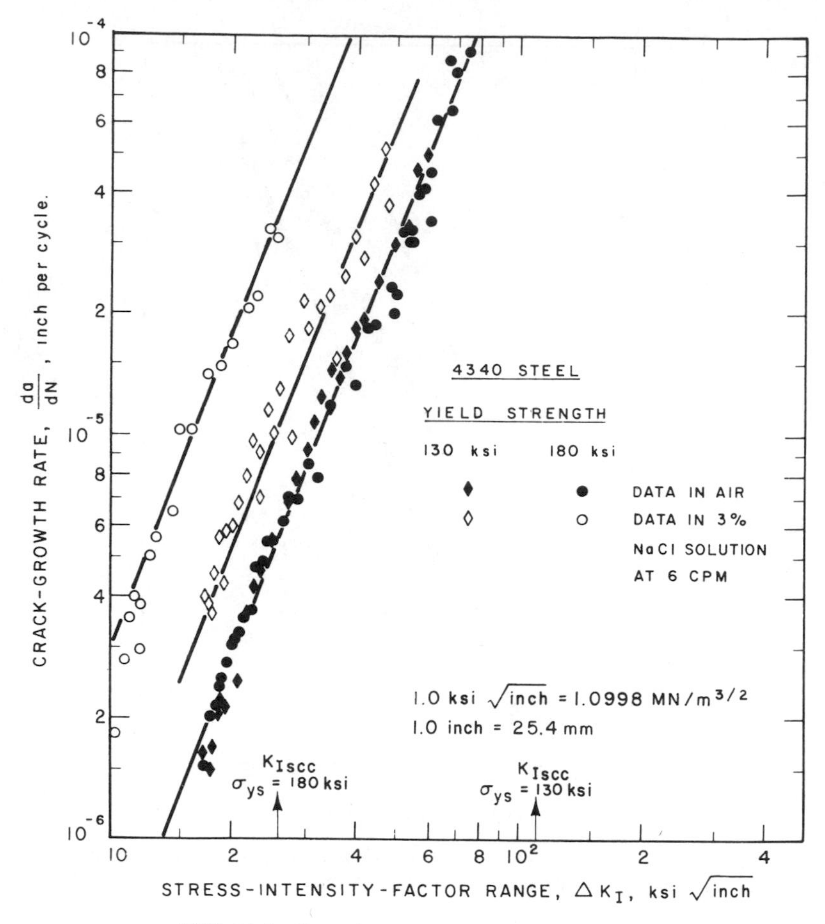

FIG. 11.6. Corrosion-fatigue-crack-growth data.

Corrosion-fatigue-crack-growth data for various low-yield-strength constructional steels have been investigated below K_{Iscc} to determine the susceptibility of these steels to aqueous environments[5].

The steels tested were A36, A588 Grade A, A588 Grade B, A514 Grade E, and A514 Grade F steels in distilled water and 3% solution of sodium chloride in distilled water. The tests were conducted under constant-amplitude and variable-amplitude random-sequence sinusoidal load fluctuations at frequencies of 60 and 12 cpm.

The data presented in Figures 11.10, 11.11, and 11.12 show that the addition of 3% (by weight) sodium chloride to distilled water had no effect on the corrosion-fatigue behavior of these steels. Similar data were obtained for the other steels that were tested.[5] The data also show that the corrosion-

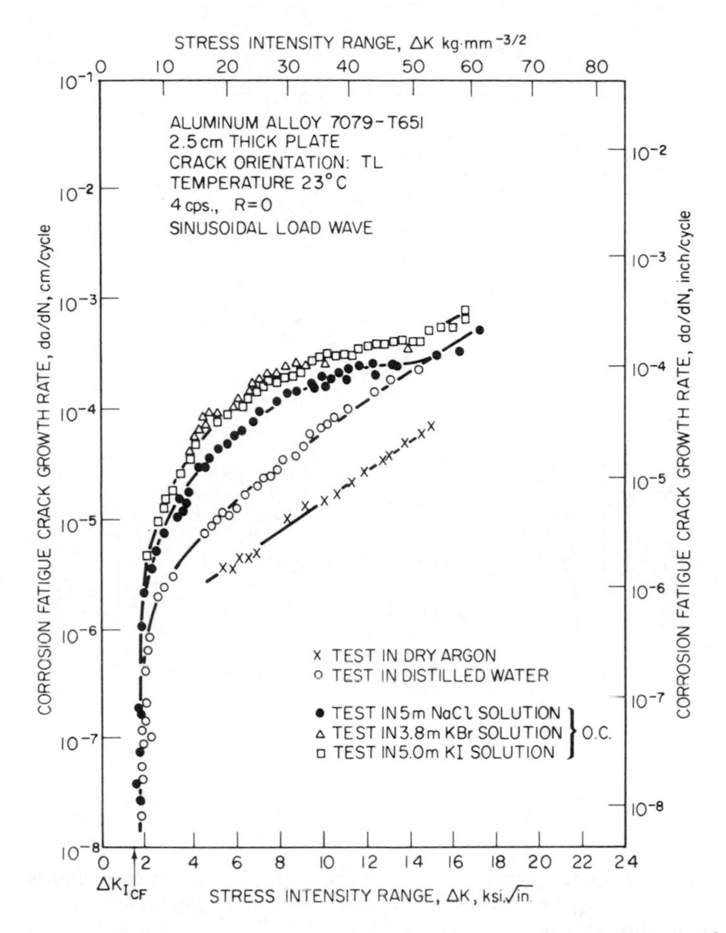

FIG. 11.7. Effect of cyclic stress intensity range on the growth rate of corrosion fatigue cracks in a high strength aluminum alloy exposed to various environments.

fatigue-crack-growth-rate behavior at 60 cpm under sinusoidal loads and under square-wave loads is essentially identical. The scatter in test results obtained in a single specimen under corrosion-fatigue conditions was equal to or greater than under fatigue conditions. The increase in scatter was caused by the general corrosion of the specimen surfaces, which decreased the accuracy for determining the exact location of the crack tip. Considering the scatter caused by the general corrosion of the specimen surfaces and the inherent scatter observed in fatigue-crack-growth data, the data presented in Figures 11.10 through 11.12 indicate that, at 60 cpm, distilled water and 3% solution of sodium chloride in distilled water had negligible effect on the

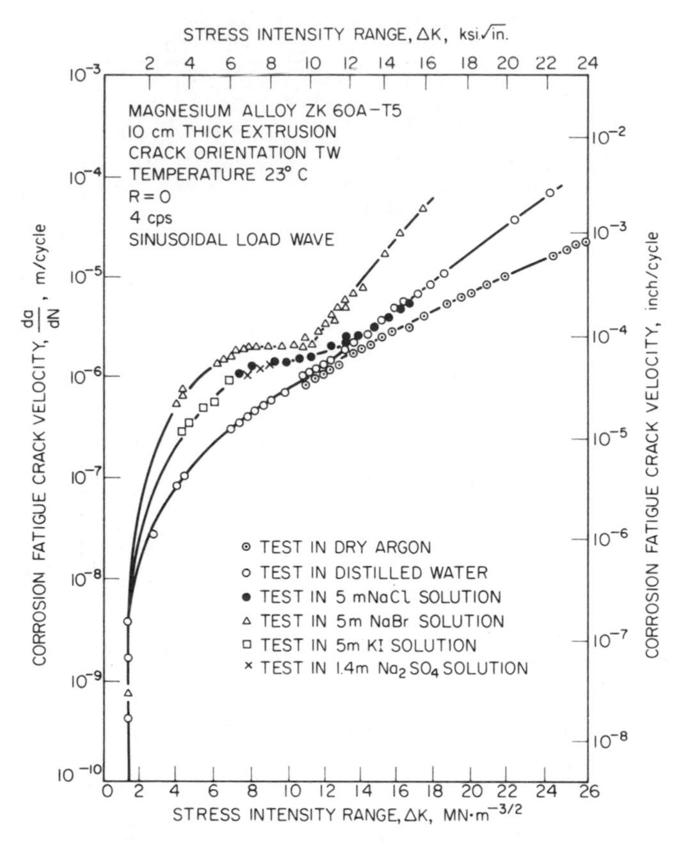

FIG. 11.8. Effect of cyclic stress intensity range and various environments on the growth rate of corrosion fatigue cracks in a high strength magnesium alloy.

rate of growth of fatigue cracks in the constructional steels investigated. Corrosion-fatigue data obtained by testing these steels in 3% solution of sodium chloride at a stress ratio, R, of 0.5 were identical to those obtained at $R = 0.1$, and corrosion-fatigue data obtained for five different heats of A588 steel were also identical.[5]

Corrosion-fatigue data obtained by testing duplicate specimens of A36, A588 Grade A, and A514 Grade F steels in 3% solution of sodium chloride at 12 cpm under constant-amplitude sinusoidal loading are presented in Figures 11.13 through 11.15.[5] Similar data were obtained for A588 Grade B and A514 Grade E steels.[5] Superimposed on these figures is the upper bound of data scatter obtained by testing these steels at 60 cpm under constant-amplitude and variable-amplitude random-sequence loading in distilled water and in 3% solution of sodium chloride. The long duration of the cor-

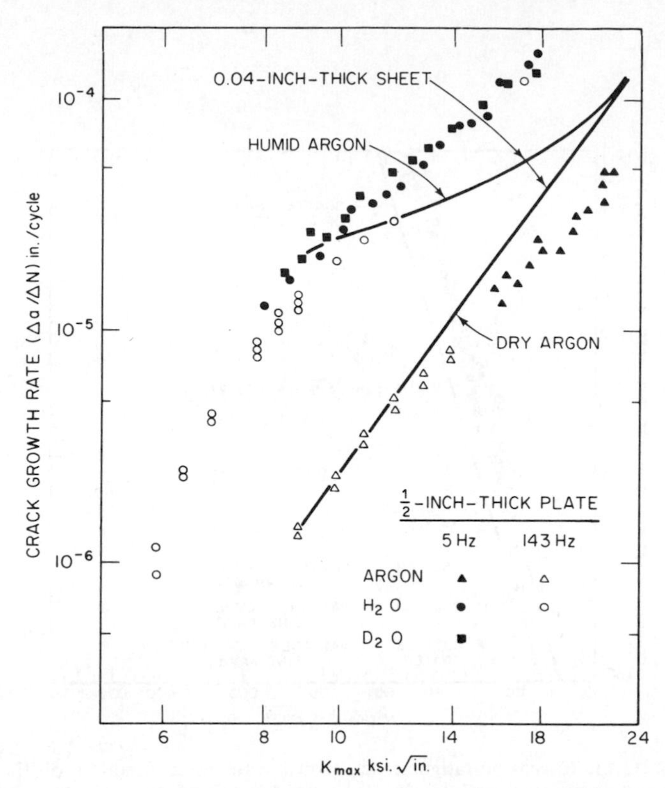

FIG. 11.9. The effect of water and water vapor on the rate of fatigue crack growth in 7075 aluminum alloy.

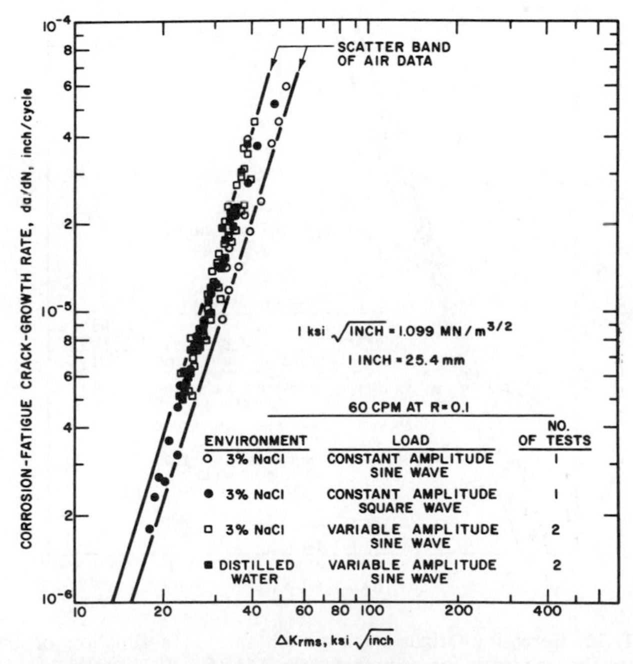

FIG. 11.10. Corrosion-fatigue-crack-growth rate as a function of the root-mean-square stress-intensity factor for A36 steel.

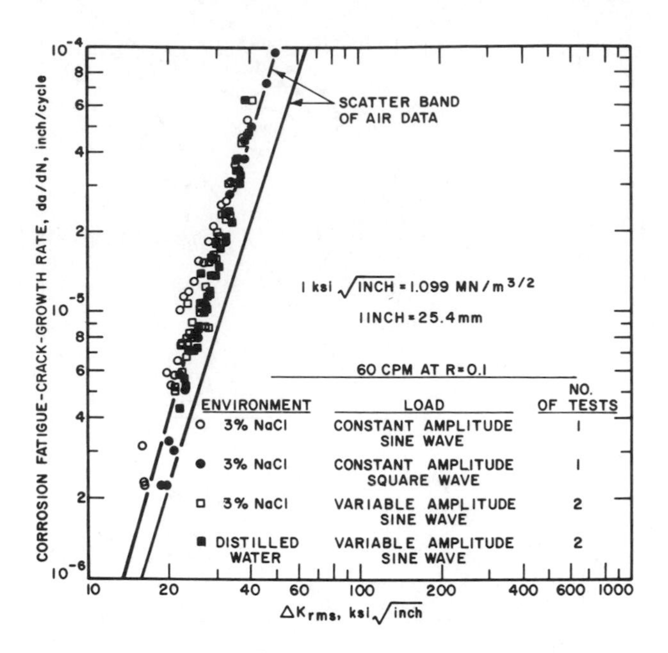

FIG. 11.11. Corrosion-fatigue-crack-growth rate as a function of the root-mean-square stress-intensity factor for A588 Grade A steel.

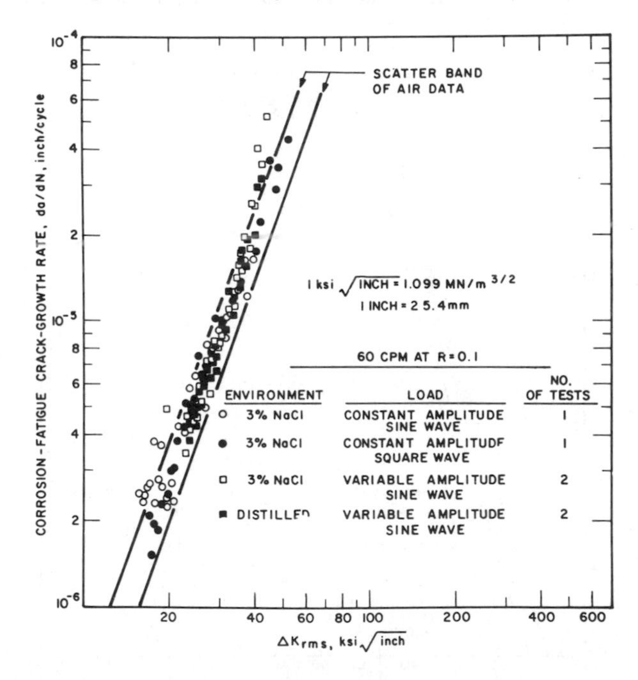

FIG. 11.12. Corrosion-fatigue-crack-growth rate as a function of the root-mean-square stress-intensity factor for A514 Grade F steel.

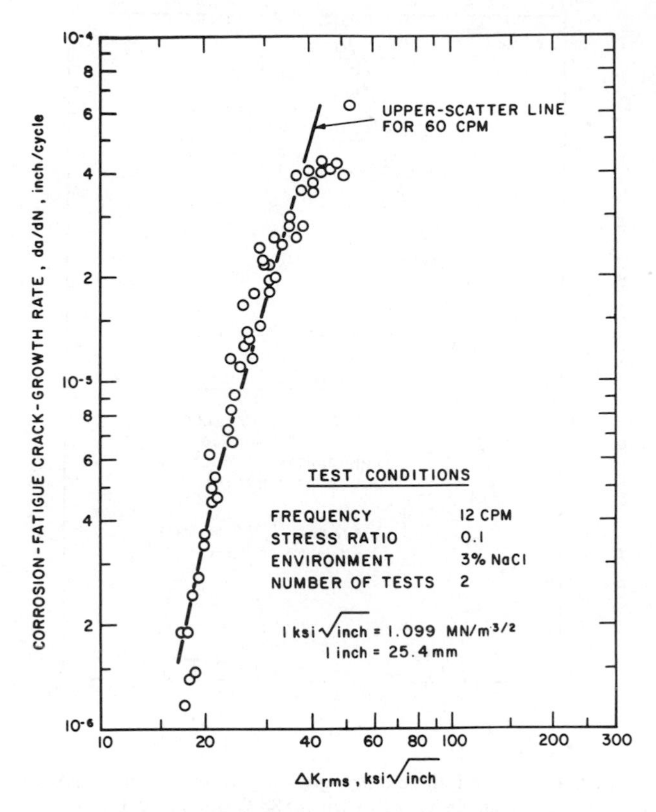

FIG. 11.13. Corrosion-fatigue-crack-growth rate as a function of the root-mean-square stress-intensity factor for A36 steel.

rosion-fatigue tests at 12 cpm caused extensive surface corrosion that resulted in greater data scatter than obtained in tests at 60 cpm. The corrosion-fatigue data presented in Figures 11.13 through 11.15 show that the rate of crack growth for the constructional steels tested at 12 cpm and at stress-intensity-factor fluctuations greater than about 15 ksi$\sqrt{\text{in.}}$ (16.5 $\text{MNm}^{-3/2}$) was equal to or slightly greater than that observed at 60 cpm.

11.3.1. Corrosion Fatigue at Low ΔK_I Values

Very limited corrosion-fatigue-crack-growth data have been obtained at low ΔK_I values.[5,12] The data presented in Figure 11.16[12] suggest that the threshold stress-intensity factor, ΔK_{th}, for fatigue crack growth for the steel is not affected by the aqueous test environment. Similar data suggest that ΔK_{th} for 7079-T651 aluminum alloy (Figure 11.7)[8] and for magnesium alloy ZK60A-T5 (Figure 11.8)[8] is essentially unaffected by various aqueous environments. Unfortunately, to establish the ΔK_{th} within a reasonable test time,

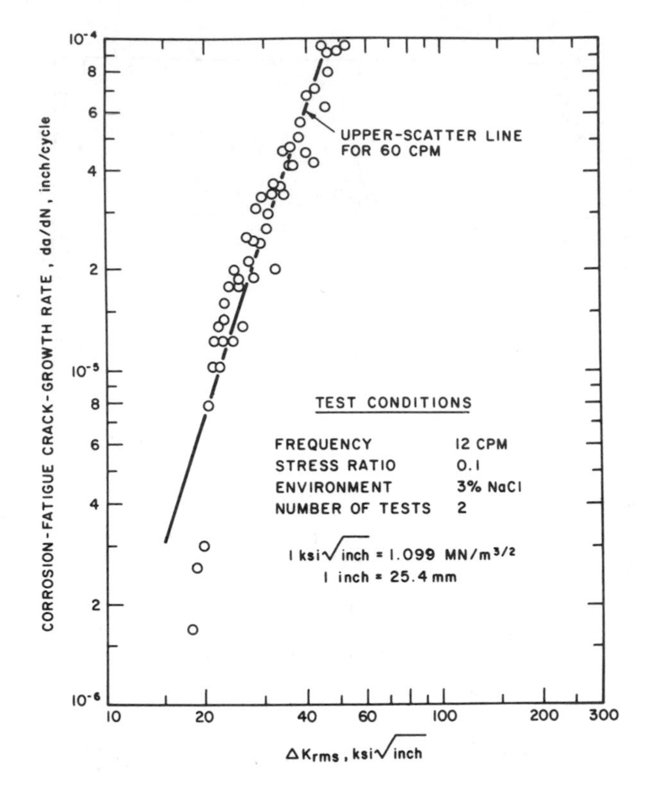

FIG. 11.14. Corrosion-fatigue-crack-growth rate as a function of the root-mean-square stress-intensity factor for A588 Grade A steel.

these data were obtained at very high cyclic frequencies. Because at high cyclic frequencies the environment has negligible effect on the rate of fatigue-crack growth below K_{Iscc}, the invariance of ΔK_{th} at high cyclic frequencies in aggressive environments should not be surprising.

Limited data related to the effect of cyclic frequency on ΔK_{th} in aqueous environments has been obtained for various constructional steels.[5] The corrosion-fatigue-crack-growth-rate data for A36 steel tested at 12 cpm indicated that the rate of crack growth decreased significantly at ΔK_I values less than 20 ksi$\sqrt{\text{in.}}$ (22.0 MNm$^{-3/2}$) (Figure 11.17). Similar behavior was observed in the A588 Grade A, A588 Grade B, A514 Grade E, and A514 Grade F steels tested. To verify this observation, one specimen of A514 Grade E and one specimen of A514 Grade F were tested at 12 cpm in 3% sodium chloride solution under cyclic-load fluctuations corresponding to ΔK_I of about 11 ksi$\sqrt{\text{in.}}$ (12.1 MNm$^{-3/2}$). The test results are presented in Figures 11.18 and 11.19. The data show that a corrosion-fatigue-crack-growth-rate threshold, ΔK_{th}, does exist in A514 steels at a value of the stress-

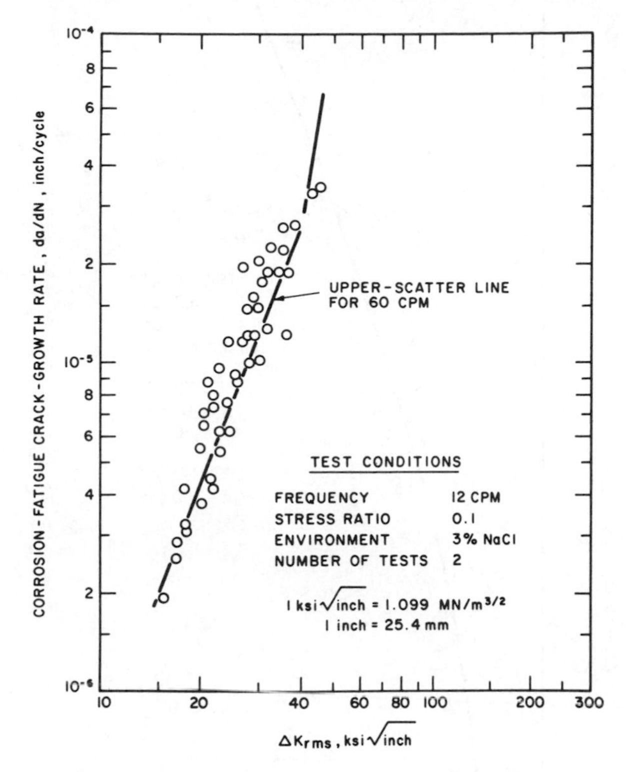

FIG. 11.15. Corrosion-fatigue-crack-growth rate as a function of the root-mean-square stress-intensity factor for A514 Grade F steel.

intensity-factor fluctuation below which corrosion-fatigue cracks do not propagate at 12 cpm in the environment-steel system tested. The value of ΔK_{th} in A514 steels tested at 12 cpm in 3% sodium chloride solution was twice as large as the value of 5.5 ksi$\sqrt{\text{in.}}$ (6.0 $MNm^{-3/2}$) for room-temperature air.[26]

Bucci and Donald[27] observed that the environmental ΔK_{th} in a 200-grade maraging steel forging was higher than the ΔK_{th} in air and that the "salt water appears to produce an inhibitive effect on fatigue cracking at the very low ΔK levels." Paris et al.[12] also observed that the threshold ΔK of ASTM A533 Grade B, Class 1 steel in distilled water was greater than that established in room-temperature air. "Since the distilled water retardation of very low crack extension rates was a somewhat surprising result, an additional specimen was tested for which a distilled water environment was provided to the crack tip and its surroundings only after an initial slow crack extension rate had been established in room air. Upon application of the distilled water, the rate of crack growth decreased from those initially obtained in air."[12]

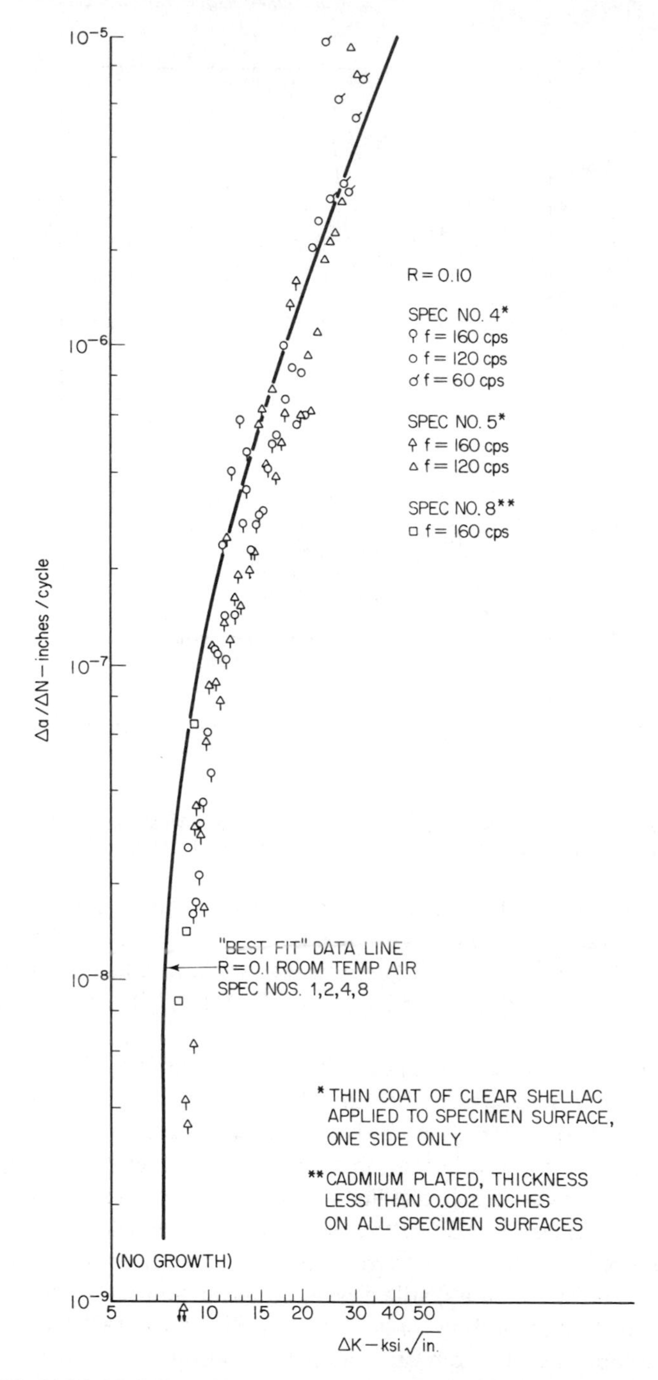

FIG. 11.16. Fatigue-crack-propagation rates of ASTM A533 B-1 steel in distilled water at ambient temperature, $R = 0.10$.

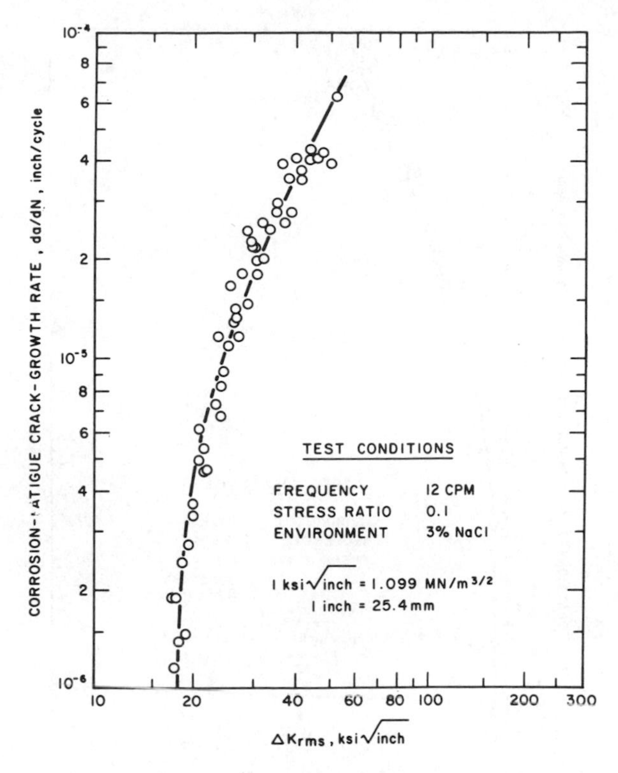

FIG. 11.17. Corrosion-fatigue-crack-growth rate as a function of the root-mean square stress-intensity factor for A36 steel.

These observations by Bucci and Donald and by Paris et al. confirm the preceeding observations.

In an air environment, the fatigue ΔK_I threshold, ΔK_{th}, in various steels tested at a stress ratio of 0.1 is independent of cyclic-load frequency and is equal to about 5.5 ksi$\sqrt{\text{in.}}$ Because hostile environmental effects decrease with increased cyclic-load frequency, the corrosion-fatigue ΔK_{th} at very high cyclic-load frequencies would have a value close to that of fatigue in air. A K_{Iscc} test can be considered a corrosion-fatigue test at extremely low cyclic-load frequency. In such tests, the rate of crack growth at a stress-intensity-factor fluctuation that is slightly lower than K_{Iscc} is, by definition, equal to zero. Consequently, at very low cyclic-load frequencies, ΔK_{th} is equal to K_{Iscc}. Hence, the value of the environmental ΔK_{th} at intermediate cyclic-load frequencies must be greater than 5.5 ksi$\sqrt{\text{in.}}$ and less than the value of K_{Iscc} for the environment-steel system under consideration. The test results show that, at 12 cpm, the environmental ΔK_{th} for A514 steels in 3% solution of sodium chloride in distilled water was equal to about 11 ksi$\sqrt{\text{in.}}$ Based on the preceding observations, a schematic representation of the corrosion-

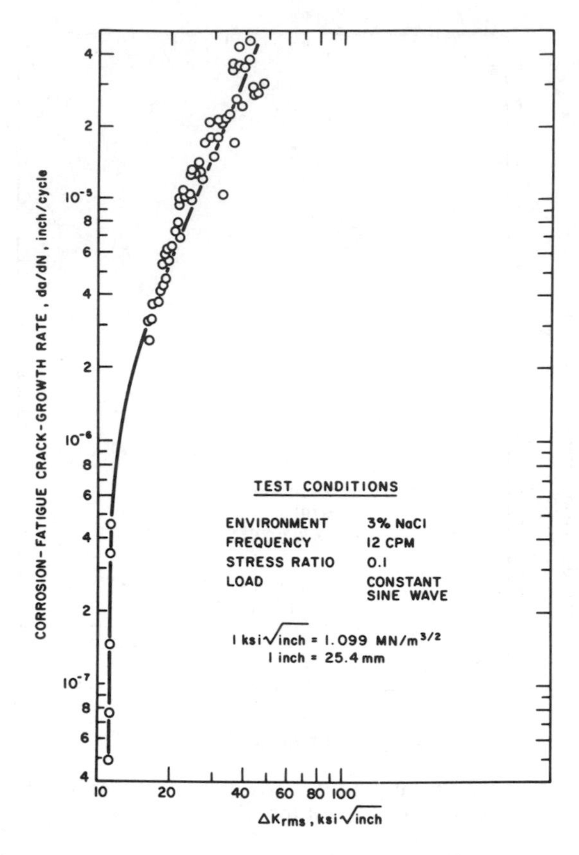

FIG. 11.18. Corrosion-fatigue-crack-growth rate as a function of the root-mean-square stress-intensity factor for A514 Grade E steel.

fatigue behavior of steels subjected to different cyclic-load frequencies has been constructed (Figure 11.20). This figure is an oversimplification of a very complex phenomenon.

11.3.2. Corrosion-Fatigue Behavior Under Alternate Wet and Dry Conditions

The preceding discussion presents the corrosion-fatigue behavior of steels under complete immersion conditions. Under actual operating conditions, some structures are subjected to alternate wet and dry environmental conditions. The effects of alternate wet and dry environmental conditions on steels were simulated by using a specimen of A514 steel.[5] First, the corrosion-fatigue-crack-growth rate in the specimen was established at 12 cpm under full immersion conditions in 3% solution of sodium chloride in distilled

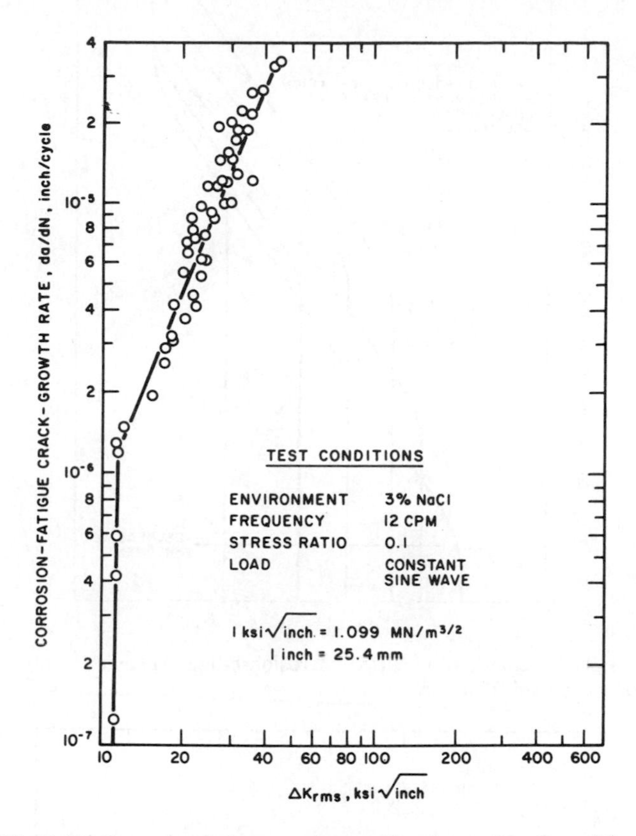

FIG. 11.19. Corrosion-fatigue-crack-growth rate as a function of the root-mean-square stress-intensity factor for A514 Grade F steel.

water. Then the specimen was removed from the environmental bath and left overnight to dry in room-temperature air. Finally, the specimen was replaced in 3% sodium chloride solution, and the corrosion-fatigue-crack-growth behavior was again measured at 12 cpm under full immersion conditions and under the same load fluctuations applied prior to drying the specimen. The data obtained from this interrupted test (Figure 11.21) indicate the existence of nonsteady-state crack growth represented by a severe retardation of the rate of corrosion-fatigue-crack growth and a corresponding substantial increase in the life of the specimen. The data show that the life of the specimen was *doubled* because 115,000 cycles were required to reestablish the steady-state crack-growth-rate behavior. Similar retardation behavior and a concomitant increase in the useful life of specimens tested under alternate wet and dry conditions have been observed by Miller et al.[11] in AISI 4340 steel tempered at 200°F (93°C).

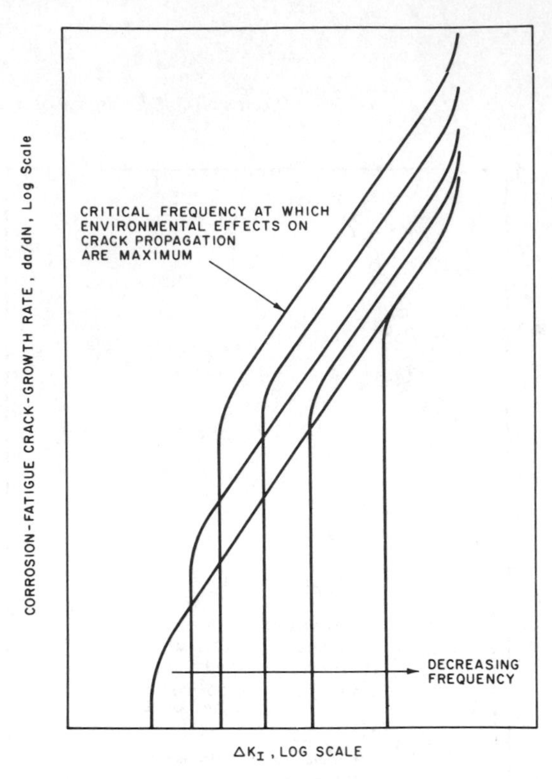

FIG. 11.20. Schematic of idealized corrosion-fatigue behavior as a function of cyclic-load frequency.

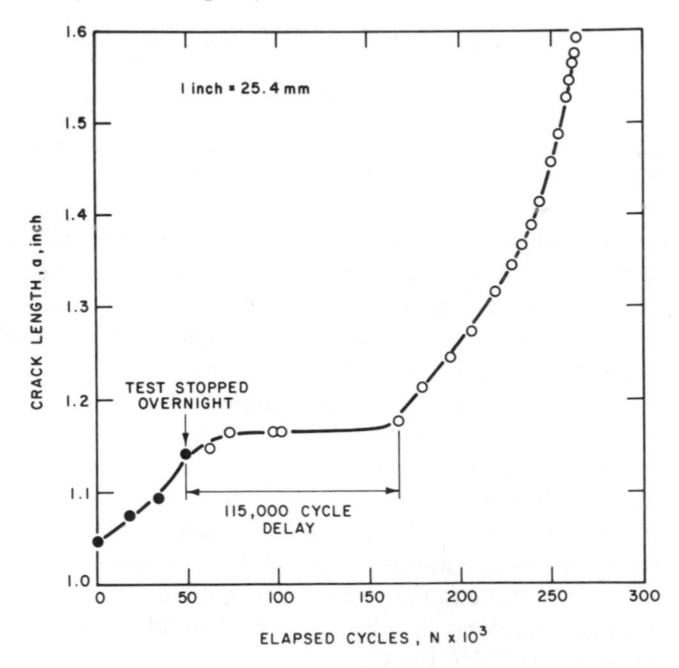

FIG. 11.21. Retardation of corrosion-fatigue-crack-growth rate under wet and dry environmental conditions for A514 Grade F steel.

11.3.3. Effect of Cyclic-Stress Wave Form

Available data indicate that the environmental effects on the rate of fatigue-crack growth in corrosion fatigue below K_{Iscc} may be highly dependent on the shape of the cyclic-stress wave.[4] This dependence is illustrated by the difference between the fatigue-crack-growth rate data for 12Ni-5Cr-3Mo steel in a room-temperature air environment (Figure 11.22) and in a 3% solution of sodium chloride (Figure 11.23) under sinusoidal loading, triangular loading, and square loading at 6 cpm. The tests were conducted on identical specimens in the same bulk environment and at the same maximum and minimum loads. The effect of the cyclic wave on the corrosion-fatigue behavior below K_{Iscc} was obtained from direct comparison between the crack-growth rates per cycle at a constant value of ΔK_{I}.

The data presented in Figure 11.22 show that the fatigue-crack-growth rates in room-temperature air environment are identical under various stress fluctuations and are independent of frequency. The data in Figure 11.23

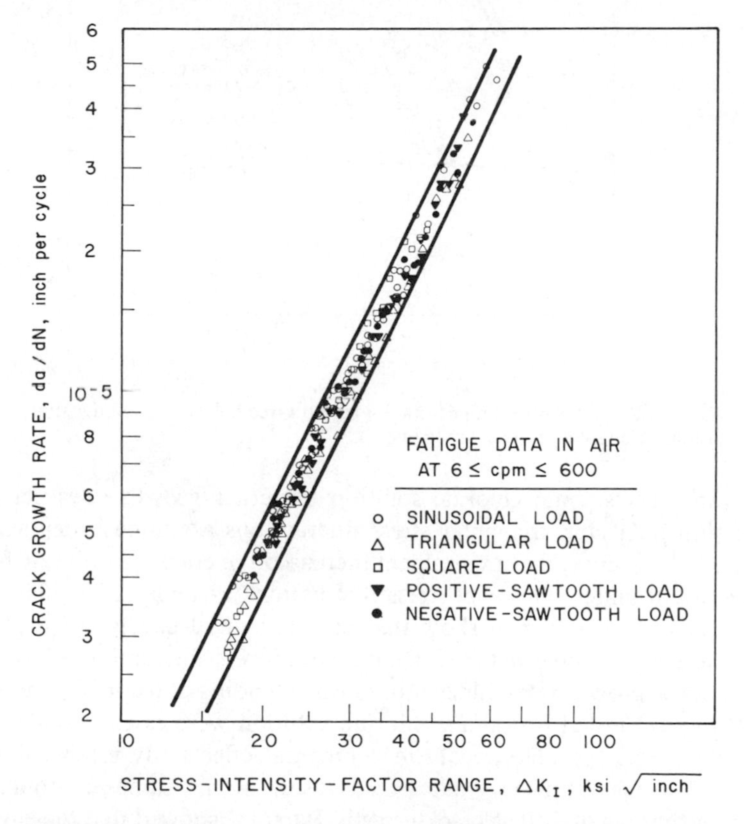

FIG. 11.22. Fatigue-crack-growth rates in 12Ni-5Cr-3Mo steel under various cyclic-stress fluctuations with different stress-time profiles.

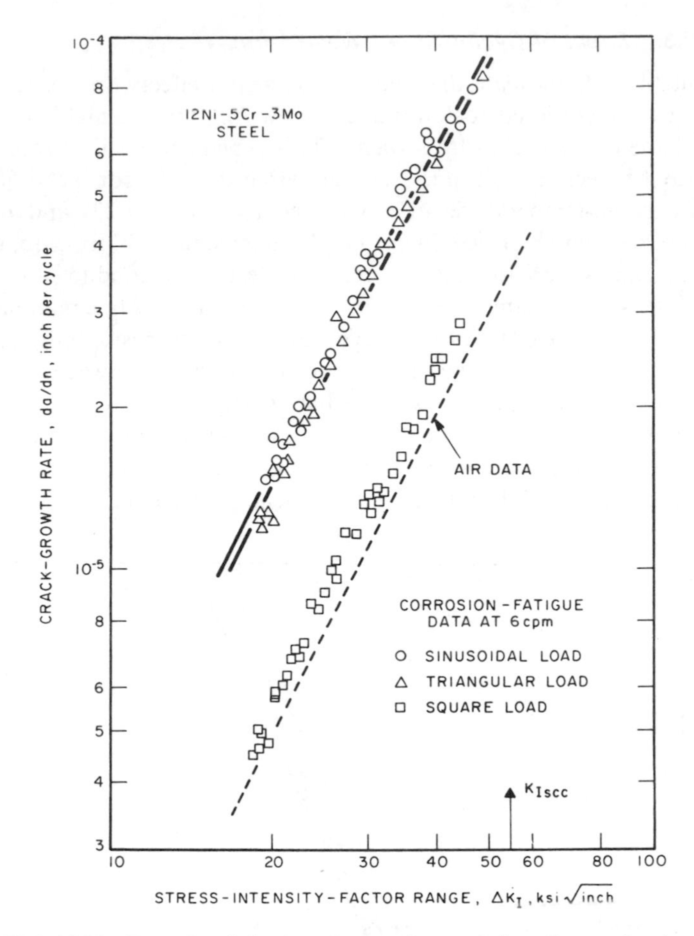

FIG. 11.23. Corrosion-fatigue-crack-growth rates below K_{Iscc} under sinusoidal, triangular, and square loads.

show that in a sodium chloride solution the crack-growth rates per cycle under sinusoidal and triangular stress fluctuations are almost identical. At a constant frequency, the environment increased the crack-growth rate by the same amount under sinusoidal stress fluctuations as under triangular stress fluctuations. The data also show that environmental effects are negligible when the steel is subjected to a square-wave stress fluctuation. Corrosion-fatigue-crack-growth rates under square-wave loading at 6 cpm for the 12Ni-5Cr-3Mo steel tested in sodium chloride solution were essentially the same as they were in the absence of environmental effects. By establishing the sinusoidal cyclic frequency that would result in the same environmental effects on the rate of fatigue crack growth, Barsom[4] showed that the environmental damage below K_{Iscc} occurred only during transient loading. Corrosion-fatigue data obtained below K_{Iscc} by using square waves having different dwell times at maximum and minimum loads also showed no environmental effects at constant tensile stresses.

Fatigue-crack-growth data for an aluminum alloy (DTD 5070A) tested in air at 60 cpm showed no difference in the rate of growth under sinusoidal, square, and pulsed waveforms.[28] On the other hand, fatigue-crack-growth data for 7075-T6 aluminum alloy tested in salt water at 6 cpm under sinusoidal, triangular, and square wave forms showed a behavior very similar to that presented for 12Ni-5Cr-3Mo steel in salt water.[14] The difference between the results obtained for the DTD 5070A aluminum alloy in air and those obtained for 7075-T6 and for 12Ni-5Cr-3Mo steel in salt water may be due to the relative immunity of the alloy to the air environment, to the high cyclic frequency, or to a difference in the mechanism of corrosion fatigue.

11.3.4. Environmental Effects During Transient Loading

Corrosion-fatigue-crack-growth test results for 12Ni-5Cr-3Mo maraging steel in 3% sodium chloride solution under sinusoidal, triangular, and square-wave loading showed that environmental effects in the environment-material investigated are significant only during the transient-loading portion of each cyclic-load fluctuation.[4] The difference between the effects of the sodium chloride solution on the rate of fatigue-crack growth during increasing and decreasing plastic deformation in the vicinity of the crack tip was investigated by using test results obtained for specimens subjected to various triangular cyclic-load fluctuations.

The effects of the environment during increasing plastic deformation in the vicinity of the crack tip were separated from the effects during decreasing plastic deformation by studying the differences in the corrosion-fatigue-crack-growth rate obtained under positive-sawtooth (⩘) and under negative-sawtooth (⩗) cyclic-load fluctuations.

The data presented in Figure 11.22[4] show that the rates of fatigue-crack growth in a room-temperature air environment under various cyclic-stress fluctuations are not affected by the form of the cyclic-stress fluctuations. Consequently, differences in the rates of corrosion-fatigue-crack growth among triangular waves, positive-sawtooth waves, and negative-sawtooth waves can be attributed primarily to variations in the interaction between plastically deformed metal at the crack tip and the surrounding environment. These variations result from differences in the pattern of stress fluctuations during each cycle.

The corrosion-fatigue-crack-growth-rate data for 12Ni-5Cr-3Mo maraging steel tested in 3% sodium chloride solution under various cyclic stress fluctuations are presented in Figure 11.24[4]. The data show that the corrosion-fatigue-crack-growth rates determined with the negative-sawtooth wave and with the square wave are essentially the same as the fatigue-crack-growth

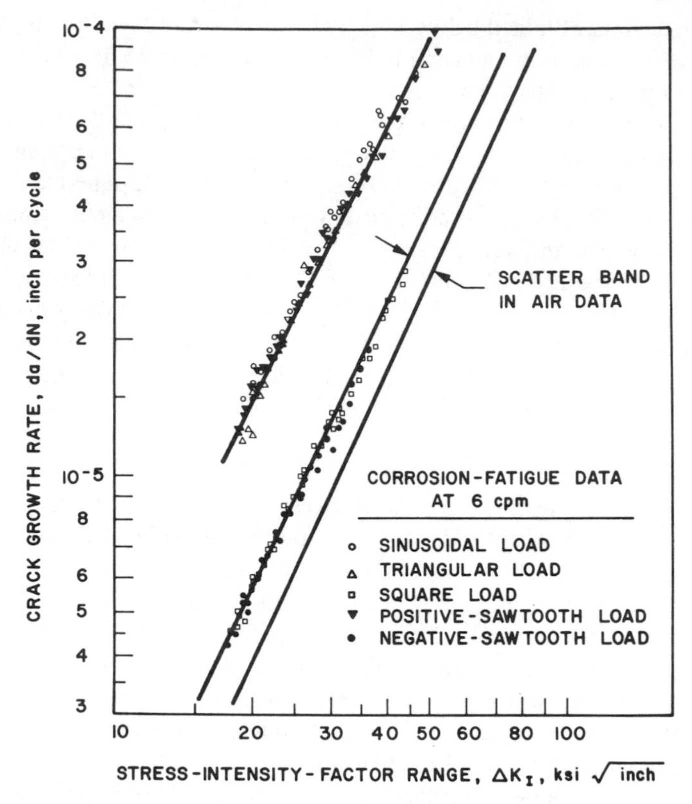

FIG. 11.24 Corrosion-fatigue-crack-growth rates in 12Ni-5Cr-3Mo steel in 3% solution of sodium chloride under various cyclic-stress fluctuations with different stress-time profiles.

rate determined in air. The corrosive effect therefore depends on the wave form. The negative-sawtooth wave and the square wave show no corrosive effect. The corrosion-fatigue-crack-growth rates determined with sinusoidal, triangular, and positive-sawtooth cyclic-stress fluctuations are identical but are three times higher than the fatigue-crack-growth rate determined in air. Thus, the environment increased the fatigue-crack-growth rate significantly.

Because the corrosive effect increased the rate of fatigue-crack growth below K_{Iscc} by the same amount with the triangular wave as with the positive-sawtooth wave, the corrosive processes in the environment-material system investigated were operative *only* while the tensile stresses in the vicinity of the crack tip were increasing. This conclusion is supported by (1) corrosion-fatigue data obtained with the negative-sawtooth wave, which showed no corrosive effect while the tensile stresses were decreasing, and (2) corrosion-fatigue data obtained with the square wave which showed no corrosive effect at constant tensile stresses.

Fatigue-crack-growth data for 7075-T6 aluminum alloys tested in salt water at 6 cpm under sinusoidal, square, positive-sawtooth, and negative-sawtooth loading showed a behavior very similar to that presented for the 12Ni-5Cr-3Mo maraging steel in salt water.[14] However, fatigue-crack-growth data for 7075-T651 aluminum alloy tested at 105 cpm in distilled water showed no significant difference in the rate of growth under positive- and negative-sawtooth loadings.[29] Further work is necessary to resolve the apparent discrepancy in the conclusions relating to the effect of wave form on the corrosion-fatigue behavior for high-strength aluminum alloys in water environments.

11.3.5. Corrosion-Fatigue Crack Growth Above K_{Iscc}

The rate of corrosion-fatigue crack growth above K_{Iscc} should be greater than the rate below K_{Iscc} because whenever the magnitude of the maximum stress-intensity-factor value in a given cycle becomes greater than K_{Iscc}, the rate of corrosion-fatigue-crack growth should be accelerated by stress-corrosion cracking. Based on the assumption that, above K_{Iscc}, the fatigue-crack growth in an aggressive environment is enhanced by the same magnitude as that for crack growth under sustained loads, Wei and Landes[15] hypothesized a linear summation model to predict the corrosion-fatigue behavior above K_{Iscc} for a high-strength steel. The model considers the corrosion fatigue-crack-growth rate above K_{Iscc} to be the sum of the rate of fatigue-crack growth in an inert reference environment and an environmental component that is computed from the load profile and sustained-load crack-growth data obtained in an identical aggressive environment. Wei and Landes were able to satisfactorily predict the rates of corrosion-fatigue-crack growth for ultra-high-strength steels tested in dehumidified hydrogen (Figure 11.25), distilled water, and water-vapor environments and for a Ti-8Al-1V-1Mo alloy tested in distilled water and sodium chloride solution. This model can be used to approximate the published rates of corrosion-fatigue-crack growth above K_{Iscc} for ultra-high-strength steels.[10,30-32]

The preceding linear-summation model has limited applicability and should not be used indiscriminately to predict the corrosion-fatigue behavior above K_{Iscc}. It can be considered, to a first-order approximation, to be applicable to environment-material systems in which the material is highly susceptible to the environment such as high-strength steels having yield strengths greater than about 200 ksi tested in water. The rates of crack growth under sustained loads in these environment-material systems are usually orders of magnitude greater than under cyclic loads in the absence of the environment. The model neglects the effects of frequency and wave form on the rate of corrosion-fatigue-crack growth below K_{Iscc}. Thus, except possibly for square-wave loading, the model is of questionable validity for most environment-material systems. It should be noted that the rate of stress-

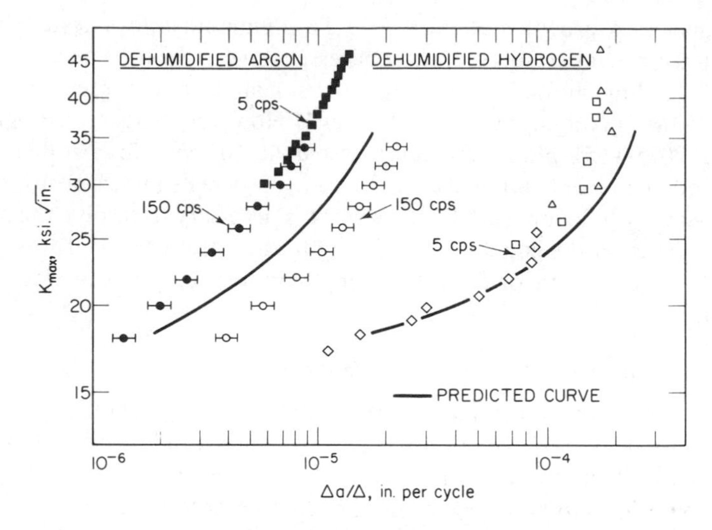

FIG. 11.25. Fatigue crack growth in 18 Ni (250) maraging steels tested in dehumidified argon and hydrogen (Ref. 15).

corrosion cracking for most environment-material systems is very slow when the applied K_I value is only slightly higher than the value for K_{Iscc}. Consequently, the corrosion-fatigue behavior below K_{Iscc} for these environment-material systems remains unaltered even when the maximum value for K_I in a given cycle becomes greater than K_{Iscc}.[33]

REFERENCES

1. *Corrosion Fatigue: Chemistry, Mechanics, and Microstructure*, International Corrosion Conference Series, Vol. NACE-2, National Association of Corrosion Engineers, Houston, 1972.
2. H. H. Johnson and P. C. Paris, "Sub-critical Flaw Growth," *Journal of Engineering Fracture Mechanics, 1*, No. 3, June 1968.
3. W. G. Clark, Jr., "The Fatigue Crack Growth Rate Properties of Type 403 Stainless Steel in Marine Turbine Environments," *Corrosion Problems in Energy Conversion and Generation*, edited by C. S. Tedman, Jr., Electrochemical Chemical Society, Princeton, N.J., 1974.
4. J. M. Barsom, "Effect of Cyclic-Stress Form on Corrosion-Fatigue Crack Propagation Below K_{Iscc} in a High-Yield-Strength Steel," in *Corrosion Fatigue: Chemistry, Mechanics, and Microstructure*, International Corrosion Conference Series, Vol. NACE-2, National Association of Corrosion Engineers, Houston, 1972.
5. J. M. Barsom and S. R. Novak, "Subcritical Crack Growth in Steel Bridge Members," Final Report, *NCHRP 12-14*, National Cooperative Highway Research Program, Washington, Sept. 1974.

6. T. W. CROOKER and E. A. LANGE, "The Influence of Salt Water on Fatigue Crack Growth in High Strength Structural Steels," *ASTM STP 462*, American Society for Testing and Materials, Philadelphia, 1970.

7. F. J. BRADSHAW and C. WHEELER, "The Influence of Gaseous Environment and Fatigue Frequency on the Growth of Fatigue Cracks in Some Aluminum Alloys," *International Journal of Fracture Mechanics*, *5*, No. 4, Dec. 1969.

8. M. O. SPEIDEL, M. J. BLACKBURN, T. R. BECK, and J. A. FEENEY, "Corrosion-Fatigue and Stress-Corrosion Crack Growth in High-Strength Aluminum Alloys, Magnesium Alloys, and Titanium Alloys, Exposed to Aqueous Solutions," *Corrosion Fatigue: Chemistry, Mechanics and Microstructure*, International Corrosion Conference Series, Vol. NACE-2, National Association of Corrosion Engineers, Houston, 1972.

9. D. A. MEYN, "An Analysis of Frequency and Amplitude Effects on Corrosion Fatigue Crack Propagation in Ti-8Al-1Mo-1V," *Metallurgical Transactions*, *2*, 1971.

10. J. P. GALLAGHER, "Corrosion Fatigue Crack Growth Behavior Above and Below K_{Iscc}," *NRL Report 7064*, Naval Research Laboratory, Washington, D.C., May 28, 1970.

11. G. A. MILLER, S. J. HUDAK, and R. P. WEI, "The Influence of Loading Variables on Environment-Enhanced Fatigue-Crack Growth in High Strength Steels," *Journal of Testing and Evaluation*, *1*, No. 6, 1973.

12. P. C. PARIS, R. J. BUCCI, E. T. WESSEL, W. G. CLARK, and T. R. MAGER, "Extensive Study of Low Fatigue-Crack-Growth Rates in A533 and A508 Steels," *ASTM STP 513*, American Society for Testing and Materials, Philadelphia, 1972.

13. R. BUCCI, "Environment Enhanced Fatigue and Stress Corrosion Cracking of a Titanium Alloy Plus a Simple Model for Assessment of Environmental Influence of Fatigue Behavior," Ph. D. Dissertation, Lehigh University, Bethlehem, Pa., 1970.

14. R. J. SELINES and R. M. PELLOUX, *Effect of Cyclic Stress Wave Form on Corrosion Fatigue Crack Propagation in Al-Zn-Mg Alloys*, Department of Metallurgy and Materials Science Report, Massachusetts Institute of Technology, Cambridge, Mass., 1972.

15. R. P. WEI and J. D. LANDES, "Correlation Between Sustained-Load and Fatigue-Crack Growth in High-Strength Steels," *Materials Research and Standards, MTRSA*, *9*, No. 7, July 1969.

16. J. M. BARSOM, "Investigation of Subcritical Crack Propagation," Ph.D. Dissertation, University of Pittsburgh, Pittsburgh, 1969.

17. J. M. BARSOM, "Corrosion-Fatigue Crack Propagation Below K_{Iscc}," *Journal of Engineering Fracture Mechanics*, *3*, No. 1, July 1971.

18. T. W. CROOKER and E. A. LANGE, "Corrosion Fatigue Crack Propagation of Some New High Strength Structural Steels," *Journal of Basic Engineering, Transactions, ASME*, 91, 1969.

19. J. M. BARSOM, J. F. SOVAK, and E. J. IMHOF, JR., "Corrosion-Fatigue Crack Propagation Below K_{Iscc} in Four High-Yield-Strength Steels," *Applied Research Laboratory Report 89.021-024(3)*, U.S. Steel Corporation, Dec. 14, 1970 (available from the Defense Documentation Center), Arlington, Va.

20. E. J. IMHOF and J. M. BARSOM, "Fatigue and Corrosion-Fatigue Crack Growth of 4340 Steel at Various Yield Strengths," *ASTM STP 536*, American Society for Testing and Materials, Philadelphia, 1973.
21. J. A. FEENEY, J. C. MCMILLAN, and R. P. WEI, "Environmental Fatigue Crack Propagation of Aluminum Alloys at Low Stress Intensity Levels," *Metallurgical Transactions, 1*, June 1970.
22. A. HARTMAN, "On the Effect of Oxygen and Water Vapor on the Propagation of Fatigue Cracks in 2024-TB3 Alclad Sheet," *International Journal of Fracture Mechanics, 1*, No. 3, Sept. 1965.
23. F. J. BRADSHAW and C. WHEELER, "The Effect of Environment on Fatigue Crack Growth in Aluminum and Some Aluminum Alloys," *Applied Materials Research, 5*, No. 2, 1966.
24. R. P. WEI, "Fatigue-Crack Propagation in a High-Strength Aluminum Alloy," *International Journal of Fracture Mechanics, 4*, No. 2, June 1968.
25. R. P. WEI and J. D. LANDES, "The Effect of D_2O on Fatigue Crack Propagation in a High Strength Aluminum Alloy," *International Journal of Fracture Mechanics, 5*, 1969.
26. J. M. BARSOM, "Fatigue Behavior of Pressure-Vessel Steels," *Welding Research Council (WRC) Bulletin*, 194, May 1974.
27. R. J. BUCCI and J. K. DONALD, *Fatigue and Fracture Investigation of a 200 Grade Maraging Steel Forging*, Del Research Corporation Report, Hellertown, Pa., Oct. 1972.
28. F. J. BRADSHAW, N. J. F. GUNN, and C. WHEELER, "An Experiment on the Effect of Fatigue Wave Form on Crack Propagation in an Aluminum Alloy," *Technical Memorandum MAT93*, Royal Aircraft Establishment, Farnborough, Hants, England, July 1970.
29. S. J. HUDAK and R. P. WEI, Comments on paper by J. M. Barsom, "Effect of Cyclic Stress Form on Corrosion Fatigue Crack Propagation Below K_{Iscc} in a High Yield Strength Steel," *Corrosion Fatigue: Chemistry, Mechanics, and Microstructure*, International Corrosion Conference Series, Vol. NACE-2, National Association of Corrosion Engineers, Houston, 1972.
30. E. P. DAHLBERG, "Fatigue Crack Propagation in High Strength 4340 Steel in Humid Air," *Transactions: American Society for Metals, 58*, 1965.
31. W. A. VAN DER SLUYS, "Effect of Repeated Loading and Moisture on the Fracture Toughness of SAE 4340 Steel," *Journal of Basic Engineering, Transactions, ASME, 87*, 1965.
32. W. A. VAN DER SLUYS, "The Effect of Moisture on Slow Crack Growth in Thin Sheets of SAE 4340 Steel Under Static and Repeated Loading," *Journal of Basic Engineering, Transactions, ASME, 89*, 1967.
33. J. P. GALLAGHER and R. P. WEI, "Corrosion Fatigue Crack Propagation Behavior in Steels," *Corrosion Fatigue: Chemistry, Mechanics, and Microstructure*, International Corrosion Conference Series, Vol. NACE-2, National Association of Corrosion Engineers, Houston, 1972.

12

Fracture Criteria

12.1. Introduction

A fracture criterion is a standard against which the expected fracture behavior of a structure can be judged. In general terms, fracture criteria are related to the three levels of fracture performance, namely plane strain, elastic-plastic, or fully plastic, as shown in Figure 12.1. Although it would appear desirable always to specify fully plastic behavior, this is rarely done because it is almost always unnecessary as well as being economically unfeasible in most cases. Furthermore it is unsound engineering because good design should be an optimization of satisfactory safe structural performance and economic considerations.

For most structural applications, some level of elastic-plastic behavior at the service temperature and loading rate is a satisfactory fracture criterion. While there may be some cases where fully plastic behavior is necessary (e.g. large dynamic loadings such as submarines being subjected to depth charges) or where plane-strain behavior can be tolerated (e.g. certain short-life aerospace applications where the loading and fabrication can be precisely controlled), for the majority of large complex structures (bridges, ships, pressure vessels, offshore drilling rigs, etc.), some level of elastic-plastic behavior is appropriate. The question becomes, What level of elastic-plastic behavior is required and how can this level of performance be ensured? The purpose of this chapter is to develop a rational engineering approach to answering this question.

Unfortunately the selection of a fracture criterion is often quite arbitrary and based on service experience for other types of structures that may have no relation to the particular structure an engineer may be designing. Also, because fracture criterion is only a part of a fracture control plan, selection of a fracture criterion alone, without considering the other factors involved in fracture control, (e.g. fatigue) will not necessarily result in a safe structure. An example of the use of a fracture criterion developed for one application

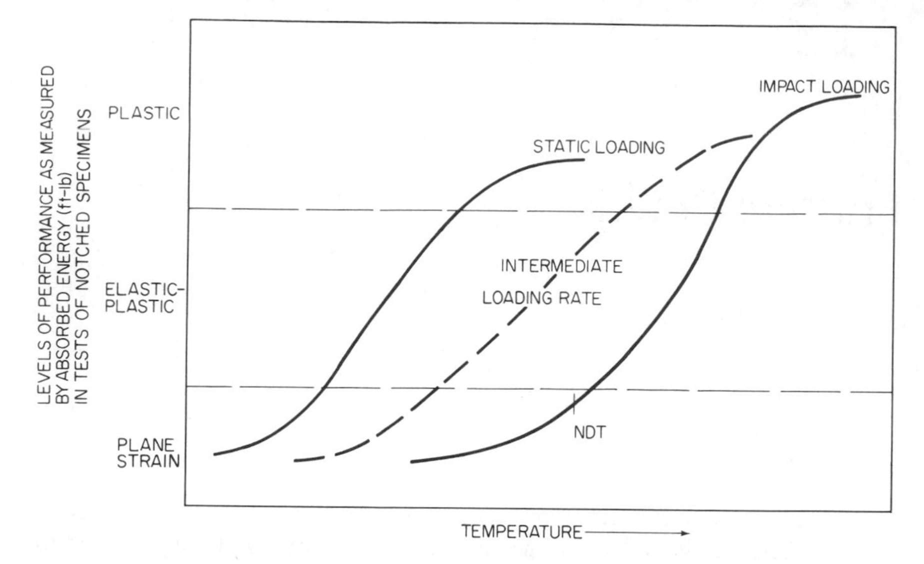

FIG. 12.1. Schematic showing relation between notch-toughness test results and levels of structural performance of various loading rates.

and yet which is widely used in many other situations is the 15-ft-lb CVN impact criterion at the minimum service temperature, which was established based on the World War II ship failures. This criterion has been widely used for various types of structures, even though the material, service conditions, structural redundancy, etc., may be considerably different from that of the World War II ships for which the criterion was established.

Criteria selection should be based on a very careful study of the particular requirements for a particular structure. In this chapter we shall describe the various factors involved in the development of a criterion, which include

1. The service conditions (loadings, temperature, loading rate, etc.) to which the structure will be subjected.
2. The significance of loading rate on performance of the structural materials to be used.
3. The desired level of performance in the structure.
4. The consequences of failure.

In the following chapter (Chapter 13) we shall present examples of various criteria that have been used for different types of structures as well as other criteria that have been proposed for various types of structures. The most widely used of these criteria is the transition-temperature approach, and various uses of this approach, as well as some of its limitations, will be described.

It should be emphasized that there can be no single *best* criterion for all structures because optimum design should involve economic considerations as well as technical ones. That is, selecting a fracture criteria that is far more conservative than necessary, based on the possible consequences of failure, is unsound engineering. Furthermore, although a study of all aspects of a fracture-control plan, i.e., design, fabrication, inspection, operation, etc., is necessary to ensure the safety and reliability of structures (as will be described in Chapter 14), *the fundamental decision in designing to prevent brittle fracture in any structure is the development of suitable fracture criteria for the structural materials to be used in the structure.* Thus the establishment of the proper fracture criteria for a given structural application should be the basis of subsequent material selection, structural design, fabrication procedures, and inspection requirements.

12.2. General Levels of Performance

The primary design criterion for most large structures such as bridges, pressure vessels, ships, etc., is still based on strength and stability requirements such that nominal elastic behavior is obtained under conditions of maximum loading. Usually the strength and stability criteria are achieved by limiting the maximum design stress to some percentage of the yield strength. In many cases, *fracture toughness also* is an important design criterion, and yet specifying a notch-toughness criterion is much more difficult, primarily because

1. Establishing the specific level of required notch toughness (i.e., the required CVN or K_{Ic} value at a particular test temperature and rate of loading) is costly and time consuming and is a subject *unfamiliar* to engineers.
2. There is no well-recognized single "best" approach. Therefore, different experts will have different opinions as to the "best" approach, although the science of fracture mechanics is helping to overcome this difficulty.
3. The cost of structural materials increases with increasing levels of inherent notch toughness. Thus economic considerations must be included when establishing any toughness criterion.

Materials that have extremely high levels of notch toughness under even the most severe service conditions (earthquakes, ice movement, dynamic loading, etc.) are available, and the designer can always specify that these materials be used in critical locations within a structure. However, because the cost of structural materials generally increases with their ability to perform satisfactorily under more severe operating conditions, a designer generally does not wish to specify arbitrarily more notch toughness than is required for the specific application. In the same sense, a designer does not specify the use of a material with a very high yield strength for a compression member if the design is such that the critical buckling stress is very low. In the former case, the excessive notch toughness is unnecessary, and in the latter case, the excessive yield strength is unnecessary. Both cases are examples of unsound engineering.

The problem of establishing specific fracture-toughness requirements that are not excessive but are still adequate for normal service conditions is a long-standing one for engineers. However, by using concepts of fracture mechanics, rational fracture criteria can be established for fracture control in different types of structures.

Previously, the maximum allowable flaw size in a member has been shown to be related to the notch toughness and yield strength of a structural material as follows:

$$a = C\left(\frac{K_c,\ K_{\mathrm{Ic}},\ \text{or}\ K_{\mathrm{Id}}}{\sigma_{ys}\ \text{or}\ \sigma_{yd}}\right)^2$$

For conditions of maximum constraint (plane strain), such as might occur in thick plates or in regions of high constraint, the flaw size becomes proportional to $(K_{\mathrm{Ic}}/\sigma_{ys})^2$ or $(K_{\mathrm{Id}}/\sigma_{yd})^2$, where both the toughness and yield strength should be measured at the service temperature and loading rate of the structure.

Thus, the $K_{\mathrm{Ic}}/\sigma_{ys}$ ratio or $K_{\mathrm{Id}}/\sigma_{yd}$ ratio (either in the valid plane-strain region or extrapolated into the elastic-plastic region) becomes a good index for measuring the relative toughness of structural materials. For most structural applications it is desirable that the structure tolerate large flaws without fracturing; therefore the use of materials with high $K_{\mathrm{Ic}}/\sigma_{ys}$ or $K_{\mathrm{Id}}/\sigma_{yd}$ ratios (i.e., elastic-plastic behavior) is a desirable condition.

The basic question in establishing a fracture criterion for large structures becomes, where complete initial inspection for cracks and continuous monitoring of crack growth through the life of a structure are not always possible, practical, or economical, how high must this ratio be for a particular structural material to ensure satisfactory and safe performance in large complex structures?

No simple answer exists because the engineer must take into account such factors as the design life of the structure, consequences of a failure in a structural member, redundancy of load path, probability of overloads, and fabrication and material cost. However, fracture mechanics can provide a conservative engineering approach to rationally evaluate this question. It is assumed that discontinuities will initiate and propagate in most large complex structures and that local yield-stress loading and plane-strain conditions may exist in parts of the structure (although the use of thin plates tends to minimize the possibility of plane-strain behavior). Therefore, the $K_{\mathrm{Ic}}/\sigma_{ys}$ ratio obtained at the appropriate loading rate for materials used in a particular structure [either by direct measurement (Chapters 3 and 4) or by empirical correlation with auxiliary fracture-toughness tests (Chapter 6)] is one of the primary material parameters that defines the relative safety of a structure against brittle fracture.

If a structure is loaded "slowly" ($\sim 10^{-5}$ in./in./sec), the K_{Ic}/σ_{ys} ratio is the controlling toughness parameter. If, however, the structure is loaded "dynamically" ($\sim 10^{1}$ in./in./sec or impact loading), the K_{Id}/σ_{yd} ratio is the controlling parameter. At intermediate loading rates the ratio of K_{Ic} and σ_{ys} determined at the appropriate loading rate is the controlling parameter. Definitions and test conditions for each of these ratios are as follows:

1. K_{Ic}: critical plane-strain stress-intensity factor under conditions of static loading as described in ASTM Test Method E-399—Standard Method of Test for Plane Strain Fracture Toughness of Metallic Materials (Chapter 3).
2. σ_{ys}: static tensile yield strength obtained in "slow" tension test as described in ASTM Test Method E-8—Standard Methods of Tension Testing of Metallic Materials.
3. K_{Id}: critical plane-strain stress-intensity factor as measured by "dynamic" or "impact" tests (Chapter 3). The test specimen is similar to a K_{Ic} test specimen but is loaded rapidly. There is no standardized test procedure.
4. σ_{yd}: dynamic tensile yield strength obtained in "rapid" tension test at loading rates comparable to those obtained in K_{Id} tests. This value is difficult to obtain experimentally, as described in Chapter 3. A good engineering approximation for σ_{yd} under loading rates comparable to impact conditions based on experimental results of structural steels is

$$\sigma_{yd} = \sigma_{ys} + (20 \text{ to } 30 \text{ ksi})$$

Because high constraint (thick plates—plane-strain conditions) at the tip of a crack can lead to premature fracture, the engineer should strive for the lowest possible degree of constraint (thin plates—plane-stress conditions) at the tip of a crack. To help the engineer establish a satisfactory level of constraint for various structural applications (as measured by K_{Ic}/σ_{ys} or K_{Id}/σ_{yd}) the following three general levels of material behavior or performance can be established using fracture-mechanics terminology:

1. ELASTIC PLANE-STRAIN BEHAVIOR (K_{Ic}/σ_{ys} or $K_{Id}/\sigma_{yd} < \sqrt{t/2.5}$)

Structural materials whose toughness and plate thickness is such that K_{Ic}/σ_{ys} or K_{Id}/σ_{yd} is less than about $\sqrt{t/2.5}$ exhibit elastic plane-strain behavior and generally fracture in a brittle manner. These materials usually are not used for most structural applications because of the high level of constraint at the tip of a crack and the rather small critical crack sizes at design stress levels. Fortunately most structural materials have toughness levels such that they do *not* exhibit plane-strain behavior at service temperatures, service loading rates, and common structural sizes normally used. However, very thick plates or plates used to form complex geometries where the constraint can be very high may be susceptible to brittle fractures even though the inherent notch toughness as measured by small-scale laboratory tests appears satisfactory. Still, there are some situations where this level of behavior can be used successfully (such as for rail steels). In these situations parameters other than fracture toughness (such as stress state, and fatigue-crack initiation and propagation) control the behavior of the structure.

2. PLANE-STRESS OR MIXED-MODE BEHAVIOR (ELASTIC-PLASTIC) [$\sqrt{t}/2.5 < K_{Ic}/\sigma_{ys}$ or $K_{Id}/\sigma_{yd} < \alpha\sqrt{t}$ (where α may be as high as 2 or 3)]

Structural materials whose toughness levels are such that they are in the above range generally exhibit elastic-plastic fractures with varying amounts of yielding prior to fracture. The tolerable flaw sizes at fracture vary considerably but can be fairly large. Fracture is usually preceded by the formation of large plastic zones ahead of the crack. Most structures are built of materials that exhibit some level of elastic-plastic behavior at service temperatures and loading rates, and thus K_{Ic} or K_{Id} values are difficult to obtain and cannot be measured directly unless very large specimens are used.

3. GENERAL YIELDING (PLASTIC) [K_{Ic}/σ_{ys} or $K_{Id}/\sigma_{yd} > \alpha\sqrt{t}$ (where α may be as high as 2 or 3)]

Structural materials falling in this range usually exhibit ductile plastic fractures preceded by large deformation. Obviously, K_{Ic} or K_{Id} values cannot be measured. This type of behavior is very desirable in structures and represents considerable notch toughness. However, this level of toughness is rarely necessary and thus is usually not specified, except for unusual cases such as submarine hulls or perhaps nuclear structures.

Photographs comparing the general fracture appearance of these three levels of behavior are presented in Figures 12.2, 12.3, and 12.4. *Plane-strain* behavior refers to fracture under elastic stresses with little or no shear-lip development and is essentially brittle. A fracture surface typical of this behavior is shown in Figure 12.2. *Plastic* behavior refers to ductile failure under general yielding conditions accompanied, usually, with the development of very large shear-lip development. A fracture surface typical of this behavior is shown in Figure 12.4. The transition between these two extremes is the *elastic-plastic* region, which is also referred to as the mixed-mode region. A fracture surface typical of this behavior is shown in Figure 12.3. However, fracture-surface appearance by itself, is neither necessary nor sufficient to indicate the mode of fracture. Ductile fractures can propagate in the elastic-plastic or plastic regions with little or no shear-lip development when the geometry of the detail prevents the development of 45° shear planes.

For structural steels, these three levels of performance are usually described in terms of the transition from brittle to ductile behavior as measured by various types of notch-toughness tests. This general transition in

FIG. 12.2. Photograph of plane-strain fracture surface.

fracture behavior was related to the three levels of behavior schematically in Figure 12.1. For static loading, the transition region occurs at lower temperatures than for impact (or dynamic) loading, for those structural materials that exhibit a transition behavior. (This general behavior was discussed in Chapter 4.) Thus, for those structures that are subjected to static loading, a static transition curve (i.e., K_{Ic} test specimens) should be used to predict the level of performance at the service temperature. For structures subjected to impact or dynamic loading, the impact transition curve (i.e., K_{Id} test specimens) should be used to predict the level of performance at the service temperature. For structures subjected to some intermediate loading rate, an intermediate loading-rate transition curve should be used to predict the level of performance at the service temperature. Because the actual loading

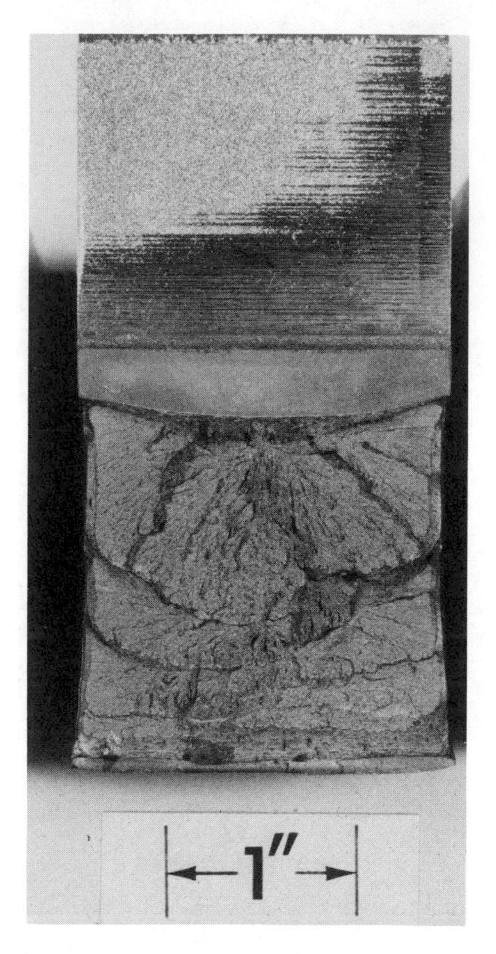

FIG. 12.3. Photograph of elastic-plastic (mixed-mode) fracture surface.

rates for many structures are not well known, the impact loading curve (Figure 12.1) is often used to predict the service performance of structures. However, this may be unduly conservative and often does not properly model the service behavior for many types of structures that are loaded statically or at intermediate loading rates, such as bridges.

Figures 12.5 and 12.6 show the three general levels of performance for typical structural steels as measured by various types of notch-toughness tests. It should be emphasized that although the upper limit of plane-strain behavior is reasonably well established as K_{Ic}/σ_{ys} (or K_{Id}/σ_{yd}) $= \sqrt{t/2.5}$, the boundary between elastic-plastic and plastic behavior *is not well established.* The boundaries shown in Figures 12.5 and 12.6 are general guidelines based on the authors' experience in which the beginning of the plastic region is

FIG. 12.4. Photograph of plastic fracture surface.

defined as that region where the transition curves begin to level off. Also, the beginning of plastic behavior can be defined as K_{Ic}/σ_{ys} or $K_{Id}/\sigma_{yd} = \alpha\sqrt{t}$ (where α may be as high as 2 or 3), although this value cannot be measured using existing standardized test methods.

For structural materials that exhibit either elastic-plastic or plastic levels of toughness, there are no standardized test methods available for measuring critical stress-intensity factors (i.e., some type of K_c value). ASTM Committee E-24 is developing test procedures for elastic-plastic behavior in terms of fracture-mechanics concepts that can be used to calculate critical crack sizes in the same manner as K_{Ic}. These procedures are described in Chapter 16.

Because most structural materials exhibit non-plane strain behavior (i.e., either elastic-plastic or plastic behavior as shown in Figure 12.1) at service

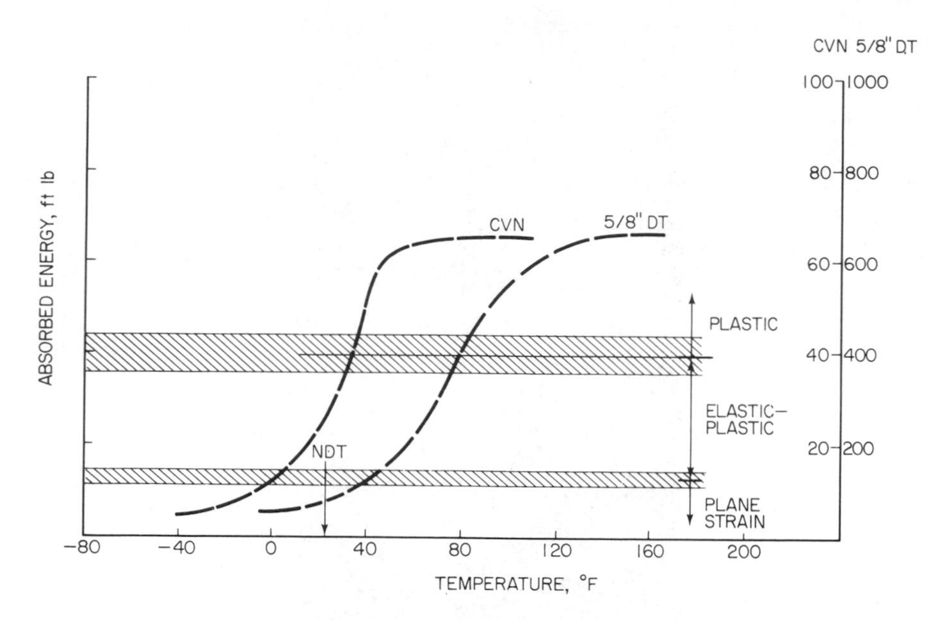

FIG. 12.5. Relation among plane-strain, elastic-plastic, and plastic levels of performance for an ABS-C steel as measured by NDT, CVN, and DT test results. *(Note: Ranges are approximate.)*

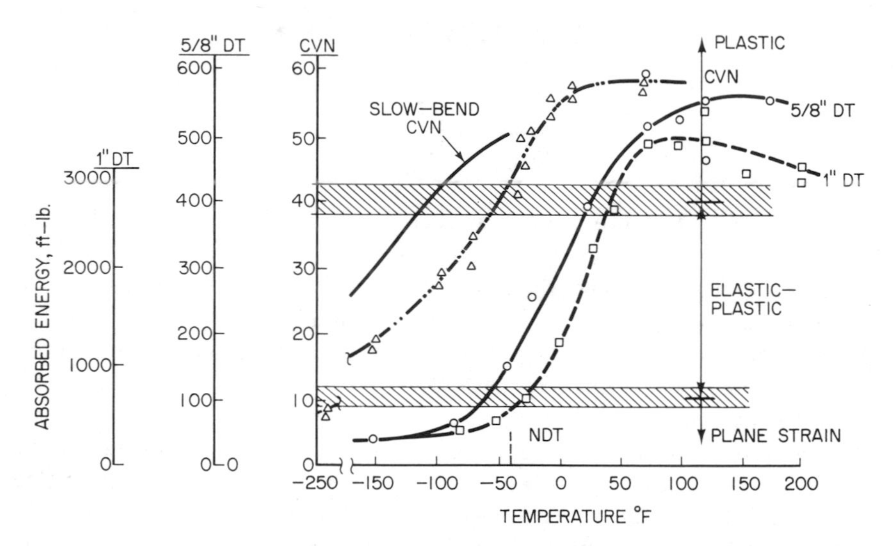

FIG. 12.6. Relation among plane-strain, elastic-plastic, and plastic levels of performance for an A517 steel as measured by various notch-toughness tests.

temperatures and loading rates, K_{Ic} values are difficult to obtain for most structural materials at service temperatures and loading rates. This fact represents a difficult situation when material properties must be specified (although it is a highly desirable one from the viewpoint that plane-strain brittle fractures are rare). Therefore, for those situations where it is desired to specify a minimum level of notch toughness that exceeds plane-strain behavior (which is usually the case), auxiliary test methods (such as described in Chapter 6) must be used to ensure that the structural material exhibits non-plane-strain behavior at the service temperature. Various examples of criteria that are used to ensure non-plane-strain behavior are presented in Chapter 13.

12.3. Significance of Loading Rate

For materials that exhibit loading-rate or strain-rate effects, such as structural steels having yield strengths less than about 140 ksi, the loading rate at a given temperature can affect the notch toughness significantly. Ideally, the loading rate in the particular notch-toughness test used to evaluate the behavior of a structural material should be essentially equal to the loading rate in the structure being analyzed. Toughness criteria can then be established by using either static or dynamic (or intermediate) toughness test results for a particular structure.

As discussed in Chapter 5, by knowing a specific K_{Ic} value for a particular structural material the critical crack size can be calculated for a given design stress level. If the structure is loaded statically, this size crack can be tolerated at all temperatures above an extremely low one, as shown schematically in Figure 12.7. If the structure is loaded dynamically, however, the minimum temperature at which this size crack can be tolerated would be much higher, possibly above the minimum service temperature.

As a specific example for a structural steel commonly used in shipbuilding, test results for an ABS, Class C steel are presented in Figure 12.8. Assume that a fracture criterion stating that $K_{Ic}/\sigma_{ys} \geq 0.6$ is required for a particular application.

The test results presented in Figure 12.8 show that if the actual structure is loaded *statically* ($\dot{\epsilon} \approx 5 \times 10^{-5}$/sec) this level of performance will be obtained at all service temperatures above −220°F. If the structure is loaded *dynamically* ($\dot{\epsilon} \approx 20$/sec), however, this level of performance ($K_{Id}/\sigma_{yd} \geq 0.6$) will be obtained at all service temperatures above −20°F. (For an intermediate loading rate ($\dot{\epsilon} \approx 10^{-1}$/sec), this level of performance will be obtained at all temperatures above −120°F.) Thus, the *actual service loading rate* has a very large influence on the toughness criterion specified, i.e., either static, dynamic, or intermediate.

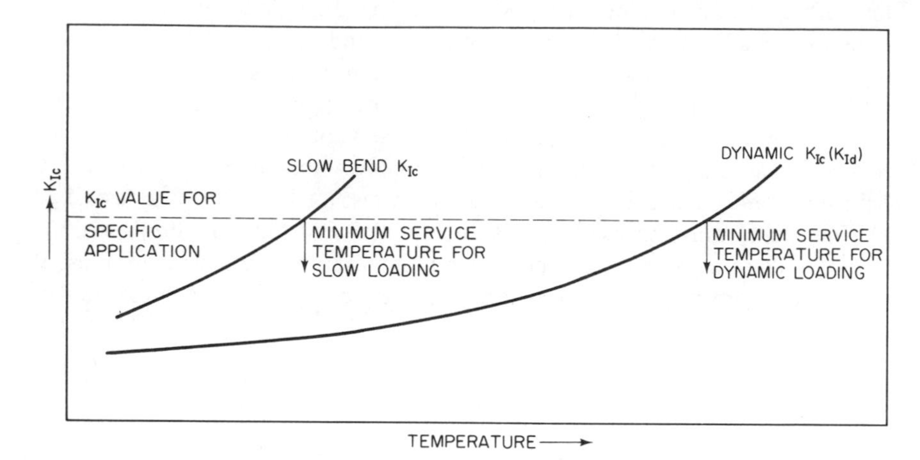

FIG. 12.7. Schematic showing effect of temperature and loading rate on K_{Ic}.

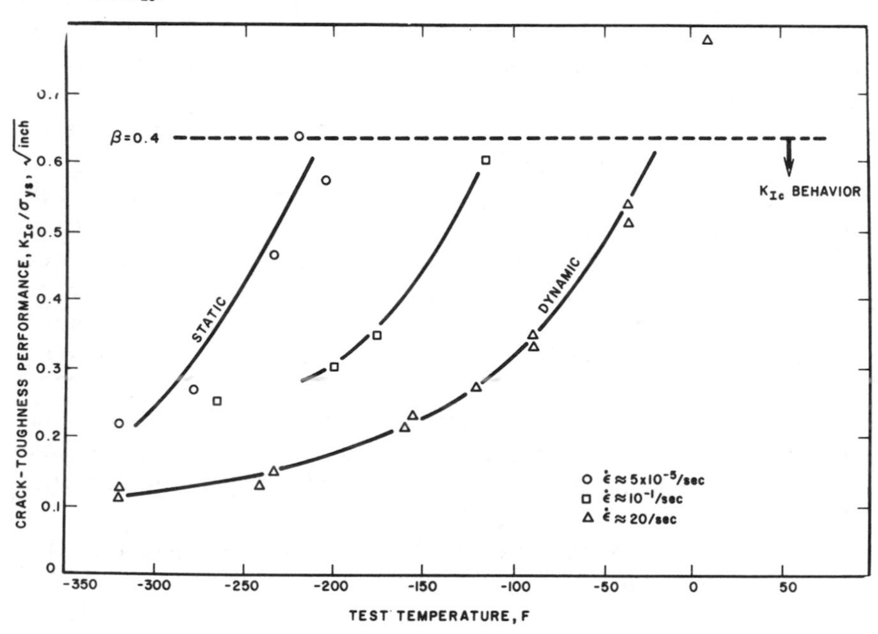

FIG. 12.8. Crack-toughness performance for ABS-C steel showing effect of loading rate.

Assume that this ABS-C steel is to be used in a low-temperature application at 0°F and that the possibility exists that a through-thickness crack is present in the vicinity of a weld, where the local stress is of yield-point magnitude. This situation is shown schematically in Figure 12.9, where the

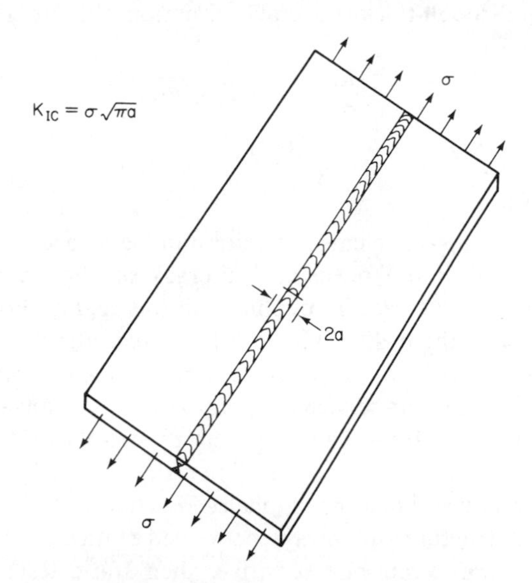

FIG. 12.9. Schematic showing flaw in welded plate.

critical crack size, $2a$, is

$$K_{Ic} = \sigma\sqrt{\pi a}$$

For

$$\frac{K_{Ic}}{\sigma_{ys}} \sim 0.6 = \sqrt{\pi a}$$

$$a \simeq \frac{(0.6)^2}{\pi} \simeq 0.11 \text{ in.}$$

Therefore $2a = 0.22$ in.

If the structure is loaded slowly, this size flaw could be tolerated at extremely low temperatures, i.e., −220°F, and the possibility of a brittle fracture at normal operating temperatures (0°F) is almost nonexistent. However, if the structure is loaded dynamically, this same size flaw can be tolerated only at temperatures above −20°F, and the possibility of a brittle fracture at a service temperature of only 0°F becomes much greater.

Note that above a K_{Id}/σ_{yd} value of about 0.6, which occurs close to above the NDT temperature of a structural steel, the relative toughness levels increase quite rapidly with temperature. This transition from plane-strain behavior to elastic-plastic and plastic behavior is the basis for many fracture criteria developed using the "transition-temperature approach." At a K_{Ic}/σ_{ys} or K_{Id}/σ_{yd} ratio of about 0.9, the minimum service temperatures for the ABS-C steel described in Figure 12.8 are approximately −180°F, −80°F, and +20°F for static, intermediate, and dynamic loading rates, respectively.

For the same through-thickness crack situation (Figure 12.9), the critical crack size is

$$\frac{K_{Ic}}{\sigma_{ys}} \sim 0.9 = \sqrt{\pi a}$$

$$a \simeq \frac{(0.9)^2}{\pi} \simeq 0.26$$

and $2a \sim 0.52$ in.

Thus, for only a 40°F change in minimum service temperature above a K_{Ic}/σ_{ys} or K_{Id}/σ_{yd} ratio of 0.6, the critical crack size has more than doubled because of the rapid increase in toughness in this region. For dynamic loading, this would be about 40°F above NDT. Accordingly, the temperature range through which the K_{Ic}/σ_{ys} or K_{Id}/σ_{yd} ratio increases rapidly can be and has been used as a limiting service temperature, i.e., a transition temperature where the transition is from plane-strain to elastic-plastic behavior.

The importance of using a test specimen that "models" the loading rate in the actual structure should be emphasized. That is, if the designer can be certain that his structure will always be loaded statically, or that the structural material is *not* strain-rate sensitive, then static tests can be used to predict the service behavior. Conversely, if impact loading is the proper loading rate then an impact test should be used to predict the service behavior.

Because of the large difference in behavior of structural materials that are strain-rate-sensitive, there are two widely differing schools of thought regarding the possibility of dynamic loading of structures.

The first school of thought assumes that there can be highly localized regions where the notch toughness is low in all large structures. Examples of these regions might be microcracks in weld metal, a grain-coarsened region in the heat-affected zone of a weldment, arc strikes on the base plate, nonmetallic inclusions, etc. These highly localized regions of low notch toughness are assumed to initiate microcracks under statically applied loads so that the surrounding material is suddenly presented with a moving dynamic (but small) crack. Thus, even though the applied structural loading rate may be slow, this school of thought believes that the crucial loading rate is that of the crack-tip pop-in. This sudden separation of a few metal grains ($<\sim 0.01$ in. in size) is assumed to create dynamic loading rates which apply thereafter in controlling crack extension. This means that the dynamic toughness (K_{Id}/σ_{yd}) always controls fracture extension, irrespective of initiation conditions. Accordingly, this leads to the conclusion that the K_{Id}/σ_{yd} parameter *always* should be used to establish the behavior for all structures regardless of the measured structural loading rates. This is often referred to as designing for crack arrest behavior.

The second school of thought believes that for those structures where the measured loading rates are slow, the static K_{Ic}/σ_{ys} ratio is the controlling notch-toughness parameter. For those structures loaded at some intermediate loading rate (which applies to many structures such as bridges, ships, and off-shore rigs) an "intermediate" (K_{Ic}/σ_{ys}) ratio is believed to be the controlling notch-toughness parameter. The latter viewpoint helps to explain the fact that many structures whose dynamic toughness (K_{Id}/σ_{yd}) is relatively low have been performing quite satisfactorily because the actual structural loading rate is slow or, at most, intermediate.

A realistic appraisal of these two schools of thought leads to the following observation. While it is true that there may exist highly localized regions of low toughness in complex structures, it seems hard to visualize that a dynamic stress field can be created by the localized extension of a small microcrack. A more realistic mechanism of creating a localized dynamic stress field under nominally static loading would appear to result from the sudden separation of a secondary stiffener, gusset plate, etc. This sudden separation might change the local stress distribution over a reasonably large area rather suddenly and could create a large dynamic stress field surrounding a small moving crack that would be necessary for the dynamic properties (K_{Id}/σ_{yd}) to control.

Accordingly, whereas it would be conservative always to design on the basis of possible dynamic loading (use of K_{Id}/σ_{yd} toughness values at the service temperature), it does not seem to be necessary unless extremely conservative toughness values are desired for a single load-path member. That is, if the failure of a single member can lead to collapse of the entire structure, then a conservative approach would be to design for the possibility of a dynamic crack; i.e., use K_{Id} values at the minimum service temperature rather than K_{Ic} values. Another approach to ensure the safety and reliability of such structures can be achieved by decreasing the maximum stress and stress fluctuation levels. This approach would result in a significant increase in the fatigue life of the structural component and in tolerating larger cracks prior to failure. Thus, factors *other than* just material toughness are important in fracture control, as discussed in Chapter 14. For multiple-load-path structures, such as most bridges, the use of static K_{Ic}/σ_{ys} or intermediate $K_{Ic(int)}/\sigma_{ys(int)}$ as the controlling toughness parameter appears quite realistic.

For example, the recent AASHTO material-toughness requirements for A36 bridge steels having service temperatures down to 0°F require that the steel plates exhibit 15-ft-lb CVN impact energy at +70°F. Because bridges have been shown to be loaded at an intermediate rate of loading, this requirement is designed to ensure non-plane-strain behavior beginning at about −50°F, well *below* the minimum service temperature of 0°F (the development of this criterion is presented in Chapter 13). Thus (1) because of the intermediate loading rate to which bridges are subjected, (2) because the loadings are reasonably well known, (3) because bridges are usually highly redundant structures (multiple-load paths) such that failure of a single member rarely leads to failure of the entire structure (except for the Pt. Pleasant

Bridge, which was *not* a multiple-load-path structure), and (4) because of the conservative AASHTO fatigue requirements for bridge details, these toughness requirements appear to be quite satisfactory. Based on satisfactory service experience and the fact that most bridges *are* multiple-load-path structures so that failure of a single member probably will not lead to failure of the entire structure, the consequences of a failure in a single member are probably minor—generally only the cost of repair or replacement of a portion of the structure. Tests conducted on nonredundant, welded bridge details indicated that the AASHTO toughness requirements were adequate even when the details were subjected to the total design fatigue life, the maximum design stress, an operating temperature of −30°F, and the maximum expected loading rate. (Discussed in Chapter 15.)

12.4. Consequences of Failure

Although rarely stated as such, the basis of structural design of large complex structures is an attempt to optimize the desired performance requirements relative to cost considerations (including materials, design, and fabrication) so that the probability of failure (and its economic consequences) is low. Generally, the primary criterion is the requirement that the structure support its own weight plus any applied loads and still have the nominal stresses less than either the tensile yield strength (to prevent excessive deformation) or the critical buckling stress (to prevent premature buckling).

Although brittle fractures can occur in riveted or bolted structures, the evolution of welded construction with its emphasis on monolithic structural members has led to the desirability of including some kind of fracture criterion for most structures, in addition to the strength and buckling criteria already in existence. That is, if a fracture initiates in a welded structure, there usually is a continuous path for crack extension. However, in riveted or bolted structures, which generally consist of many individual plates or shapes, there is no continuous path for crack extension. Any cracks that may extend are arrested as soon as they traverse a single plate or shape. Thus, there is a large difference in the possible fracture behavior of welded structures, compared with either riveted or bolted structures.

Failure of most engineering structures is caused by the initiation and propagation of cracks to critical dimensions. Because crack initiation and propagation for different structures occur under different stress and environmental conditions, no single fracture criterion or set of criteria should be used for all types of structures. Most criteria are developed for particular structures based on extensive service experience and thus are valid only for a particular design, fabrication method, and service use. Furthermore, the criteria usually are not formalized and often may consist only of a requirement that the structural material meet a certain ASTM standard, which may not include any type of fracture-toughness criterion. Furthermore, as modifications are made in the design, fabrication, inspection, and operating conditions of the structure as a result of service experience, any fracture criteria that exist often are not changed.

However, one of the biggest reasons that no single fracture criterion should be applied uniformly to the design of different types of structures is the fact that the *consequences of structural failure* are vastly different for different types of structures. For example, a recently proposed fracture criterion for steels used in seagoing ship hull structures is that the NDT temperature be 0°F, for a minimum service temperature of 30°F. This criterion was based on the assumption, which may be too severe, that ships are subjected to full impact loading. The excellent service experience of this type of structure is such that even this requirement is considered by some to be too conservative. Nonetheless, in view of the consequences of failure of a ship, i.e., either in terms of loss of life or of cargo, this criterion appears reasonable, if the assumption that ships are loaded dynamically is true.

In contrast, the fracture criterion for the steels to be used in the hull structure of stationary but floating nuclear power plants inside a protective breakwater is that the NDT be −30°F, for a minimum service temperature of +30°F. Thus, for *less severe loading* (because the platforms are stationary), the steels in the floating nuclear power plants are required to have an NDT temperature 60°F *below* their service temperature, compared with seagoing ship steels whose NDT temperature is proposed to be 30°F below their service temperatures (and whose existing NDT temperature is about equal to the service temperature). However, for the hull structure of floating nuclear power plants, where service experience is nonexistent and the occurrence of a brittle fracture might result in the development of an entire industry being drastically curtailed, the fracture criterion is extremely conservative because of the consequences of failure, even though the design of the stationary floating hull structure is similar to that of seagoing ship hull structures.

Another example of consequences of failure is the *lack* of a necessity for specifying a fracture criterion for a piece of earth-moving equipment where the consequences of failure of a structural member may be loss of the use of the equipment for a short time until the part can be replaced. If the consequences of failure are minor, then specification of a toughness requirement that might increase the cost of each piece of equipment significantly may be unwarranted.

Thus, the consequences of failure should be a major consideration when determining

1. The need for some kind of fracture criteria, and
2. The level of performance (plane-strain, elastic-plastic, or fully plastic) to be established by the toughness criteria.

Each class or type of structure must be carefully evaluated and the consequences of failure factored into the selected fracture criterion. In fact, in his 1971 AWS Adams Memorial Lecture, Pellini has stated that "one should not use a design criterion in excess of real requirements because this results in specifications of lower NDT and therefore, increased costs." Needless to say, determining "real requirements," i.e., balancing safety and reliability against economic considerations (which if too great may lead to the structure not being built), is a difficult task. Examples of existing fracture criteria (Chapter 13) illustrate this point.

12.5. Safe-Life Versus Fail-Safe

Safe-life design is a design philosophy usually applied to structural components that are subjected to fatigue loading and whose failure may directly or indirectly cause failure of a structure. A service life, either in numbers of cycles of operation or in time, is established. The life of the component then is predicted by the best analytical and experimental techniques available, and the component is removed from service at or before its predicted life so that the possibility of structural failure is eliminated. For this design philosophy to be reliable, the engineer must know the loading conditions throughout the life of the structure and the material response to these loads, and the material must behave in service as it does in any testing program used to establish the material response. If all these conditions are not known exactly, then a fairly large factor of safety must be used to account for the unknown variables. Often, the safe-life design philosophy results in a very safe structure because of the large factors of safety used, although the structure may be very uneconomical. Fixed structures, such as bridges and buildings, generally fall into this category.

Another design philosophy is the fail-safe approach, which is a design philosophy applied to components loaded in fatigue (although the philosophy is similar for statically loaded structures). In this approach, appropriate safeguards are incorporated into the design of a structure so that even if a local failure occurs, the structure is safe—i.e., catastrophic failure of the structure does not occur even if a member fails. Thus the structure can be repaired in an orderly manner.

The fail-safe design philosophy need not be limited to structural members loaded in fatigue. Members in which brittle fractures are possible or in which stress-corrosion cracks may occur can also be designed using a fail-safe philosophy. Several examples of fail-safe design applications are given below.

1. Multiple Load-Path Members

If several parallel load-carrying members are used (multiple load paths), such as multiple-stringer bridges, then the failure of one member does not necessarily lead to overall failure of the structure because the load is carried by adjacent members. This was the case for the Kings Bridge in Australia where one of four members failed during one winter and the remaining three girders carried the entire load for almost a year before the failure of a second girder, which then led to collapse of the entire bridge.

The amount of load redistribution cannot be too severe; otherwise failure can still occur. This was the case in the Pt. Pleasant Bridge failure in which two I-bar members carried the load and the brittle fracture of one led immediately to an extremely severe eccentricity and overload, resulting in immediate failure of the other member and subsequent collapse of the entire structure.

2. Use of Crack Arresters

By placing crack arresters or crack stoppers at various locations within a structure, excessive crack growth (either by brittle fracture, fatigue, or stress-corrosion) is arrested before the structure fails. This concept is widely used in the aircraft industry where bonded doubler plates or crack arresters are attached to the skin of the aircraft. Thus, if a crack occurs in the aircraft skin, it cannot grow beyond the first doubler plate it intersects and is thus arrested, as shown in Figure 12.10. Loss of pressure may occur in the aircraft, but the structure is still safe.

This philosophy is also used in ship hull design where either riveted out-of-plane crack arresters (Figure 12.11) or welded in-plane crack arresters (Figure 12.12) are used. Although the riveted crack arresters have been shown to be satisfactory from a technical viewpoint, they do not appear to be so

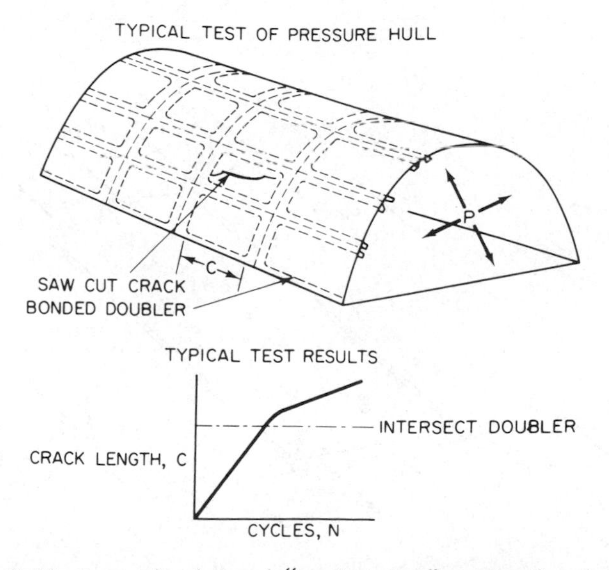

FIG. 12.10. Conventional aircraft "crack-stopper" construction showing tear-stopper straps in the form of internal bonded-skin doublers.

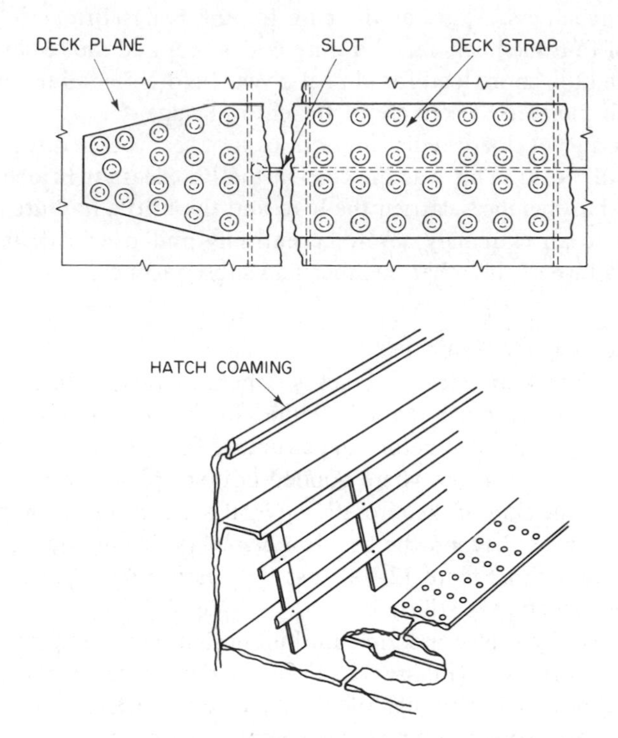

FIG. 12.11. Typical geometry of riveted crack arrester.

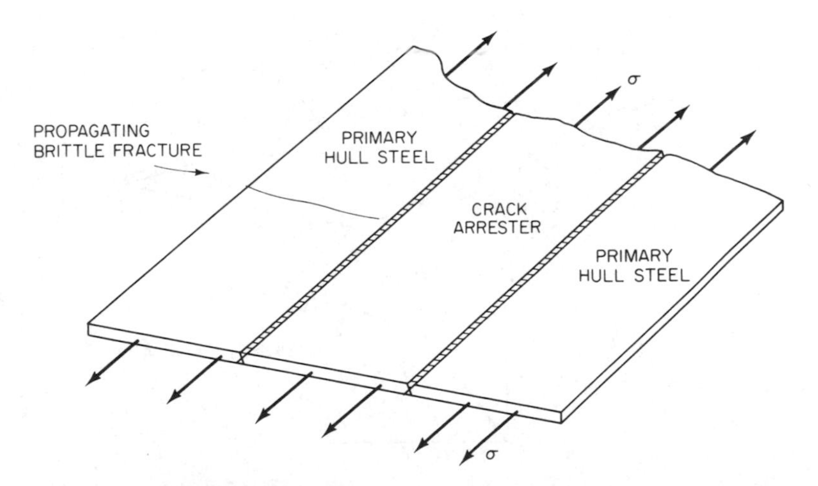

FIG. 12.12. Typical geometry of in-plane crack arrester.

widely used now as they were in the 1940s to 1950s because the use of riveted construction in combination with welding may result in a longer construction period. Furthermore the overall decline of riveted construction in the 1960s and 1970s has lowered the availability of riveters.

Welded in-plane crack arresters are used in welded steel ship hulls as integral load-carrying components in conjunction with the primary hull structure (Figure 12.13). However, the arresters are usually made of materials with a much higher level of notch toughness than the basic material used in the hull structure. For very large cracks, this type of crack arrester may not be too effective unless the arrester plates are extremely wide, because of the large movements of the ends of a ship and the large stored elastic energy that continues to drive the crack through the crack arresters.

Out-of-plane crack arresters may also be very practical because their configuration resembles the girders and stiffening members commonly used in structures. An example of this type geometry is presented in Figure 12.14, showing a wide-flange beam welded as an integral part of the primary structure. Out-of-plane arresters such as this would appear to have several advantages compared with in-plane crack arresters.

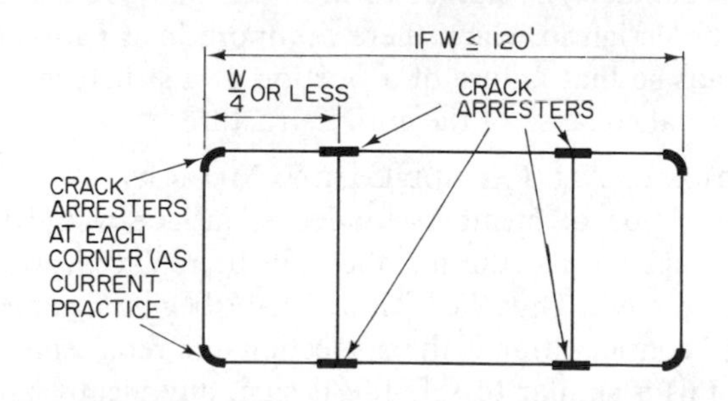

TYPICAL CROSS SECTION OF SHIP HULL STRUCTURES

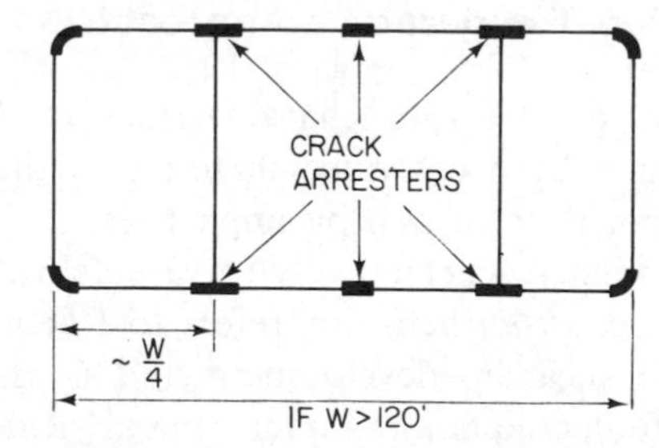

FIG. 12.13. General locations of crack arresters in hull section.

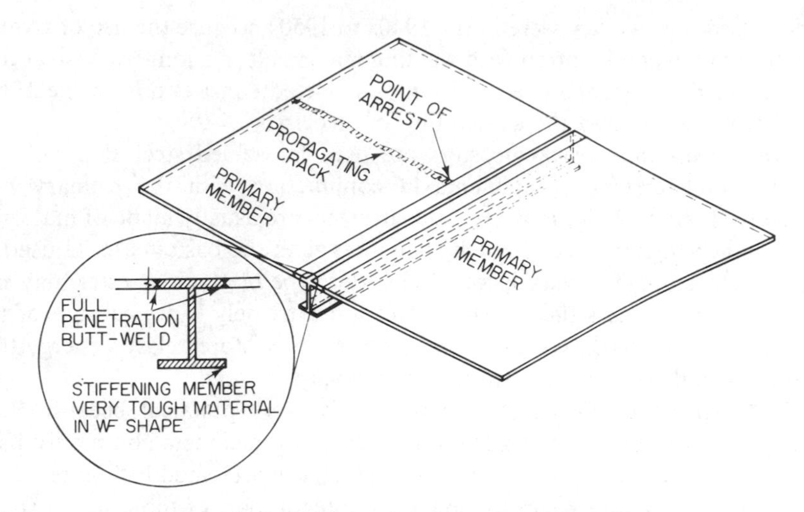

FIG. 12.14. Schematic showing out-of-plane crack arrester.

3. REDUNDANCY

Structural redundancy is similar to multiple-load paths and is a primary mode of fail-safe design in which there are more load paths than necessary to carry the loads so that failure of a portion of a structure does not necessarily result in total collapse of the entire structure.

4. INSPECTION OF ALL FATIGUE-LOADED MEMBERS

Periodic inspection of members loaded in fatigue would result in early detection of a crack so that the member can be replaced prior to complete collapse of the structure. Thus the "failure" is the beginning of a fatigue crack, and the "safety" consideration is the inspection and replacement of the member. Although this is similar to safe-life design, any member that has a safe life presumably would not need to be inspected because, by definition and design, fatigue-crack growth leading to failure should not occur during the life of the member.

12.6. Transition-Temperature Approach

Traditionally, the fracture characteristics of low- and intermediate-strength steels have been described in terms of the transition from brittle to ductile behavior as measured by impact tests. This transition in fracture behavior can be related schematically to various fracture states, as shown in Figure 12.15. *Plane-strain* behavior refers to fracture under elastic stresses with little or no shear-lip development and is essentially brittle. *Plastic* behavior refers to ductile failure under general yielding conditions with very large shear-lip development. The transition between these two extremes is the

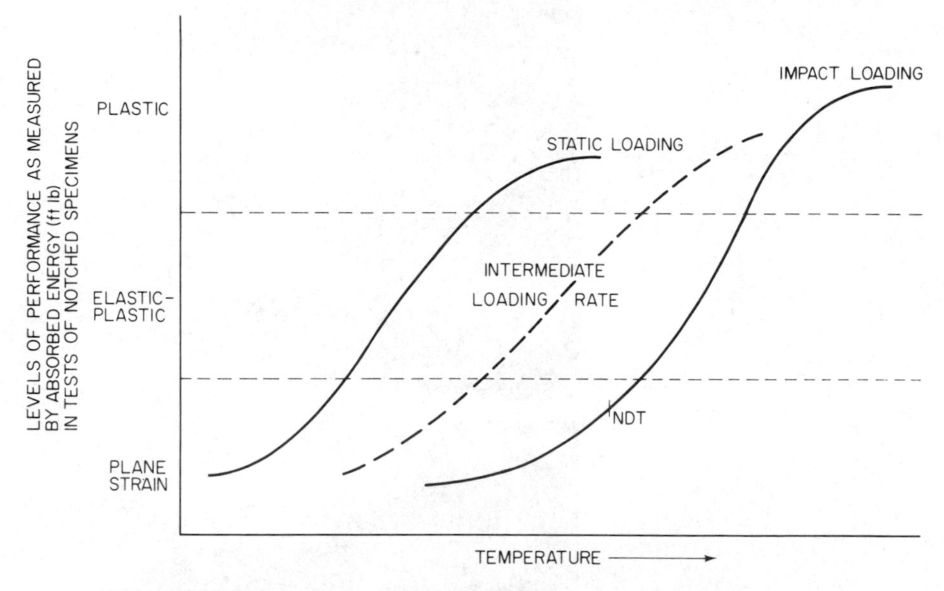

FIG. 12.15. Schematic showing relation between notch-toughness test results and levels of structural performance for various loading rates.

elastic-plastic region, which is also referred to as the mixed-mode region. Typical fracture surfaces showing each of these three regions were shown in Figures 12.2, 12.3, and 12.4. Overall views of test specimens showing the change from no deformation (elastic behavior) to large deformation (general yielding) are shown in Figure 12.16.

For static loading, the transition region occurs at lower temperatures than for impact (dynamic) loading, depending on the yield strength of the steel. Thus, for structures subjected to static loading, the static transition curve should be used to predict the level of performance at the service temperature. For structures subjected to impact or dynamic rates of loading, the impact transition curve should be used to predict the level of performance at the service temperature. For structures subjected to some intermediate loading rate, an intermediate loading-rate transition curve should be used to predict the level of performance at the service temperature. If the loading rate for a particular type of structure is not well defined, and the consequences of failure are such that a fracture will be extremely harmful, a conservative approach is to use the impact loading curve to predict the service performance. As noted in Figure 12.15, the nil-ductility transition (NDT) temperature is close to the upper limit of plane-strain conditions under conditions of *impact* loading.

After establishing the loading rate for a particular structure, and the corresponding loading rate for the test specimen to be used, the next question

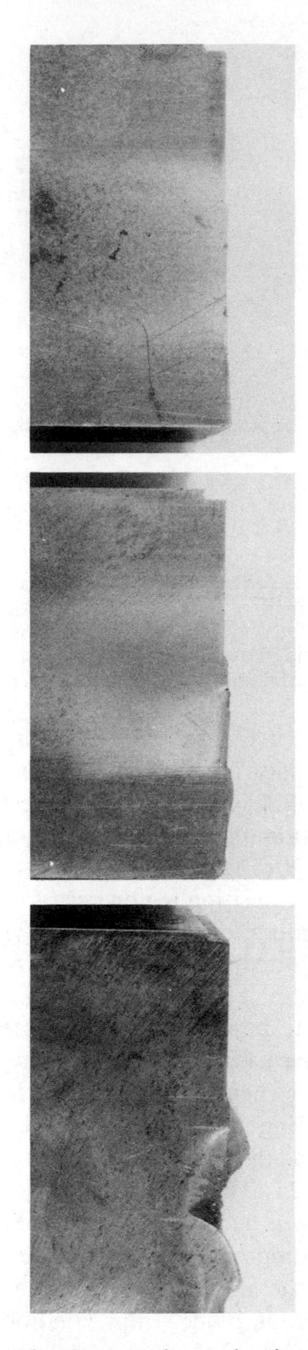

FIG. 12.16. Photographs of test specimens showing side view of no deformation to considerable deformation.

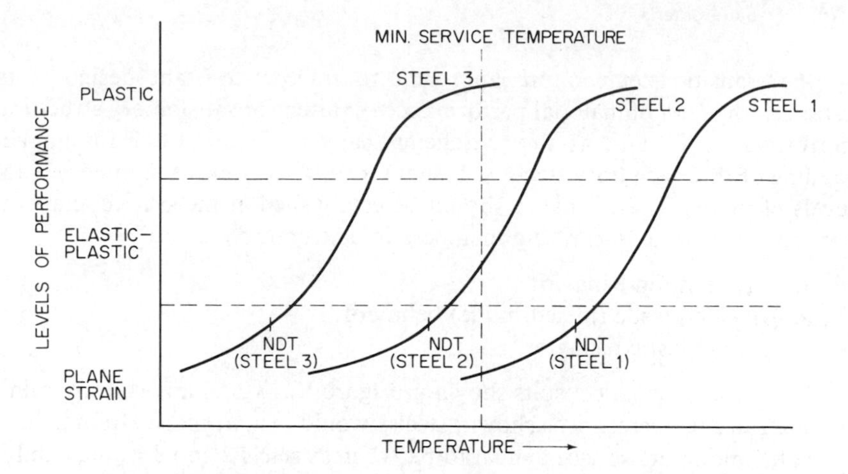

FIG. 12.17. Schematic showing relation between level of performance as measured by impact tests and NDT for three arbitrary steels.

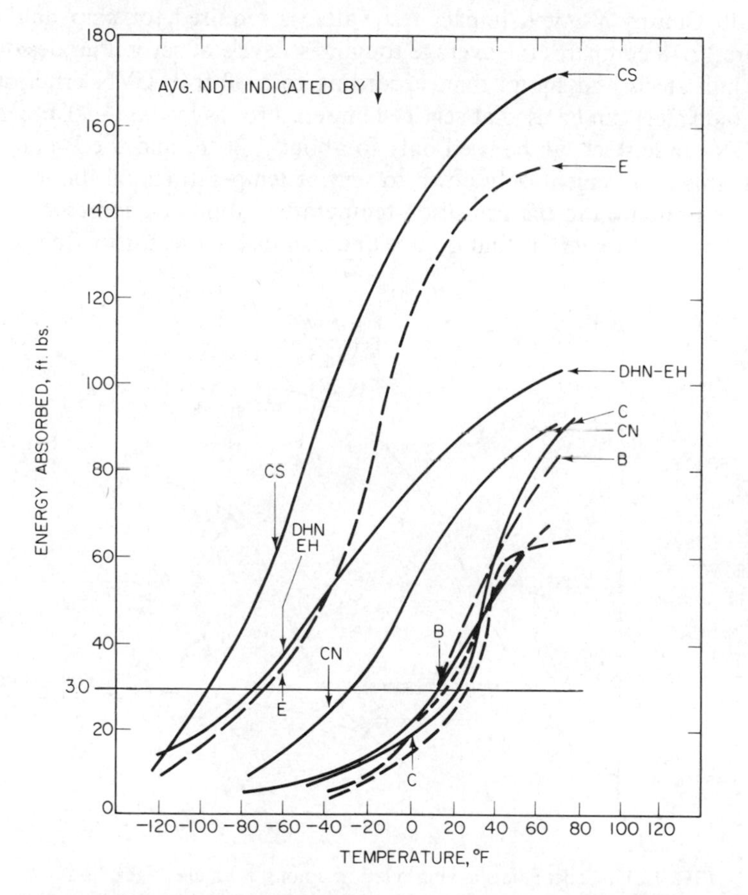

FIG. 12.18. Comparison of minimum service temperature of ship steels using an arbitrary criterion of 30 ft-lb.

in the transition-temperature approach to fracture-resistant design is to establish the *level* of material performance required for satisfactory structural performance. That is, as shown schematically in Figure 12.17 for impact loading of three arbitrary steels 1, 2, and 3, one of the following three general levels of material performance should be established at the service temperature for primary load-carrying members in a structure:

a. Plane-strain behavior.
b. Elastic-plastic (mixed-mode) behavior.
c. Fully plastic behavior.

Using the schematic results shown in Figure 12.17, and an arbitrary minimum service temperature as shown, steel 1 would exhibit plane-strain behavior at the minimum service temperature, whereas steels 2 and 3 would exhibit elastic-plastic and fully plastic behavior, respectively.

As an example of the transition-temperature approach assume that a 30 ft-lb Charpy V-notch impact test value is required for ship hull steels. Figure 12.18 compares the average toughness levels of several grades of ABS ship hull steels and shows that, according to a 30-ft-lb CVN criterion, the CS-grade steel can be used at service temperatures as low as −90°F, whereas the CN-grade steel can be used only to about −30°F, and the B-grade steel meets this requirement only down to service temperatures of about +10°F.

One limitation to the transition-temperature approach sometimes occurs with the use of materials that do not undergo distinct transition-temperature

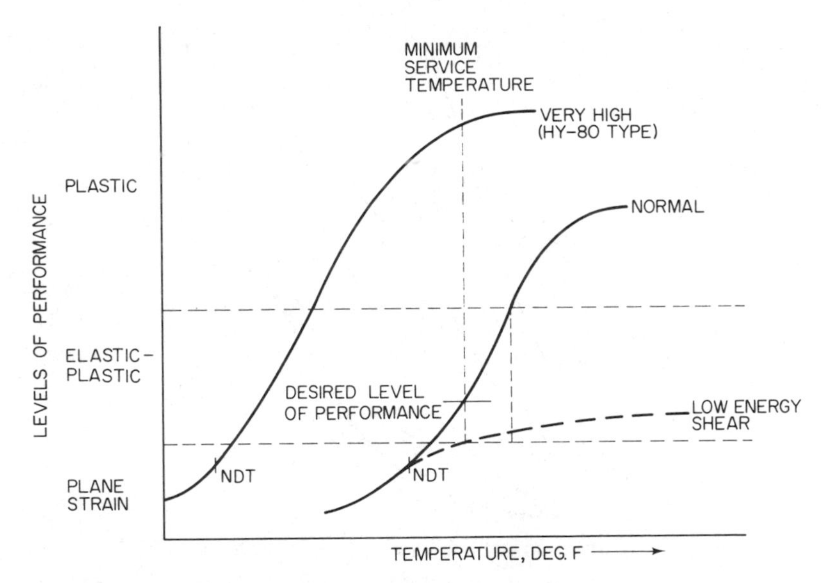

FIG. 12.19. Schematic showing relation among normal-, high-, and low-energy shear levels of performance as measured by impact tests.

behavior or with materials that exhibit a low-energy shear behavior. Figure 12.19 shows the relationship of low-energy performance to normal behavior and to very high-level toughness behavior (such as obtained in an HY-80-type steel used for military applications).

Low-energy shear behavior usually does not occur in low- to intermediate-strength structural steels ($\sigma_{ys} \leq 100$ ksi) but sometimes is found in high-strength steels ($\sigma_{ys} > 100$ ksi). For example, if the desired level of performance is as shown in Figure 12.19, a material exhibiting low-energy shear behavior may never achieve this level of performance at *any* temperature. For these high-strength materials, the through-thickness yielding, the leak-before-burst, and the ratio analysis diagram (RAD) methods as described in Chapter 13 are very useful to establish fracture criteria.

13

Examples of Various Notch-Toughness Criteria

13.1. Introduction

A notch-toughness criterion or a fracture criterion is defined as an established standard against which the expected fracture behavior of a structural material can be judged. This established standard is defined in terms of the levels of performance (plane-strain, elastic-plastic, and plastic) described in Chapter 12. Each of these three levels of performance can be related to the general expected structural behavior in actual structures.

However, a general fracture criterion must be translated into some *specific fracture test requirement* that ensures the engineer he is obtaining the desired level of performance. For example, a general requirement that a structural material exhibit elastic-plastic behavior at service temperatures is a general criterion that is useful to the engineer. However, because of ambiguity and differences in opinion, this general criterion must be made specific in terms of a fracture test specimen and specified value. An *example* of a specific notch-toughness criterion might be the requirement that a low level of elastic-plastic behavior is desired and that the *specification* should be that "all structural steels must exhibit 19 ft-lb of energy absorption as measured in a longitudinal Charpy V-notch impact test specimen tested at 32°F." Hopefully, this particular criterion would have been based on sufficient laboratory results, service experience, and fracture-mechanics analysis to ensure the desired structural behavior. The criterion would then be specified for purchase of materials and quality control.

As a result of several dramatic structural failures in the late 1960s and early 1970s, as well as a growing concern with the overall reliability and safety of structures, many specifications are now beginning to include specific minimum toughness requirements. This trend is expected to grow as regulatory governmental agencies become increasingly active in the development of mandatory rational fracture prevention criteria. Recent examples are the ASME Nuclear Code, AASHTO Material Toughness

Requirements, and the floating nuclear power plant hull structure-toughness requirements imposed by the U.S. Coast Guard and the Nuclear Regulatory Commission. These examples are described in Chapter 15.

There are two general parts to a fracture criterion:

1. THE GENERAL TEST SPECIMENS TO CATEGORIZE THE MATERIAL BEHAVIOR

Throughout the years, various fracture criteria have been specified using notch-toughness tests such as CVN impact, NDT, dynamic tear, and, more recently, the fracture-mechanics test specimens described in Chapter 3 which are used to measure K_{Ic} and K_{Id}. Test specimens currently used as research tools and expected to be used more extensively in the future are J_{Ic}, COD, and R-curve specimens, as described in Chapter 16. The test specimen used for a particular application should be that one which most closely models the actual structural behavior. However, selection of the general test specimen to use is often based on past experience as well as economics and convenience of testing rather than on the basis of the test specimen that most closely models the actual structure.

2. THE SPECIFIC NOTCH-TOUGHNESS VALUE OR VALUES

The second and more difficult part of establishing a fracture criterion is the selection of the specific level of performance in a particular test specimen in terms of measurable values for material selection and quality control.

The specified values in any criterion should be an optimization of both safe structural performance and cost and are always subject to considerable differences in opinion between knowledgeable engineers. Accordingly, one of the main objectives of this textbook is to provide some rational guidelines for the engineer to follow in establishing toughness criteria for various structural applications.

In the remainder of Chapter 13 we shall present examples of the use of different test specimens to establish guidelines as well as criteria for material selection. In some cases, the guidelines are translated into specific test specimen requirements for a particular type of structure. It should be emphasized that there is no single *best* test specimen and corresponding fracture criterion for all types of structures. Thus the engineer must select that criterion which most closely meets the needs of his particular structure. While it is true that the use of K_{Ic} or K_{Id} test specimens (Chapter 3) and the corresponding stress–flaw-size relations as described in Chapter 5 *does* provide the engineer with the most quantitative relationships available, many structural materials have sufficiently high levels of notch toughness at service temperatures and loading rates so that these values cannot be used *directly* in establishing criteria—hence, the widespread use of the various correlations described in Chapter 6 between K_{Ic} and K_{Id} test results and other notch-toughness tests (such as CVN, DT, NDT, etc.) to establish some of the existing fracture criteria described in the following sections.

13.2. Original 15-ft-lb CVN Impact Criterion for Ship Steels

Although occasional brittle fractures were reported in various types of structures (both welded and riveted) prior to the 1940s,[1,2] it was not until the rapid expansion in all-welded ship construction during the early 1940s that brittle fracture became a well-recognized structural problem. During the early 1940s, over 2,500 Liberty ships, 500 T-2 tankers, and 400 Victory ships were constructed as a result of World War II. Because the basic designs of each of these three types of ships (Liberty, Victory, and tankers) were similar, it was possible to analyze any structural difficulties on a statistical basis.

The first of the Liberty-type ships were placed in service near the end of 1941, and by January 1943 there were ten major fractures in the hull structures of the ships that were in service at that time. Numerous additional failures throughout the next few years led to the establishment of an investigative board to conduct an investigation into the design and methods of construction of welded steel merchant ships. In 1946, this board made its final report[3] to the Secretary of the Navy.

The role of materials, welding, design, fabrication, and inspection is described in various extensive reviews of this problem,[3-10] and the interested reader is referred to these documents. Of interest in this particular section is the work leading to the development of the 15-ft-lb notch-toughness criterion, which is still widely used for many other types of structures.

In the development of the 15-ft-lb criterion, samples of steel were collected from approximately 100 fractured ships and submitted to the National Bureau of Standards for examination and tests. This particular study resulted in the collection of an extremely complete body of data relative to the failure of large welded ship hull structures. The plates from fractured ships were divided into three groups:

1. Those plates in which fractures originated, called *source plates*.
2. Those plates through which the crack passed, called *through plates*.
3. Those plates in which a fracture stopped, called *end plates*.

An analysis of the Charpy V-notch impact test results showing the frequency distribution of the 15-ft-lb Charpy V-notch transition temperature of the source, through, and end plates is presented in Figure 13.1. These results show that the *source* plates had a high average 15-ft-lb transition temperature, about 95°F. The through plates had a more normal distribution of transtion temperatures, and the average 15-ft-lb transition temperature was lower, i.e., about 65°F. The *end* plates had the lowest average 15-ft-lb

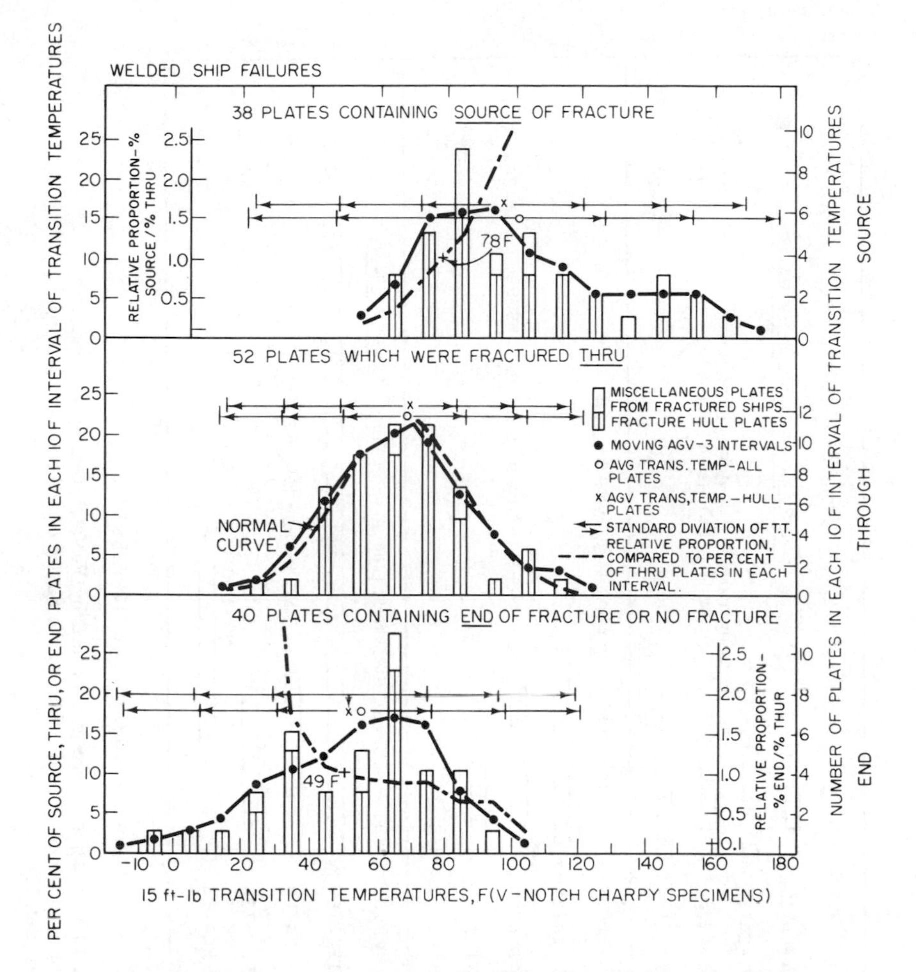

FIG. 13.1. Frequency distribution of 15-ft-lb V-Notch Charpy transition temperatures for fractured ship plates (Ref. 10).

transition temperature (about 50°F) and a distribution with a long tail at lower transition temperatures.

Parker[10] points out that "the character of these distribution curves was not wholly unanticipated; one would suspect that the through plates would be most representative of all ship plates and hence might tend toward a normal distribution, whereas the *source* and *end* plates were *selected* for their role in the fracturing of the ship. A factor in this selection is the notch toughness of the plate as measured in the Charpy test. The overlapping of the *source* and *end* plate distributions with the through plate distribution can be explained by factors involved in the fracturing in addition to the notch toughness of the plate. For example, a plate in the through fracture category

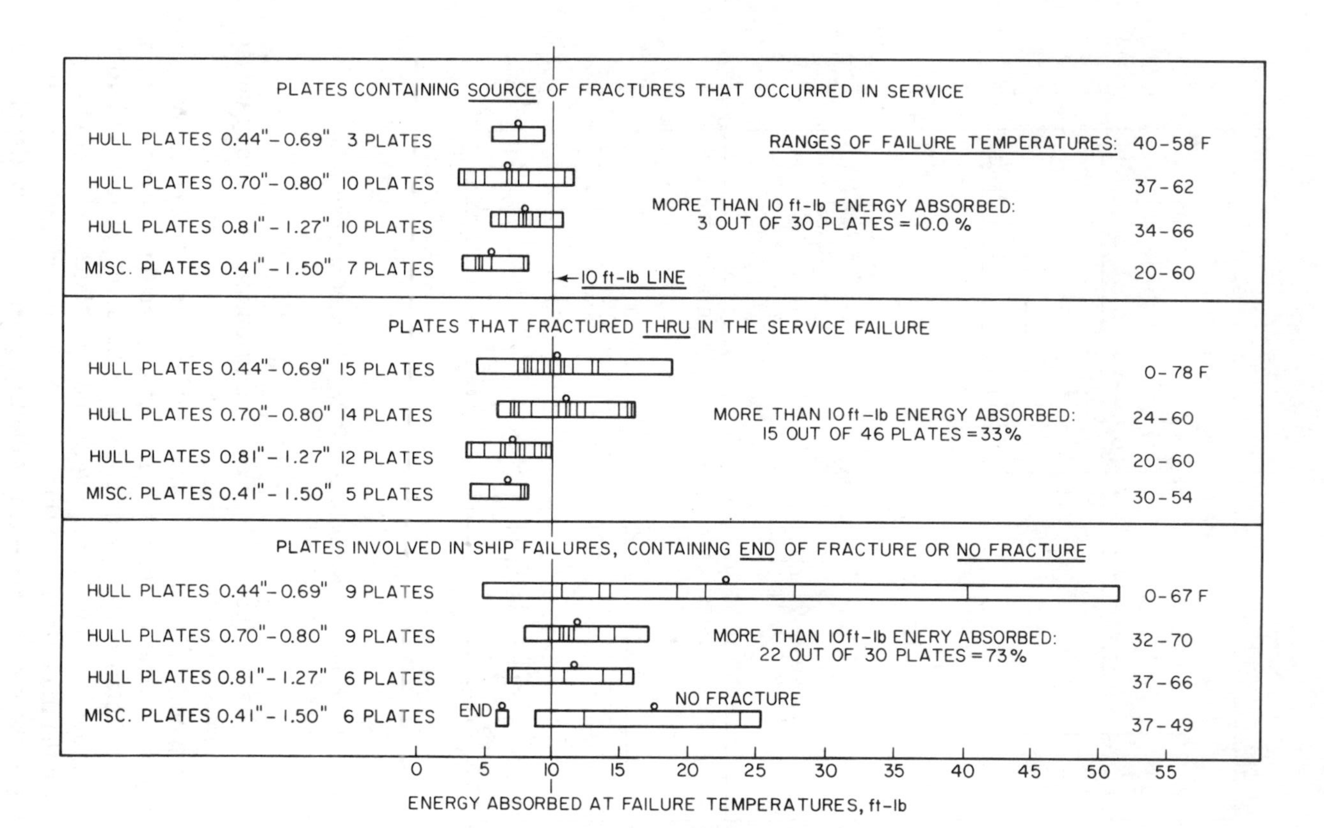

FIG. 13.2. Relation of energy absorbed by V-Notch Charpy specimens at the temperature of ship failure to the nature of the fractures in ship plates (Ref. 10).

having a transition temperature between about 60°F and 90°F [Figure 13.1] might have been a fracture *source* plate under more severe stress conditions such as those in the region of a notch; under less severe conditions of average stress it might have been an *end* plate. Even though factors other than notch toughness contributed to the selection of a plate for its role in fracturing, it should be pointed out that statistical analyses have indicated that the differences in transition temperature and energy at failure temperature between the *source* and *through* and *through* and *end* plates are not due to chance."

The extremely low probability of the differences in Charpy properties of plates in the three fracture categories being due to chance permitted the development of several criteria that are very important to engineers. Figure 13.2 shows that only 10% of the *source* plates absorbed more than 10 ft-lb in the V-notch Charpy test at the failure temperature; the highest value encountered in this category was 11.4 ft-lb. At the other extreme, 73% of the *end* plates absorbed more than 10 ft-lb at the failure temperature. It was therefore concluded, based on these data and for steels of this quality, that in the large ship hull structures brittle fractures are not likely to initiate in a plate that absorbs more than 10 ft-lb in a Charpy V-notch test conducted at the anticipated operating temperature.

Thus, a slightly higher level of performance, namely the *15-ft-lb* transition temperature as measured with a CVN impact test specimen, was selected as a fracture criterion on the basis of actual service behavior of a large number of similar-type ship hull structures. Since the establishment of this relation between CVN values and service behavior in ship hulls, the 15-ft-lb transition temperature has been a widely used fracture criterion, even though it was developed only for a particular type of steel and a particular class of structures, namely ship hulls.

Fortunately for the engineering profession and the general public safety, similar statistical correlations between test results and service failures do not exist for any other class of structures because there have not been such a large number of failures in any other type of structure. However, the difficulty of obtaining service experience creates a problem for the design engineer in establishing toughness criteria for new types of structures.

13.3. Loading-Rate Shift Criterion

In general, the notch toughness of most structural steels increases with increasing temperature and decreasing loading rate. The effect of temperature is well known and has led to the transition-temperature approach to designing to prevent fracture. However, the effect of loading rate may be equally as important, not only in designing to prevent fracture but in understanding the satisfactory behavior of many existing structures built from structural materials that have low levels of *impact* toughness at their service temperatures.

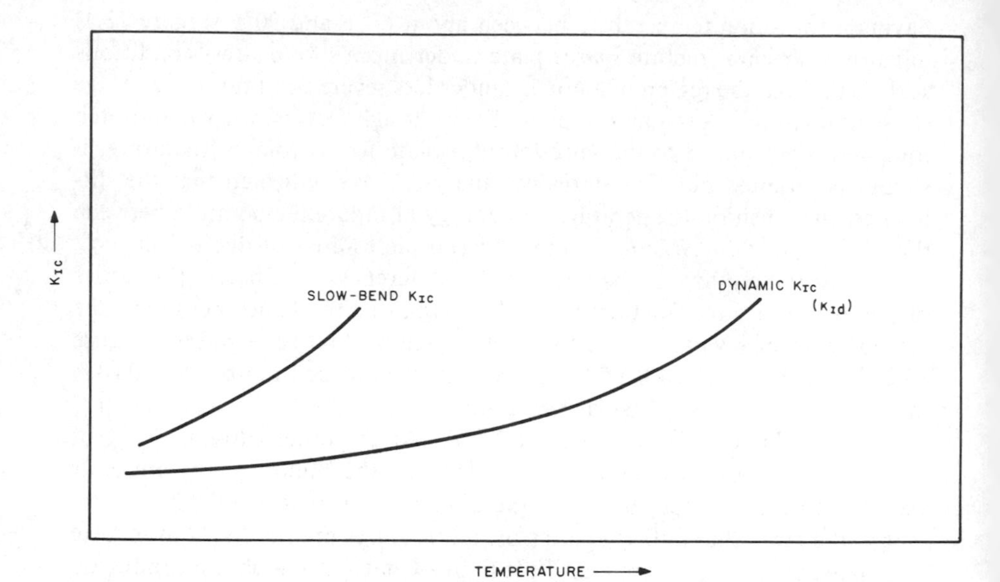

FIG. 13.3. Schematic showing effect of temperature and loading rate on K_{Ic}.

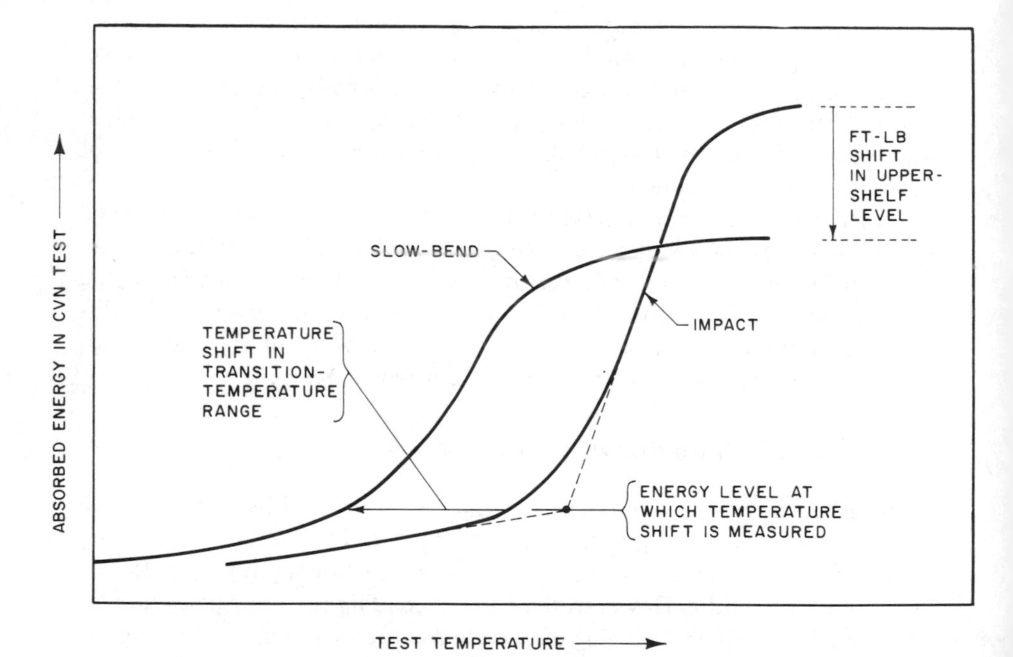

FIG. 13.4. Schematic representation of shift in CVN transition temperature and upper shelf level due to strain rate.

The general effects of temperature and loading rate on K_{Ic} and Charpy V-notch behavior are shown schematically in Figures 13.3 and 13.4, respectively. The toughness of most structural steels tested at a constant loading rate undergoes a significant increase with increasing temperature. Data obtained for various steels demonstrate that a true K_{Ic} temperature transition exists that is independent of specimen geometry. That is, the rate of increase of K_{Ic} with temperature does not remain constant but increases markedly above a given test temperature, as described in Chapter 4. The general effect of a slow loading rate, compared with impact loading rates, is to shift the fracture-toughness curve to lower temperatures, regardless of the test specimen used.

Examples of this shift in behavior with loading rate are presented in Figures 13.5 and 13.6 for an A36 grade structural steel and an A572 Grade

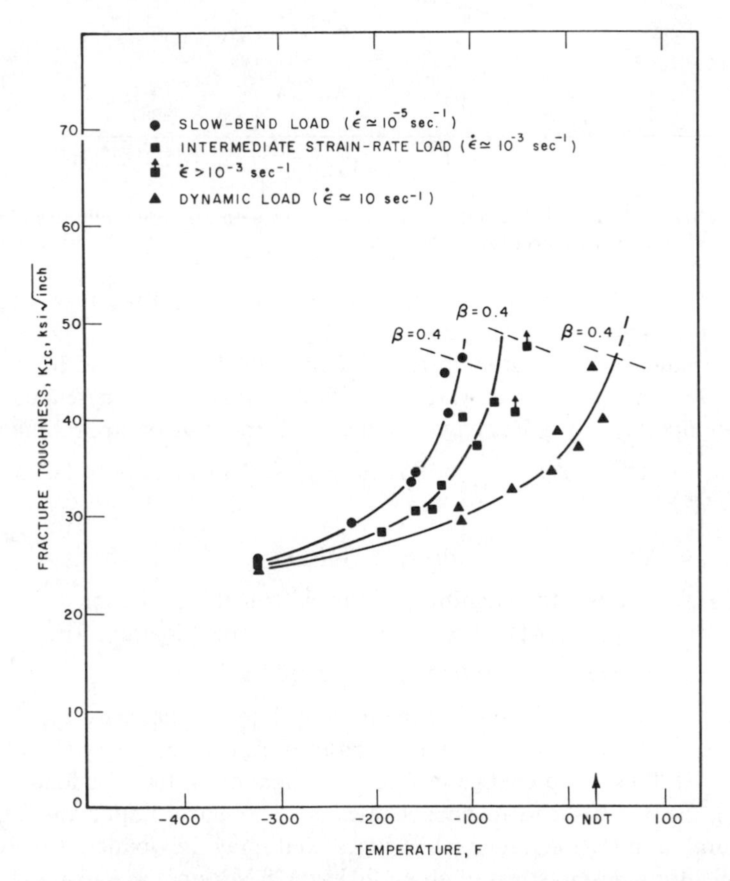

FIG. 13.5. Effect of temperature and strain rate on fracture toughness of A36 steel.

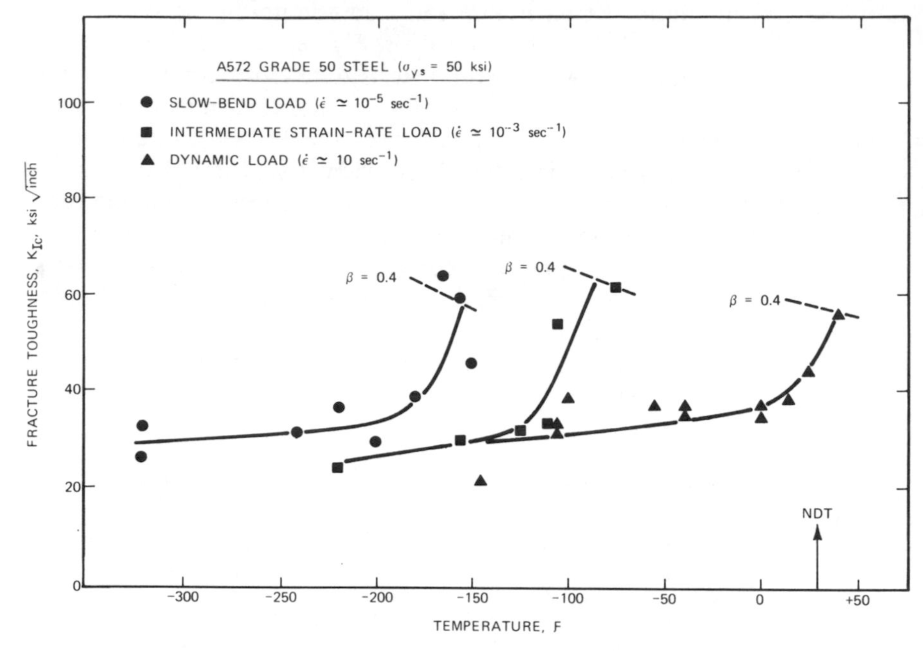

FIG. 13.6. Effect of temperature and strain rate on fracture toughness of A572 Grade 50 steel (σ_{ys} = 50 ksi).

50 structural steel, respectively. Additional examples for other structural steels were presented in Chapter 4.

The magnitude of the temperature shift between slow-bend loading and impact loading in steels of various yield strengths has been related to the room-temperature yield strength of the steel and can be approximated by

$$T_{\text{shift}} = 215 - 1.5\sigma_{ys} \qquad \text{for } 36 \text{ ksi} < \sigma_{ys} < 140 \text{ ksi } (248 \text{ to } 965 \text{ MN/m}^2)$$

and

$$T_{\text{shift}} = 0.0 \qquad \text{for } \sigma_{ys} > 140 \text{ ksi}$$

where T_{shift} = absolute magnitude of the shift in the transition temperature between slow-bend loading and impact loading, °F,

σ_{ys} = *room-temperature* yield strength, ksi.

Because of this shift, increasing the loading rate can decrease the fracture-toughness value at a particular temperature for steels having yield strengths less than 140 ksi. The change in fracture-toughness values for loading rates varying from slow-bend to impact rates is particularly important for those structural applications such as bridges that may be loaded slowly. For example, for a design stress of about 20 ksi (138 MN/m²), a K_{Ic} of 40 ksi$\sqrt{\text{in.}}$ (44 MN/m$^{3/2}$) for a given steel would correspond to the tolerance of a through-thickness crack of approximately 3 in. (76 mm). If a structure were loaded statically, this size crack could be tolerated at extremely low temperatures. If the structure were loaded dynamically, however, the temperature at which this size crack could cause failure would be much higher.

If the loading rates of a particular type of structure, such as bridges, are closer to those of slow-bend loading than to impact loading, a considerable difference in the actual behavior of these structures would be expected, compared with the predicted results using impact test results. Thus, not only should the effects of temperature and geometry (plate thickness) on structural steels be established (for several plate thicknesses), but more importantly the maximum nominal loading rates that will occur in a structure under operating conditions should be established.

As a specific example of the use of the loading-rate shift as a fracture criterion, assume that a structure is loaded at a slow-loading rate of 10^{-5} in./in./sec and that the fracture toughness of the material is as shown in Figure 13.6. If stress–crack size calculations show that a K_{Ic} value of about

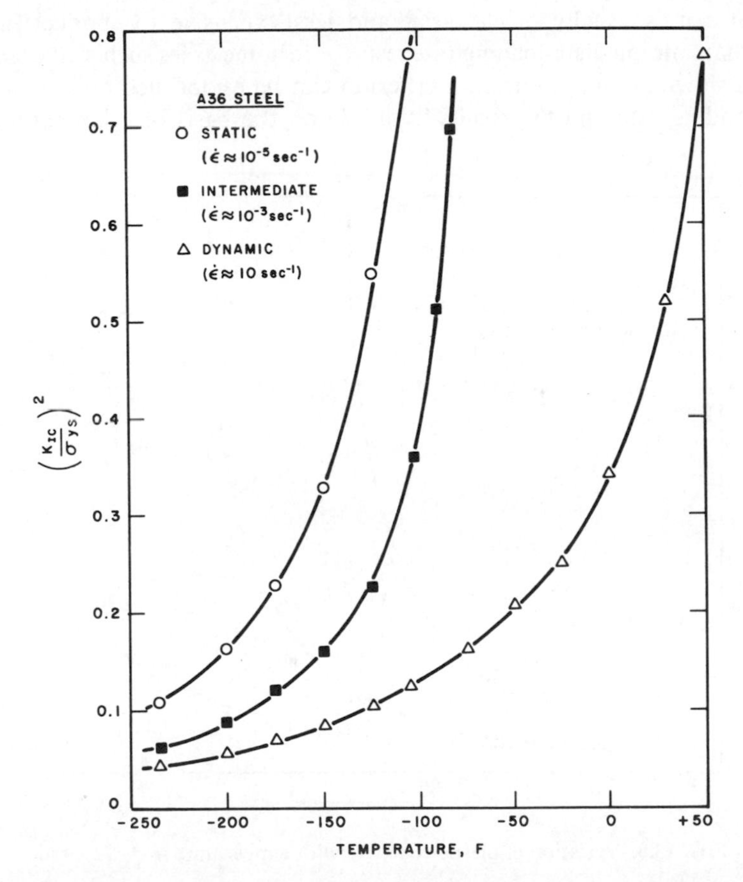

FIG. 13.7. Variation of plastic-zone size with temperature in A36 steel.

60 ksi$\sqrt{\text{in.}}$ would ensure satisfactory structural performance, the results presented in Figure 13.6 show that this behavior can be obtained at about +40°F *dynamically* ($\dot{\epsilon} \simeq 10$ in./in./sec), whereas this behavior can be obtained at about −90°F at an *intermediate loading* rate and at about −150°F for a *slow-loading* rate.

Figures 13.7 and 13.8 show the variation of the parameter $(K_{Ic}/\sigma_{ys})^2$, which is proportional to the plastic-zone size, as a function of temperature in the A36 and A572 Grade 50 steels investigated. Each data point in these figures represents the ratio of the critical plane-strain stress-intensity factor, K_{Ic} (or K_{Id}), at a given temperature and strain rate to the yield strength, σ_{ys} or σ_{yd} at the same temperature and strain rate. The magnitude of the plastic zone as a function of test temperature appears to increase asymptotically. This behavior indicates that valid K_{Ic} or K_{Id} values obtained at a given strain rate cannot be measured above a given temperature regardless of specimen size. This temperature is established, primarily, by the inherent metallurgical properties of the steel at the strain rate of interest.

Since it is usually much easier and less expensive to conduct impact tests than intermediate-loading rate tests (or in some cases such as the Charpy test, a slow-loading rate test), a criterion can be established on the basis of one loading rate and the results "shifted" on the basis of a laboratory test

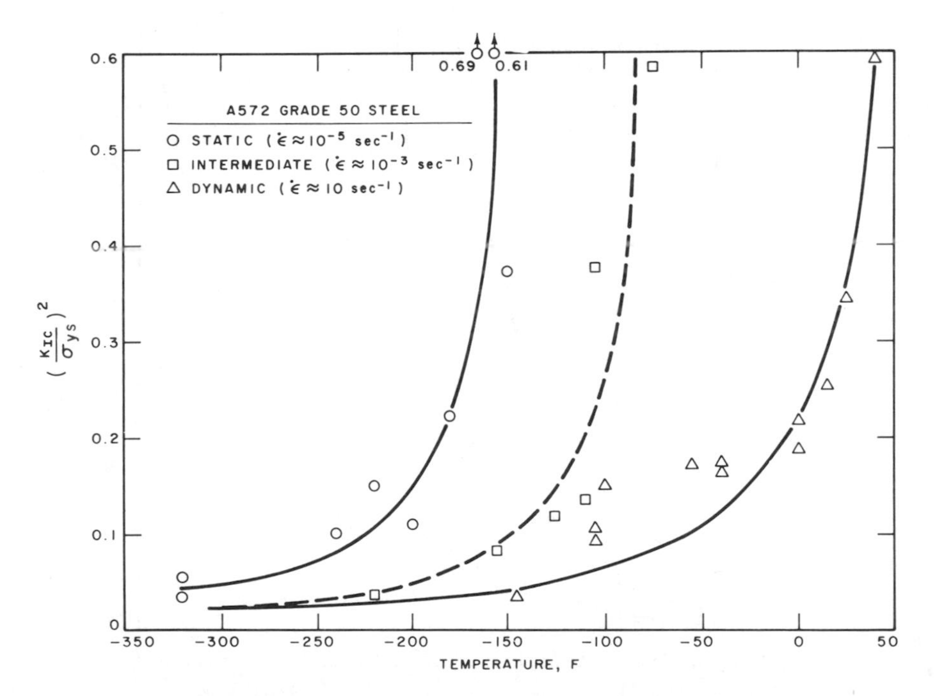

FIG. 13.8. Variation of plastic-zone size with temperature in A572 Grade 50 steel.

conducted at a different loading rate. The recently developed American Association of State Highway and Transportation officials (AASHTO) material-toughness requirements were based on this line of reasoning. The specific development of this criterion for bridge steels is presented in Chapter 15. It should be emphasized that this criterion can be used only with those materials that exhibit a shift in transition behavior with changes in loading rate. However, the magnitude of this shift can be considerable and helps to explain why many structures have operated successfully at service temperatures well below their impact transition temperature.

13.4. Fracture Analysis Diagram (FAD)

The fracture analysis diagram (FAD) was developed by Pellini of the Naval Research Laboratory to provide design guidelines or criteria for structural steels that exhibit a transition temperature. The general diagram is presented in Figure 13.9. In his 1971 AWS Adams Lecture on Principles of Fracture-Safe Design,[11,12] Pellini described the use of the fracture analysis diagram and defined four critical transition-temperature-range reference points which also serve as "design" points:

1. *NDT (nil ductility transition) reference point.* Restricting the service temperature to slightly above the NDT provides fracture initiation protection for the most common types of service failures. These involve fractures which are initiated due to small cracks subjected to yield stress loading levels under conditions of dynamic loading. Note that the NDT reference point (as well as all the other reference points) assumes that the structure is subjected to dynamic loading conditions, either from external loading or from local "loading" by fracturing of embrittled regions such as microcracks or welding arc strikes.
2. *NDT to FTE (fracture transistion elastic) midrange reference point.* Restricting the service temperature to above NDT + 30°F (NDT + 17°C), i.e., the midrange of the NDT to FTE region, provides fracture arrest protection if the nominal stress level does not exceed $0.5\sigma_{ys}$.
3. *FTE reference point.* Restricting the service temperature to above the FTE provides fracture arrest protection if the nominal stresses do not exceed yield level.
4. *FTP (fracture transition plastic) reference point.* Restricting the service temperature to above the FTP ensures that only fully ductile fracture is possible even under conditions of dynamic loading.

The degree of protection against fracture initiation due to crack size and stress combinations is increased dramatically in the NDT to FTE region.

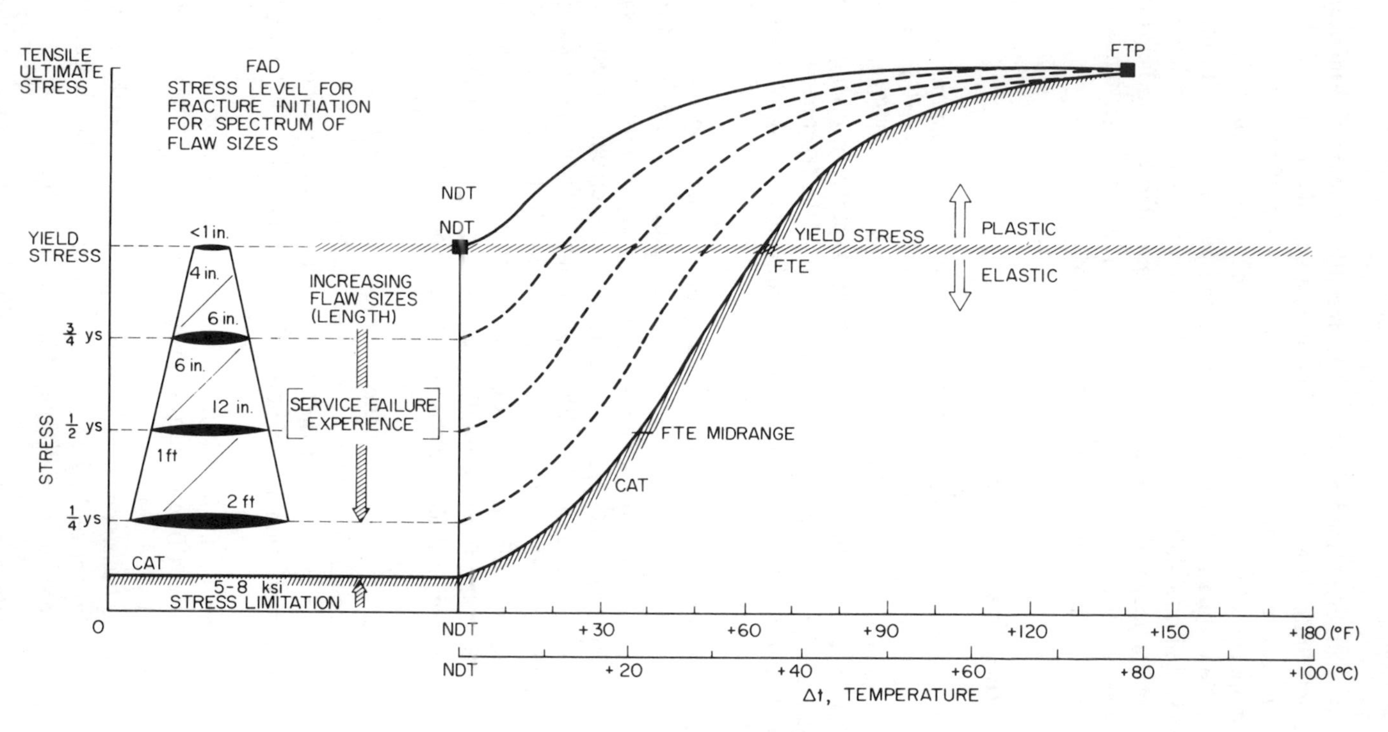

FIG. 13.9. Fracture analysis diagram (FAD) (Ref. 11).

The assignment of subdesign points to this narrow temperature region would require exacting definitions of temperature, crack size, and stress. This observation is made to emphasize that finer distinctions than the described 30°F (17°C) four-design-points sectioning of the FAD are not required for most engineering purposes. Thus, the large increases in fracture resistance, obtained by successive 30°F (17°C) temperature increments above the NDT, reduce the problem of fracture-safe design to a straightforward temperature reference system, if dynamic loading is assumed.

The choice of steel is dictated by the following factors:

1. The lowest service temperature.
2. The design reference point criterion chosen, i.e., NDT, NDT to FTE midrange, FTE, or FTP.

Pellini's statement that "one should not use a design criterion in excess of real requirements because this results in specifications of lower NDT and therefore, increased costs" is, of course, the real problem in establishing realistic toughness criteria, as discussed in Chapter 12.

Recently Pellini has noted[11] that the FAD transition-temperature range is expanded significantly by an increase in mechanical constraint, e.g., plate thickness. As shown schematically in Figure 13.10, the temperature span between the NDT and the FTE is expanded in the order of 60°–80°F (33°–45°C) by increases in section size from 1 to 2 in. to 6 to 12 in. in thickness. Full shelf energy ductility (FTP) is estimated to be attained in the order of 60°–80°F (33°–45°C) above the FTE temperature of these thick sections. Because of the limited statistical data and the recognition of metallurgical variations within a given plate, it is not feasible to generalize the relationships to a finer degree.

Pellini[11] states that the major items of interest are

1. "There is no effect of section size for the small flaw curve of the FAD. The instability conditions for small flaws are controlled by the flaw size and not the section size. For example, a small flaw of a few tenths of an inch does not recognize that it is located in a 1 or 12 in. thick section—both are semi-infinite with respect to the flaw size.
2. "The instabilities of very large flaws are influenced by section size because increasing size provides additional constraint for the large flaws. Thus, there is a moderate shift of the large flaw size transition curves to higher temperatures, as indicated by the expanded FAD.
3. "The rise of the CAT (crack arrest temperature) curves for small and thick sections starts commonly at the NDT temperature; however, the rate of change in risk is more gradual for the thick sections. This is a consequence of the shift of the FTE to a higher relative temperature.

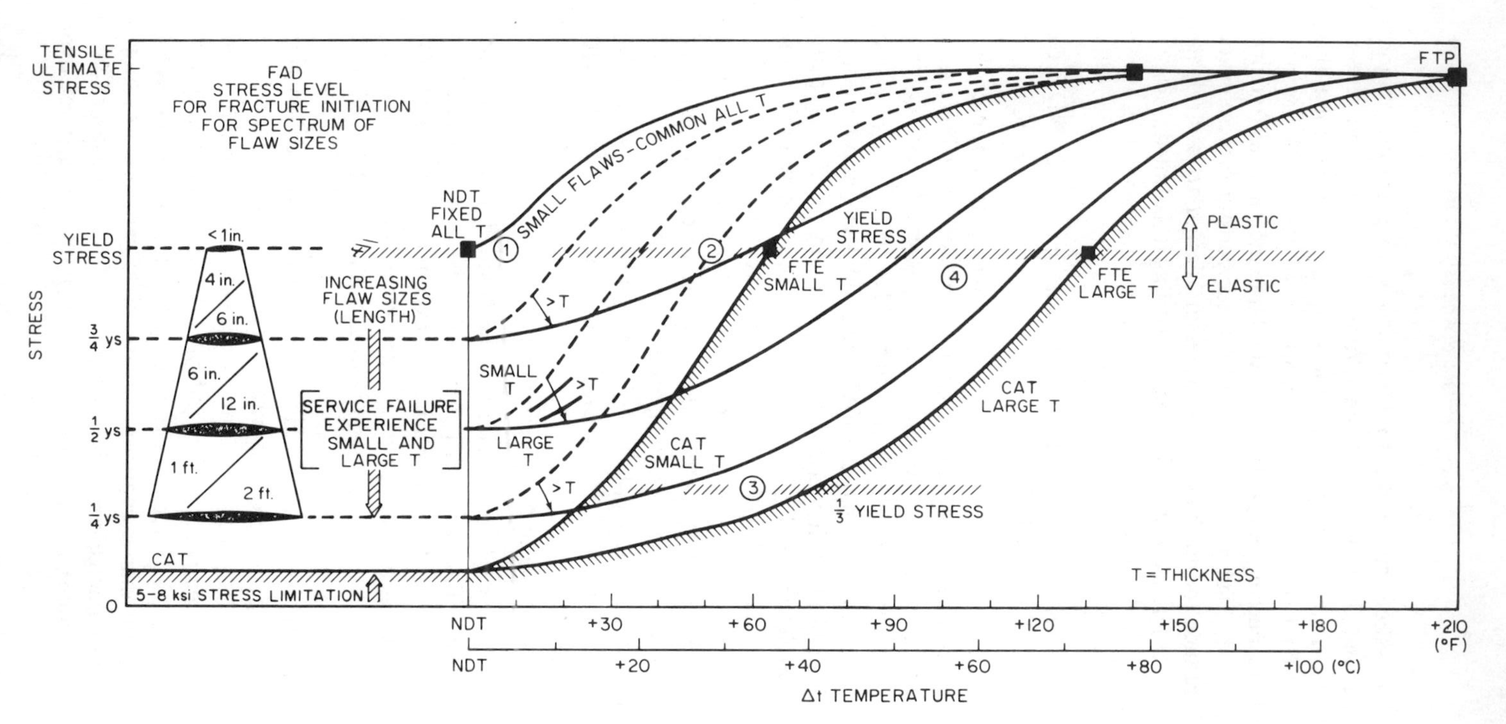

FIG. 13.10. FAD expansion to include the effects of very large section size. Fracture safe performances with respect to unstable fracture is indicated for temperatures in excess of the limits defined for the engineering design points noted as (1) small flaws subject to yield stress levels, (2) intermediate sized flaws subjected to yield stress levels, (3) very large flaws subjected to nominal design stresses, and (4) very large flaws subjected to yield stress levels. (Ref. 11.)

"If uncertainties of stress levels, temperature, etc., are considered, it will be evident that practical adjustments of the FAD in excess of 30°F (17°C), due to section size, are required only for special situations. These involve very large flaws located in very thick sections which are loaded to relatively high levels of stress, i.e., huge structures, very large defects, and high stress levels.

"For most structural applications, e.g., ships, bridges, earthmoving equipment, etc., the use of the *expanded* FAD diagram is unnecessary."

In summary, the FAD criterion is based on establishing a level of performance related to NDT (nil-ductility transition), FTE (fracture transition elastic), or FTP (fracture transition plastic) for impact loading conditions.

13.5. Through-Thickness Yielding Criterion

The through-thickness yielding criterion for structural steels is based qualitatively on two observations.[13] First, increasing the design stress in a particular application (which generally requires a higher yield strength material) results in more stored elastic energy in a structure, which means that the fracture toughness of the steel should also be increased to have the same degree of safety against fracture as a structure with a lower working stress. Second, increasing plate thickness promotes a more severe state of stress, namely plane strain. Thus, a higher level of toughness is required to obtain the same level of performance in thick plates as would be obtained in thin plates.

By using concepts of linear-elastic fracture mechanics[14,15] a quantitative approach to the development of toughness requirements is based on the requirement that in the presence of a large sharp crack in a large plate, through-thickness yielding should occur before fracture. Specifically, the requirement is based on the ratio of plate thickness to plastic-zone size ahead of a large sharp flaw.

Although fracture-mechanics toughness requirements preferably should be established on the basis of K_{Ic} values, the development of toughness requirements using Charpy V-notch impact values would be more useful to the materials engineer because of the difficulty in obtaining K_{Ic} values.

From a qualitative viewpoint, the effect of plate thickness on the fracture toughness of steel plates tested at room temperature has been generally established.[14,15] As the plate thickness is decreased, the state of stress changes from plane strain to plane stress, and ductile fractures generally occur along 45° planes through the thickness. Except for very brittle materials, failure is usually preceded by through-thickness yielding and is not catastrophic. Conversely, in thick plates the state of stress is generally plane strain, and fractures usually occur normal to the direction of loading. Except for ductile materials, through-thickness yielding does not occur prior to fracture, and failure may be unstable. The transition in state-of-stress from plane strain to plane stress is responsible for a large increase in toughness and is quite desirable. Thus, if plane-stress behavior can be assured, structural members will fail only when plastically overloaded.

The behavior of most structural members is somewhere between the two limiting conditions of plane stress and plane strain. Hahn and Rosenfield[16,17] have shown that in terms of either through-thickness strain or crack-opening displacement, there is a significant increase in the rate at which through-thickness deformation occurs when the following relation exists:

$$\left(\frac{K_{Ic}}{\sigma_{ys}}\right)^2 \frac{1}{t} \geq 1 \tag{13.1}$$

where K_{Ic} = plane-strain stress-intensity factor, ksi$\sqrt{\text{in.}}$,

σ_{ys} = yield strength, ksi.

Brown and Srawley,[15] as well as ASTM Committee E-24 on Fracture Testing of Metals,[18] have indicated that the following relation must be satisfied to ensure plane-strain behavior:

$$\left(\frac{K_{Ic}}{\sigma_{ys}}\right)^2 \frac{1}{t} \leq 0.40 \tag{13.2}$$

Detailed analysis of numerous K_{Ic} test results as well as other experimental evidence[16,17,19,20,21] indicates that plane-strain conditions may exist at somewhat smaller plate thicknesses than required by Equation (13.2). However, assume that Equation (13.1) represents an upper bound for plane-strain behavior and may be used to define the condition at which considerable through-thickness yielding begins to occur. This type of behavior is desirable in structural applications and can be used as a criterion to obtain satisfactory performance in structures where through-thickness yielding can occur, such as in large thin plates that contain through-thickness cracks and in which prevention of fracture is an important consideration.

Equation (13.1), which is based on Hahn and Rosenfield's experimental observation of the plastic-zone size in silicon steel, is similar to Irwin's $\beta_I = 1.0$ criterion for a part-through crack model.[22] In this model, Irwin proposed that the minimum K_{Ic} value for a "leak-before-break" criterion be as follows:

$$\beta_I = \frac{1}{t}\left(\frac{K_{Ic}}{\sigma_{ys}}\right)^2 \simeq 1.5 \tag{13.3}$$

Therefore

$$t \simeq \frac{1}{1.5}\left(\frac{K_{Ic}}{\sigma_{ys}}\right)^2 \tag{13.4}$$

Thus, for through-thickness yielding to occur in the presence of a large sharp crack in a large plate, a fracture-toughness criterion based on Equation (13.1) appears reasonable on the basis of both experimental and theoretical considerations. This condition is probably conservative and there are many design applications for which this type of performance is not required. However, it is a desirable goal in establishing toughness requirements for large thin plates that may contain through-thickness cracks and are used in critical applications.

Rearranging Equation (13.1), a toughness criterion for steels to obtain through-thickness yielding before fracture can be developed in terms of yield strength and plate thickness as follows:

$$K_{Ic} \geq \sigma_{ys}\sqrt{t} \qquad \text{for } t \leq 2 \text{ in.} \tag{13.5}$$

From this relation, the K_{Ic} values required for through-thickness yielding before fracture at any temperature were developed for steels with various yield strength levels and plate thicknesses. These values, shown in Table 13.1

TABLE 13.1. K_{Ic} Values Required for Through-Thickness Yielding Before Fracture

Yield Strength, σ_{ys} *(ksi)*	*Plate Thickness,* *t (in.)*	K_{Ic} *(ksi$\sqrt{in.}$)*
40	$\frac{1}{2}$	28
	1	40
	2	57
60	$\frac{1}{2}$	42
	1	60
	2	85
80	$\frac{1}{2}$	57
	1	80
	2	113
100	$\frac{1}{2}$	71
	1	100
	2	141
120	$\frac{1}{2}$	85
	1	120
	2	170
140	$\frac{1}{2}$	99
	1	140
	2	198
160	$\frac{1}{2}$	113
	1	160
	2	226
180	$\frac{1}{2}$	127
	1	180
	2	255
200	$\frac{1}{2}$	141
	1	200
	2	283

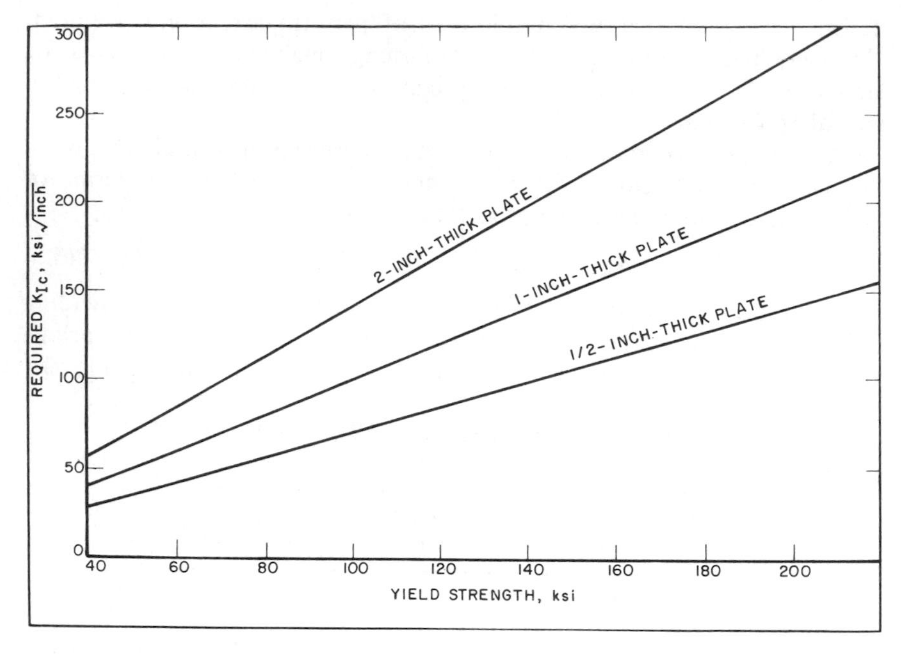

FIG. 13.11. K_{Ic} required to satisfy through-thickness yielding before fracture criterion.

and plotted in Figure 13.11, demonstrate the marked dependence of toughness on yield strength and plate thickness if through-thickness yielding is to precede fracture. That is, the desired toughness increases linearly with yield strength and with the square root of plate thickness.

Service experience would indicate that the required toughness on the basis of Equation (13.5) may be excessive for thick plates. Until investigations of very thick plates establish the necessary toughness requirements, the use of Equation (13.5) should be limited to plates less than 2 in. thick.

It is generally not possible to determine K_{Ic} when the plate has sufficient toughness according to the criterion described above. According to the ASTM recommended practice,[18] the plate thickness for valid results should be

$$\left(\frac{K_{Ic}}{\sigma_{ys}}\right)^2 \frac{1}{t} < 0.40 \tag{13.2}$$

Rearranging terms, the maximum valid K_{Ic} that can be measured by current practice is

$$K_{Ic} = 0.63\sigma_{ys}\sqrt{t}$$

However, the required K_{Ic} according to the through-thickness yielding criterion is

$$K_{Ic} = 1.0\sigma_{ys}\sqrt{t}$$

Having the required toughness greater than the maximum toughness that can currently be measured means that the specimen should behave in a somewhat ductile manner in the test and that elastic plane-strain behavior will not be obtained in materials that meet the criterion.

Although a K_{Ic} value is in principle the best criterion for evaluating the fracture toughness of steels, it is expensive and often impossible to obtain in the plate thicknesses used. The development of empirical correlations between K_{Ic} and CVN values (Chapter 6) can be used to establish similar through-thickness toughness guidelines for steels using Charpy V-notch impact test results.

The correlation described in Chapter 6 between slow-bend K_{Ic} test results and room-temperature CVN impact results for high-strength steels having yield strengths greater than 100 ksi is widely used. This correlation is limited to upper "shelf" CVN values where the loading rate or notch acuity do not affect the fracture behavior of high-strength steels significantly.

The room-temperature CVN-K_{Ic} correlation presented in Chapter 6 in terms of K_{Ic}, σ_{ys}, and CVN is as follows:

$$\left(\frac{K_{Ic}}{\sigma_{ys}}\right)^2 = \frac{5}{\sigma_{ys}}\left(\text{CVN} - \frac{\sigma_{ys}}{20}\right) \tag{13.6}$$

Combining Equations (13.5) and (13.6) results in the following relation for the minimum CVN impact energy required to obtain through-thickness yielding before fracture in large thin plates that contain through-thickness cracks:

$$\text{CVN} = \frac{\sigma_{ys}}{5}(t + 0.25) \tag{13.7}$$

From this relation, the CVN values required for through-thickness yielding before fracture were established for various yield strength levels and plate thicknesses (Table 13.2).

Because Equation (13.7) is based on steels having yield strengths greater than 100 ksi, it is applicable primarily for establishing toughness guidelines for steels in this strength range. In Table 13.3, the required K_{Ic} and CVN values are compared with the experimentally obtained K_{Ic} and CVN values for various high-strength steels.

The CVN impact values of HY-80 steel plates are generally well above the 36-ft-lb requirement shown in Table 13.3. Explosion-bulge test results have demonstrated that this steel exhibits extremely large plastic strains prior to failure. Thus, HY-80 steel has significantly more toughness than is required to satisfy the proposed criterion. A517-F steel plates generally exhibit more than the 25-ft-lb requirement shown in Table 13.3 and have been used successfully in many structural applications for which the presently proposed toughness requirements would apply.

Although HY-130, HY-140(T), and the 10Ni-Cr-Mo-Co steel have not been used in many service applications, the required values shown in Table

TABLE 13.2. Charpy V-Notch Upper Shelf Values Required for Through-Thickness Yielding Before Fracture

Yield Strength, σ_{ys} *(ksi)*	*Plate Thickness,* *t (in.)*	$CVN = \frac{\sigma_{ys}}{5}(t + 0.25)$ *(ft-lb)*
100	½	15
	1	25
	2	45
120	½	18
	1	30
	2	54
140	½	21
	1	35
	2	63
160	½	24
	1	40
	2	72
180	½	27
	1	45
	2	81
200	½	30
	1	50
	2	90

TABLE 13.3. Comparison of Required Toughness Values for Through-Thickness Yielding and Measured Values for Selected Steels

			Required Values		*Measured Values*†	
Steel	*Minimum Yield Strength, 0.2% Offset (ksi)*	*Plate Thickness* (in.)*	K_{Ic} *(ksi* $\sqrt{in.}$*)*	*Charpy V-Notch Energy Absorption at +80°F (ft-lb)*	K_{Ic}‡ *(ksi* $\sqrt{in.}$*)*	*Charpy V-Notch Energy Absorption at +80°F (ft-lb)*
HY-80	80	2	113	36	—	100
A517-F	100	1	100	25	170	62
4147	140	½	99	21	109	26
HY-130	130	2	184	59	246	89
HY-140(T)	140	1	140	35	250	85
10Ni-Cr-Mo-Co	180	2	254	81	300	130
18Ni Maraging (250 grade)	240	¾	207	48	87	16

*Plate thicknesses chosen are representative of thicknesses normally used in these steels.
†These values are laboratory data on individual plates.
‡Apparent K_{Ic}.

13.3 appear to be desirable objectives on the basis of the general belief[23] that the required toughness should increase with increasing yield strength. In addition, these steels were developed for critical applications. Severely notched pressure vessels of 1-inch thick HY-140(T) steel have been tested to failure[24]. These vessels had CVN values well above the 35-ft-lb requirement shown in Table 13.3, and, even though the vessels had 20-in.-long deep surface cracks, all vessels exhibited through-thickness yielding in the vicinity of the crack before final failure. All fractures exhibited 100% shear.

Notched pressure-vessel tests conducted on 4147 steel have also exhibited through-thickness yielding before fracture. One vessel that contained a 1.6-in.-long through-thickness crack (three times the wall thickness) failed at 90% of the nominal yield stress, after exhibiting through-thickness yielding at the notch. This steel had a CVN value of 25 ft-lb, which was above the required value of 21 ft-lb shown in Table 13.3.

The toughness requirements for the 18Ni maraging (250-grade) steel appear quite extreme. It should be noted, however, that a 260-in. missile-motor case which failed during hydrotest[25] (Chapter 5) did not have the required toughness values shown in Table 13.3. This does not mean that this steel (or similar steels) cannot be used successfully in structural applications in which the fabrication and inspection procedures are such that the crack size is smaller than the critical crack size. Many similar vessels have performed satisfactorily even though they did not exhibit the levels of toughness shown in Table 13.2. Successful testing of maraging steel vessels demonstrates that through-thickness yielding is not required provided that fabrication and inspection preclude cracks greater than the critical size. Vessels with levels of toughness that do not satisfy the proposed criterion can be designed using fracture mechanics by calculating permissible stress-crack size relations (Chapter 5). Failure of the 260-in. missile-motor case demonstrates the consequences of inadequate fabrication and inspection if the toughness levels suggested in Equation (13.7) cannot be obtained.

The criteria proposed in Equations (13.5) and (13.7) are strongly dependent on plate thickness, and thus for thicker plates and higher-strength steels extremely high K_{Ic} and CVN values are required. However, the condition that through-thickness yielding must occur in the presence of a large sharp crack (fatigue crack) is conservative for many applications. The criteria presented in Equations (13.5) and (13.7) are guidelines for fracture-toughness levels of medium- and high-strength steels. It should be emphasized that because of careful fabrication and inspection procedures, there are many design applications for which this type of performance is not required and that these toughness requirements are intended only for those applications in which through-thickness yielding in the presence of a large sharp crack is required before fracture. Moreover, the through-thickness yielding criterion is based on the development of through-thickness plastic deformation at 45° planes in the region of a through-thickness crack tip in a plate having large in-plane dimensions, consequently, this criterion should not be used for geometries where the development of such deformation is very difficult (e.g. part-through cracks).

13.6. Leak-Before-Break

The leak-before-break criterion was proposed by Irwin et al.[26] as a means of estimating the necessary toughness of pressure-vessel steels so that a surface crack could grow through the wall and the vessel "leak" before fracturing. That is, the critical crack size at the design stress level of a material meeting this criterion would be greater than the wall thickness of the vessel so that the mode of failure would be leaking (which would be relatively easy to detect and repair) rather than catastrophic fracture.

Figure 13.12 shows schematically how such a surface crack might grow through the wall into a through-thickness crack having a length approximately equal to $2B$. Thus the leak-before-break criterion assumes that a crack of twice the wall thickness in length should be stable at a stress equal to the nominal design stress.

The value of the general stress intensity, K_{I}, for a through crack in a large plate (Figure 13.13) is

$$K_{\mathrm{I}}^2 = \frac{\pi \sigma^2 a}{1 - \frac{1}{2}(\sigma/\sigma_{ys})^2}$$

where $2a$ = crack length,

σ = tensile stress normal to the crack,

σ_{ys} = yield strength.

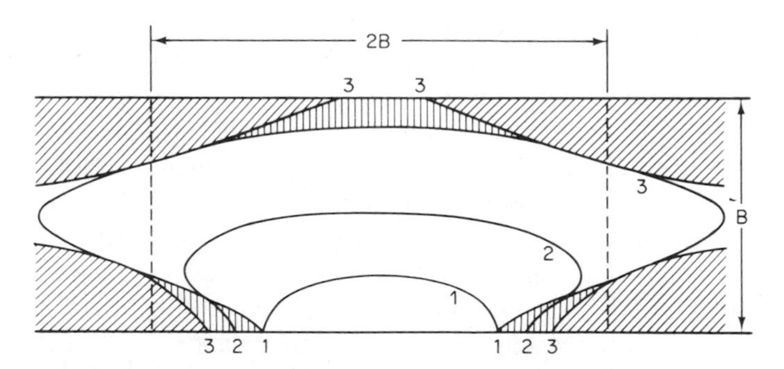

FIG. 13.12. Spreading of a part—through crack to critical size (line 3) for the short-crack failure models. Lines 1, 2, 3 are assumed to represent crack edge positions during increase of crack size. Shaded regions are shear-lip (vertical shading) or potential shear-lip (slant shading).

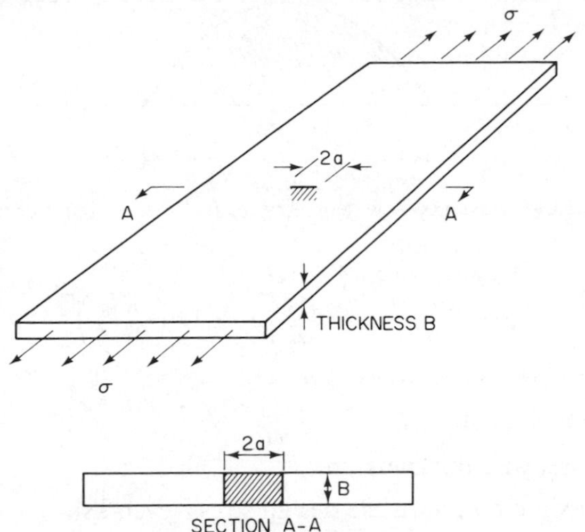

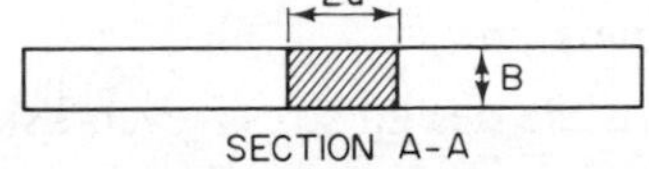

FINAL THROUGH-THICKNESS CRACK
$2a \cong 2B$

FIG. 13.13. Final dimensions of through-thickness crack.

(Note that for low values of design stress, σ, this cxpression reduces to $K_{\text{I}} = \sigma\sqrt{\pi a}$.)

At fracture, $K_{\text{I}} = K_c$ (assuming plane-stress behavior) and

$$K_c^2 = \frac{\pi\sigma^2 a}{1 - \frac{1}{2}(\sigma/\sigma_{ys})^2}$$

Because standard material properties are usually obtained in terms of K_{Ic}, the following relation between K_c and K_{Ic} is used to establish the leak-before-break criterion in terms of K_{Ic}:

$$K_c^2 = K_{\text{Ic}}^2(1 + 1.4\beta_{\text{Ic}}^2)$$

where

$$\beta_{\text{Ic}} = \frac{1}{B}\left(\frac{K_{\text{Ic}}}{\sigma_{ys}}\right)^2, \qquad \text{a dimensionless parameter.}$$

Thus, substituting for K_c and β_{Ic}, the following general relation is obtained:

$$\frac{\pi\sigma^2 a}{1 - \frac{1}{2}(\sigma/\sigma_{ys})^2} = K_{\text{Ic}}^2\left[1 + 1.4\left(\frac{K_{\text{Ic}}^2}{B\sigma_{ys}^2}\right)^2\right]$$

In the leak-before-break criterion the depth of the surface crack, a, is set equal to the plate thickness, B (Figure 13.14), and we obtain

$$\frac{\pi\sigma^2 B}{1 - \frac{1}{2}(\sigma/\sigma_{ys})^2} = K_{\text{Ic}}^2\left[1 + 1.4\left(\frac{K_{\text{Ic}}^4}{B^2\sigma_{ys}^4}\right)\right]$$

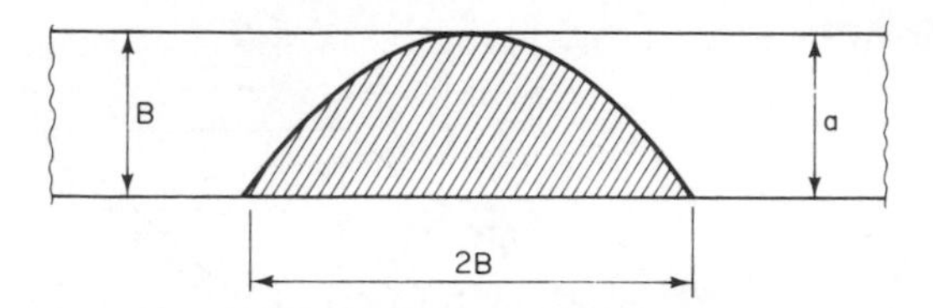

FIG. 13.14. Assumed flaw geometry for leak-before-burst criterion.

or

$$\frac{\pi\sigma^2}{1 - \frac{1}{2}(\sigma/\sigma_{ys})^2} = \frac{K_{Ic}^2}{B}\left[1 + 1.4\left(\frac{K_{Ic}^4}{B^2\sigma_{ys}^4}\right)\right]$$

where σ = nominal design stress, ksi,

σ_{ys} = yield strength, ksi,

B = vessel wall thickness, in.,

σ = maximum permissible design stress level, ksi,

K_{Ic} = plane-strain crack toughness (ksi$\sqrt{\text{in.}}$) required to satisfy the leak-before-break criterion for a material with a particular σ_{ys}, a vessel with wall thickness B, and design stress σ.

In this expression σ, σ_{ys}, and K_{Ic} are the design stress, yield strength, and material toughness at the particular service temperature and loading rate.

Irwin et al.[26] have calculated various values of the nominal design stress that will satisfy the above leak-before-break criterion for three structural steels, namely A212B, A302B, and HY-80 steel. In their example, they estimated K_{Id} values based on the NDT temperature plus either 60° or 120°F for each of the three steels. The basic input data are presented in Table 13.4. Using these estimates of material properties they selected three thickness values (1, 2.5, and 5.0 in.) and calculated the allowable dynamic design

TABLE 13.4. Actual and Estimated Material Properties Used by Irwin et al.[26] in Examples of Leak-Before-Break Criterion

Actual Test Values			*Estimated Material Properties*					
			NDT +60°F			*NDT +120°F*		
Steel	σ_{ys} *+70°F (ksi)*	*NDT (°F)*	*NDT +60°F (°F)*	σ_{yd} *(ksi)*	K_{Id} *(ksi$\sqrt{in.}$)*	*NDT +120°F (°F)*	σ_{yd} *(ksi)*	K_{Id} *(ksi$\sqrt{in.}$)*
A212B	36	20	80	55.9	51.1	140	50.1	57.0
A302B	62	0	60	82.7	74.4	120	77.8	79.1
HY-80	82	−150	−90	122	116	−30	113	126

TABLE 13.5. Allowable Design Stresses to Satisfy Leak-Before-Break Criterion for Three Steels Having Cracks Equal to Wall Thickness

Steel	*Estimated Dynamic Values*	*B (in.)*	β_{Ic}	$\frac{\sigma}{\sigma_{yd}}$	*Allowable Design Stress, σ_d, to Satisfy the Leak-Before-Break Criterion (ksi)*
A212	At 80°F	1	0.84	0.65	36
	$\sigma_{yd} = 55.9$ ksi	2.5	0.33	0.34	19
	$K_{Id} = 51.1$ ksi$\sqrt{\text{in.}}$	5	0.17	0.23	13
	At 140°F	1	1.29	1.10	55 (greater than σ_{yd})
	$\sigma_{yd} = 50.1$ ksi	2.5	0.52	0.45	23
	$K_{Id} = 57.0$ ksi$\sqrt{\text{in.}}$	5	0.26	0.29	15
A302B	At 60°F	1	0.81	0.63	52
	$\sigma_{yd} = 82.7$ ksi	2.5	0.32	0.33	28
	$K_{Id} = 74.4$ ksi$\sqrt{\text{in.}}$	5	0.16	0.23	19
	At 120°F	1	1.03	0.76	59
	$\sigma_{yd} = 77.8$ ksi	2.5	0.41	0.39	29
	$K_{Id} = 79.1$ ksi$\sqrt{\text{in.}}$	5	0.21	0.26	20
HY-80	At −90°F	1	0.912	0.69	84
	$\sigma_{yd} = 122$ ksi	2.5	0.364	0.36	44
	$K_{Id} = 116$ ksi$\sqrt{\text{in.}}$	5	0.182	0.24	30
	At −30°F	1	1.25	0.88	99
	$\sigma_{yd} = 113$ ksi	2.5	0.50	0.44	50
	$K_{Id} = 126$ ksi$\sqrt{\text{in.}}$	5	0.25	0.29	33

stress that would satisfy the leak-before-break criterion for each of the three wall thicknesses chosen for the example. That is, the critical crack size for the 1-in.-thick vessel is 1 in. and for the 5-in.-thick vessel is 5 in. Obviously, because the toughness levels are the same for all vessel thicknesses studied, the allowable design stress should decrease significantly with increasing vessel thickness.

The results of their calculations are presented in Table 13.5 and show the allowable design stress to satisfy the leak-before-break criterion. Note that the general trend in each case is to decrease the allowable design stress by a factor of about 3 as the wall thickness is increased from 1 to 5 in. The same general trend for static design stresses would be expected if K_{Ic} values could be obtained for these steels. K_{Id} values were used apparently because it was easier to extrapolate these values from the K_{Id} at NDT.

Because of the uncertainties in stress level at intersections, because of the effects of residual stress, and as a general conservative approach, the leak-before-break criterion can also be established assuming that the nominal stress, σ, is equal to the yield stress, σ_{ys}. Thus, for $\sigma = \sigma_{ys}$, the general criterion reduces to

$$\frac{\pi\sigma_{ys}^2}{1-\frac{1}{2}(\sigma_{ys}/\sigma_{ys})^2}=\frac{K_{Ic}^2}{B}\left[1+1.4\left(\frac{K_{Ic}^4}{B^2\sigma_{ys}^4}\right)\right]$$

$$2\pi\sigma_{ys}^2=\frac{K_{Ic}^2}{B}+(1.4)\frac{K_{Ic}^6}{B^3\sigma_{ys}^4}$$

or

$$\frac{1.4K_{Ic}^6}{B^3\sigma_{ys}^4}+\frac{K_{Ic}^2}{B}=2\pi\sigma_{ys}^2$$

As an example of the use of this criterion, the engineer must first select the nominal yield strength steel that he wishes to use, then determine the wall thicknesses (these two factors might be established on the basis of a general strength criterion to withstand a given internal pressure), and, finally, select the required minimum toughness level necessary to meet the criterion. *Then*, from the steels available, he must select that one or ones which meet the criterion. Final material selection would be on the basis of the above, plus other criteria, such as cost, fabrication, etc.

The required K_{Ic} values that will satisfy the leak-before-break criterion

TABLE 13.6. K_{Ic} Values Required to Satisfy Leak-Before-Break Criterion for Yield Strength Loading

Material Yield Strength and Assumed Applied Stress (ksi)	*Vessel Thickness, B (in.)*	*Required K_{Ic} (ksi$\sqrt{in.}$)*
40	$\frac{1}{2}$	35
	1	50
	2	70
	4	100
80	$\frac{1}{2}$	70
	1	100
	2	140
	4	195
120	$\frac{1}{2}$	105
	1	145
	2	210
	4	295
160	$\frac{1}{2}$	145
	1	195
	2	280
	4	325
200	$\frac{1}{2}$	180
	1	245
	2	345
	4	490

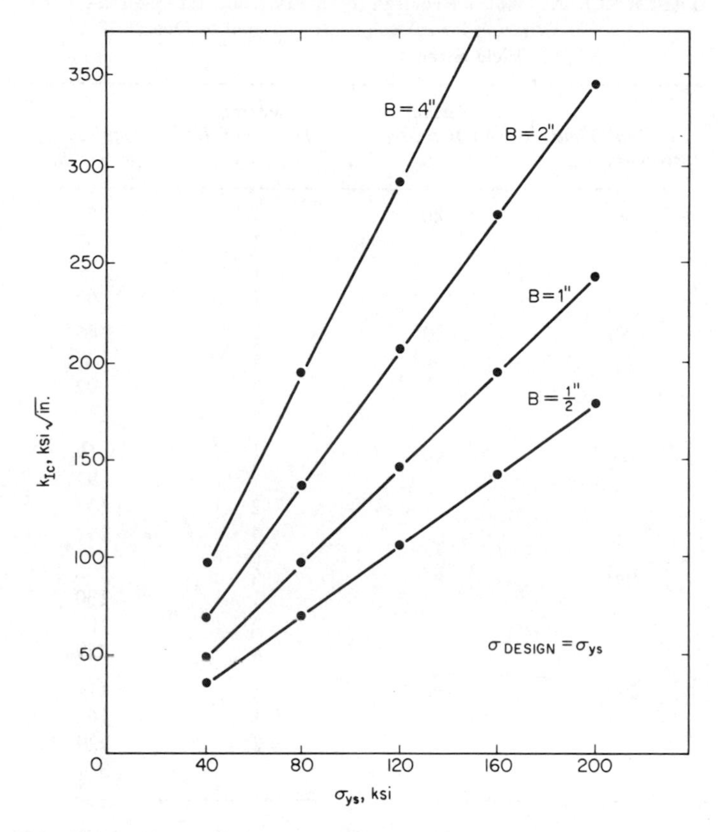

FIG. 13.15. Effect of yield strength and thickness on K_{Ic} required to satisfy leak-before-burst criterion ($\sigma_{\text{design}} = \sigma_{ys}$).

at yield strength levels ranging from 40 to 200 ksi and for wall thicknesses ranging from $\frac{1}{2}$ to 4 in. are presented in Table 13.6 and Figure 13.15. These results illustrate the significant effect of an increase in thickness or yield strength when selecting materials to satisfy the leak-before-break criterion. Thus this criterion can be used to ensure a high level of notch toughness (elastic-plastic behavior). The required K_{Ic} values for $\sigma_{\text{design}} = \sigma_{ys}/2$ are presented in Table 13.7 and are plotted in Figure 13.16, showing the effect of design stress level on the required K_{Ic} values.

13.7. Ratio Analysis Diagram (RAD) Criterion

Pellini developed the ratio analysis diagram (RAD) to describe the general decrease in upper-shelf-level toughness of structural materials with increasing yield strength level.[12] This decrease in upper-shelf-level toughness

TABLE 13.7. K_{Ic} Values Required to Satisfy Leak-Before-Break Criterion for a Design Stress Equal to One-Half the Yield Strength

Material Yield Strength (ksi)	*Assumed Applied Stress (ksi)*	*Vessel Thickness, B (in.)*	*Required K_{Ic} (ksi$\sqrt{in.}$)*
40	20	$\frac{1}{2}$	23
		1	32
		2	46
		4	65
80	40	$\frac{1}{2}$	46
		1	65
		2	92
		4	130
120	60	$\frac{1}{2}$	69
		1	97
		2	137
		4	194
160	80	$\frac{1}{2}$	92
		1	130
		2	183
		4	259
200	100	$\frac{1}{2}$	114
		1	162
		2	229
		4	324

for different materials having increasing strength levels is analogous to the decrease in notch toughness of a single material with decreasing temperature. Hence there are two general transitions in notch-toughness behavior—the temperature transition and the strength transition, as shown schematically in Figure 13.17. (There is also a thickness or constraint transition from plane stress to plane strain as described in Chapter 4, but this is not included in this figure.) The vertical scale in Figure 13.17 represents a measure of the ability of a steel to absorb energy in a notched bar impact test, e.g., CVN, DT, etc. One of the horizontal axes defines the transition-temperature features of a *particular* material, whereas the other axis defines the strength-transition features of *different* materials. The shaded region at the bottom of the surface follows the contour of the toe region of the DT test temperature lower-transition range. Accordingly, it indexes the course of the temperature-induced plane-strain regime for which K_{Ic} or K_{Id} values can be measured. Since Pellini determined the linear-elastic transition features using

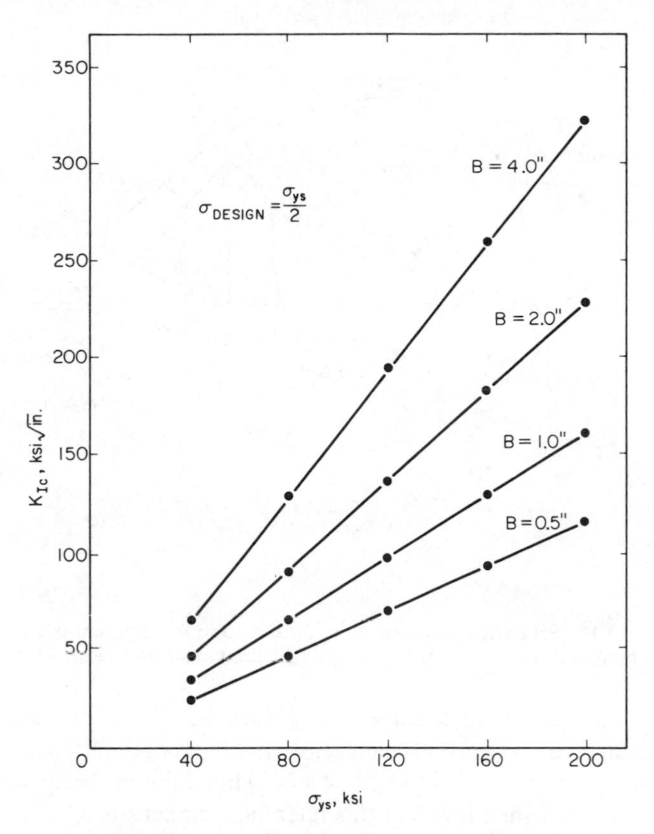

FIG. 13.16. Effect of yield strength and thickness on K_{Ic} required to satisfy leak- before- burst criterion ($\sigma_{design} = \sigma_{ys}/2$).

1-in.-thick DT test specimens, the plane-strain limit relates to a section of small size, as indicated by the notation "plane-strain—small *B*." The temperature limits of plane strain for large section sizes should be located at approximately 60°–80°F (33°–45°C) higher temperatures if the metallurgical quality is retained for the large section size.

The appearance of the DT test specimen fracture in the plane-strain region is flat and devoid of shear lips. The designation of "mixed-mode *fB*" signifies that the nature, in the region of the three-dimensional surface which is not shaded, involves flat central regions with surface shear lips. This is elastic-plastic behavior. The relative fraction of flat and slant fracture is a function of thickness. The shaded region at the top of the surface indicates very high levels of fracture toughness and fractures of ductile (plastic) features.

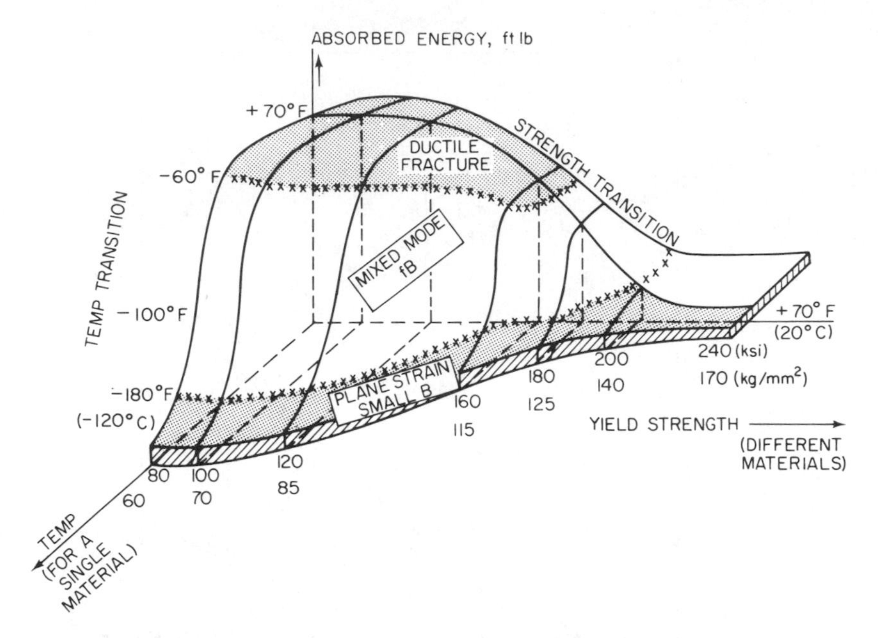

FIG. 13.17. Schematic showing strength transition for different materials and temperature transition for various different materials (Ref. 11).

A slice through the three-dimensional diagram of Figure 13.17 at 70°F (20°C) indicates the effects of increasing yield strength on the shelf level fracture characteristics of different steels. This follows because all these steels have attained shelf levels at this reference temperature.

Based on a summary of numerous upper-shelf-level test results, Pellini established the ratio analysis diagram (RAD) as a system to index structural performance as approximated by DT or explosion-bulge test results to fracture mechanics. Correlations between K_{Ic}, CVN, and DT test results were used to develop this diagram. This coupling of structural behavior as measured by CVN or DT test values and fracture-mechanics K_{Ic} values (which can be used to describe stress–flaw size trade-offs) is shown in Figure 13.18 as the ratio analysis diagram.

The technological limit line describes the locus of materials having the highest available toughness level at a particular period in time. Thus, as metallurgical improvements are made in the materials field, this "upper-bound" limit line continually shifts to the right. That is, improvements in materials technology lead to improved toughness levels at higher yield strength levels. At present, structural steels above about 180-ksi yield strength exhibit a rapid decrease in toughness with increasing strength level. Obviously steels are used that do not have this optimum toughness, and they can be significantly below the "upper-bound" technological limit line.

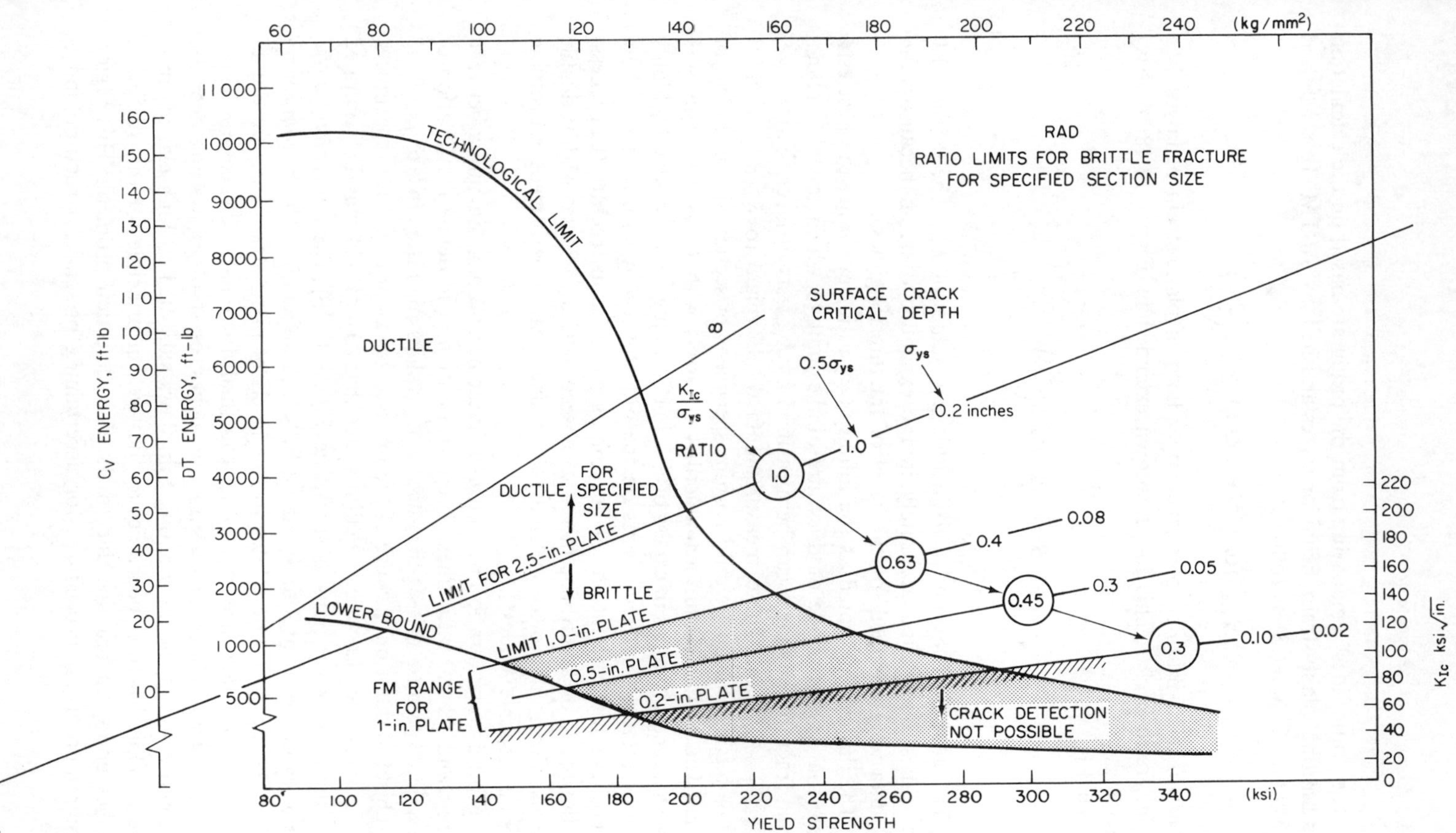

FIG. 13.18. Ratio analysis diagram (RAD) (Ref. 11).

The "ratio lines" emanating from the origin are based on the calculation for limiting plane-strain behavior as established by ASTM Test Method E-399 (see Chapter 3) as follows:

$$B_{\min} \text{ (for plane strain)} \geq 2.5\left(\frac{K_{Ic}}{\sigma_{ys}}\right)^2$$

Thus for a 2.5-in.-thick plate the upper limit of plane-strain behavior is a K_{Ic}/σ_{ys} ratio of 1.0. This line is shown extended in Figure 13.18 for easy reference.

That is,

$$B = 2.5 = 2.5\left(\frac{K_{Ic}}{\sigma_{ys}}\right)^2$$

$$2.5 = 2.5(1.0)^2$$

The diagram is expanded to include low values of K_{Ic}. The 0.5 and 1.0 ratio lines are located accurately in the expanded diagram because valid K_{Ic} data were obtained for this level of fracture toughness.

The dashed line noted as the infinity (∞) ratio represents the best estimate that can be made at this time of the limit to which unstable plane-strain fracture toughness can be measured by K_{Ic} tests of large section size, i.e., $B = 10$–12 in. These measurements would relate to K_{Ic}/σ_{ys} ratios of 2.0 or higher and thus would require specimens of section size in excess of 10 in. The high metallurgical ductility of steels with CVN and 1-in. DT test shelf energy values in excess of this line is amply demonstrated by highly ductile fracture for section sizes on the order of 3–4 in. in thickness. There is no basis for expectations of developing plane-strain fracture (brittleness) as the result of additional increase of section size for such highly ductile metals. Confirming evidence that the ∞ line location is correct is evolving from large-scale K_{Ic} tests.

The 1.5 and 2.0 ratio lines must lie in the narrow gap remaining between the ∞ and the 1.0 ratio lines. There is no attempt to define the location of these lines because the significance of K_{Ic} values for ratios in excess of 1.0 is subject to question because of the inelastic behavior of the test specimens.

In his Adams Lecture,[11] Pellini states that "The RAD may be separated into three general regions as shown in [Figure 13.19]. The *top region*, above the ∞ ratio line, relates to ductile or fully plastic fracture. The *bottom region* below the 0.5 ratio line relates to low levels of plane strain fracture toughness, involving flaw sizes which are too small for reliable inspection. These are generally in the order of tenths of inches for stubby flaws and decrease to hundredths of inches for long thin flaws subjected to high elastic stress levels. While fracture mechanics applies well in principle for this region, it applies poorly in practice due to the flaw detection problem. This region requires proof-test procedures for ascertaining preservice structural integrity.

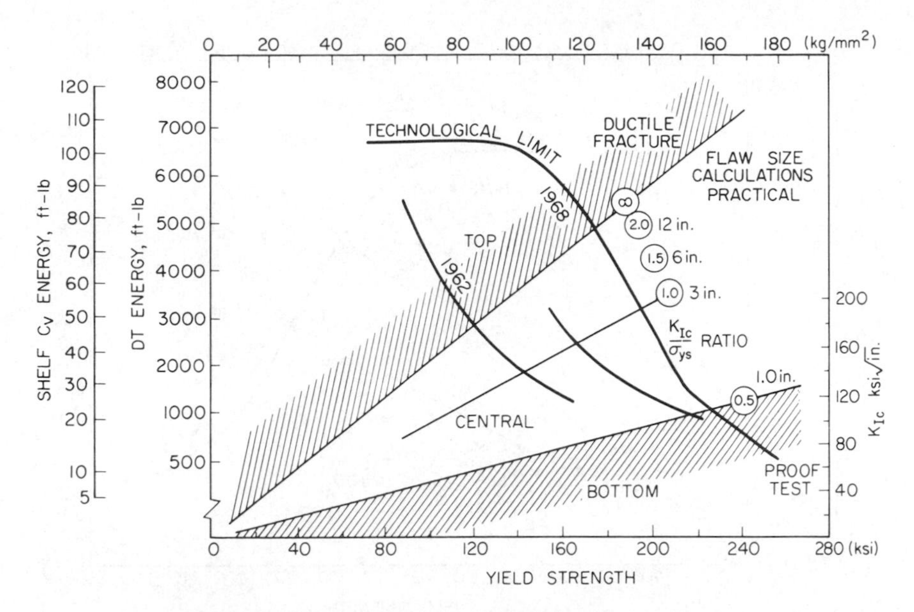

FIG. 13.19. Ratio analysis diagram divided into three regions (Ref. 11).

"The width of the remaining slice of the RAD, the *central region* for which plane strain fracture mechanics calculations appear feasible, depends on the section size. For very thick sections (to over 10 in.) the applicable region available for possible plane strain calculations is between the 0.5 and ∞ ratio lines. For section sizes in the order of 0.6 to 2.5 in. thickness, this region is bounded by the 0.5 and 1.0 ratio lines. This range represents a very thin slice through the population of the steels represented by the diagram.

"Factors which relate to the metallurgical quality of the steel become readily evident by metallurgical zoning of the RAD. The full span of fracture toughness definitions (plane strain to plastic fracture), provided by the DT test specimen energy scale and the ratio lines for the regions of plane strain fracture for different plate thicknesses, serve as a fixed mechanical reference grid. The analytical value of this grid is exploited by the superimposition of zone reference systems which partition the diagram in relation to metallurgical features.

"A simple, yet highly significant, zoning of metallurgical type is developed [in Figure 13.20] by tracing the effects of increasing strength level on the shelf transition features of various generic classes of steels. A series of metallurgical quality corridor zones, which are related to the melting and processing practices used to produce the steels, become evident. The lowest corridor zone involves relatively low-alloy commercial Q&T steels produced by conventional low-cost melting practices. The corridor is defined by the

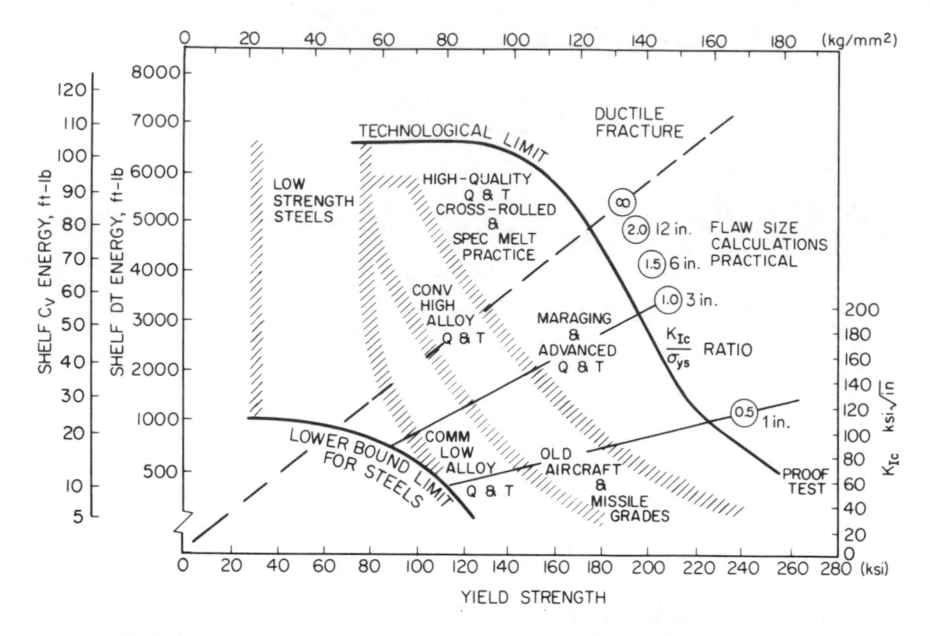

FIG. 13.20. Metallurgical zoning of the RAD which defines the general effects of melting and processing factors on the strength transition. The three corridors of strength transition relate to metallurgy quality (void-site density) which controls microfracture processes and, thereby, the macroscopic fracture toughness of the metal. The locations of generic alloy steel types are notes (Ref. 11).

strength transition of these steels to the 0.5 ratio level, as the result of heat treatment to yield strength levels in the order of 130 to 150 ksi (92 to 105 kg/mm²). Optimizations of the alloy content, coupled with improved melting and processing practices, elevate the corridors to higher levels. The strength transition to the 0.5 ratio is shifted accordingly to higher levels of yield strengths.

"Recent metallurgical investigations of high-strength steels have emphasized processing and metal purity aspects rather than purely physical metallurgy considerations of transformations. New scientific knowledge of void growth mechanisms and of the importance of void-site-density factors clearly indicates that control of metal bridge ductility can only be effective within limits. As these limits are reached by optimization of microstructures, it then is essential to obtain further increases in ductility by suppression of void initiation aspects. Void sites are provided by the presence of nonmetallic particles of microscopic size featuring noncoherent phase boundaries with the metal grains. Compared to coherent metal, grain boundaries separate easily. The extent to which such void sites can be eliminated is related to a number of factors including melting and deoxidation practices as well as the *P* and *S* impurity contents.

"[Figure 13.21] illustrates the effects of specific melting practices which relate directly to void-site-density. Air melting under conventional slags, followed by normal deoxidation practices, results in steels which have numerous nonmetallic particles, as is easily seen with low-power optical microscopes. The carbon deoxidized, vacuum induction-vacuum arc remelted (VIM-VAR) metal requires high-magnification electron microscope examination in order to locate the few nonmetallic particles that are present. Clearly, a new era in steel making has evolved which involves producing ultrahigh-purity steels at large production scale. [Figure 13.21] also notes a relocation of the Technological Limit curve of the RAD throughout the years to the position which has resulted from these metal processing improvements by the steel industry.

"The RAD corridor relationships provide a vastly improved definition of metallurgical factors. Systematization of vast amounts of data becomes feasible in easily understood form. The required directions for metal process-

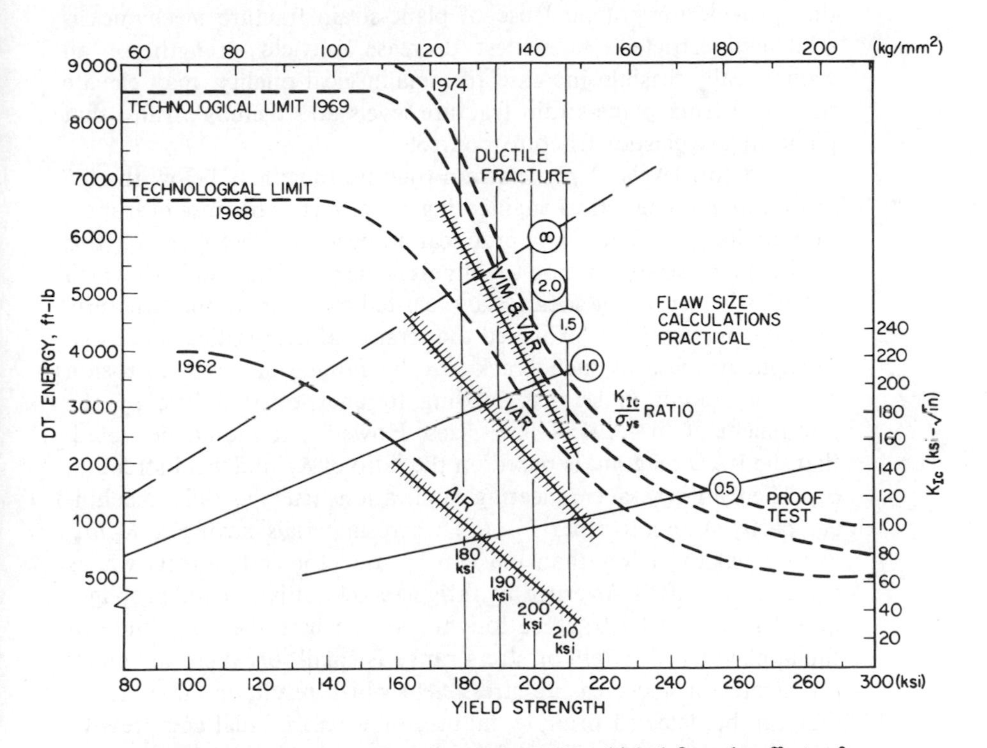

FIG. 13.21. Metallurgical zoning of the RAD which defines the effects of specific melting practices. These include air melting under slags (air), vacuum induction melting (VIM), and vacuum arc remelting (VAR); a combination of these is indicated by the VIM-VAR notation. The extension of the technological limit curve to the 1969 and 1974 positions results from the use of advanced VIM-VAR melting practices at large production scale (Ref. 11).

ing improvements are predictable from these corridor relationships. Moreover, it becomes apparent that the increased costs of high-quality processing are economically defensible, as well as technologically essential, at high strength levels. Analyses which combine economic, metallurgical, and fracture considerations over the full span of strength levels are of major interest to both users and producers of the steels."

The foregoing discussion of the RAD by Pellini has emphasized that a choice can be made between steels of brittle or ductile shelf characteristics over a wide range of yield strength levels. The integration of metallurgical and mechanical parameters provided by the RAD serves to define conditions which require the direct use of plane-strain fracture mechanics in design as well as those which preclude its use. These definitions are most important to engineering applications of fracture-safe design.

The following major points should be noted in these respects:

1. The combination of metallurgical quality and yield strength features that provide for practical use of plane-strain fracture mechanics is relatively restricted. A modest decrease in yield strength, or an economically feasible increase in metallurgical quality, may elevate the metal from plane-strain fracture levels and thereby ensure that plane-stress (plastic) fracture controls.
2. The very low levels of plane-strain fracture toughness below the 0.5 ratio restrict the practical engineering use of metals because of inspection limitations. The only practical inspection procedure for this level of plane-strain fracture toughness is the proof test method, which defines whether or not critical flaws existed by failure of the structure. Usually, no warning is provided that cracks of near-critical size exist. A slight increase in such crack size by fatigue or stress-corrosion may then occur in service, leading to unexpected failure by the attainment of the critical flaw sizes. However, it should be noted, that the RAD diagram is based on the ratio of K_{Ic} and yield strength, σ_{ys}. There are several engineering applications (such as rails, machinery parts, automotive parts—etc.) where materials having a K_{Ic}/σ_{ys} ratio equal to or less than 0.5 can be used for components whose useful design life is governed by fatigue-crack initiation and propagation rather than by fracture toughness, or where the magnitude of the applied tensile stress or stress range is small, or where the crack propagates in a decreasing stress field which results in large cracks that can be detected prior to failure, or when residual compressive stresses are induced in the surface of the component by shot-peening, carburizing, or nitriding.
3. If the structures involve relatively large plate thickness and feature relatively simple geometric configurations, crack size–stress calculations may be feasible for metals with plane-strain properties in the

range of approximately 0.8–2.0 ratios. The crack sizes involved will generally be within limits which permit reasonably reliable non-destructive inspection. The engineering reliability of this approach increases markedly with an increase in the ratio value.

4. For most engineering structures of complex design and critical applications, fracture safety is best assured by choice of metals featuring plane-stress fracture toughness. It is emphasized that the ratio value above which the plane-stress (ductile) conditions apply is a function of the section size.

In conclusion, the RAD is a means of characterizing optimum shelf levels of performance (plastic behavior) that can be expected for steels of varying strength levels. It can be used to select the highest strength level that should be used for specifying desired levels of fracture resistance. That is, if a particular K_{Ic}/σ_{ys} level of performance is desired, the RAD diagram can be used to determine the current technological limitations for steels of various strength levels. For example, if a K_{Ic}/σ_{ys} level of 1.0 is desired, the maximum strength level steel that can be considered is one having a yield strength of about 200 ksi. Also, this steel would have to be produced under optimum conditions and thus would be relatively expensive. A less expensive steel could be used to satisfy a performance requirement of $K_{Ic}/\sigma_{ys} = 1.0$ if the designer could use a steel with a yield strength of only about 150 ksi.

13.8. Summary

The preceding sections have described various notch-toughness criteria that have been proposed for use in structural applications. Obviously, one of the most desirable criteria is one which is based directly on K_{Ic} or K_{Id} values that are used to calculate stress–crack size trade-offs, as discussed in Chapter 5. However, the *direct* use of linear-elastic fracture mechanics assumes that only materials whose K_{Ic} or K_{Id} values can be determined using the test method described in Chapter 3 are used. As has been discussed before, most structural materials exhibit a *greater* level of performance than plane strain at service temperatures and loading rates. Furthermore, most *criteria* are established to ensure a level of performance greater than plane strain.

Table 13.8 summarizes the various criteria discussed above as well as their applications. Before selecting a particular criterion, the engineer should make sure that the general guidelines discussed in Chapter 12 are thoroughly understood and that the particular criterion used does indeed model his particular structure as close as possible.

The crack-opening-displacement (COD) criteria has been included in Table 13.8 as a matter of reference because it is widely used in the United Kingdom and Japan. It is based on elastic-plastic behavior, as described in Chapter 16.

TABLE 13.8. Comparison of Various Fracture Criteria

Name	*General Use*	*Level of Performance (See Figure 12.1)*	*Test Specimen Used*	*Applications*
K_{Ic} or K_{Id}	All materials	Plane strain	K_{Ic} standard (ASTM E-399) or K_{Id}—nonstandard	Linear-elastic behavior Quantitative stress–flaw size calculations Aerospace applications Very high-strength materials Low-temperature applications
15-ft-lb CVN impact	Low- and medium-strength (30–140 ksi) structural steels	Upper plane strain to lower elastic-plastic	ASTM standard (E-23); Charpy V-notch impact	Ship hull steels Bridge steels Applications where moderate level of toughness is required
Loading-Rate Shift	Structural steels; σ_{ys} = 30–140 ksi that exhibit a load-rate shift	Elastic-plastic	CVN impact (ASTM E-23) or K_{Ic} slow-bend (ASTM E-399)	AASHTO Bridge Steel Toughness Specifications; structures that are loaded at slow or intermediate loading rates
Fracture analysis diagram	All materials exhibiting transition-temperature behavior	All levels: plane strain, elastic-plastic, fully plastic	Nil-ductility transition NDT (ASTM E-208); dynamic tear—DT (Mil Standard 1601)	Transition-temperature region Applications where intermediate to high levels (elastic-plastic) of toughness are required; defines NDT, FTE, and FTP criteria
Through-thickness yielding	Structural steels	Elastic-plastic	CVN or K_{Ic}	Accounts for desired increase in toughness with increasing yield strength and plate thickness. Large thin plates with edge or through-thickness cracks.

TABLE 13.8. Comparison of Various Fracture Criteria (Continued)

Name	*General Use*	*Level of Performance (See Figure 12.1)*	*Test Specimen Used*	*Applications*
Leak-before-break	Primarily pressure vessels to ensure "leaking" before unstable crack growth	Elastic-plastic	K_{Ic} or K_c (nonstandard plane stress) *R*-curve (see Chapter 16)	Pressure vessels Design stress–flaw size calculations possible as a function of vessel thickness
Ratio analysis diagram	All materials; primarily used as a comparison of materials rather than as a criterion	All levels but primarily fully plastic	K_{Ic}, dynamic tear, or CVN (upper shelf values)	Limited to upper shelf behavior; for very high-strength materials that do not exhibit a transition behavior, this may be plane strain behavior.
The following three methods of analyzing the fracture toughness of materials are still in the research stage and are discussed in Chapter 16; for purposes of comparison, they are included in this table.				
Crack-opening displacement	All structural materials but primarily low- to medium-strength steels	Elastic-plastic to fully plastic	Crack-opening displacement bend specimen; British standard—similar to slow-bend K_{Ic} test specimen (E-399)	Widely used in United Kingdom and Japan; Alaskan Line pipe
R-Curve	Primarily high-strength sheet materials	Elastic-plastic	Oversize compact tension specimen	Primarily used in studies to measure resistance to stable crack extension; plane-stress behavior
J Integral	All structural materials; general extension of K_{Ic} into elastic-plastic behavior	Elastic-plastic (directly related to K_{Ic} in elastic region	Similar to K_{Ic} test specimens	Research extension of linear-elastic K_{Ic} behavior; quantitative stress–flaw size calculations possible

Examples of the use of these various criteria that have been developed for particular structural applications are discussed in Chapter 15.

REFERENCES

1. H. G. ACKER, "Review of Welded Ship Failures," *Ship Structure Committee Report, Serial No. SSC-63,* U.S. Coast Guard, Washington, D.C., Dec. 15, 1953.
2. D. P. BROWN, "Observations on Experience with Welded Ships," *Welding Journal,* Sept. 1952, pp. 765–782.
3. *Final Report of a Board of Investigation to Inquire into the Design and Methods of Construction of Welded Steel Merchant Vessels,* GPO, Washington, D.C., 1947.
4. F. JONASSEN, "A Resumé of the Ship Fracture Problem," *Welding Journal, Research Supplement,* June 1952, pp. 316-s–318-s.
5. M. L. WILLIAMS and G. A. ELLINGER, "Investigation of Fractured Steel Plates Removed from Welded Ships," *Ship Structure Committee Report, Serial No. NBS-1,* U.S. Coast Guard, Washington, D.C., Feb. 25, 1949.
6. M. L. WILLIAMS, M. R. MEYERSON, G. L. KLUGE, and L. R. DALE, "Investigation of Fractured Steel Plates Removed from Welded Ships," *Ship Structure Committee Report, Serial No. NBS-3,* U.S. Coast Guard, Washington, D.C., June 1, 1951.
7. M. L. WILLIAMS, "Examination and Tests of Fractured Steel Plates Removed from Welded Ships," *Ship Structure Committee Report, Serial No. NBS-4,* U.S. Coast Guard, Washington, D.C., April 2, 1953.
8. M. L. WILLIAMS and G. A. Ellinger, "Investigation of Structural Failures of Welded Ships," *Welding Journal, Research Supplement,* Oct. 1953, pp. 498-s–527-s.
9. M. L. WILLIAMS, "Analysis of Brittle Behavior in Ship Plates," *Ship Structure Committee Report, Serial No. NBS-5,* U.S. Coast Guard, Washington, D.C., Feb. 7, 1955. (Also presented in *ASTM Spec. Tech. Pub. No. 158,* Philadelphia, 1954).
10. E. R. PARKER, *Brittle Behavior of Engineering Structures,* prepared for the Ship Structure Committee, Wiley, New York, 1957.
11. W. S. PELLINI, 1971 AWS Adams Lecture: *Principles of Fracture-Safe Design,* Part I, *Welding Journal Research Supplement,* March 1971, pp. 91-s–109-s.
12. W. S. PELLINI, 1971 AWS Adams Lecture: *Principles of Fracture-Safe Design,* Part II, *Welding Journal Research Supplement,* April 1971, pp. 147-s–162-s.
13. S. T. ROLFE, J. M. BARSOM, and MAXWELL GENSAMER, "Fracture-Toughness Requirements for Steels" presented at the Offshore Technology Conference, Houston, May 18–21, 1969.
14. "Fracture Toughness Testing and Its Applications," *ASTM STP 381,* American Society for Testing and Materials, Philadelphia, 1965.

15. "Plane Strain Crack Toughness Testing," *ASTM STP 410*, American Society for Testing and Materials, Philadelphia, 1967.
16. G. T. HAHN and A. R. ROSENFIELD, "Sources of Fracture Toughness: The Relation Between K_{Ic} and the Ordinary Tensile Properties of Metals," *ASTM STP 432*, American Society for Testing and Materials, Philadelphia, 1968, pp. 5–32.
17. G. T. HAHN and A. R. ROSENFIELD, "Plastic Flow in the Locale of Notches and Cracks in Fe-3Si Steel Under Conditions Approaching Plane Strain," *Ship Structure Committee Report SSC-191*, U.S. Coast Guard, Washington Nov. 1968.
18. "Standard Method of Testing for Plane-Strain Fracture Toughness of Metals," *ASTM Standards, Part 10, E-399*, Philadelphia, 1974.
19. A. R. ROSENFIELD, P. K. DAI, and G. T. HAHN, *Proceedings of the International Conference on Fracture, Sendai, Japan, 1965.*
20. S. T. ROLFE and S. R. NOVAK, discussion of "What Does an Engineer Need to Know about Measurement of Fracture Toughness When Using High-Strength Structural Steels," ASME Metals Engineering Conference, Houston, April 2–5, 1967.
21. S. T. ROLFE and S. R. NOVAK, "Slow-Bend K_{Ic} Testing of Medium-Strength High-Toughness Steels," *ASTM STP 463*, American Society for Testing and Materials, Philadelphia, 1970.
22. G. R. IRWIN, "Relation of Crack-Toughness Measurements to Practical Applications," *ASME Paper No. 62-MET-15*, presented at Metals Engineering Conference, Cleveland, April 9–13, 1962.
23. J. H. GROSS and R. D. STOUT, "Steels for Hydrospace," *Ocean Engineering*, Vol. 1, 1969.
24. J. M. BARSOM and S. T. ROLFE, "Fatigue and Burst Analysis of HY-140(T) Steel Pressure Vessels," *Journal of Eng. for Industry*, ASME Trans., Vol 92, Series B, No. 1, Feb. 1970.
25. J. E. SRAWLEY and J. B. ESGAR, "Investigation of Hydrotest Failure of Thiokol Chemical Corporation 260-Inch-Diameter Sl-1 Motor Case," *NASA TM X-1194*, Cleveland, Jan. 1966.
26. G. R. IRWIN, J. M. KRAFFT, P. C. Paris, and A. A. WELLS, "Basic Aspects of Crack Growth and Fracture," *NRL Report 6598*, Washington, D.C. Nov. 21, 1967.

14

Fracture-Control Plans

14.1. Introduction

The objective in structural design of large complex structures such as bridges, ships, pressure vessels, aircraft, etc., is to optimize the desired performance requirements relative to cost considerations (i.e., the overall cost of materials, design, fabrication, operation, and maintainance). In other words, the purpose of engineering design is to produce a structure that will perform the operating function efficiently, economically, and safely. To achieve these objectives, engineers make predictions of service loads and conditions, calculate stresses in various structural members resulting from these loads and service conditions, and compare these stresses with the critical stresses in the particular failure modes that may lead to failure of the structure. Members are then proportioned and materials specified so that failure does not occur by any of the pertinent failure modes. Because the response to loading can be a function of the member geometry, an iterative process may be necessary.

Possible failure modes that usually are considered are

1. General yielding or excessive plastic deformation.
2. Buckling or general instability, either elastic or plastic.
3. Subcritical crack growth (fatigue, stress-corrosion, or corrosion fatigue) leading to loss of section or unstable crack growth.
4. Unstable crack extension, either ductile or brittle, leading to either partial or complete failure of a member.

Although other failure modes exist, such as corrosion or creep, the above-mentioned failure modes are the ones that usually receive the greatest attention. Furthermore, of these four failure modes, engineers usually concentrate on only the first two and assume that proper selection of materials and design stress levels will prevent the other two failure modes from occurring. This reasoning is not always true and has led to several spectacular structural failures. In good structural design, *all* possible failure modes should be considered. In this particular textbook, failure by fracture or subcritical crack growth by fatigue, stress-corrosion, or corrosion fatigue are the dominant failure modes to be considered. However, it is assumed that the engineer looks at all possible failure modes and designs structures to prevent failure by any of the possible failure modes.

In the case of brittle fracture or fatigue, many of the fracture-control guidelines that have been followed to minimize the possibility of brittle fractures in structures are familiar to structural engineers. These guidelines include the use of structural materials with good notch toughness, elimination or minimization of stress raisers, control of welding procedures, proper inspection, etc. When these general guidelines are integrated into specific requirements for a particular structure they become part of a fracture-control plan. A fracture-control plan is therefore a specific set of recommendations developed for a particular structure and should not be indiscriminantly applied to other structures.

The four basic elements of a fracture-control plan are as follows:

1. *Identification* of the factors that may contribute to the fracture of a structural member or to the failure of an entire structure. Description of service conditions and loadings.
2. *Establishment* of the relative contribution of each of these factors to a possible fracture in a member or to the failure of the structure.
3. *Determination* of the relative efficiency and trade-offs of various design methods to minimize the possibility of either fracture in a member or failure of the structure.
4. *Recommendation* of specific design considerations to ensure the safety and reliability of the structure against fracture. This would include recommendations for desired levels of material performance as well as material selection, design stress levels, fabrication, and inspection.

The development of a fracture-control plan for large complex structures, such as bridges, airplanes, ships, etc., is very difficult. Despite the difficulties, attempts to formulate a fracture-control plan for a given application, even if only partly successful, should result in a better understanding of the possible fracture characteristics of the structure under consideration.

The total useful life of a structural component is generally determined by the time necessary to initiate a crack and to propagate the crack from subcritical dimensions to the critical size. The life of the component can be prolonged by extending the crack-initiation life and/or the subcritical-crack-propagation life. Consequently, the crack-initiation, subcritical-crack-propagation, and unstable-crack-propagation (fracture) characteristics of structural materials, as well as their fracture behavior, are primary considerations in the formulation of fracture-control guidelines for structures.

Unstable crack propagation is the final stage in the useful life of a structural component subject to failure by the fracture mode. This stage is governed by the material toughness, the crack size, and the stress level. Consequently, unstable crack propagation cannot be attributed *only* to material toughness, or *only* to stress conditions, or *only* to poor fabrication, but rather to particular combinations of the above factors. However, if any of these factors are significantly different from what is usually obtained in a particular type of structure, the possibility of failure may increase markedly for most structures.

Structural materials that have adequate notch toughness to prevent brittle fractures at service temperatures and loading rates are available. However, when these structural materials are used in conjunction with inadequate design or poor fabrication, or both, the safety and reliability of a structural component cannot be guaranteed because the useful life of a structural component is governed by the time required to initiate a crack and to propagate the crack to terminal conditions (for example, leaks or unstable crack propagation). Thus, the useful life of a structural component depends on the magnitude and fluctuation of the applied stress, the magnitude of the stress-concentration factors and the size, shape and orientation of the initial discontinuity, on the stress-corrosion susceptibility, the fatigue characteristics, and the corrosion-fatigue behavior of the structural material in the environment of interest. Because most of the useful life is expended in initiating and propagating cracks at low values of the stress-intensity factor (Chapters 7–9), an increase in the fracture toughness of the steel may have a negligible effect on the useful life of a structural component whose primary mode of failure is fatigue. Despite these facts, oversimplification of failure analyses has led some to advocate the philosophy that structures should be fabricated of "forgiving materials." Such materials are characterized as having sufficient notch toughness to fracture in a ductile manner at operating temperatures and under impact loading even though the structure may be loaded only for slow or intermediate loading rates. The use of these materials is advocated to "forgive" any mistakes that may be committed by the fabricator, inspector, designer, and user. While this approach to ensure the safety and reliability of structures does work most of the time, it perpetuates a false sense of security, places an unjustifiable burden on the structural material, and unnecessarily increases the cost of the structure.

Another prevalent yet unfounded philosophy advocates that the primary cause of fractures in welded structures is the inherent inferior characteristics of the weldments. These characteristics include yield-strength residual stresses, and weld discontinuities. Advocates of this philosophy often neglect to consider environmental effects, cyclic history, stress redistribution caused by load fluctuation or proof tests, and when an obvious weld discontinuity does not exist in the vicinity of the fracture origin, the cause of failure is attributed to "microweld" defects. Although residual stress and weld discontinuities can contribute to failure, the above oversimplification can lead to an incorrect fracture analysis of a structural failure. The use of oversimplification or gross exaggerations in the analysis of failures, exemplified by the above philosophies, can lead to erroneous conclusions. Correct diagnosis and preventive action can be established only after a thorough study of *all* the pertinent parameters related to the specific problem under consideration. An integrated look at all these parameters and their synergistic effect on the safety and reliability of a structure is necessary to develop a fracture-control plan.

As pointed out earlier, the recent development of fracture mechanics has been extremely helpful in synthesizing the various elements of fracture-control plans into more unified quantitative plans than was possible previously. Specifically, fracture mechanics has shown that although numerous factors (e.g., service temperature, material toughness, design, welding, residual stresses, etc.) can contribute to brittle fractures in large welded structures, there are three primary factors that control the susceptibility of a structure to brittle fracture, namely

1. Notch toughness of a material at a particular service temperature, loading rate, and plate thickness,
2. Size of crack or discontinuity at possible locations of fracture initiation; and
3. Tensile stress level, including residual stress.

All three factors can be interrelated using concepts of fracture mechanics to predict the susceptibility of a structure to brittle fracture. When the particular combination of stress and crack size in a structure reaches the critical stress-intensity factor for a particular specimen thickness and loading rate, fracture can occur. It is the specific intent of any fracture-control plan to establish the possible ranges of K_I that might be present throughout the lifetime of a structure because of the various service loads and to ensure that the critical stress-intensity factor (K_c, K_{Ic}, K_{Id}, or other critical values such as K_{Iscc}) for the materials used is sufficiently large so that the structure will have a safe life.

One of the key questions in developing a fracture-control plan for any particular structure is how large must the degree of safety and reliability be for the particular structure in question. The degree of safety and reliability needed, sometimes referred to as the factor of safety, is often specified by a code. However, the degree of safety depends on many additional factors such as consequences of failure or redundancy and thus varies even within a generic class of structures.

Accordingly a fracture-control plan is developed only for the specific structure under consideration and can vary from one which must, in essence, provide assurance of very low probability of service failures to one which may allow for occasional failures during manufacturing or service. An example of the former situation would be a nuclear power plant structure where the consequences of a structural failure may not be tolerable. In the latter case, the consequences of failure might be minimal, and it would be more efficient and economical to periodically maintain and replace parts rather than design them so that no failures occur. An example of a situation where a service failure could be tolerated might be that of the loading bed of a dump truck where periodic inspection would indicate when plates would need to be repaired or replaced because the consequences of failure would be minor.

In commenting on fracture-control plans, Irwin[1] has noted that "For certain structures, which are similar in terms of design, fabrication method, and size, a relatively simple fracture control plan may be possible, based upon extensive past experience and a minimum adequate toughness criterion. It is to be noted that fracture control never depends solely upon maintaining a certain average toughness of the material. With the development of service experience, adjustments are usually made in the design, fabrication, inspection, and operating conditions. These adjustments tend to establish adequate fracture safety with a material quality which can be obtained reliably and without excessive cost."

14.2. Historical

Prior to about 1940, fracture-control plans or fracture-safety guidelines did not really exist in a formalized sense because the number of brittle fractures was small. Most large structures were built out of low-strength materials using thin, riveted plates with the structural members arranged so that in the event of failure of one plate the fracture was usually arrested at the riveted connections. Thus although there were some exceptions as noted in Chapter 1, most failures were not catastrophic. While it is true that the number of failures in particular structural situations was reduced by various design modifications based on experience from any service failures, the first general overall fracture-control guideline was merely to use lower design stress levels. One of the early fracture-control applications was in the boiler and pressure-vessel industry where the allowable stress was decreased continually as a certain percentage of the maximum tensile stress, thereby decreasing the number of service failures in succeeding years.

A second general guideline to fracture control was to eliminate stress concentrations as much as possible. In the early 1940s brittle fracture as a potentially serious problem for large-scale structures was brought to the designer's attention by the large number of World War II ship failures. Of the approximately 5,000 merchant ships built during World War II, over 1,000 had developed cracks of considerable size by 1946. Between 1942 and 1952, more than 200 ships had fractures classified as serious, and at least 9 tankers and 7 Liberty ships had broken completely in two as a result of brittle fractures. The majority of fractures in the Liberty ships started at the square hatch corners of square cutouts at the top of the shear strake. The design changes involving rounding and strengthening of the hatch corners, removing square cutouts of the shear strake, and adding rivets and crack arresters in various locations led to an immediate reduction in the incidence of failures. The use of crack arresters and improved workmanship in the tankers reduced the incidence of failures in these vessels. Thus the second general type of fracture-control guideline was that of design improvement to minimize stress concentrations.

The ship-failure problem called attention to general inadequacies of designing to prevent fracture primarily by just limiting stresses or minimizing stress concentrations. The World War II shipbuilding program was the first large-scale use of welding to produce monolithic structures where cracks could propagate continuously from one plate to another, in contrast to the multiple plate usage for riveted structures. While problems of fracture were apparent prior to the large-scale use of welding, there was evidence that the nature of welded structures provided for continued extension of a fracture, which can result in total failure of a structure. Previous experience with riveted structures indicated that brittle fractures were usually limited to a single plate.

The third general type of fracture control was to improve the notch toughness of materials. The first material-control guideline occurred about the late 1940s and early 1950s following extensive research on the cause of the ship failures. The particular fracture-control guideline was the observation that plates having a CVN impact energy value greater than 10–15 ft-lb at the service temperature did not exhibit complete fracture but rather arrested any propagating cracks. Studies indicated that generally the ship fractures experienced during World War II did not occur at temperatures above the 15-ft-lb Charpy V-notch impact energy transition temperature. This observation led to the development of the 15-ft-lb transition-temperature criterion as a means of fracture control. Early in the 1950s development of the nil-ductility transition- (NDT) temperature test led to another general fracture-control guideline for structural materials, namely that structural steels have an NDT temperature below their service temperature when loaded dynamically.

Since that time, and particularly during the 1960s, the development of general criteria for prevention of brittle fractures, or at least analyses of methods for designing to prevent brittle fractures, have been numerous. These included the development of the fracture analysis diagram (FAD) by the Naval Research Laboratory, the early use of fracture-mechanics concepts in the Polaris Missile Motor Case failures, and the turbine-rotor generator spin test analyses. In the 1960s and 1970s, rapid development of fracture mechanics as a research tool and eventually as an engineering tool led to the use of fracture-mechanics concepts in the development of fracture criteria for several structures such as nuclear pressure vessels, various aircraft structures, and bridges. However, the three basic elements of fracture control, i.e., (1) use lower design stress, (2) minimize stress concentrations, and (3) use materials with improved notch toughness, have long been known to engineers. Fracture mechanics' basic contribution is to make these guidelines *quantitative* and to show the relative importance of each of these elements.

Although the literature has been dominated with the use of fracture mechanics as a research tool, Irwin[1] has stated that "the practical objective of fracture mechanics is a continuing increase in the efficiency with which undesired fractures of structural components are prevented. The fracture-control plans most commonly used in the past have not possessed a high degree of efficiency. Generally the visible portions of such plans consist in statements of minimum required toughness and in statements of inspection standards. No quantitative connection between these two fracture control elements is employed in establishing these statements. The unseen parts consist mainly in adjustments of design and fabrication. After enough years of experience so that further adjustments of design and fabrication are unnecessary, the fracture control plan is complete. The plan then consists of certain minimum toughness requirements and adherence to certain inspection requirements plus the state-of-art methods of design and fabrication which fracture failure experience showed to be desirable or necessary. Proof testing has the great advantage, where employed, because much of the fracture failure experience necessary for development of the plan tends to occur in the proof test rather than in service. The use of transition temperature based measurements rather than fracture mechanics based measurements of fracture toughness is not a significant disadvantage to the reliability of such plans, once established. However, the efficiency may be low and use of fracture mechanics methods would be helpful in designing modifications toward improved efficiency."

At present there are no formalized procedures for developing fracture-control plans, although several have been developed. General procedures for the development of fracture-control plans are outlined in Section 14.3. Examples of fracture-control plans either proposed or currently being used for specific structural applications are described in Chapter 15. The remainder of Chapter 14 will be devoted to a general discussion of how fracture-control plans are developed as well as to a description of the various items that go into a fracture-control plan.

14.3. Development of a Fracture-Control Plan

In Section 14.1, the four general elements of a fracture-control plan to prevent failure by the fracture mode of failure were presented. It should be reemphasized that *all* possible failure modes, e.g., buckling, yielding, corrosion, etc., should be considered during a structural design. Textbooks on design of particular types of structures (bridges, buildings, pressure vessels, aircraft, etc.) should be used to design to prevent failures by buckling, yielding, etc., and metallurgical textbooks should be consulted to design to prevent chemical failures such as by corrosion. The purpose of this textbook is to provide technical information and design guidelines that can be used to prevent failure by fracture or subcritical crack growth leading to fracture.

The four general elements of a fracture-control plan are described in as much detail as possible, realizing that the specific details of any fracture-control plan depend on the particular structure being analyzed.

14.3.1. Identification of the Factors that May Contribute to the Fracture of a Structural Member or to the Failure of an Entire Structure. Description of Service Conditions and Loadings

The first step in all structural design is to establish the probable loads and service conditions throughout the design life of a structure. Usually the live loads are specified by codes [e.g., American Association of State Highway and Transportation Officials (AASHTO) for bridges] or performance criteria such as operating pressures in pressure vessels or payload requirements in aircraft structures. These types of loadings usually are reasonably well defined, although there are certain types of structures such as ships where the loading is not well defined. Nonetheless, the first step is to establish the probable loads.

From a fracture toughness viewpoint, a major decision is establishing the *rate* at which these loads are applied since this determines whether K_{Ic} for slow loading, or a K_{Ic} value for an intermediate loading rate, or K_{Id} for impact loading should be the controlling toughness parameter.

Wind loads, hurricane loadings, sea-wave loadings, etc., are established by field measurements in the particular location of the structure in the world. Also field measurements on similar types of structures are used to estimate the effect of these loads on structural response and behavior. Earthquake loadings for particular regions are based on experience plus various code requirements, as well as a judgment factor related to the degree of conservatism desired for a particular structure.

Repeated or fatigue loading is usually established by the particular design function of the structure. That is, bridges are subjected to fatigue by the movement of vehicles, and thus fatigue must be considered in the design of bridges, whereas the loads on buildings are primarily dead loads so fatigue is usually neglected. Fatigue loadings can be constant amplitude, such as in rotating machinery, or variable amplitude, such as in bridges or aircraft structures. Regardless of the type of loading, fatigue can result in subcritical crack growth by various means, as described in Chapters 7–11. Thus even through the initial flaw size may be small (based on quality of fabrication), the possibility of larger cracks is present when the structure is subjected to repeated loading.

Chemical or environmental factors such as corrosion, stress-corrosion, cavitation, etc., must be considered for various structures depending on the particular design function of the structure. Crack growth by stress-corrosion was described in Chapter 10.

From a fracture viewpoint, temperature can have a significant effect on the service behavior of structural members. For those structures fabricated from materials that exhibit a brittle to ductile transition in behavior (primarily the low- to medium-strength structural steels), the minimum service temperature must be established.

Quality of fabrication which generally controls the initial crack or defect size is another factor that should be established so that some estimate of possible initial flaw sizes can be made.

Obviously, the inherent notch toughness (K_{Ic} or K_{Id}) of a structural material based on the particular chemistry, heat treatment, etc., is a primary factor that may contribute to the fracture behavior of the material and is mentioned last only because the other factors are often slighted.

The point is that for each structure for which a fracture-control plan is desired, the designer should consider all possible loadings and service conditions *before* the selection of materials or allowable stresses. In this sense, the basis of structural design in all large complex structures should be an attempt to optimize the *desired performance requirements* in terms of the service loadings relative to *cost considerations* so that the probability of failure is low. Accordingly, if the failure mode is brittle fracture, then the K_{Ic}/σ_{ys}, K_c/σ_{ys}, or K_{Id}/σ_{yd} ratio (at the appropriate loading rate and plate thickness) should be selected to minimize the probability of fracture. However, if desired performance requirements are such that the overall weight of the structure must be minimized, then *another* possible "failure mode" is *nonperformance* because of excessive weight, and therefore a higher allowable design stress must be used. Because the allowable design stress is usually some percentage of the yield strength, a higher-yield-strength material usually is required.

These two requirements of (1) a high K_{Ic}/σ_{ys} ratio and (2) a high yield strength (σ_{ys}) are conflicting, and a compromise often must be reached. However, as long as this analysis is made *prior to* material selection and establishment of design or allowable stresses, the designer has a good chance of achieving a "balanced design" in which the *performance requirements* as well as the *prevention of failure requirements* both are met as economically as possible. A design example for selection of a material for a specific pressure-vessel application based both on performance and minimum weight was presented in Chapter 5 and showed the advantages of the fracture-mechanics approach to design.

Often it is not possible to change materials for a particular structure because of existing codes, past practices, or economics. In these cases, a limited fracture-control plan can still be effective by reducing the design stress levels, restricting the range of operating temperatures, improving fabrication and inspection, etc. Thus, even though it is desirable to develop complete fracture-control plans during the early design stages, it is not always possible to do so. However, considerable benefits can still be realized by limited fracture-control plans developed at later stages in the design of a structure. In all cases, the designer should identify as closely as possible all service conditions and loadings to which the structure will be subjected.

14.3.2. Establishment of the Relative Contribution of Each of These Factors to a Possible Fracture in a Member or to the Failure of the Structure

Of the many factors that can contribute to fracture in large welded structures (e.g. service loadings, fatigue, quality of fabrication, loading rate, material toughness, etc.) the science of fracture mechanics has shown that there are three primary factors that control the susceptibility of a structure to brittle fracture, namely (1) tensile stress level (σ), (2) material toughness (preferably measured in terms of K_c, K_{Ic}, or K_{Id}), and (3) crack size (a). As shown schematically in Figure 14.1, these factors can be related quantitatively using the K_I relationships developed in Chapter 2.

The contribution of the various types of loadings to the possibility of a brittle fracture occurs primarily in the calculation of the maximum value of stress which can occur in the vicinity of a crack. The calculation of these

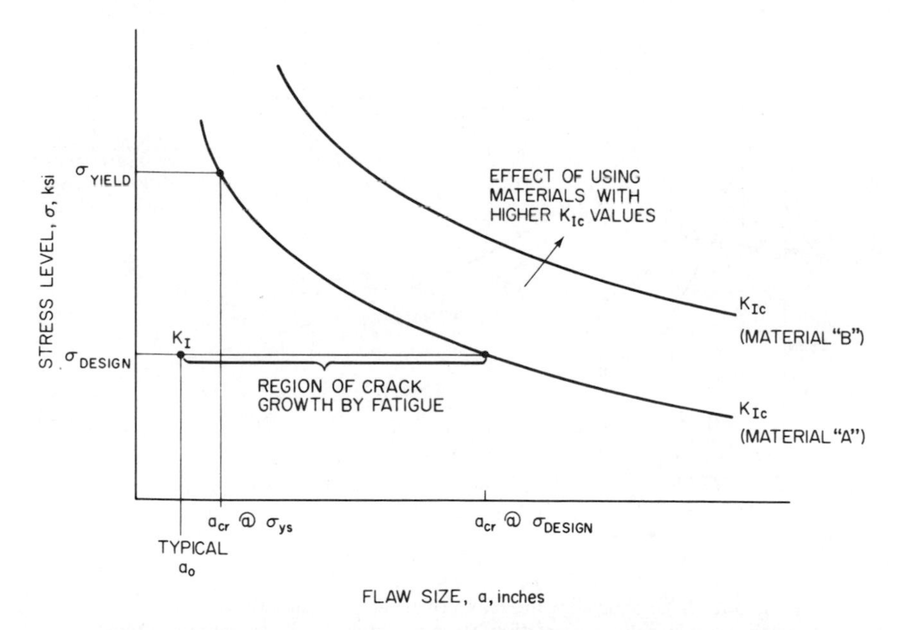

FIG. 14.1. Schematic showing relation among stress, critical flaw sizes, and material toughness.

stresses ranges from simple calculations of P/A or $M \cdot y/I$ to extremely complex elasticity solutions for various structural shapes such as plates, shells, box girders, etc. The task of calculating nominal stresses has been made easier by digital computers and new methods of stress analysis, such as the finite element method.

For welded construction, the possibility of residual stresses exists, and the local stress level can be of yield-point magnitude. In fact, it is often assumed that local yielding exists in the vicinity of stress concentrations. The ductility of structural materials is relied upon to redistribute these stresses so that premature failure does not exist. However, if the possibility of brittle fracture exists, the designer may want to assume that yield-point loading is present in portions of the structure where cracks can be present and to calculate the resistance to fracture accordingly. That is, determine the critical crack size at yield stress loading (Figure 14.1) and compare it with the maximum possible flaw size based on fabrication and inspection capabilities. However, local yielding can occur in the vicinity of stress concentrations and in regions of tensile residual stresses such as for weldments. This local yielding causes a redistribution of stresses such that during subsequent loadings those regions may exhibit a purely elastic-stress behavior.

Figure 14.2 is a schematic showing the effect of local residual stresses on crack growth as well as the effects of plane-strain or plane-stress conditions

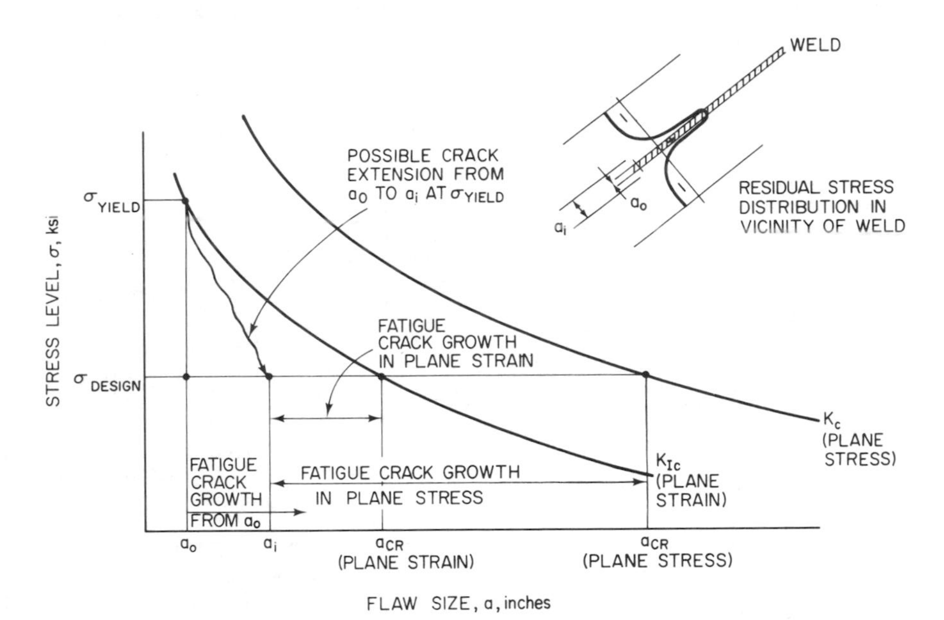

FIG. 14.2. Schematic showing effect of local residual stresses and plane-strain to plane-stress transition (loss of constraint) on fatigue crack growth.

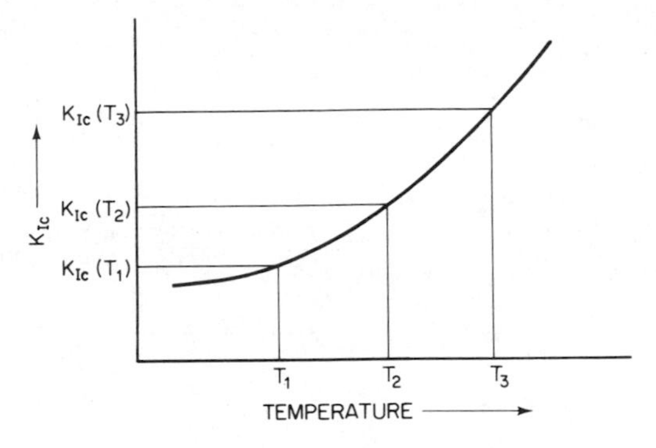

(a) EFFECT OF TEMPERATURE ON K_{Ic}

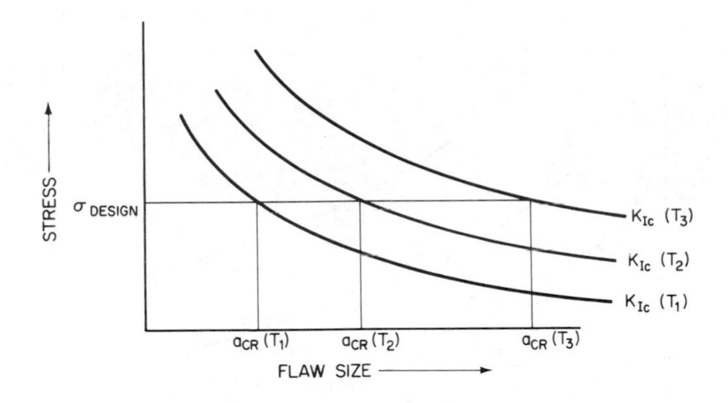

(b) EFFECT OF K_{Ic} ON CRITICAL FLAW SIZE

FIG. 14.3. Schematic showing effect of temperature on critical flaw size.

on subsequent fatigue crack growth. Figure 14.3 is a schematic showing the effect of service temperature on critical flaw size, and Figure 14.4 is a schematic showing the design use of K_{Iscc}. Referring to Figures 14.1–14.4, the following general conclusions can be made with respect to fracture control regarding the relative contribution of stress level, material toughness, and crack size:

1. In regions of high residual stress, where the actual stress can equal the yield stress over a small region, the critical crack size is computed for σ_{yield} instead of the design stress, σ_{design}. If, for the particular structural material being used, both the base metal and the weld metal are sufficiently tough (e.g. material B in Figure 14.1), the critical crack size for full yield stress loading should be satisfactory. Under fatigue loading, a crack can grow out of the residual stress zone, and the

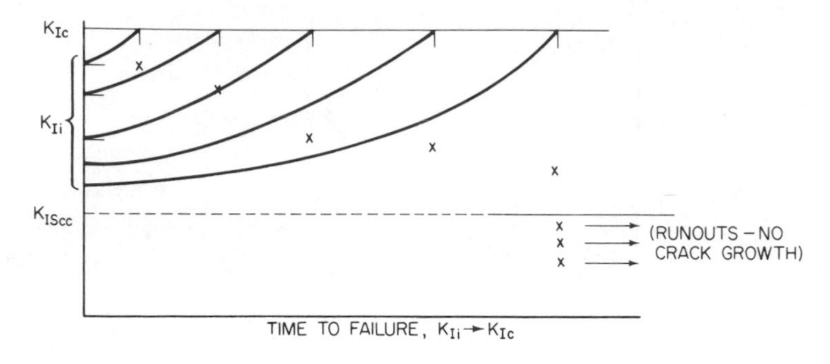

(a) RELATION BETWEEN K_{Ii}, K_{Ic}, K_{Iscc}

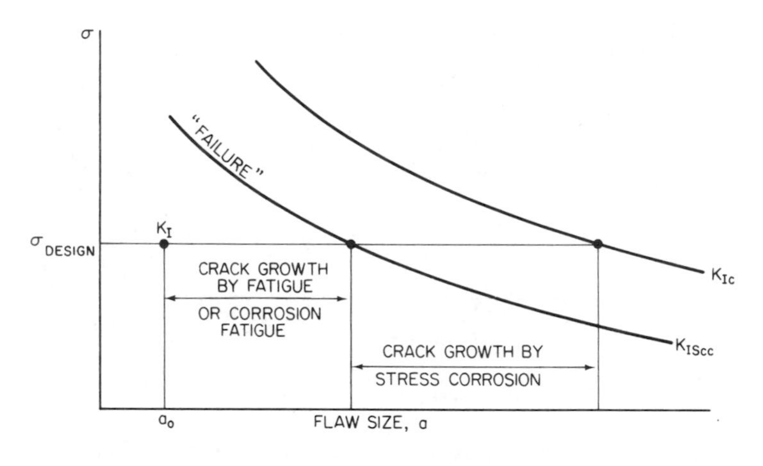

(b) RELATION BETWEEN CRACK GROWTH BY FATIGUE OR CORROSION FATIGUE AND BY STRESS CORROSION

FIG. 14.4. Schematic showing design use of K_{Iscc}.

critical crack size becomes the value at the design stress level. Note that the "critical crack size" in a particular material is dependent on the particular design stress level and therefore is *not* a material constant. However, this conservative approach does not account for redistribution and shakedown of stresses that occur during proof tests, erection, or initial loading, or for crack blunting caused by high residual stresses.

2. If the level of toughness of the material is sufficiently high, any crack which does initiate in the presence of residual stresses (a_0, Figure

14.2) should arrest quickly as soon as the crack propagates out of the region of high residual stress. However, the initial crack size for any subsequent fatigue crack growth will be fairly large (a_i, Figure 14.2).

3. For the design stress level, determine the calculated critical crack size. If it is large, compared for example to the plate thickness, subcritical crack growth (by fatigue) should lead to relaxation of the stress ahead of the crack, resulting in plane-stress or elastic-plastic behavior. For this case the critical stress-intensity factor for plane stress, K_c, will be greater than K_{Ic} or K_{Id}, which is an additional degree of conservatism (Figure 14.2).
4. Materials with low toughness values can still be used if the applied stresses under tensile loads are reduced, or if residual compressive stresses applied to the component by shot-peening or induction case hardening are used to inhibit crack initiation such as the case for gear teeth or landing gears, or if the distribution of stresses causes the initiation and propagation of cracks in a decreasing stress field, or if the crack orientation is not in a critical plane to cause unstable crack extension such as the case for "shelling" in rails.
5. The relative contribution of temperature is, of course, to establish the particular level of toughness to be used in calculating the stress–crack size trade-offs. That is, as shown in Figure 14.3(a), K_{Ic} (or K_{Id}) can vary with temperature for certain structural materials. Thus the K_{Ic} values at temperatures T_1, T_2, and T_3 are different and lead to different values of critical flaw size even though the design stress remains constant [Figure 14.3(b)].
6. The relative effect of fatigue loading is to grow a crack of initial size a_0 to a_{cr} (Figure 14.1). The number of cycles (or time in the case of stress-corrosion) required to grow an existing crack to critical size was discussed in Chapters 7–11.
7. Subcritical crack growth by stress-corrosion was discussed in Chapter 10. For materials susceptible to stress corrosion, a good design practice for sustained-load applications is to use K_{Iscc} as a limiting design curve (rather than K_{Ic}) so that "failure" is defined as the start of stress-corrosion crack growth, as shown in Figure 14.4(b). This conservative practice is followed because once stress-corrosion crack growth is started, it is only a matter of time until the K_I level reaches K_{Ic} and complete failure occurs. That is, fatigue-crack growth is usually considered to be deterministic because the number of cycles of loading can be estimated. However, the rate of stress-corrosion crack growth is extremely difficult to predict because of the large number of variables such as chemistry of the corrodent, concentration

of corrodent at the crack tip, temperature, etc. Thus, once the K_{Iscc} level is reached, a realistic design approach for sustained-load applications is to consider failure to be imminent.

Using the above guidelines, the relative contribution of material properties (K_{Ic}, K_c, K_{Id}, K_{Iscc}, da/dN), stress level (σ), and crack size (a) can be established for a given set of service conditions for a particular structure.

14.3.3. Determination of the Relative Efficiency and Trade-Offs of Various Design Methods To Minimize the Possibility of Either Fracture in a Member or Failure of the Structure

Previously, it has been established that the three primary factors that control the susceptibility of a structure to brittle fracture are

1. Tensile stress level.
2. Size of crack or severity of discontinuity.
3. Material toughness at the particular temperature, loading rate, and plate thickness.

Thus there are three general design approaches to minimizing the possibility of brittle fracture in a structural member, and each of these is directly related to the above factors:

1. Decrease design stress.
2. Minimize initial discontinuities.
3. Use materials with improved notch toughness.

Each of these design approaches has been used by engineers in various types of structures for many years. The science of fracture mechanics merely makes the process more quantitative.

For example, Figure 14.5 shows that for the same quality of fabrication and inspection as well as the same critical material toughness, K_{Ic} or K_c, reducing the design stress or stress fluctuation leads to a new margin of safety. This new margin of safety can be either a larger margin of safety against fracture or an increased fatigue life because of the possibility of larger subcritical crack growth before failure. Specific examples of this fact were presented in Chapter 5.

Figure 14.6 shows the general effect of improving the quality of fabrication and inspection while using the same design stress level and material. Finally Figure 14.7 shows the general effect of using a structural material with improved notch toughness and the same design stress and quality of fabrication. These examples lead to the general conclusion that because $K_{Ic} \simeq C\sigma\sqrt{a}$, it would be expected that increasing K_{Ic} or decreasing σ would have a greater effect on the resistance to fracture than reducing the initial crack size, a_0. Also, it is usually easier to control σ or K_{Ic} than a_0.

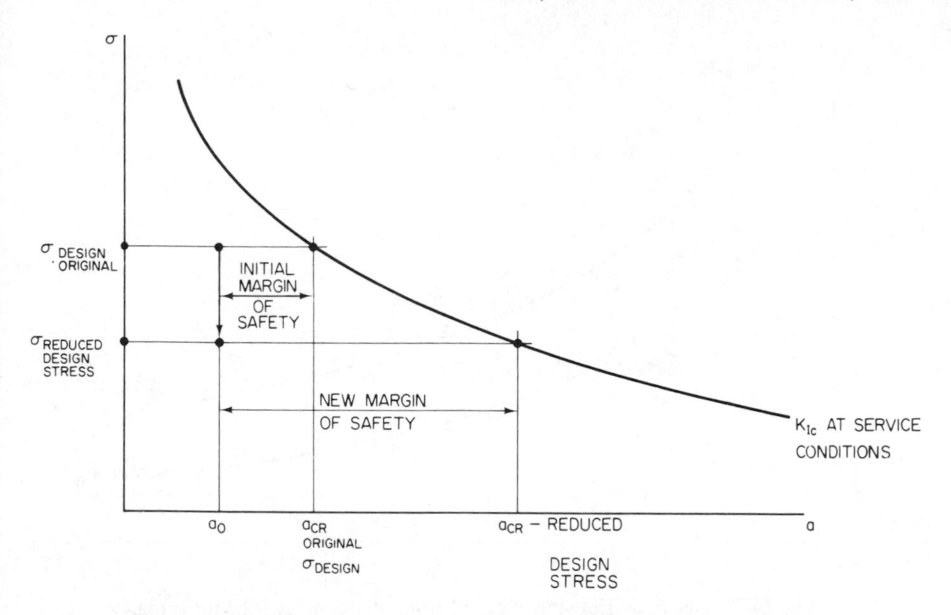

FIG. 14.5. Schematic showing effect of lowering the design stress on fracture control.

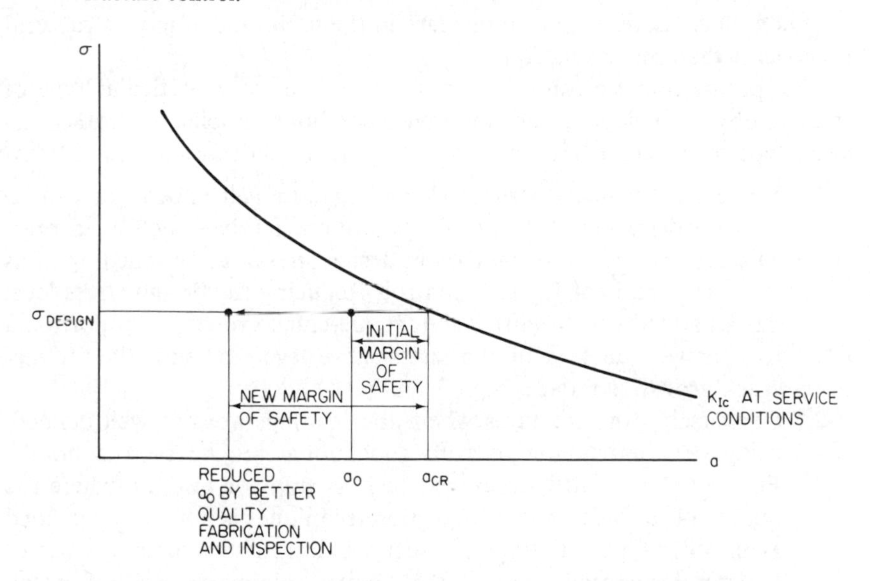

FIG. 14.6. Schematic showing effect of reducing the initial flaw size on fracture control.

However, because the primary parameter that governs the rate of subcritical crack growth is the stress-intensity factor, K_I, or the stress-intensity factor range, ΔK_I, raised to a power greater than 2, decreasing σ or $\Delta\sigma$ should

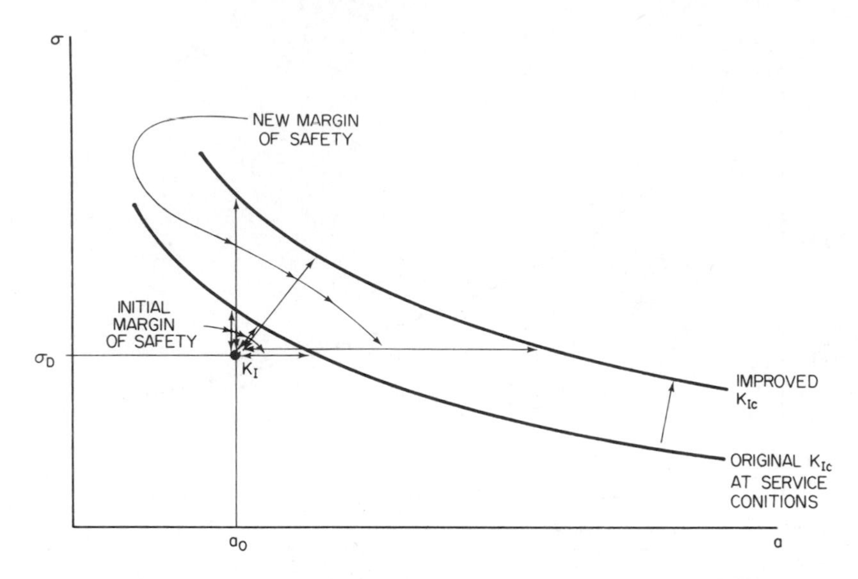

FIG. 14.7. Schematic showing effect of using material with better toughness (improved K_{Ic}) on fracture control.

result in a much more significant increase in the useful life of most structural components than increasing K_{Ic}.

The specific fracture-control approach followed for a particular type of structure obviously depends on the service conditions applicable to the particular type of structure. However, several general guidelines are as follows:

1. For cases where the structural loadings are well known, or can be controlled through the use of pressure relief valves such as in pressure vessels, moderate changes in design stress can be relied upon as a primary basis of fracture control. Reducing the design stress level has an added benefit in that the fatigue-crack growth is proportional to a power function of the stress-range level and thus the fatigue-crack growth is reduced significantly.
2. Conversely, for structures where the loadings are not well defined, using structural materials with good notch toughness is desirable.
3. For complex welded structures, such as ships or bridges, where the quality of fabrication and inspection techniques is not so well defined as in other types of structures such as aircraft structures, the use of structural materials with some known minimum level of notch toughness is very desirable.
4. If the design stress levels are well known and the quality of fabrication and inspection are able to be controlled quite closely, such as in the manufacture of small pressure vessels, then materials with relatively low notch toughness can be used safely.

5. For structures where the consequences of failure are such that failures cannot be tolerated, such as the pressure vessel in a nuclear reactor, or the hull structure of a floating nuclear power plant, or the main structure in an airplane, then *all three* factors should be controlled.

For those cases where fatigue crack growth is a consideration, such as in bridges, the total useful design life of a structural component can be estimated from the time necessary to initiate a crack and to propagate the crack from sub-critical dimensions to the critical size. The life of the component can be prolonged by extending the crack-initiation life and/or the sub-critical-crack-propagation life. In an engineering sense, the initiation stage is that region in which a very small initial discontinuity or crack grows to become a measurable propagating crack in fatigue. The sub-critical crack growth stage is that region in which a propagating fatigue crack follows one of the existing crack-growth "laws," e.g., $da/dN = A \cdot \Delta K^n$ as described in Chapter 8. The unstable crack growth stage is that region in which either fatigue crack growth is *very* rapid, or a brittle fracture occurs, or ductile tearing occurs, resulting in loss of section and failure.

The effect of each of the three primary factors that control the susceptibility of a structure to failure (tensile stress level (σ), flaw size (a), and material toughness (K_{Ic}, K_{Id}, CVN, etc.)) on the total life of a structure subjected to fatigue loading may be summarized as follows:

a. *Tensile Stress* — Large effect on life (Region I—Fig. 14.8) because the rate of fatigue crack growth is decreased significantly as the applied stress range is decreased (σ_1 compared with σ_2—Fig. 14.8). Design stress range ($\sigma_{max} - \sigma_{min}$) is the primary factor to control.

b. *Flaw Size* — Large effect on life (Region II—Fig. 14.8) because the rate of fatigue crack growth for *small* flaws is very low. Quality of fabrication and inspection is the primary factor to control.

c. *Material Toughness* —
   i) Large effect on life in moving from plane-strain behavior to elastic-plastic behavior (Region III—Fig. 14.8). The AASHTO Material Toughness Requirements insure this level of performance under intermediate rates of loading. The three levels of performance, namely plane-strain, elastic-plastic, and plastic behavior, are shown schematically in Fig. 14.9. For most structural applications, some moderate level of elastic-plastic behavior at the service temperature and loading rate constitutes a satisfactory performance criterion.
   ii) *Small effect* on life in moving from elastic-plastic

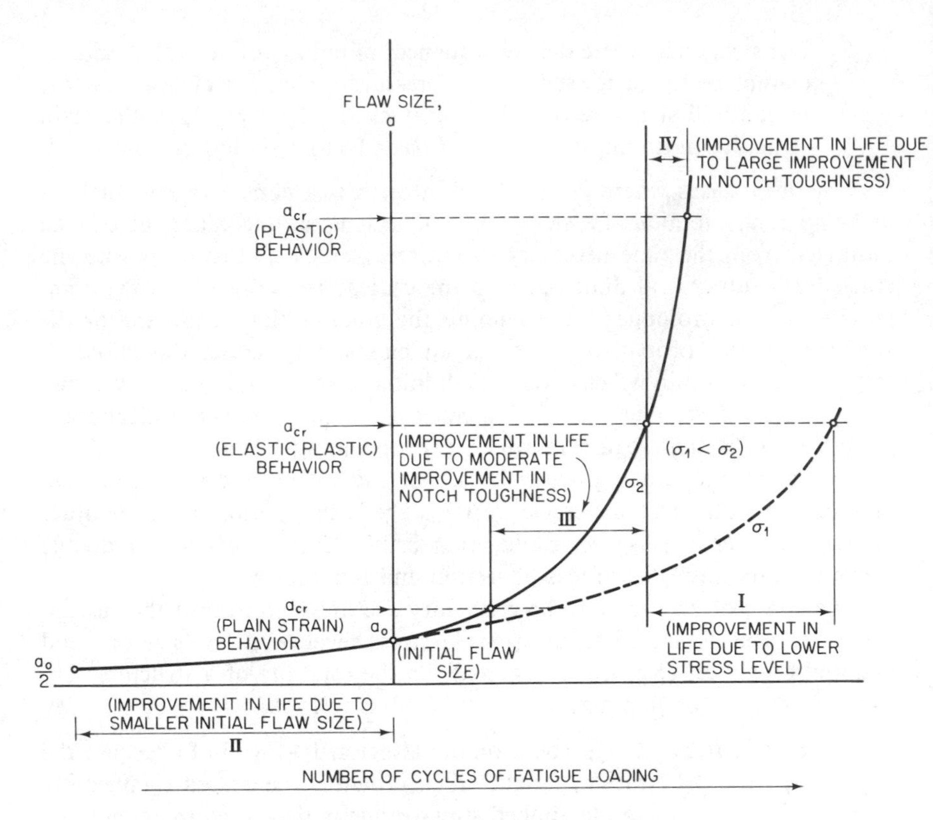

FIG. 14.8. Schematic showing effect of notch toughness, stress, and flaw size on improvement of life of a structure subjected to fatigue loading.

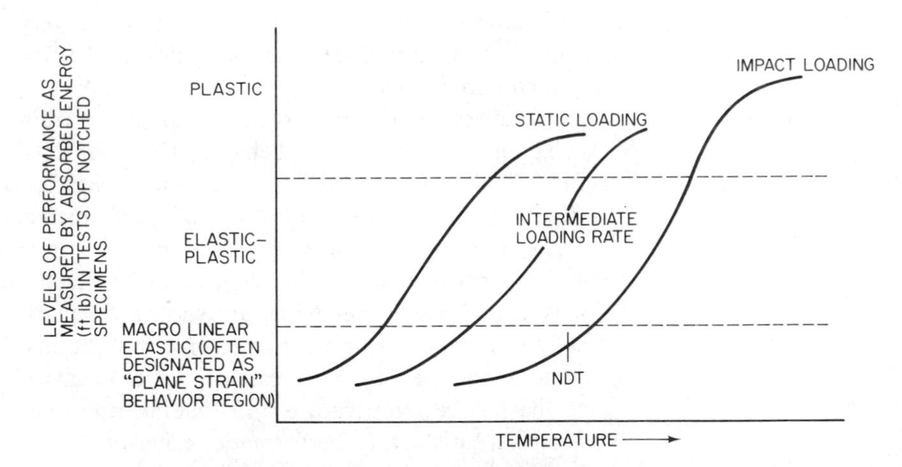

FIG. 14.9. Schematic showing relation between notch-toughness test results and levels of structural performance for various loading rates.

behavior to plastic behavior (Region IV—Fig. 14.8) because the rate of fatigue crack growth becomes so large that even if the critical crack size (a_{cr}) is doubled or even *tripled*, the effect on the remaining fatigue life is small. However, the mode of failure may change.

In summary, use of stress–crack size–toughness relations such as presented in Figures 14.1–14.8 should be used to determine the general design approach to minimize the possibility of brittle fracture in the particular type of structure being considered. It should be noted that in most structures all three methods of fracture control are usually used to some degree.

Although the above three methods of minimizing the possibility of brittle fracture (use lower design stress, better fabrication, or tougher materials) are the basic design approaches, there are other design methods which can be used which are also very effective. These methods include the following:

1. Use structural materials whose notch toughness is such that the material does not fail by brittle fracture even under the most severe operating conditions to which the structure may be subjected. The use of HY-80 steel for submarine hull structures is an example of this method. This method is an extreme use of the method described above of using materials with improved notch toughness. However, as shown in Fig. 14.8, this method is not that effective for structures subjected to fatigue loading.
2. Provide multiple-load fracture paths so that a single fracture cannot lead to complete failure. Multiple-load-path structures should not be confused with redundant structures.

A redundant structure is one in which the laws of statics are insufficient to solve for the loads and stresses and thus the structure is usually called indeterminant from an analysis viewpoint. A multiple-load-path structure is defined by the particular geometry of the members used to make up the structure. For example a simply supported single-span bridge structure is *determinant* (nonredundant) because the reactions can be determined by the laws of statics. If the geometry of this single-span determinant structure is a single box girder such that failure of the single tension flange leads to collapse of the bridge, then the structure is also a *single-load-path* structure. However, if the geometry consists of eight independent structural shapes with a concrete deck, then the structure is a *multiple-load-path* structure and is much more resistant to complete fracture than the single box girder.

Another example is a truss member composed of one structural shape (e.g., a wide-flange shape in tension) or multiple shapes (e.g., four to ten eye-bar members parallel to each other). The former is a single-load-path structure, while the latter is a multiple-load-path structure.

The distinguishing feature is whether or not, in the event of fracture of a primary structural member, the load can be transferred to and carried by other members. If so, the structure is a multiple-load-path one; if not, the structure is a single-load-path one. In this sense, multiple-load-path structures are usually more resistant to failure than single-load-path structures. For example, if a single member fails in a single-load-path structure, the entire structure may collapse, as occurred with the Silver Bridge at Point Pleasant, West Virginia.[2] Conversely, if a single member fails in a multiple-load-path structure, the entire structure may not collapse. This type of behavior was demonstrated in the failure of the Kings Bridge in Australia.[3] That is, at the instant of failure, the failed span in the Kings Bridge contained three cracked girders. One member had cracked while the girders were still in the fabrication shop. A second girder failed during the first winter the bridge was opened to traffic, a full 12 months before the failure of the bridge. Failure of a third girder led to final failure, although architectural concrete sidewalls (which added to the multiple-load paths of the overall structure) prevented complete collapse. Similar examples of the importance of multiple-load paths (as well as adequate inspection) can be shown by analysis of individual failures of other structures during the past several years.

Therefore, assuming that the load can be transferred and carried by other members, the factor of safety applicable to toughness and applied stress should be greater for single-load-path structures than for multiple-load-path structures. If redistribution of load occurs suddenly or if the crack-driving force is not restricted, the fracture behavior of the structure will be governed, primarily, by the dynamic stability of the structure. Under these conditions, fracture toughness above a given value and the design stresses may have negligible effect on the fracture behavior of the structure. The arrest of unstable crack propagation in these cases must be based on a fully dynamic theory that considers the dynamic behavior of the structure and of the propagating crack. Unfortunately, fully dynamic theories that can be applied to engineering structures, such as bridges, ships, off-shore rigs etc., are yet to be developed. The best method to minimize the consequences of failure under these conditions is by proper selection of crack arresters.

Fatigue-crack propagation in multiple-load-path structures may occur essentially under constant deflection, which corresponds to a decreasing stress-field intensity, rather than under constant load. Thus, cracks propagating in multiple-load-path structures may eventually arrest and, although individual structural components will have to be replaced or repaired, complete failure of the structure is not expected to occur so long as redistribution of load can occur.

3. Provide crack arresters so that in the event that a crack should initiate, it will be arrested before catastrophic failure occurs. Crack arresters or a fail-safe philosophy (i.e., in the event of "failure" of a member, the structure is still "safe") have been used extensively in the aircraft industry as well as the shipbuilding industry. A properly designed crack arrest system must satisfy four basic requirements, namely,

a. Be fabricated from structural materials with a high level of notch toughness,
b. Have an effective local geometry,
c. Be located properly within the overall geometry of the structure, and
d. Be able to act as an energy-absorbing system or deformation-restricting system in cases such as gas-pressurized line pipes where the high rate of loading continually deforms the structure.

4. An additional aspect of fracture control deals with fatigue-crack growth. If fatigue-crack growth is possible, the same general design methods described above for fracture control also apply because of the correspondence between a fracture-control plan based on restriction of fatigue-crack initiation, fatigue-crack propagation, and fracture toughness of materials and a fracture-control plan based on fabrication, inspection, design, and fracture toughness of materials. The fact that fatigue-crack initiation, fatigue-crack propagation, and fracture toughness are functions of the stress-intensity fluctuation, ΔK (as described in Chapters 7–11), and of the critical stress-intensity factor, K_{Ic}—which are in turn related to the applied nominal stress (or stress fluctuation), the crack size, and the structural configuration—demonstrates that a fracture-control plan for various structural applications in which fatigue is a consideration depends on the same factors in which just fracture is a consideration. This type of fracture control was described in Fig. 14.8.

5. Although not really a design method to prevent brittle fracture, the fact that many structures are loaded at slow to intermediate loading rates where their notch toughness is quite satisfactory even though their notch toughness is considered to be very low on the basis of *impact* loading-rate tests leads to an understanding of why there are so few brittle fractures in older structures. The recent understanding of the *loading-rate* shift (Chapters 4, 12, and 13) leads to this conclusion. Thus, if the structure can be designed such that it is loaded *slowly*, so that the controlling toughness parameter is K_{Ic} rather than K_{Id}, the possibility of fracture is reduced considerably. Thus, control of the loading rate is a very effective method of fracture control.

14.3.4. Recommendation of Specific Design Considerations To Ensure the Safety and Reliability Against Fracture. This Would Include Recommendations for Desired Levels of Material Performance as Well as Material Selection, Design Stress Levels, Fabrication, and Inspection

This step in the development of a control plan should be the easiest if the preceding steps have all been followed and all technical and performance factors have been properly considered. However, it is usually the most difficult because it involves making the difficult decision of just how much fracture control can be justified *economically*.

Although not widely stated as such, the general nature of structural design of all types of large complex structures is such that the possibility of failure does exist, particularly when the possibility of overloads or natural loads such as earthquakes must be considered. The goal of good sound engineering design is to optimize structural performance consistent with economic considerations such that the probability of structural failure is minimized. To *completely* eliminate the possibility of structural failure (where failure can be yielding, instability, brittle fracture, etc.) is essentially impossible if structures are to perform their design function economically. For the yielding and general buckling modes of failure, sufficient experience has been acquired so that the designer can usually prevent these types of failure quite economically, and the various codes and specifications reflect this fact. However, this is not always true for the catastrophic mode of brittle fracture. Thus the very real question of just how much notch toughness is necessary *economically* to perform satisfactorily *functionally* is very difficult to answer.

All factors related to toughness criteria (as described in Chapters 12 and 13) must be considered, and an economic decision must be made, based on technical input obtained regarding the level of performance to be specified. Once this level of performance (plane-strain, elastic-plastic, fully plastic) has been established for the service loadings and conditions, then some material-toughness property must be specified. Even if this toughness level can be specified directly in terms of K_{Ic}, or K_{Id} (which is usually not the case unless plane-strain behavior is specified), setting material specifications on the basis of K_{Ic} or K_{Id} values is essentially prohibitive, both from an economic and a technical viewpoint. The K_{Ic} test is just too complex and too expensive to conduct on a routine quality-control basis, and there is no standardized K_{Id} test. Hence some auxiliary test specimen must be used, based on the various correlations described in Chapter 6.

Specific examples of fracture criteria and material specifications for various structures are described in Chapter 15, and although they are based on concepts of fracture mechanics, *all* the material specifications are in terms of fracture toughness tests *other than* K_{Ic} or K_{Id}.

14.4. Comprehensive Fracture-Control Plans—George R. Irwin

Dr. Irwin has prepared the following general comments on fracture-control plans,[1] which are reprinted in their entirety.

"For certain structures, which are similar in terms of design, fabrication method, and size, a relatively simple fracture control plan may be possible, based upon extensive past experience and a minimum adequate toughness criterion. It is to be noted that fracture control never depends solely upon maintaining a certain average toughness of the material. With the development of service experience, adjustments are usually made in the design, fabrication, inspection, and operating conditions. These adjustments tend to establish adequate fracture safety with a material quality which can be obtained reliably and without excessive cost. A fair statement of the basic philosophy of fracture control for such structures might be as follows. Given that the material possesses strength properties within the specified limits, and given that the fracture toughness lies above a certain minimum requirement (Nil-Ductility Temperature, Fracture Appearance Transition Temperature, Plane-Strain or Plane-Stress Crack Toughness), then it is assumed that past experience indicates well enough how to manage design, fabrication, and inspection so that fracture failures in service occur only in small tolerable numbers.

"With the currently increasing use of new materials, new fabrication techniques, and novel designs of increased efficiency, the preceding simple fracture control philosophy has tended to become increasingly inadequate. The primary reason is the lack of suitable past experience and the increased cost of paying for this experience in terms of service fracture failures. Indeed, modern technology is beginning to exhibit situations with space vehicles, jumbo-jet commercial airplanes, and nuclear power plants for which not even *one* service fracture failure would be regarded as acceptable without consequences of disaster proportions. Consideration must be given, therefore, to comprehensive plans for fracture control such as one might need in order to provide assurance of zero service fracture failures. A review of the fracture control aspects of a comprehensive plan may be advantageous even for applications such that the required degree of fracture control is moderate. One reason for this would be that an understanding of how to minimize manufacturing costs in a rational way is assisted when we assemble all of the elements which contribute to product quality and examine their relative effectiveness and cost. In the present case, the quality aspect of interest is the degree of safety from service fractures.

"After these introductory comments, it is necessary to point out that the concept termed comprehensive fracture control plan is quite recent and cannot yet be supported with completely developed illustrations. We know in a general way how to establish plans for fracture control in advance of extensive service trials. However, until a number of comprehensive fracture control plans have been formulated and are available for study, detailed recommendations to guide the development of such plans for selected critical structures cannot be given.

"The available illustrative examples of fracture control planning which might be helpful are those for which a large number of the elements contributing to fracture control are known. At least in terms of openly available information, these examples are incomplete in the sense that the fracture control elements require collection, re-examination with regard to relative efficiency, and careful study with regard to adequacy and optimization. Substantial amounts of information relative to fracture control are available in the case of heavy rotating components for large steam turbine generators, components of commercial jet airplanes (fuselage, wings, landing gear, certain control devices), thick-walled containment vessels for BW and PW cooled nuclear reactors, large diameter underground gas transmission pipelines, and pressure vessel components carried in space vehicles. Certain critical fracture control aspects of these illustrative examples are as follows.

A. Large Steam Turbine Generator Rotors and Turbine Fans
   1. Vacuum de-gassing in the ladle to reduce and scatter inclusions and to eliminate hydrogen.
   2. Careful ultra-sonic inspection of regions closest to center of rotation.
   3. Enhanced plane-strain fracture toughness.

B. Jet Airplanes
   1. Crack arrest design features of the fuselage.
   2. Fracture toughness of metals used for beams and skin surfaces subjected to tension.
   3. Strength tests of models.
   4. Periodic re-inspection.

C. Nuclear Reactor Containment Vessels
   1. Quality uniformity of vacuum degassed steel.
   2. Careful inspection and control of welding.
   3. Uniformity of stainless cladding.
   4. Proof testing.
   5. Investigations of low cycle fatigue crack growth at nozzle corners and of cracking hazard from thermal shock.

D. Gas Transmission Pipelines
   1. Adequate toughness to prevent long running cracks.
   2. High-stress-level, in-place, hydrotesting.
   3. Corrosion protection.

E. Spacecraft Pressure Vessels
   1. Surface finishing of welds so as to enhance visibility of flaws.
   2. Heat treatment to remove residual stress and produce adequate toughness within given limits of strength.
   3. Adjustment of hydrotesting to assure adequate life relative to stable crack growth in service.

"In a large manufacturing facility, the inter-group cooperation necessary to achieve successful fracture control on the basis of a comprehensive fracture control plan may require special attention. In general, the comprehensive plan will contain various elements pertaining to Design, Materials, Fabrication, Inspection, and Service Operation. These elements should be directly or indirectly related to fracture testing information. However, the coordination of the entire plan to insure its effectiveness is not a priori a simple task. The following outline lists certain fracture control tasks under functional headings which might, in some organizations, imply separate divisions or departments.

I. Design
   A. Stress distribution information.
   B. Flaw tolerance of regions of largest fracture hazard due to stress.
   C. Estimates of stable crack growth for typical periods of serivce.
   D. Recommendation of safe operating conditions for specified intervals between inspection.

II. Materials
   A. Strength properties and fracture properties.
      σ_{YS}, σ_{UTS}, K_{Ic}, K_c.
      K_{Iscc} for selected environments.
      da/dN for selected levels of ΔK and environments.
   B. Recommended heat treatments.
   C. Recommended welding methods.

III. Fabrication
   A. Inspections prior to final fabrication.
   B. Inspections based upon fabrication control.
   C. Control of residual stress, grain coarsening, grain direction.
   D. Development or protection of suitable strength and fracture properties.
   E. Maintain fabrication records.

IV. Inspection
   A. Inspections prior to final fabrication.
   B. Inspections based upon fabrication control.
   C. Direct inspection for defects using appropriate non-destructive evaluation (NDE) techniques.
   D. Proof testing.
   E. Estimates of largest crack-like defect sizes.

V. Operations
   A. Control of stress level and stress fluctuations in service.
   B. Maintain corrosion protection.
   C. Periodic in-service inspections.

"From the above outline, one can see that efficient operation of a comprehensive fracture control plan requires a large amount of inter-group coordination. If a complete avoidance of fracture failure is the goal of the plan, this goal cannot be assured if the elements of the fracture control plan are supplied by different divisions or groups in a voluntary or independent way. It would appear suitable to establish a special fracture control group for coordination purposes. Such a group might be expected to develop and operate checking procedures for the purpose of assuring that all elements of the plan are conducted in a way suitable for their purpose. Other tasks might be to study and improve the fracture control plan and to supply suitable justifications, where necessary, of the adequacy of the plan."

REFERENCES

1. GEORGE R. IRWIN, private communication.
2. Point Pleasant Bridge, National Transportation Safety Board, "Collapse of U.S. 35 Highway Bridge, Point Pleasant, West Virginia, December 15, 1967," *Report No. NTSB*-HAR-71-1, Washington, D.C., 1971.
3. R. B. MADISON and G. R. IRWIN, "Fracture Analysis of Kings Bridge, Melbourne," *Journal of the Structural Division, ASCE, 97*, No. ST9, Sept. 1971.

15

Examples of Fracture-Control Plans

15.1. Introduction

Only a few fracture-control plans have been formulated for nonproprietary structural applications, and these are usually materials-oriented. That is, the primary recommendation of many fracture-control plans is one of specifying the material toughness for preexisting design stress levels and fabrication techniques.

Nonetheless, a review of several fracture-control plans (some of which were only proposed as of the publication of this textbook) should provide the engineer with an insight into the factors involved in developing them, realizing, of course, that the specific factors in developing a fracture control plan depend on the particular structure being analyzed. Examples of fracture-control plans and/or toughness criteria are presented for the following types of structures:

1. Section 15.2: Pressure Vessel Research Committee (PVRC) Recommendations on Toughness Requirements for Ferritic Materials in Nuclear Power Plant Components.
2. Section 15.3: American Association of State Highway and Transportation Officials (AASHTO) Material Toughness Requirements for Steel Bridges.
3. Section 15.4: Proposed Fracture-Control Guidelines for Welded Steel Ship Hulls.
4. Section 15.5: Proposed Fracture-Control Plan for Floating Nuclear Plant Platforms.
5. Section 15.6: Fracture-Control and Structural Integrity in Military Aircraft Structures.

15.2. Pressure Vessel Research Committee (PVRC) Recommendations on Toughness Requirements for Ferritic Materials in Nuclear Power Plant Components.

In January 1971, PVRC formed a task group to review current knowledge and prepare recommendations on toughness requirements for ferritic materials in nuclear power plant components. The recommendations were requested from PVRC by the ASME Boiler and Pressure Vessel Committee for their use in considering any revisions to the requirements of the ASME Code Section III—Nuclear Power Plant Components.

This task group undertook to recommend, on the basis of current knowledge, criteria for ferritic-material-toughness requirements for pressure-retaining components of the reactor coolant pressure boundary operating below 700°F. These criteria, when used in addition to the stress limits allowed by the ASME code, should permit the establishment of safe procedures for operating nuclear reactor components under normal, upset, and testing conditions; emergency and faulted conditions should be considered on a case basis.

The following condensation is based on *Welding Research Council Bulletin No. 175*, "PVRC Recommendations on Toughness Requirements for Ferritic Materials, by the PVRC Ad Hoc Task Group on Toughness Requirements, August 1972."

The development of design procedures to prevent the occurrence of unexpected and sudden failures in engineering structures and components has been the subject of intensive activity for many years. Much progress has been attained and many of the factors which need to be considered for a reliable design procedure are now known.

It is almost self-evident that the presence of flaws, defects, and other sources of high stress concentration in structural members can decrease the load-carrying capability. However, for engineering design analysis purposes, this general observation must be made quantitative and must include the following:

1. The relation of the strength reduction to the size and geometry of the flaw.
2. The important material properties which determine the magnitude of the strength reduction.
3. The influence of the geometry of the structural member and the kind and the rate of applied loading.
4. The effect of nonapplied loads such as residual stresses from welding.

For applications involving structural grades of carbon and alloy steels, an engineering design procedure based on the "transition-temperature" concept, largely developed in the 1940s, has been widely adopted and used with very good success in preventing the occurrence of unexpected failures of the brittle fracture type. In essence, this procedure is based on the marked change in fracture behavior and characteristics from "low" to "high" values with increasing temperature when certain varieties of test specimens are used. Specifically, this transition behavior is best exhibited when the specimen contains a sharp notch or a crack. The Charpy V-notch and the drop weight tests are among the most commonly used specimens for this purpose. The resulting transition temperatures can be defined for the Charpy specimen in terms of an energy level, the amount of deformation, or the fracture appearance. For the drop weight test, the nil-ductility temperature (NDT) is defined in ASTM Specification E-208 and is the temperature above which the specimen will sustain a specific amount of deformation without cracking to either edge.

The transition-temperature procedure is applied in design by permitting loading on the structure only at temperatures higher than the transition temperature by a specified temperature increment. The increment to be used is determined by one or a combination of several methods: (1) correlation with service experience, (2) correlation with model tests, and (3) engineering judgment. The transition-temperature procedure has been developed, in a more formal and organized sense, into the fracture analysis diagram (see Chapter 13) through the substantial efforts of Pellini and his associates. This procedure was also the implicit basis of material-toughness requirements and operational criteria contained in Section III of the ASME Boiler and Pressure Vessel Code Prior to the Summer of 1972.

Although the transition-temperature procedure has the virtues of simplicity and successful experience, it does not inherently have the capability of quantitatively treating some of the items mentioned earlier. The linear-elastic fracture-mechanics concept has these capabilities to a greater degree. This concept is based on an elastic analysis of the stresses in the neighborhood of the tip of a sharp crack in solids. Thus, it is essentially a stress-analysis-based approach, as described in Chapter 2.

The basic analysis assumes elastic behavior of the stresses in the body, including the region around the crack tip. It is found that the stress distribution near the crack tip is always the same and that the stress magnitudes all depend on a single quantity termed the *stress-intensity factor*, generally designated as K. For loadings which produce an opening mode of displacement between the crack surfaces, the stress-intensity factor is further designated as K_I.

Equations are available for the calculation of the value of K_I in terms of the applied load and the crack size for various combinations of crack geometries, structural member dimensions and shapes, and type of applied loading.

All these equations have an identical general form, namely

$$K_I = C\sigma\sqrt{\pi a} \tag{15.1}$$

where σ = nominal (gross section) tensile stress perpendicular to the plane of the crack at the location of the crack, psi,

a = characteristic crack dimension, such as crack depth for surface cracks, in.,

C = nondimensional constant whose value depends on the crack geometry, the ratio of the crack size to the size of the structural member, and the type of loading (tension, bending, etc).

The basic premise of linear-elastic fracture mechanics is that unstable propagation of an existing flaw will occur when the value of K_I attains a critical value designated as K_{Ic} (Chapter 5). The K_{Ic}, generally called the fracture toughness of the material, is a temperature-dependent material property (Chapter 4). The implementation of the fracture-mechanics concept as a fracture-control design procedure then consists of two essential steps:

1. Determine the K_{Ic} properties of the material using suitable test specimens and conditions.
2. Determine the actual or anticipated flaw size in the structural component and calculate the limiting value of stress which will keep the value of K_I in the component less than K_{Ic}. A safety factor may be applied to the stress, and a safety margin may also be incorporated in the flaw size by choosing a reference flaw size considerably larger than the actual or anticipated maximum flaw size.

In the case of structural carbon and alloy steel materials, it has been found that fracture-toughness properties are dependent on temperature and on the loading rate imposed on the flaw. The loading-rate effect leads to several categories of fracture toughness values as follows:

K_{Ic} = static initiation fracture toughness obtained under slow loading conditions.

K_{Id} = dynamic initiation fracture toughness obtained under fast or rapidly applied loading rates.

K_{Ia} = crack arrest fracture toughness obtained from the value of K_I under conditions where a rapidly propagating fracture is arrested within a test specimen.

In structural steels, experimental evidence shows that K_{Id} and K_{Ia} generally are less than K_{Ic} at a given temperature. Later in this section, a quantity designed as K_{IR}, the reference value of the fracture toughness, is used; it is the lower bound of the available K_{Ic}, K_{Id}, and K_{Ia} values for the pertinent materials.

Experimental investigations of the temperature effect on the fracture toughness of carbon and alloy steels of the lower- and intermediate-yield-strength grades show that K_{Ic}, K_{Id}, and K_{Ia} all exhibit a sharp increase with temperature over a relatively small temperature range. Further, analytical studies and correlations have indicated that this temperature range can be related to the transition temperature of each material item as determined by the Charpy and/or drop weight tests. A specific example of this kind of correlation is the indexing of the temperature dependence of K_{IR} to the drop weight nil-ductility transition (NDT) temperature. Other empirical correlations between fracture toughness (K_{Ic}, K_{Id}) and Charpy test values (foot-pounds) have also been developed.

The foregoing discussion has been a capsule summary of the transition-temperature and fracture-mechanics concepts which form useful bases for a fracture-control design procedure for nuclear power plant components. In summary, the toughness requirements and design criteria developed in the PVRC document make use of both concepts. Fracture mechanics provides the formulas and equations for defining a relation between the type and magnitude of the applied stresses, the crack-tip stress-intensity factor, and the size and geometry of the postulated flaw. The transition-temperature approach provides methods for estimating the material's fracture toughness and for verifying the temperature dependence of the fracture toughness.

It should be recognized that although extremely useful, fracture mechanics does not provide a complete solution for all fracture-control design situations. For example, extension into conditions involving large plastic deformation is not adequately possible now. Considerable effort is now in progress to overcome these limitations, and this effort can be expected to be the source of more complete methods in the future, as described in Chapter 16.

There are *four major regions of reactor coolant boundary requiring criteria*, as follows:

1. Vessel shell and head regions remote from geometric discontinuities. These may be subjected to sufficient radiation to affect the mechanical and toughness properties. Representative ferritic materials for the pressure-retaining components are listed in Table 15.1.
2. Pumps and valve bodies and vessel regions at or near nozzels and flanges not subjected to significant radiation.
3. Piping.
4. Bolting.

In this condensation, only region 1 will be considered. For the other regions, reference is made to the original fracture-control plan, *Welding Research Council Bulletin No. 175.*

TABLE 15.1. Ferritic Materials for Pressure-Retaining Components of the Reactor Coolant Pressure Boundary Operating Below 700°F

Component	*Section Thickness (in.)*
Reactor shell	
SA-533, Grade B, Class 1 plate	4 to 12
SA-508, Class 2 forging	4 to 12
SA-508, Class 3 forging	4 to 12
Reactor nozzle and flange	
SA-508, Class 2 forging	2 to 20
SA-508, Class 3 forging	2 to 20
SA-508, Class 1 forging	3 to 4
Steam generator	
SA-533, Grade B, Class 1 plate	7 to 9
SA-533, Grade B, Class 2 plate	7 to 9
SA-516, Grade 70 plate	4 to 5
SA-508, Class 2 forging	6 to 24
SA-508, Class 1 forging	3 to 5
SA-216, WCC casting	4 to 6
Pressurizer	
SA-533, Grade B, Class 1 plate	4 to 5
SA-533, Grade B, Class 2 plate	4 to 5
SA-533, Grade A, Class 1 plate	3 to 4
SA-516, Grade 70 plate	5 to 7
SA-508, Class 2 forging	2 to 5
Piping	
SA-516, Grade 70 plate	3 to 5
SA-106, Grade B pipe	To $1\frac{1}{2}$
SA-106, Grade C pipe	To 3
SA-53, Grade B pipe	To 2
SA-333, Grade 6 pipe	To 2
SA-155, KCF pipe	To 2
SA-105, Grade I forging	To 1
SA-105, Grade II forging	$1\frac{1}{2}$ to 5
SA-182, F1 forging	$1\frac{1}{2}$ to 5
Bolting	
SA-540, Grade B23 and B24 bar (4340)	To 8
SA-320, Grade L43 bar (4340)	To 8
SA-193, Grade B7 bar (4140, 4142, 4145)	$\frac{3}{4}$ to 2
Pumps and valves	
SA-216, WCB casting	$\frac{1}{2}$ to 10

Note: The listed materials and section thicknesses are typical for current practice. Other materials and thicknesses may be used.

15.2.1. *Allowable K_{IR} Values*

Figure 15.1 is a curve which shows the relationship that can be conservatively expected between the critical (or reference) stress-intensity factor,

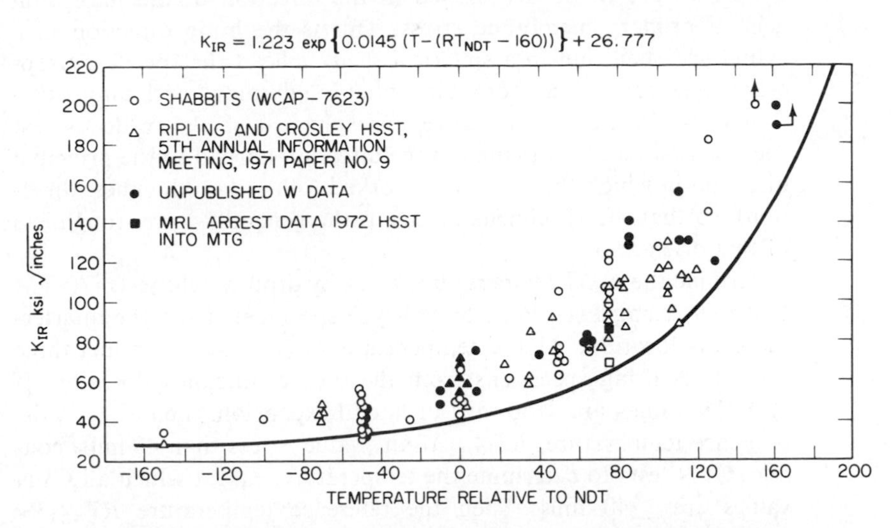

FIG. 15.1. Derivation of curve of reference stress-intensity factor (K_{IR}).

K_{IR}, and a temperature which is related to the reference nil-ductility temperature, RT_{NDT}. This curve is based on the lower bound of dynamic and crack arrest K_{Id} and K_{Ia} values measured as a function of temperature on specimens of SA-533 B-1, SA-508-2, and SA-508-3 steels. The data used for the derivation of Figure 15.1 are described in a later section on derivation of the K_{IR} curve. No available data points for static, dynamic, or arrest tests fall below the curve. An analytical approximation to the curve in Figure 15.1 is

$$K_{IR} = 1.223 \exp\{0.0145[T - (RT_{NDT} - 160)]\} + 26.777$$

Curves similar to Figure 15.1 are needed for other materials, but the data are not available in sufficient quantity. In the interim, until data become available, it is recommended and considered adequately conservative to use the curve of Figure 15.1 for steels with specified minimum yield strengths up to and including 50 ksi, but not for steels with higher specified yield strengths.

15.2.2. Testing Requirements and Radiation-Induced Changes

The following acceptance tests for materials in shell or head regions remote from discontinuities are recommended:

1. Charpy V-notch tests are required for materials having a nominal section thickness greater than $\frac{1}{2}$ in. Drop weight tests are required when geometrically possible.

2. Properties are to be determined in the direction of the maximum general primary membrane stress. This is the hoop direction in a cylindrical shell, but in a spherical shell or head the specified properties are required in both tangential-longitudinal and tangential-transverse directions. Therefore, for spherical shells or heads test specimens should be oriented in the direction normal to the principal direction in which the metal was worked (other than thickness direction), so that the specimens represent the generally lower toughness of that orientation.
3. Determine the NDT temperature, T_{NDT}, by drop weight tests (ASTM E-208-69) using P-1-, P-2-, or P-3-type specimens from the quarter-thickness location. At the temperature $T_{\text{NDT}} + 60°\text{F}$ conduct three CVN tests using specimens from the quarter-thickness location. If all CVN values are $\geq$40 mils of lateral expansion, then T_{NDT} is the reference temperature, RT_{NDT}. If any value is less than 40 mils, conduct CVN tests to determine the temperature, T_{40}, at which all CVN values are $\geq$40 mils. Then the reference temperature $RT_{\text{NDT}} = T_{40} - 60°\text{F}$. Thus RT_{NDT} is the higher of T_{NDT} or $T_{40} - 60°\text{F}$. An alternative requirement of 35 mils of lateral expansion, but not less than 50 ft-lb absorbed energy, would also seem reasonable. If the material is to be subjected to significant radiation, data from surveillance specimens must be obtained and used as required by specifications of ASTM E-185-70, "Recommended Practice for Surveillance Tests on Structural Materials in Nuclear Reactors" now undergoing revision. (Also see test 5 below.)
4. Apply the procedure to specimens representing base material, heat-affected zone and weld metal.
5. The impact properties described in test 3 must be met throughout the life of the component. Since neutron radiation increases the transition temperature and reduces the fracture toughness of ferritic steels, determination must be made of the margin between initial and end-of-life properties to provide assurance that the required properties will be obtained at the quarter-thickness location throughout life. Further, since the degree of radiation-induced change varies widely depending on the neutron fluence, the temperature at the point of irradiation, and the relative sensitivity to radiation of the particular steel component, a systematic surveillance program as required by ASTM E-185-70 (as revised) is essential. Maintenance of the required properties in the materials of lowest (limiting) end-of-life toughness, from those in test 4 above, and with due consideration of the variable factors noted above must be verified by surveillance specimens. The radiation-induced temperature change ΔT_{NDT} and the maintenance of the required impact properties at

$RT_{NDT} + 60°F$ at the quarter-thickness location is to be established by tests of surveillance specimens conducted in accordance with the methods of ASTM. It is not feasible to name a specific fluence above which a surveillance program is mandatory. Available data show, however, that if the fluence at the inner wall is less than 10^{18}_{nvt} ($>1\ M_{ev}$) no significant radiation damage is to be expected. For higher values of end-of-life fluence the omission of a surveillance program should be justified by showing that for the particular lots of base and weld metal being used the reactor vessel shell will not become more limiting than other parts of the vessel.

The recommended tests and acceptance standards given above were based on the following considerations:

1. The use of both drop weight and CVN tests gives protection against the possibility of errors in the conducting of tests or the reporting of test results.
2. The CVN requirements are given in terms of *lateral expansion* rather than absorbed energy because this provides protection from variation in yield strength from initial heat treatment and the change in yield strength produced by irradiation.
3. The requirement of 40 mils of lateral expansion or, alternatively, 35 mils and 50 ft-lb at $RT_{NDT} + 60°F$ throughout the life of the component provides assurance of adequate fracture toughness at upper shelf temperatures.
4. The CVN test at $T_{NDT} + 60°F$ serves to weed out nontypical materials such as those which might have a low transition temperature but an abnormally low energy absorption on the upper shelf.

15.2.3. Reference Flaw Size

After the K_{IR} which the material is capable of providing has been established, the next step in the procedure for obtaining allowable loading is the choice of a postulated defect size. For normal and upset conditions, as defined in the ASME code (Section III, NB-3113) in shell or head regions remote from discontinuities, a very conservative defect size is recommended. It consists of a sharp surface flaw, perpendicular to the direction of maximum stress, having a depth of one fourth of the section thickness over most of the thickness range of interest. Since the nondestructive test procedures used are not strongly dependent on thickness above 12 in. and below 4 in., the postulated depth is assumed to be constant beyond those values, as shown in Figure 15.2. The assumed shape of the defect is semielliptic, with a length six times its depth. Consideration was also given in some cases to the more conservative crack of infinite length. This assumed quarter-thickness defect

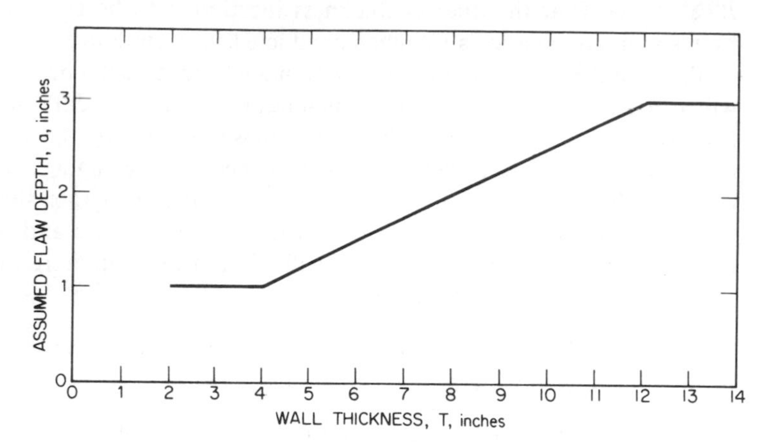

FIG. 15.2. Reference flaw depth.

provided the rationale for using the quarter-thickness location for test specimens and the same location for the radiation damage calculations.

If the ASME code's 2% radiography standard is taken at face value, it may be noted that the postulated defect is 12.5 times as large in linear dimensions as the code-allowable dimensions and thus has over 150 times the area. It is not safe to assume that no defects larger than those that the code allows will ever occur, but it does seem reasonable to assume that with the combination of examinations (generally radiography) required by Section III of the ASME code and the volumetric examination (generally ultrasonic mapping) required by Section XI, there is a very low probability that a defect larger than about four times that allowed by the code will escape detection. The postulated defect is about three times as large in linear dimensions and has about ten times the area of even that conservative value.

Other defect shapes and locations might have been chosen to develop the recommended criteria. Buried defects are more apt to escape detection than surface defects, but when not too near the surface they produce only about half the K_I value for a given size, so the use of a buried defect would not be as conservative. Surface defects longer than six times their depth would produce higher K_I values, but even if an infinitely long defect had been postulated, the recommended properties and calculation methods would still provide protection against a defect having a depth one sixth instead of one fourth of the section thickness.

In summary, the postulation of the large defect size, the use of dynamic test data to give the lower-bound toughness value, and the safety factors (discussed below) on the allowable stresses combine to provide a large degree of conservatism in the safety and reliability of nuclear vessels.

15.2.4. Allowable Loading

The final step in the calculation of allowable loading is the calculation of the applied K_I and comparison with the available K_{IR} value. In a vessel shell or head remote from discontinuities, the significant loadings are (1) general primary membrane stress due to pressure and (2) thermal stress due to thermal gradient through the thickness during startup and shutdown. Effects of residual stress are not included in the recommended procedure because

1. Peak values in a postweld heat-treated component usually are less than 20% of the yield strength.
2. Service stresses and radiation effects both tend to reduce residual stresses during the life of the component.
3. Conservatisms throughout the whole recommended procedure and the safety factors applied appear to be ample to cover any incalculable adverse effects.

Therefore, the procedure for calculating the allowable system pressure during startup consists of adding the K_I corresponding to the membrane tension produced by pressure to the K_I produced by thermal gradient at a desired startup and shutdown rate and requiring that the sum of these K_I values not exceed the available K_{IR} of Figure 15.1 at each temperature. Methods for calculating these two K_I values for the assumed maximum credible defect are described in the PVRC document. These methods result in the curves shown in Figures 15.3 and 15.4. The K_I produced by pressure is

$$K_{I\ \text{pressure}} = [M_t] \times [\text{general primary membrane stress}]$$

Figure 15.3 shows M_t as a function of wall thickness. The four lines shown for the M values are to be used for stress values 0.1, 0.5, 0.7, and 1.0 times the material yield strength. The M values for other stress levels can be interpolated linearly with sufficient accuracy since they are not strongly affected by the stress level. For calculated stresses above yield level, linear-elastic fracture mechanics is not applicable.

The K_I produced by thermal gradient is

$$K_{I\ \text{therm.}} = [M_\theta] \times [\text{temperature difference through wall}]$$

Figure 15.4 shows M_θ as a function of wall thickness, based on the following assumptions:

1. An assumed shape of the temperature gradient.
2. The shutdown starts after a steady-state temperature has been attained.
3. The rate of change of temperature is of the magnitude associated with startup and shutdown, i.e., less than about 100°F/hr. The result would be overly conservative if applied to rapid temperature changes.

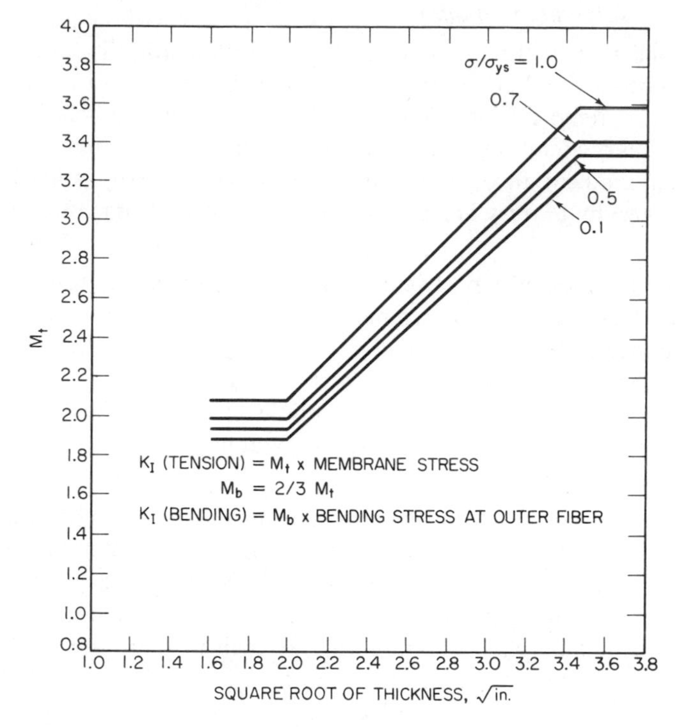

FIG. 15.3. M_t and M_b versus the square root of the wall thickness for semielliptical surface flaw $\frac{1}{4}$T deep and $1\frac{1}{2}$T long.

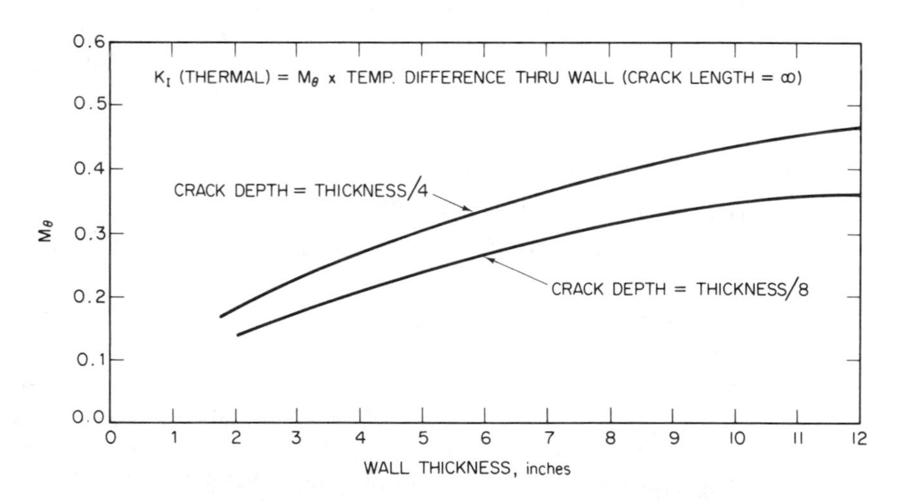

FIG. 15.4. Stress-intensity factor caused by thermal stress.

4. The postulated defect has a depth one fourth of the wall thickness and infinite length. (The values of M_θ for $\frac{1}{8}$-thickness crack depth are also shown in Figure 15.4.)

For some purposes, Figure 15.5 may be found more convenient to use than Figure 15.4. It is based on a radius-to-thickness ratio of 10 and uses cooling rate directly, thus avoiding the need to calculate the temperature difference across the wall.

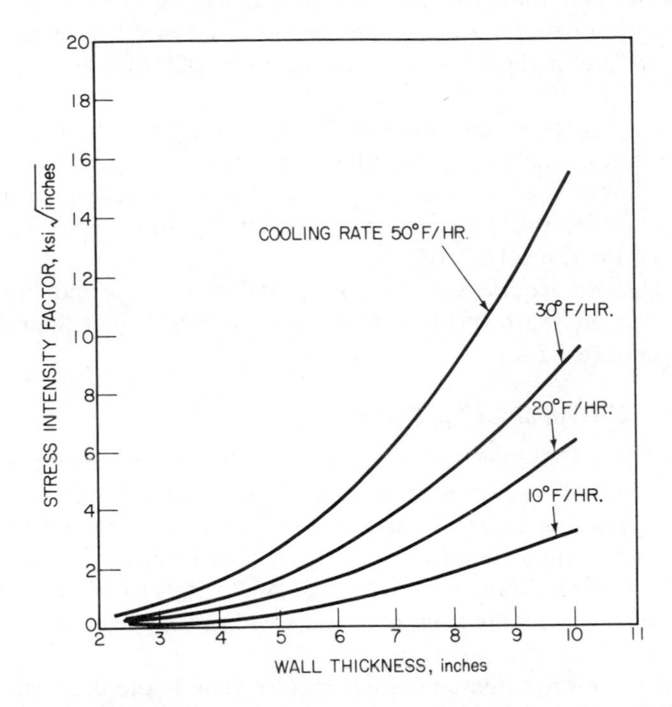

FIG. 15.5. Stress-intensity factor caused by thermal stress for cylinders with radius/thickness = 10.

The governing operating condition is most apt to be *shutdown* because decreasing temperature produces tensile stress in the portion of the reactor vessel wall which is subjected to the greatest radiation damage.

The requirement for startup, shutdown, and normal and upset operating conditions is that, throughout the life of the component at each temperature, the following inequality holds:

$$K_{\mathrm{I\,pressure}} + K_{\mathrm{I\,therm.}} \leq K_{\mathrm{IR}} \qquad \text{from Figure 15.1}$$

15.2.5. Safety Factors

The methods described here for the calculation of allowable loadings do not include any safety factors beyond the use of conservative assumptions for the available K_{IR} and the defect size. The application of additional "factors of safety" is a function of the code-writing organizations rather than the research organizations such as PVRC, since the choice of safety factors depends on the chosen relationship between economics and conservatism and on the consequences of failure. If additional conservatism is desired, it is suggested that in the equation for normal operating conditions $K_{I\,pressure}$ be multiplied by a factor in the range 1.5–2.0. Due to its secondary and self-relieving nature, the factor for $K_{I\,therm.}$ can be smaller. In addition, the values given here for $K_{I\,pressure}$ are based on a crack length six times its depth. Therefore, it should be adequate to use a safety factor of 1.0–1.3 on $K_{I\,therm.}$.

During hydrotest of the vessel and/or system, there is no thermal stress and there is less uncertainty regarding the stress values. Also, the consequences of failure are less severe. Therefore, less conservatism is justifiable, and the suggested safety factor for the system hydrotest after installation should be in the range 1.25–1.5.

The suggested requirement for a shop hydrotest of an individual component is that the test be made at a temperature 60°F higher than the reference temperature RT_{NDT}.

15.2.6. Derivation of K_{IR} Curve

In general, tests to determine K_{Ic}, K_{Id}, and K_{Ia} are expensive and time-consuming, but the advantages of using linear-elastic fracture mechanics for fracture analysis are so strong that many ways have been developed to estimate critical stress-intensity factors from data from other tests. Enough data exist on A533, Grade B, Class 1, and A508 steels to relate actual measured K_{Ic} values to results from other tests with a very high degree of confidence.

One of the more commonly used fracture tests is the drop weight test. The NDT determined from the drop weight test represents the upper temperature where brittle fracture occurs under defined conditions of strain and flaw size. Because conditions and geometry are defined, the critical stress-intensity factor can be calculated, and this has been done by Irwin and others. The relationship between dynamic yield strength (σ_{yd}) and K_{Id} resulting from this work is

$$K_{Id} = (\text{factor}) \times \sigma_{yd}$$

where the factor is between 0.5 and 0.78 (Chapter 6).

The development of the K_{IR} versus temperature curve (where temperature is relative to the NDT) was done by plotting all known data—K_{Id}, K_{Ia}, and low-temperature K_{Ic} versus the temperature relative to the drop weight NDT of the same plate or forging. A lower-bound curve was drawn and the values at NDT (i.e., $T - \text{NDT} = 0$) checked against the analytical relationships discussed above. The value of K_{IR} at NDT on this curve is equal to

$0.55\,\sigma_{yd}$ of the steels tested. The resulting curve, therefore, represents a very conservative assumption as to the critical stress-intensity versus temperature properties of materials similar to those tested, as related to the measured NDT.

Figure 15.1 shows the K_{IR} versus NDT adjusted temperature curve, on which are plotted all pertinent data available at present. Only K_{Id} and K_{Ia} data are shown, because K_{Ic} values are invariably much higher, so they do not affect the lower bound.

To facilitate computer calculations, the equation representing this curve is also shown, and actual values for reference are as follows:

$$K_{IR} = 26.777 + 1.223 \exp\{0.0145[T - (\text{NDT} - 160)]\}$$

RT_{NDT}	K_{IR} (ksi$\sqrt{\text{in.}}$)
NDT − 160	28.00
NDT − 140	28.41
NDT − 120	28.96
NDT − 100	29.7
NDT − 80	30.68
NDT − 60	31.99
NDT − 40	33.74
NDT − 20	36.08
NDT	39.21
NDT + 20	43.39
NDT + 40	48.97
NDT + 60	56.44
NDT + 80	66.41
NDT + 100	79.73
NDT + 120	97.54
NDT + 140	121.33
NDT + 160	153.13
NDT + 180	195.61

15.3. American Association of State Highway and Transportation Officials (AASHTO) Material Toughness Requirements for Steel Bridges

15.3.1. Introduction

Steel bridges have an excellent service record extending over millions of operational years. However, the collapse of the Point Pleasant Bridge[1] in the United States as well as other localized fractures in bridges has led to an increasing concern about the possibility of catastrophic fractures in steel bridges. This concern is accentuated by a growing awareness by material scientists and design engineers of the use of fracture-mechanics methodology in other applications and the increasing emphases on safety and reliability of structures.

Most bridge structures in existence perform safely and reliably. The safety and reliability of these structures have been achieved by improving the weak links that were observed during the life of each component in the system. Present specifications on material, design, and fabrication generally are based on correlations with service experience. The comparatively few service failures in steel bridges indicate that steel properties, design, and fabrication procedures used for bridges basically are satisfactory. However, service failures that have occurred in the late 1960s and early 1970s indicated that some modifications in practices were needed. The identification of the specific modifications needed required a thorough study of material properties, design, fabrication, inspection, erection, and service conditions. Fracture-mechanics methodology was of considerable help in this endeavor.

Based on considerable research and study, including an industry-sponsored research program,[2,3] the Committee on Bridges and Structures of the American Association of State Highway and Transportation Officials (AASHTO) adopted in 1973 the Charpy V-notch (CVN) impact-toughness requirements for bridge steels shown in Table 15.2. These requirements for primary tension members of bridge steels were based primarily on the information presented in the subsequent sections.

15.3.2. Fracture Control in Bridges

A well-conceived fracture-control plan recognizes that the performance of a structure or a structural component is governed not only by material properties but also by the design, fabrication, erection, inspection, and final use of the structure. The objective in developing fracture-toughness specifications for a structure should be to establish the necessary and sufficient toughness values that ensure the adequacy of the material for the intended application.

Analysis of available bridge test results[4-6] indicates that the maximum loading rates observed in bridges are closer to slow-bend loading rates than to impact loading rates. The loading times in bridges are greater than 1 sec, which corresponds to a strain rate of less than 10^{-3} sec^{-1} on the elastic-plastic boundary in the vicinity of the crack tip. Thus, a strain rate of 10^{-3} sec^{-1} can be used as a conservative measure of the maximum strain rate for bridges.

The K_{Ic} data obtained by testing A36 and A572 Grade 50 steels at a strain rate of 10^{-3} sec^{-1} indicate that fracture does not occur under plane-strain conditions when the test temperature is greater than about -80°F (-62°C) (Figures 15.6 and 15.7). The fracture-toughness transition region (the temperature region bounded by the lower end of the fracture-toughness transition temperature and the highest temperature at which fracture does

TABLE 15.2. AASHTO Notch-Toughness Specifications for Bridge Steels

ASTM Designation	*Thickness (in.)*	*CVN Impact Value, ft lb*		
		*Zone 1**	*Zone 2**	*Zone 3**
A36		15 @ 70°F	15 @ 40°F	15 @ 10°F
A572	Up to 4 in. mechanically fastened	15 @ 70°F	15 @ 40°F	15 @ 10°F
	Up to 2 in. welded	15 @ 70°F	15 @ 40°F	15 @ 10°F
A440		15 @ 70°F	15 @ 40°F	15 @ 10°F
A441		15 @ 70°F	15 @ 40°F	15 @ 10°F
A242		15 @ 70°F	15 @ 40°F	15 @ 10°F
A588**	Up to 4 in. mechanically fastened	15 @ 70°F	15 @ 40°F	15 @ 10°F
	Up to 2 in. welded	15 @ 70°F	15 @ 40°F	15 @ 10°F
	Over 2 in. welded	20 @ 70°F	20 @ 40°F	20 @ 10°F
A514	Up to 4 in. mechanically fastened	25 @ 30°F	25 @ 0°F	25 @ −30°F
	Up to $2\frac{1}{2}$ in. welded	25 @ 30°F	25 @ 0°F	25 @ −30°F
	Between $2\frac{1}{2}$–4 in. welded	35 @ 30°F	35 @ 0°F	35 @ −30°F

*Zone 1: minimum service temperature 0°F and above.
Zone 2: minimum service temperature from −1° to −30°F.
Zone 3: minimum service temperature from −31° to −60°F.
**If the yield point of the material exceeds 65 ksi, the temperature for the CVN value for acceptability shall be reduced by 15°F for each increment of 10 ksi above 65 ksi.

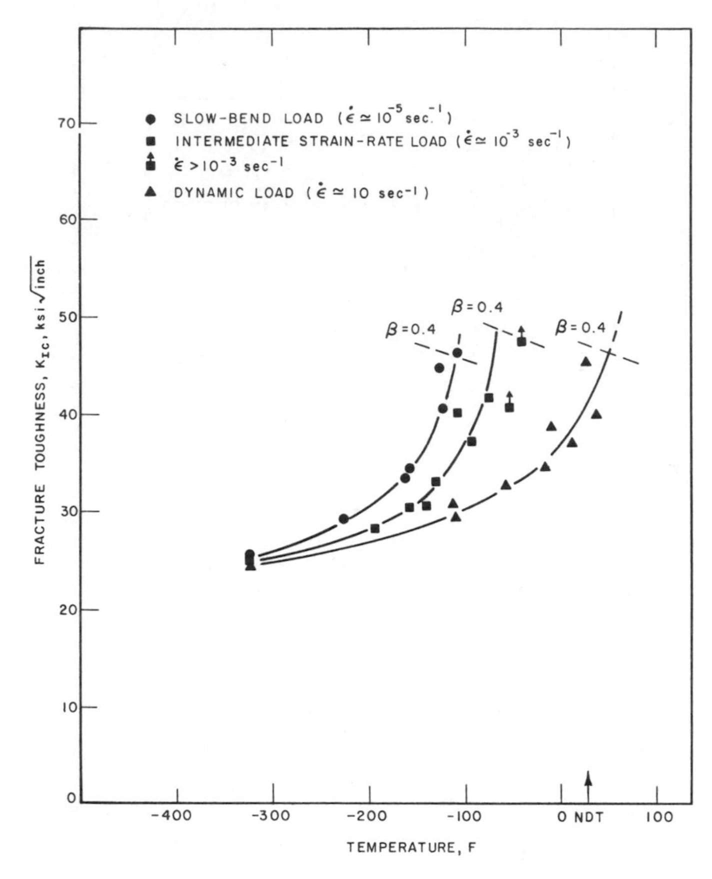

FIG. 15.6. Effect of temperature and strain rate on fracture toughness of A36 steel.

not occur under plane-strain conditions) for the steels investigated is on the order of 50°F (28°C). A fracture-toughness transition temperature of −130°F (−90°C) corresponded to a non-plane-strain fracture behavior at −80°F (−62°C) for these steels tested at the intermediate strain rate of 10^{-3} sec^{-1}.

One of the requirements used in the development of the AASHTO toughness requirements was that fracture of a 50-ksi (345-MN/m^2) yield strength steel subjected to the intermediate strain rate does not occur under plane-strain conditions. This requirement should be satisfied when the fracture-toughness transition temperature of the steel under the intermediate strain rate occurs at about −80°F. Because K_{Ic} tests are expensive and difficult to conduct, and because of the apparent correspondence between K_{Ic} test results and CVN test restults, the CVN test was selected as the

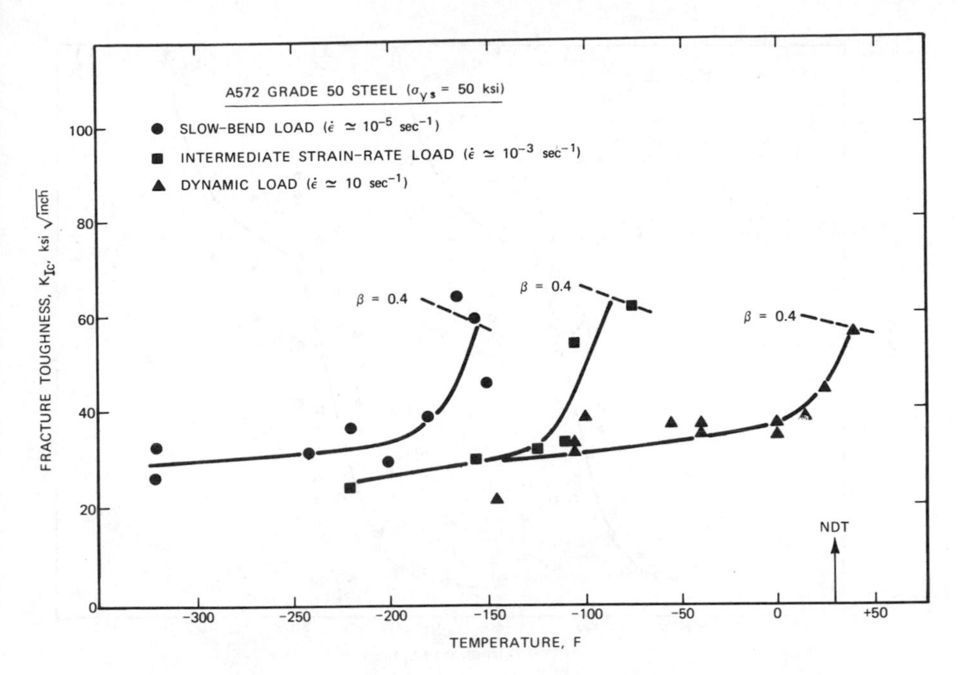

FIG. 15.7. Effect of temperature and strain rate on fracture toughness of A572 Grade 50 steel (σ_{ys} = 50 ksi).

reference test for the AASHTO fracture-toughness requirements. The fracture-toughness transition temperature is the temperature at which the fracture toughness of the steel begins to increase rapidly from plane-strain behavior to fully ductile behavior. The CVN test results (Figure 15.8) show that this transition behavior at 15 ft-lb under the intermediate rate of loading at −80°F should ensure a non-plane-strain fracture behavior at the minimum operating temperature of −30°F (−34°C) for a 50-ksi yield strength steel.

Although the intermediate-loading-rate test is the test that more properly describes the expected service performance of bridge steels, the impact-loading-rate test is much easier to conduct and analyze and is less expensive than an intermediate-loading-rate CVN test. Consequently, the difference in fracture-toughness behavior at the two strain rates was used to develop the toughness values in terms of the impact test rather than an intermediate-loading-rate test. The temperature shift between the CVN (and K_{Ic}) curves of a 50-ksi yield strength steel tested at a strain rate of 10^{-3} sec^{-1} and at an impact strain rate of 10 sec^{-1} was on the order of 120°F (67°C) (Figure 15.8). Consequently, the requirement of a 15-ft-lb CVN impact value at +40°F (4°C) corresponds to a 15-ft-lb CVN value under intermediate strain rate at −80°F (−62°C), which in turn corresponds to a non-plane-strain fracture behavior at the assumed minimum operating temperature of −30°F (−34°C).

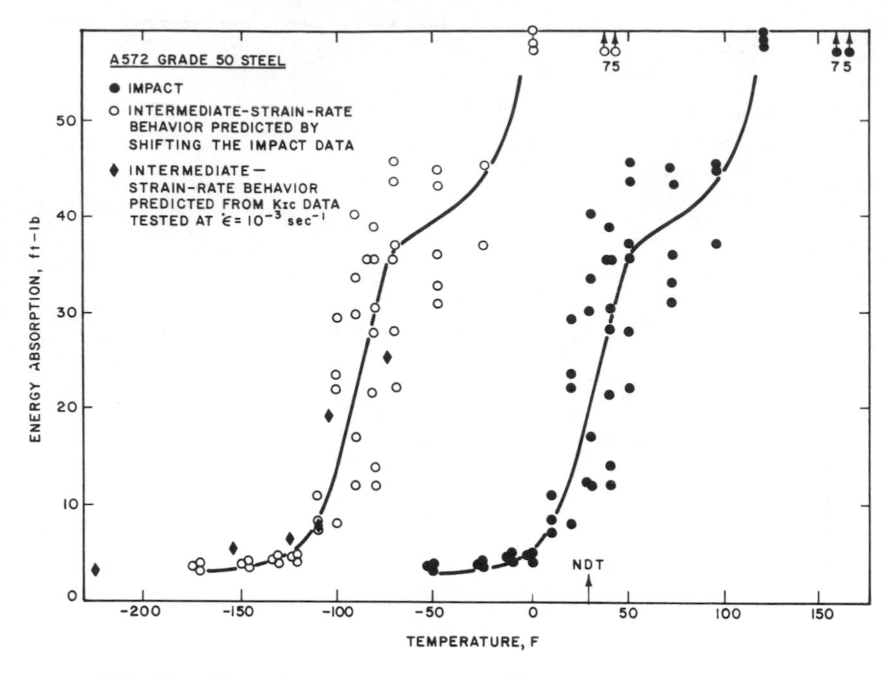

FIG. 15.8. Charpy V-notch energy-absorption behavior for impact loading and intermediate strain-rate loading of standard CVN specimens.

Thus, a CVN fracture-toughness requirement of 15 ft-lb at +40°F was imposed on all primary tension members of 50-ksi yield strength steels for bridge applications. This same requirement was also imposed on all primary tension members of bridge steels having yield strengths less than 50 ksi [242, A440, and A441 steels in thicknesses greater than 0.75 in. (19 mm)], which is a conservative requirement for these steels.

The 15-ft-lb CVN impact-toughness requirement at +40°F for steels of 50-ksi yield strength or less was based on a −30°F minimum operating temperature. The preceding procedure can be used to develop toughness requirements for any minimum operating temperature. The resulting toughness requirement for 50-ksi yield strength steels is 15 ft-lb at a test temperature that is 70°F (39°C) higher than the specified minimum operating temperature. Thus the CVN test temperatures and the minimum operating temperatures are linearly related. To minimize the proliferation of a variety of testing temperatures, and the resulting problems in the design and fabrication of steel bridges, the variable testing temperatures were comprehended by establishing three zones of service temperatures and providing temperatures and CVN values for each zone. The three zones of service temperatures and the corresponding test temperatures for bridge steels for 50-ksi yield strength or less are presented in Table 15.2.

The general relations between service temperatures and test temperatures for A36 steel satisfying the requirements of each of the three service-temperature zones are shown in Figure 15.9. These results show that, because of the loading-rate shift, CVN-toughness levels greater than 15-ft-lb are expected at intermediate loading rates approximately 70°F below the impact testing temperature. In terms of the NDT temperature measured using drop weight test specimens under impact loading, the minimum service temperature is approximately 70°F below NDT.

The specifications of the American Society for Testing and Materials (ASTM) for A572 Grade 50 and A588 steels require a minimum yield strength value of 50 ksi. Consequently, these steels may have yield strengths that are significantly higher than 50 ksi. The data in Figure 15.10 show that the magnitude of the temperature shift between static and impact loading rates decreased with increased yield strength. The magnitude of the decrease in the temperature shift is about 15°F (8°C) for every 10-ksi (69-MN/m²) increase in yield strength. To ensure the same fracture behavior for A572 Grade 50, A440, A441, A242, and A588 steels having yield strengths significantly greater than 50 ksi, the CVN test temperature was decreased incre-

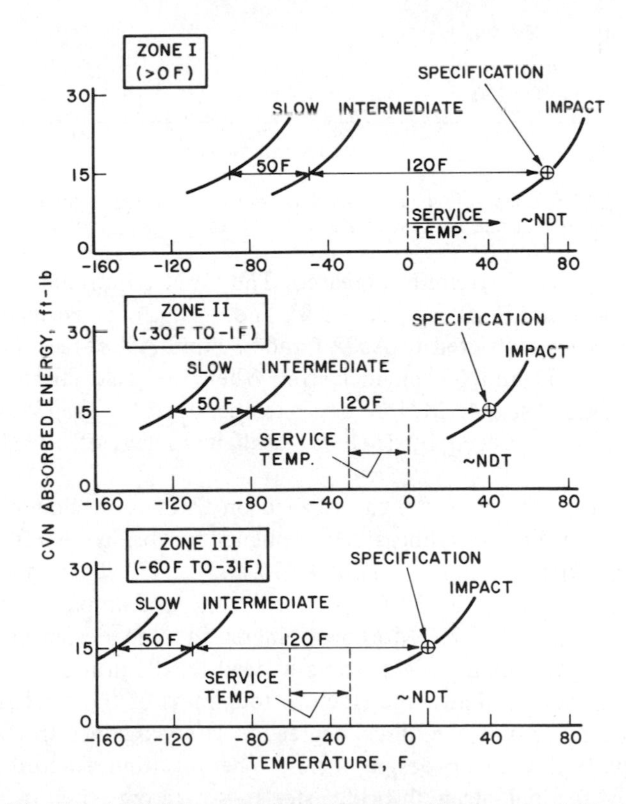

FIG. 15.9. AASHTO material-toughness specifications for A36 steel.

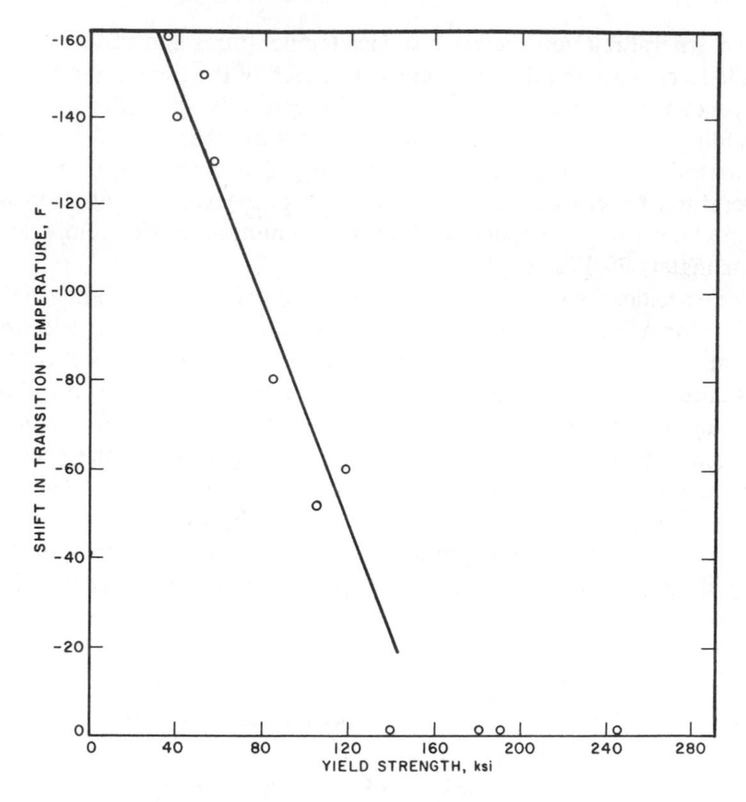

FIG. 15.10. Effect of yield strength on shift in transition temperature between impact and slow-bend K_{Ic} or CVN tests.

mentally as the yield strength increased. The CVN requirements of 15 ft-lb at +70°, +40°, and +10°F (+21°, +4°, and −12°C) for zones 1, 2, and 3, respectively, were restricted to A572 Grade 50 and A588 steels having yield points between 50 and 65 ksi, inclusive. When the yield strength of these steels exceeds 65 ksi (448 MN/m²), the temperature for the CVN value for acceptability was reduced by 15°F for each increment of 10 ksi above 65 ksi.

The above philosophy, which is based on fracture-mechanics concepts, was used to develop toughness requirements for bridge steels of 100-ksi (689-MN/m²) yield strength (ASTM A514 and A517). These steels show a temperature shift of 60°F (33°C) between static and impact loading rates (Figures 15.11 and 15.12). Moreover, increasing the design stress (which generally requires a higher-yield-strength steel) results in more stored elastic energy in a structure. Thus, the fracture toughness of the steel should also be increased to ensure the same degree of safety against fracture as the structure with the lower design stress. The resulting fracture-toughness requirements for high-strength bridge steels also are presented in Table 15.2.

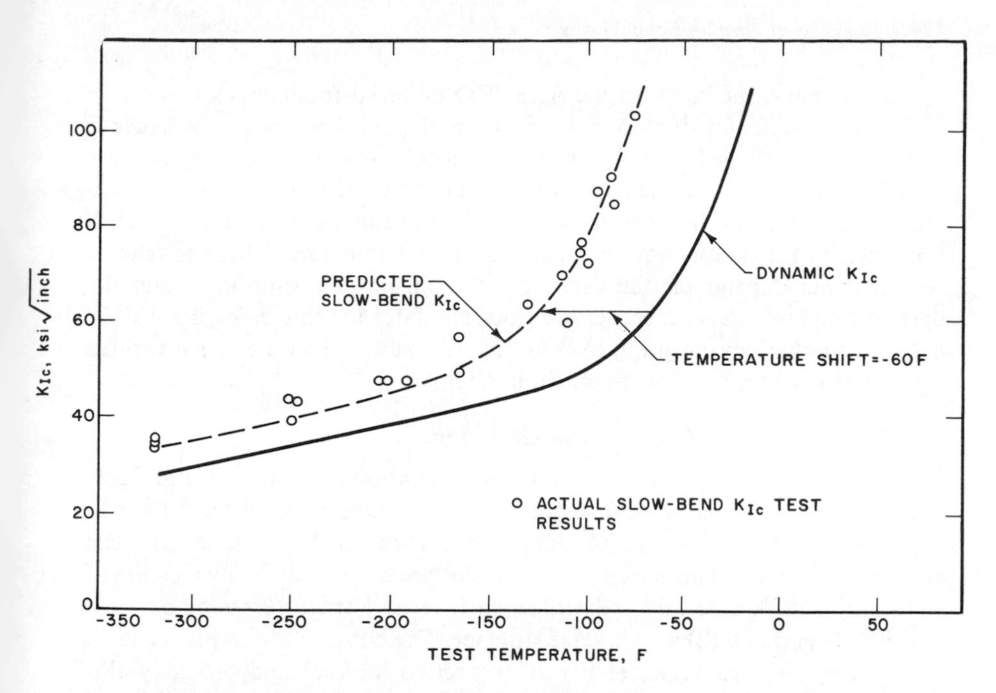

FIG. 15.11. Use of CVN test results to predict the effect of loading rate on K_{Ic} for A517-F steel.

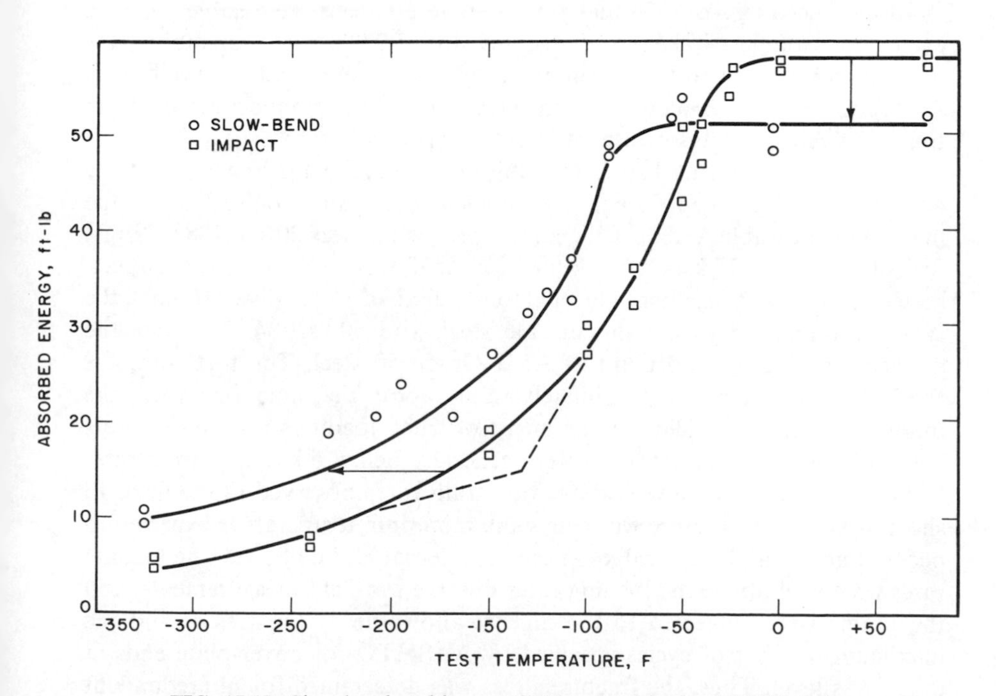

FIG. 15.12. Slow-bend and impact CVN test results for A517-F steel.

In summary, the basis for the AASHTO material-toughness specification is fracture mechanics. However, because the desired level of performance is outside the range of linear-elastic fracture-mechanics behavior and because toughness values cannot be currently measured by existing fracture-mechanics tests, correlations between K_{Ic} and CVN test results were used to establish the material-toughness requirements shown in Table 15.2. These toughness requirements depend on the existence of a strain-rate shift to obtain the desired toughness levels at a service (intermediate) loading rate 70°F (21°C) below the actual specification loading rate (impact), which has been verified by several investigators, as described in Chapter 4.

15.3.3. Fracture Tests of Welded Beams

To verify the adequacy of the toughness requirements presented in Table 15.2, beam specimens of A36 steel and A572 Grade 50 steel were used to study the fracture behavior of simulated bridge members under extreme service conditions. The specimens were designed to include two common structural details that adversely affect fatigue and fracture strength: (1) a cover-plate end and (2) a transverse stiffener. The cover-plate end is the most severe common structural detail with respect to fatigue[7] and probably also fracture.[5] A detailed description of the beam specimens and an analysis of the test data have been presented elsewhere.[8]

Beam specimens of A36 and A572 Grade 50 steels were subjected to a cyclic stress range of 21 ksi (145 MG/m^2) for 100,000 cycles or more. This loading corresponds to the maximum allowable fatigue loading specified by AASHTO[9] for cover-plate ends in either steel but is much more severe than the cyclic loadings measured in actual bridges.[4]

After the specimens had been subjected to cyclic loading, some were cooled to −30°F (−34°C) and then loaded under an impulse load to the maximum allowable AASHTO bending stress, which was 20 ksi (138 MN/m^2) for A36 steel and 27 ksi (186 MN/m^2) for A572 Grade 50 steel. If the specimen did not fail, it was then subjected to impulses of 36 ksi (248 MN/m^2), the specified minimum yield point of A36 steel, and 50 ksi (345 MN/m^2), the specified minimum yield point of A572 Grade 50 steel. The total time for the 20-ksi impulse was approximately 1 sec—about the same time as for the impulses observed in field measurements of truck loadings.[4] Since the stress levels in the field measurements were generally below 6 ksi, the strain rates in the test impulse were well above the strain rates observed in the field. In short, the test temperature was below the minimum temperature expected to occur in actual highway bridges in the continental United States, the loading rates were well above the loading rates observed in field measurements, and the beams were subjected to the highest allowable cyclic stresses and the maximum number of cycles specified by AASHTO for cover-plate ends at that stress level. Thus, the fracture stress was determined for nonredundant (single-load-path) bridge members containing the *most severe* common structural detail and for the *most severe combination* of temperature, strain rate, and prior fatigue loading that could reasonably be expected to occur in actual highway bridges in the continental United States.

All the simulated bridge members tested contained fatigue cracks at the end of the cover plate which had initiated and propagated when the specimens were subjected to the full allowable fatigue loadings permitted by AASHTO. However, these simulated highway-bridge members had sufficient fracture toughness under the most severe test temperatures and strain rates to prevent premature fracture when the members were subjected to the maximum design stress permitted by AASHTO at that time—20 ksi for A36 steel and 27 ksi for A572 Grade 50 steel. Specifically, the nominal stresses at actual fracture (which exclude residual stresses and stress intensification caused by stress concentrations) were equal to about 34 ksi (234 MN/m^2) and 46 ksi (317 MN/m^2) for the A36 and the A572 Grade 50 steels, respectively; these values are close to the specified minimum yield point of the respective steels. Thus, the simulated highway-bridge members fractured at stress levels about 70% higher than the maximum design stress permitted by AASHTO at that time.

Under the extreme prior fatigue-loading conditions, which far exceeded those usually encountered in actual bridges, cracks developed at the cover-plate ends and ranged in depth from 0.03 to 0.375 in. (0.76 to 9.5 mm). Fatigue-life estimates suggest that the useful fatigue life of the beams had been almost exhausted. In service, members with cracks about 0.4 in. (10.2 mm) deep would soon fail by fatigue regardless of the level of fracture toughness of the steel. Thus, an increase in fracture toughness above the values measured in the steels investigated would not significantly improve the service performance of such bridge members as discussed in Chapter 14.

The above discussion should not imply that crack sizes significantly larger than 0.4 in. cannot exist in actual bridge components. That such large cracks can be tolerated is explained by the redundancy built into the design of many bridges, by the propagation of fatigue cracks in a decreasing stress field, by the small stresses applied to actual bridges compared with the maximum design stresses, and by the extremely low probability that the severest detail, lowest temperature, and highest strain rate will occur simultaneously.

In summary, the test results suggest that a beam with an end-welded cover plate subjected to 100,000 cycles of an applied 21-ksi (145-MN/m^2) stress range would be expected to contain a crack about 0.4 in. (10.2 mm) deep and with a length equal to the width of the cover plate. The results of the fracture tests showed that these cracks would not result in fracture of the beams under the maximum allowable design stress, even when the beams are subjected to the maximum loading rate that occurs in bridges and the minimum operating temperature. The test results also indicated that a substantial increase in the fracture toughness of the steel would have resulted in a negligible increase in the useful life of the tested beams.

15.4. Proposed Fracture-Control Guidelines for Welded Ship Hulls

15.4.1. General

The problem of brittle fractures in ships was first recognized as a serious problem in the early 1940s when there were numerous brittle fractures in the World War II Liberty and Victory Ships. While it is true that since the 1940s considerable research has led to various changes in design, fabrication, and materials so that the incidence of brittle fractures in welded ship hulls has been reduced markedly,[10] brittle fractures continue to occur in welded ship hulls fabricated with ordinary-strength steels. With the use of higher-strength steels, there is a definite concern that brittle fractures may occur in these steels also. Because there were no specific fracture-control guidelines or overall toughness criteria available for the practicing naval architect to specify in designing welded steel ship hulls of all strength levels, the Ship Structure Committee initiated an investigation in 1972 to provide rational fracture-control guidelines consistent with economic realities which, when implemented, would minimize the probability of brittle fractures in welded ship hulls. Although the guidelines developed[11] were primarily material-oriented, this does not relieve the naval architect of responsibility for good ship design but recognizes the fundamental importance of using good-quality structural steels in large complex welded structures.

15.4.2. Service Conditions for Ship Hulls

A review of current practice of designing ship hulls indicates that the actual loadings are not well known.[12,13] Therefore, general rules of proportioning the cross section of ships have been developed, primarily on the basis of experience. Recent developments in analytical techniques and actual measurements of ship loadings have led to improvements in the understanding of the structural behavior of ships.[14] However, the design of ship hulls is primarily an empirical proportioning based on satisfactory past experience rather than a systematic analytical design, and therefore calculated design stresses for specific sea states are rarely found.

Strain measurements on actual ships have indicated that the maximum vertical wave-bending-stress *excursion* (peak-to-trough) ever measured is about 24 ksi (165 MN/m^2). Also the maximum bending stress for slender cargo lines is about 10 ksi (69 MN/m^2) and for bigger ships such as tankers and bulk carriers, about 14 ksi (97 MN/m^2).[13,15] Therefore, 14 ksi (97

MN/m^2) appears to be a reasonable maximum *nominal* stress level in ship hulls. Although this stress is less than one-half the yield stress of most ship hull steels, the local stress level at stress concentrations can reach the yield strength level, particularly when the additional effects of residual stress are considered. Furthermore, because actual loading rates for ship hulls are not known, the following fracture criterion was based on the very conservative assumption that ships can be loaded under impact conditions, i.e., the loads can be applied rapidly enough so that the dynamic yield stress is reached. This is a crucial assumption and is still under investigation. It should be emphasized that the proposed material toughness requirements described in this section would be changed significantly if an intermediate loading rate (such as described in Section 15.3.2 for Bridges), were assumed for ship hull structures (rather than a dynamic loading rate).

The actual loading rate for ship hulls is probably between the limits of "static" loading (strain rate approximately 10^{-5} sec^{-1}) and dynamic or impact loading (strain rate approximately 10 sec^{-1}). However, in view of the general service conditions to which ships are subjected, the fact that ships are essentially single-load path structures rather than multiple-load path structures such as most bridges, and the lack of information on specific loading rates, the conservative assumption that ships are loaded dynamically was made.

Studies[16] have shown that ships operate at temperatures less than 32°F (0°C) only about 3% of the time (Figure 15.13). Therefore, a design service temperature of 32°F (0°C) for welded steel ship hulls appears realistic. For special applications, such as icebreakers, the design service temperature should be lower. For the present fracture-control guidelines, it was assumed that critical regions in welded ship hulls can be subjected to impact loadings at 32°F (0°C) and thus, the dynamic fracture parameters, K_{Id}/σ_{yd}, rather than static fracture parameters, K_{Ic}/σ_{ys}, were used. If subsequent investigations should show that the maximum loading rates are somewhat slower, then it may be possible to modify the proposed toughness requirements, by using, for example, the strain-rate shift procedures discussed in Section 15.3.

15.4.3. Required Performance Characteristics

Previously, it has been shown that brittle fractures occur because of particular combinations of material toughness, crack size, and tensile stresses. If this basic principle is combined with the realistic fact that the stress level in critical parts of a ship hull can reach yield stress magnitude and that discontinuities can be present in the hull, the naval architect is faced with three possible solutions to prevent catastrophic brittle fractures in ships:[17]

1. Develop multiple-load paths within the hull so that failure of any one part of the cross section does not lead to total failure of the ship.

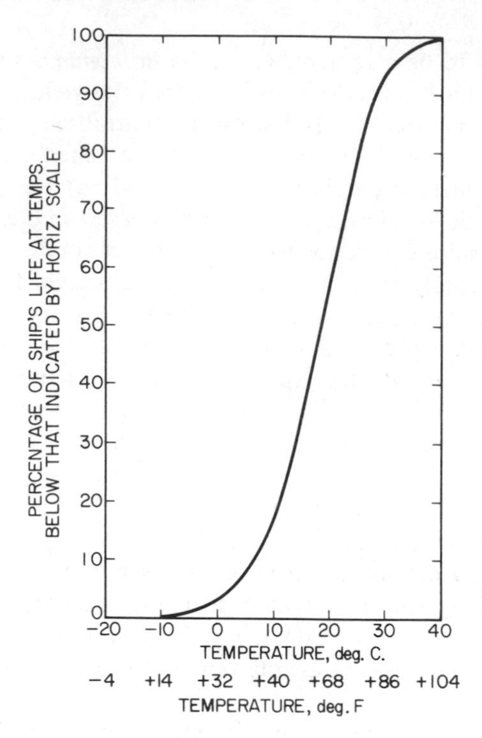

FIG. 15.13. Distribution of service temperature for ships (Ref. 16).

Although this solution is satisfactory for other types of welded structures such as stringer-type bridges with concrete decks, it does not appear to be feasible for monolithic welded steel ship hulls.

2. Use extremely notch-tough steels so that no brittle fractures can initiate or propagate, even at very high stress levels. Although this solution would eliminate the problem of brittle fracture in welded steel ship hulls, it is economically unfeasible because such extreme levels of notch toughness actually are not required. Furthermore even notch-tough materials can fail under cyclic loads or if the loading is severe enough.
3. Provide a fail-safe design using steels with moderate levels of notch toughness in combination with properly designed crack arresters, so that even if a crack initiates, it will be arrested before catastrophic failure occurs.

The fundamental problem in a realistic fracture-control plan for welded ship hulls is to optimize the above possible performance criteria with cost considerations so that the probability of complete structural failure due to brittle fracture in welded ship hulls is very low. In that sense, the toughness criterion proposed was an attempt to optimize satisfactory performance with reasonable cost, following a fail-safe philosophy.

Thus, the third solution, namely the use of steels and weldments with moderate levels of notch toughness combined with properly designed crack arresters, is recommended as a fracture criterion for welded ship hulls.

In line with this general fracture-control plan, the following items are noted:

1. As has been well documented during the past 30 years, the definite possibility of brittle fracture in welded ship hulls exists because welded ship hulls are complex structures that can be subjected to local yield-point stresses at temperatures as low as 32°F (0°C). In these guidelines it was assumed that the loading rate of ships was dynamic because of (a) lack of sufficient data on loading rates and (b) the fact that many welded ships are single-load-path structures. Obviously this is the most conservative assumption regarding loading rate that could be made.
2. Because of current limitations in fabrication practice and inspection at shipyards, the probability exists that large discontinuities may be present at some time during the life of welded ship hulls. Even with improvements in control of welding quality during fabrication, some discontinuities will still be present prior to the service life of the structure, and fatigue loading may cause these discontinuities to grow in size during the life of the structure. Unfortunately, unlike bridges, there are very limited data on the fracture behavior for ship hull details or on the number and magnitude of stress fluctuations that would be applied to ship hulls during their design life. Thus, it was assumed that discontinuities are present in all welded ship hulls.
3. The naval architect generally does not have absolute control over the fabrication of a welded ship hull. Thus, he should establish material and design controls during the design process that are adequate to prevent the occurrence of brittle fractures in welded ship hulls. Although the designer tries to avoid details that act as stress raisers, this is practically an impossible task in large complex welded structures. Hence, the emphasis in this fracture-control plan is on the choice of proper materials (toughness specifications for steels and weldments) and design (proper use of crack arresters), even though quality fabrication and inspection of welds are extremely important.
4. Although specifying solely the metallurgy and manufacturing process, including composition, deoxidization practice, heat treatment, etc., has been used as one method of controlling the level of notch tough-

ness in a steel, the only method of measuring the actual toughness of a steel is a toughness test. A direct measure of toughness would appear to be better for the user because he is ultimately concerned with the performance of the steel or weldment, and this performance can best be determined by a notch-toughness test. Also a specification based on a notch-toughness test would appear to be more equitable for steelmakers in that it leaves them some latitude to adopt the process best suited to their particular operation in satisfying the toughness requirement. However, a toughness test does have the disadvantage in that a test value pertains to only one location in a plate, whereas *proper* processing control *should* pertain to the entire plate. However, because this may not always be true, a toughness test is *no less* effective as an indication of the service performance of the entire plate.

5. Because of the difficulties in conducting a toughness test on a composite weldment, existing ABS rules[18] specify that five sets of impact specimens be taken during welding procedure qualification testing for weldments used for very low-temperature service. The notches for the specimens are located at the center line of the weld, on the fusion line, and in the heat-affected zone, 0.039 in. (1 mm), 0.118 in. (3 mm), and 0.197 in. (5 mm) from the fusion line. For weld qualification tests it may be desirable to follow this practice.

The specific requirements to implement these fail-safe fracture-control guidelines consist of (1) establishing a satisfactory level of notch toughness in the steels and weldments and (2) development of properly designed crack arresters. It should be reemphasized that improper fabrication or operation can still lead to structural failure regardless of the level of notch toughness. Thus good-quality welding, inspection, and operating procedures should be followed.

15.4.4. Development of Toughness Requirement for Main-Stress Regions

In general, the primary load-carrying members of steel ship structures are the plate members within the center 0.4L of the hull that comprise the upper deck, bottom shell, side plating, and longitudinal bulkheads. Because these members are the primary load-carrying members, material-toughness requirements should be specified for them. Although stiffeners can also be primary load-carrying members, they are not connected to each other, and thus failure of one stiffener should not lead to failure of adjacent stiffeners. Therefore they need not be subject to the proposed criteria.

Stresses in a ship hull vary from extreme levels in the upper deck and bottom shell to essentially zero at the neutral axis, as indicated in Figure 15.14,

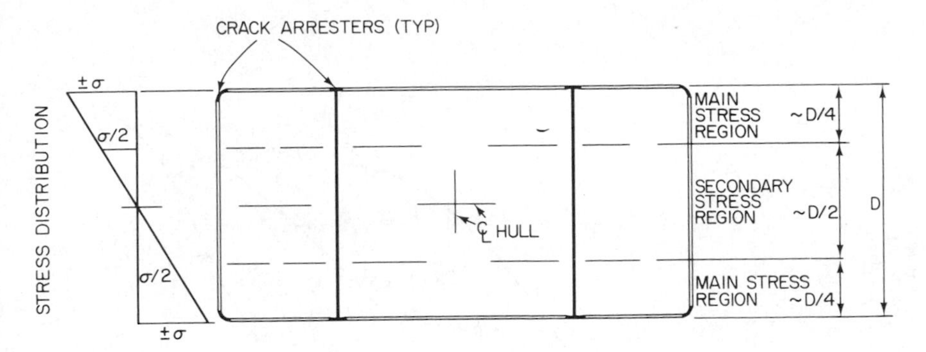

FIG. 15.14. Schematic cross section showing primary load-carrying members in main- and secondary-stress regions.

which illustrates an idealized stress distribution in the section. The critical crack size for a given material is influenced by the nominal tensile stress level. Therefore because stresses in the main-stress regions (Figure 15.14) can reach critical levels, the material performance characteristics of the primary load-carrying plate members in these areas should be specified by a toughness requirement. Stresses in the secondary-stress region are somewhat lower, and for primary load-carrying plate members in this area a less stringent toughness requirement is proposed.

Traditionally, the fracture characteristics of low- and intermediate-strength steels have been described in terms of the transition from brittle to ductile behavior as measured by impact tests. This transition in fracture behavior can be related schematically to various fracture states, as shown in Figure 15.15. *Plane-strain* behavior refers to fracture under elastic stresses generally with little or no shear lip development and is essentially brittle. *Plastic* behavior refers to ductile failure under general yielding conditions generally with very large shear lip development. The transition between these two extremes is the *elastic-plastic* region, which is also referred to as the mixed-mode region.

For static loading, the transition region occurs at lower temperatures than for impact (or dynamic) loading, depending on the yield strength of the steel. Thus, for structures subjected to static loading, the static transition curve should be used to predict the level of performance at the service temperature. For structures subjected to impact loading, the impact transition curve should be used to predict the level of performance at the service temperature. For structures subjected to some intermediate loading rate, an intermediate loading-rate transition curve should be used to predict the level of performance at the service temperature. Because the actual loading rates for ship hulls are not well defined, and to be conservative, the impact loading curve (Figure 15.15) is used to predict the service performance of ship hull

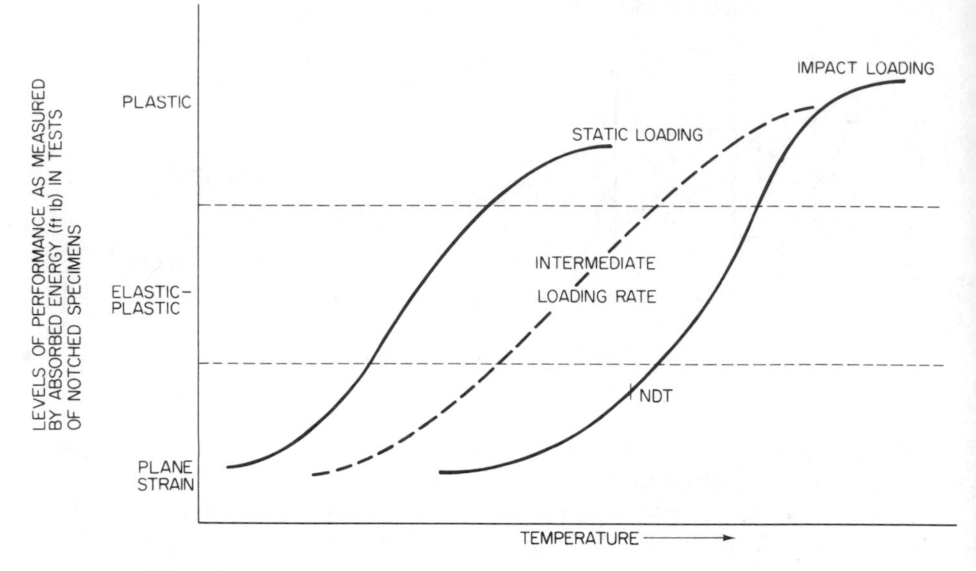

FIG. 15.15. Schematic showing relation between notch-toughness test results and levels of structural performance for various loading rates.

steels. As noted in Figure 15.15, the nil-ductility transition (NDT) temperature generally defines the beginning of a significant increase in fracture toughness under conditions of impact loading.

A fundamental question to be resolved regarding a fracture criterion for welded ship hull steels is, What level of material performance should be required for satisfactory performance in a ship hull consistent with realistic economic and safety considerations? That is, as shown schematically in Figure 15.16 for an assumed impact loading, one of the following three general levels of material performance must be established at the service temperature for the steels that are primary load-carrying members:

1. Plane-strain behavior: Use steel 1, Figure 15.16.
2. Elastic-plastic behavior: Use steel 2, Figure 15.16.
3. Fully plastic behavior: Use steel 3, Figure 15.16.

Although fully plastic behavior would be a very desirable level of performance for ship hull steels, it is not necessary or even economically feasible. A reasonable level of elastic-plastic behavior (steel 2, Figure 15.16) should be satisfactory to prevent initiation and propagation of brittle fractures under normal service conditions. (If fractures do initiate, they should not lead to catastrophic failure of a ship so long as properly designed crack arresters are used.) Specifying that the NDT temperature of all steels and weldments used in primary load-carrying members in the center 0.4L of ships be equal to or less than 0°F [32°F below the minimum service tem-

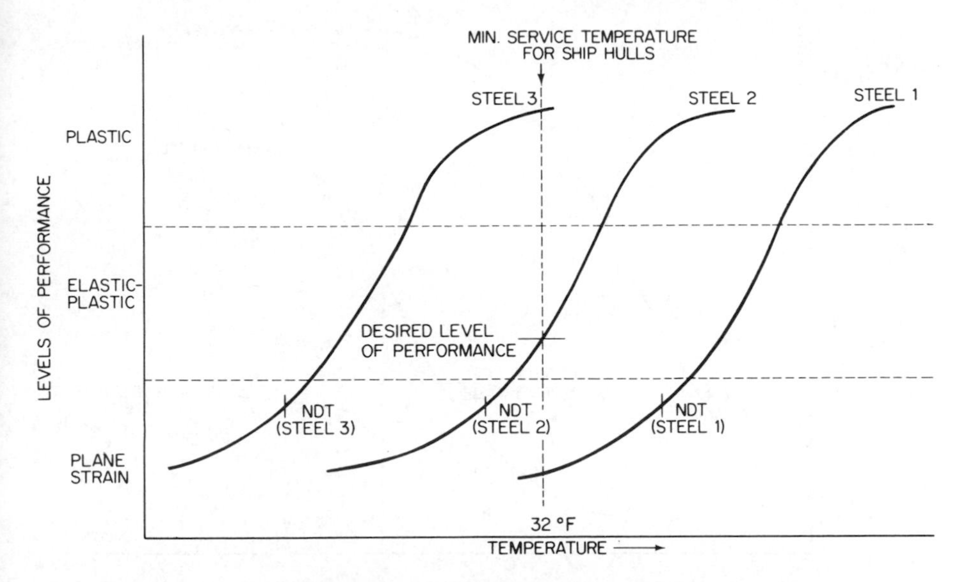

FIG. 15.16. Schematic showing relation between level of performance as measured by impact tests and NDT for three arbitrary steels.

perature] should establish the desired performance level under conditions of dynamic loading.

Consequently, the primary material specification in an overall fracture-control plan for welded steel ship hulls was that all steels and weldments used in primary load-carrying plate members in the main-stress regions of ships have a maximum NDT of 0°F (−18°C) as measured by ASTM Test Method E-208-69.[19]

Although necessary, this primary NDT requirement alone is not completely sufficient, since an additional toughness requirement is desirable to ensure that the resistance to fracture of the steels and weldments whose NDT is 0°F (−18°C) (or lower) is actually satisfactory at 32°F (0°C). That is, this additional requirement is necessary to ensure that materials follow the general performance level shown in Figure 15.16 rather than exhibit a low-energy shear behavior. Figure 15.17 shows the relationship of low-energy performance to normal behavior and very high-level behavior (HY-80-type behavior for military applications). Low-energy shear behavior usually does not occur in low-strength steels but is sometimes found in high-strength steels. Thus the additional toughness requirement is desirable to eliminate the possibility of low-energy shear failures, primarily for the higher-strength steels.

In terms of fracture-mechanics concepts, the critical dynamic toughness, K_{Id}, is approximately equal to $0.6\sigma_{yd}$ at NDT, where σ_{yd} is the dynamic

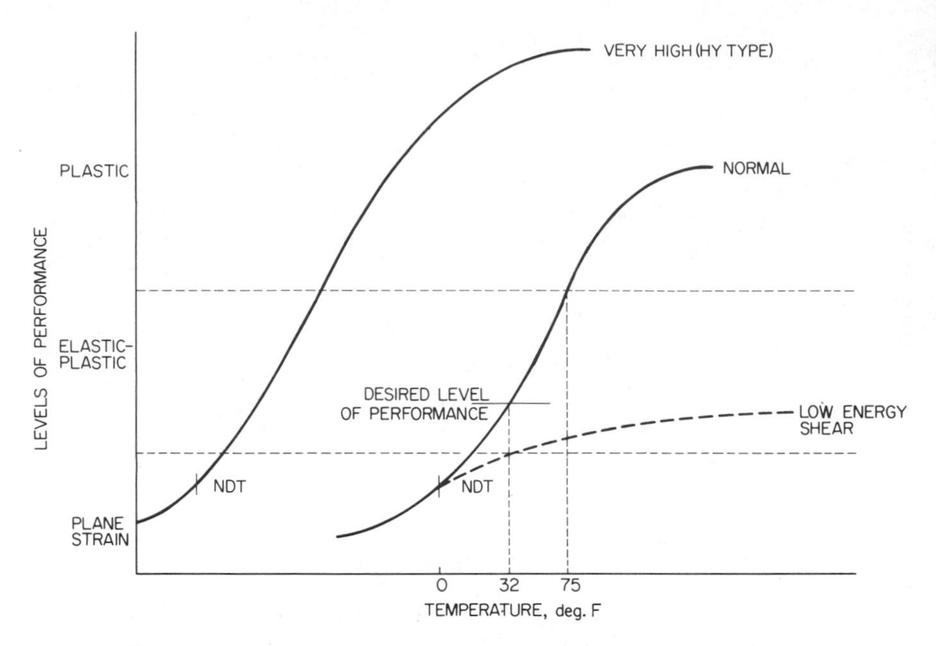

FIG. 15.17. Schematic showing relation between normal-, high-, and low-energy shear levels of performance as measured by impact tests.

yield strength of the material. Thus for the ship hull materials that satisfy the criterion that NDT be equal to or less than 0°F (−18°C),

$$\frac{K_{\mathrm{Id}}}{\sigma_{yd}} \geq 0.6 \qquad \text{at } 0°\text{F } (-18°\text{C})$$

At the minimum service temperature of 32°F (0°C),

$$\frac{K_{\mathrm{Id}}}{\sigma_{yd}} \text{ is estimated to be about } 0.9$$

because of the rapid increase in K_{Id} with temperature in the transition-temperature region. Although the value of 0.9 cannot be established theoretically, experimental results for various steels,[20] including ABS-C and ASTM A517 steels (Figures 15.18 and 15.19), indicate that this is a realistic value.

It should be emphasized that although concepts of fracture mechanics have been used to develop an auxiliary toughness requirement that $K_{\mathrm{Id}}/\sigma_{yd} \sim 0.9$ [for 1-in.-thick (25.4-mm) plates], materials satisfying this criterion will exhibit elastic-plastic, *non*-plane-strain behavior. Therefore, this toughness level *cannot* be measured using existing state-of-the-art fracture-mechanics tests as specified by ASTM[21] (Chapter 3). That is, for 1-in.-thick (25.4-mm) plates, the upper limit of dynamic plane-strain behavior is

$$1.0 = 2.5\left(\frac{K_{\mathrm{Id}}}{\sigma_{yd}}\right)^2$$

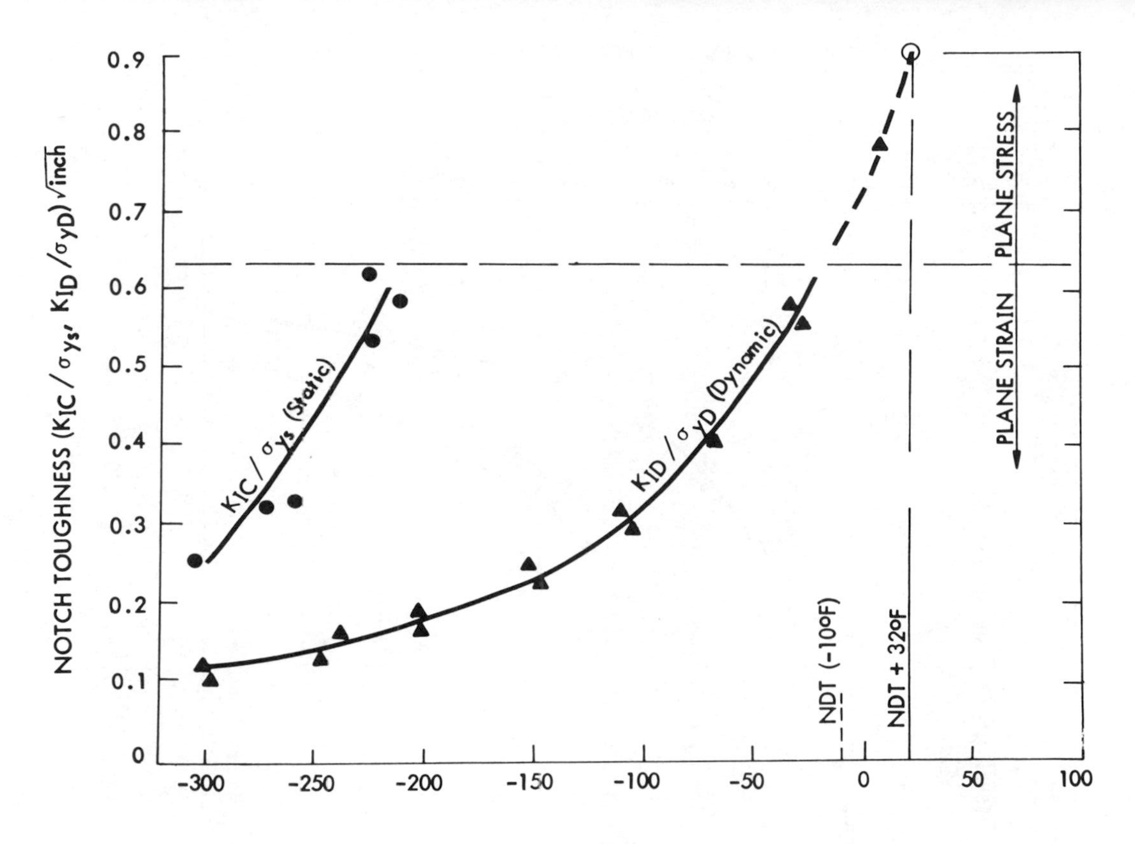

FIG. 15.18. Crack-toughness performance for ABS-C steel.

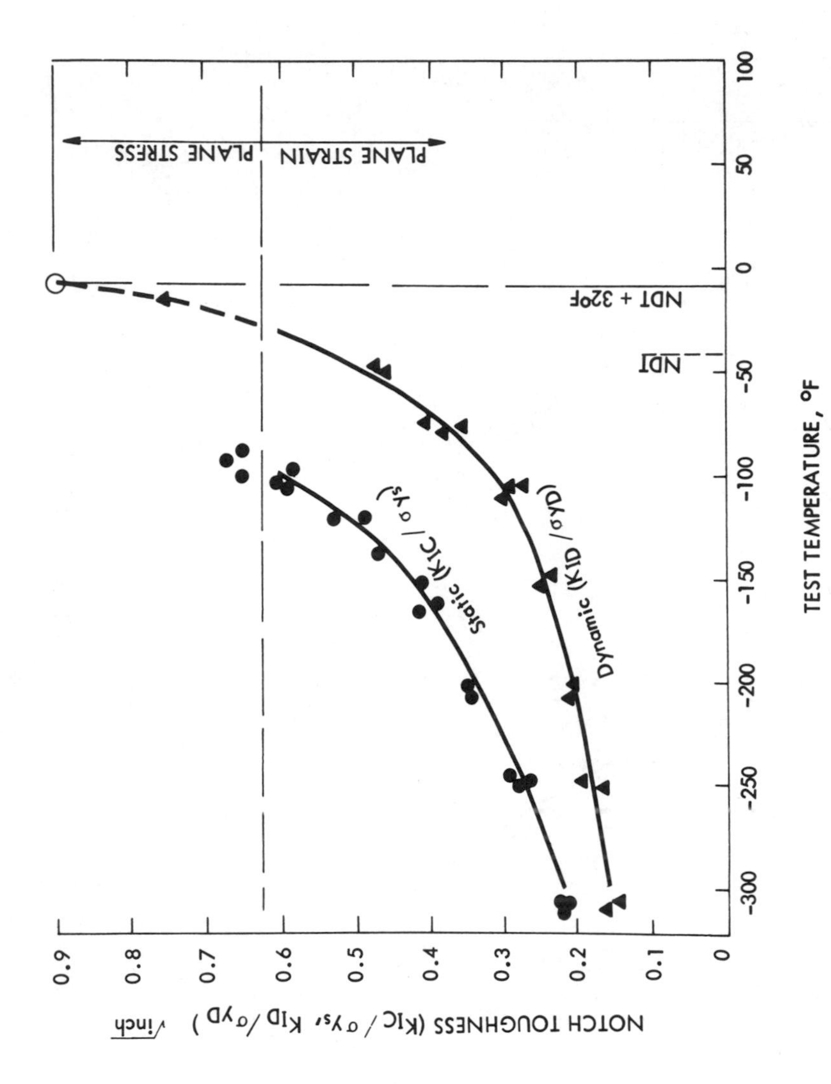

FIG. 15.19. Crack-toughness performance for A517-F steel.

or

$$\frac{K_{\mathrm{Id}}}{\sigma_{yd}} = 0.63$$

Thus NDT (where $K_{\mathrm{Id}}/\sigma_{yd} \simeq 0.6$) is close to the upper limit of dynamic plane-strain behavior for 1-in.-thick (25.4-mm) plates.

At 32°F (0°C), $K_{\mathrm{Id}}/\sigma_{yd}$ is specified in this proposed criterion to be 0.9, which is beyond the limits of dynamic plane-strain behavior for 1-in.-thick (25.4-mm) plates.

For 2-in.-thick (50.8-mm) plates,

$$2.0 = 2.5\left(\frac{K_{\mathrm{Id}}}{\sigma_{yd}}\right)^2$$

or $K_{\mathrm{Id}}/\sigma_{yd} = 0.89$ is the limit of dynamic plane-strain behavior. Thus, a 2-in.-thick (50.8-mm) plate, loaded dynamically to the full yield stress of a material in the presence of a sharp crack at 32°F (0°C), would be at the limit of dynamic plane-strain behavior. Because the probability of all these factors occurring simultaneously is minimal, the requirement that $K_{\mathrm{Id}}/\sigma_{yd} \geq 0.9$ appears to be satisfactory for all thicknesses of plate 2 in. (50.8 mm) or less. However, the required toughness levels for plates thicker than 2 in. (50.8 mm) should be increased.

Using concepts of fracture mechanics, as well as engineering experience, the following observations can be made regarding the level of performance at 32°F (0°C) for steels and weldments that satisfy the *primary* toughness requirement of NDT $\leq$ 0°F (−18°C) and the *auxiliary* toughness requirement that $K_{\mathrm{Id}}/\sigma_{yd} \geq 0.9$ at 32°F (0°C):

1. The start of the transition from brittle to ductile behavior will begin *below* the minimum service temperature of 32°F (0°C). Therefore, at the minimum service temperature, the materials will exhibit some level of elastic-plastic non-plane-strain behavior in the presence of a sharp crack under dynamic loading.
2. Although not specified in the proposed toughness requirement, the materials will exhibit some percentage of fibrous fracture appearance at 32°F (0°C). Service experience has shown that fracture appearance is an effective indicator of the resistance to brittle fracture. Thus, this criterion is consistent with service experience of ship hulls.
3. Although precise stress–flaw size calculations cannot be made for material exhibiting elastic-plastic behavior, estimates of critical crack sizes for 40-ksi (276-MN/m^2) yield strength steels can be made as follows:
   a. For $K_{\mathrm{Id}} \geq 0.9\sigma_{yd}$ and a nominal stress of 14 ksi (97 MN/m^2) the critical crack size at 32°F (0°C) is estimated to be 8–10 in. (203–254 mm), as shown in Figure 15.20.

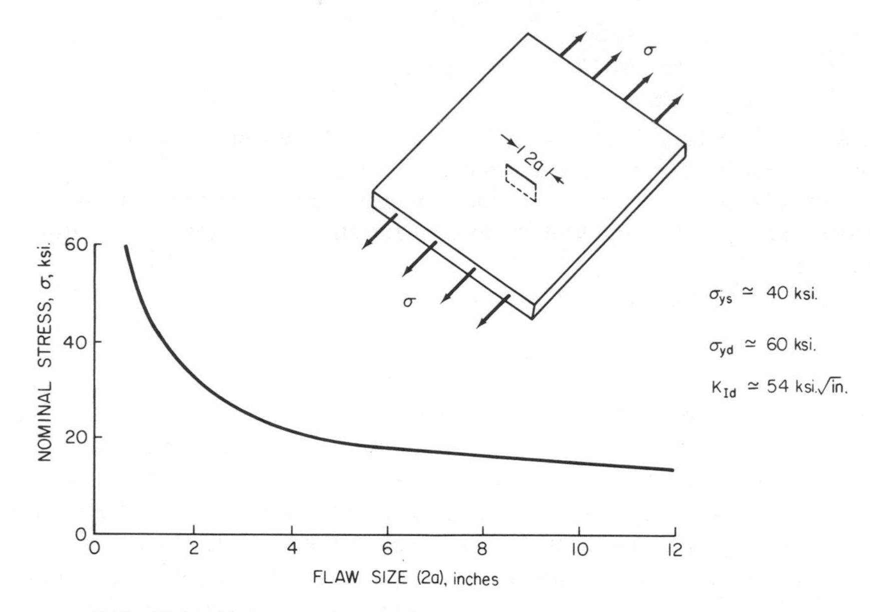

FIG. 15.20. Estimate of stress-flaw size relation for ABS steel with $K_{Id}/\sigma_{yd} = 0.9$.

b. For one of the largest stress ranges (peak-to-trough) ever recorded in ships, i.e., about 24 ksi (165 MN/m²), the critical crack size is estimated to be 3 in. (76 mm).

c. For the worst possible case of dynamic loading of yield-point magnitude, the dynamic critical crack size is estimated to be about $\frac{1}{2}$ in. (12.7 mm).

Ideally, the *auxiliary* toughnesss requirement that $K_{Id}/\sigma_{yd} \geq 0.9$ at 32°F (0°C) should be established by conducting a K_{Id} test at 32°F (0°C). Unfortunately, no inexpensive standard K_{Id} test specimen exists. Furthermore, research test procedures to obtain K_{Id} values directly are currently too complex for use in specifications (Chapter 3). Thus some other test specimen must be used to ensure that $K_{Id}/\sigma_{yd} \geq 0.9$ at 32°F (0°C).

The test specimen should be loaded dynamically, easy to use, and standardized, and the results should be readily interpretable. In addition, the specimen should have a sharp notch to approximate closely the sharp crack conditions that exist in large complex welded structures such as welded ship hulls. Finally, the test specimen should be as large as practical because of the effect of constraint on the fracture behavior of structural steels.

After careful consideration of which of the various fracture test specimens (e.g., CVN, pre-cracked CVN, Crack-Opening Displacement-COD, DT, and K_{Id}) would be most applicable to the particular requirement for welded ship hulls, the $\frac{5}{8}$-in.-thick (15.9-mm) dynamic tear (DT) test specimen[22] was recommended as the auxiliary test specimen.

For the ship hull steel application, the DT test specimen satisfied all the above requirements better than any other test specimen. The DT test is an impact test (high loading rate) that has a sharp pressed notch with residual tensile stresses (thus the strain concentration is larger than for machined notches or possibly a fatigue crack). The beginning of the elastic-plastic transition occurs just above NDT, as shown in Figures 15.21, 15.22, and 15.23, for representative ABS-B, ABS-C, and A517 steels, respectively. Thus the DT test specimen results may be related to the NDT values for ship steels.

For the plate thicknesses normally used in ship hull construction [less than 2 in. (50.8 mm) thick], thickness has a second-order effect on the toughness behavior in the transition-temperature region compared with the first-order effects of loading rate and notch acuity. Increasing the load rate of fatigue-cracked steel specimens raises the transition temperature, as shown in Figures 15.18 and 15.19. Increasing the notch acuity (from that in a machined CVN specimen to that in a pressed-notch DT specimen) also raises the beginning of the transition-temperature range, as shown in Figures 15.21–15.23. The second-order effect of thickness [namely the very small change in transition behavior between $\frac{5}{8}$- (15.9-mm) and 1-in.-thick (25.4-mm) DT specimens] is shown in Figure 15.21–15.23. There are *larger* changes in transition temperature for much thicker plates [e.g., 3- to 12-in.-thick (76- to 305-mm) plates used in thick-walled pressure vessels], but for the ship hull application [plates less than 2 in. (50.8 mm) thick], the effects of specimen thickness are second-order and are ignored.

Therefore, although it would be technically more desirable to use full-thickness DT specimens to specify the behavior of ship steels, only the $\frac{5}{8}$-in.-thick (15.9-mm) DT specimen was recommended because the practical aspects of testing the $\frac{5}{8}$-in.-thick (15.9-mm) DT specimen far outweighed the disadvantage of having to use a less than full-plate thickness test specimen. The $\frac{5}{8}$-in. (15.9-mm) DT specimen has recently been standardized (MIL Standard 1601[22]) and can be conducted in existing NDT-type falling weight test machines or in relatively small pendulum-type machines.

For the above reasons, the DT test was recommended as the auxiliary test specimen to be used to ensure that elastic-plastic behavior is actually being obtained in steels and weldments for welded ship hulls even though CVN impact test results currently are widely used as reference values for predicting the behavior of ship steels.

After having selected the DT test specimen as the auxiliary test specimen, the next step was to establish the DT value at 32°F (0°C) that will ensure a K_{Id}/σ_{yd} ratio of 0.9 so that the desired level of elastic-plastic behavior is obtained for all steels and weldments. Because there are no direct theoretical

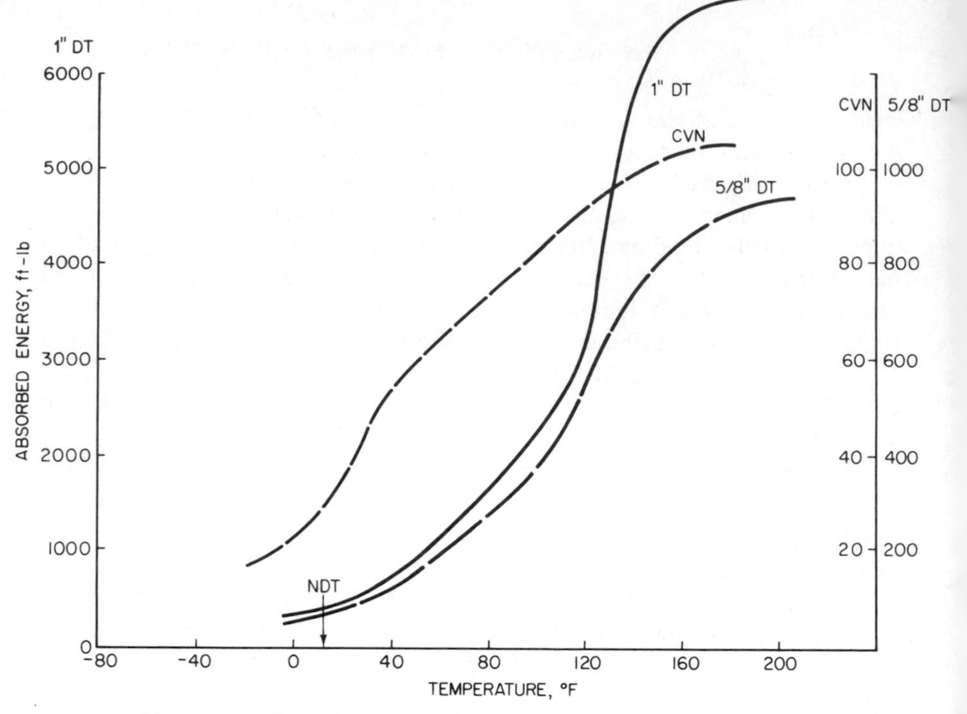

FIG. 15.21. Relation between NDT, CVN, and DT test results for ABS-B steel.

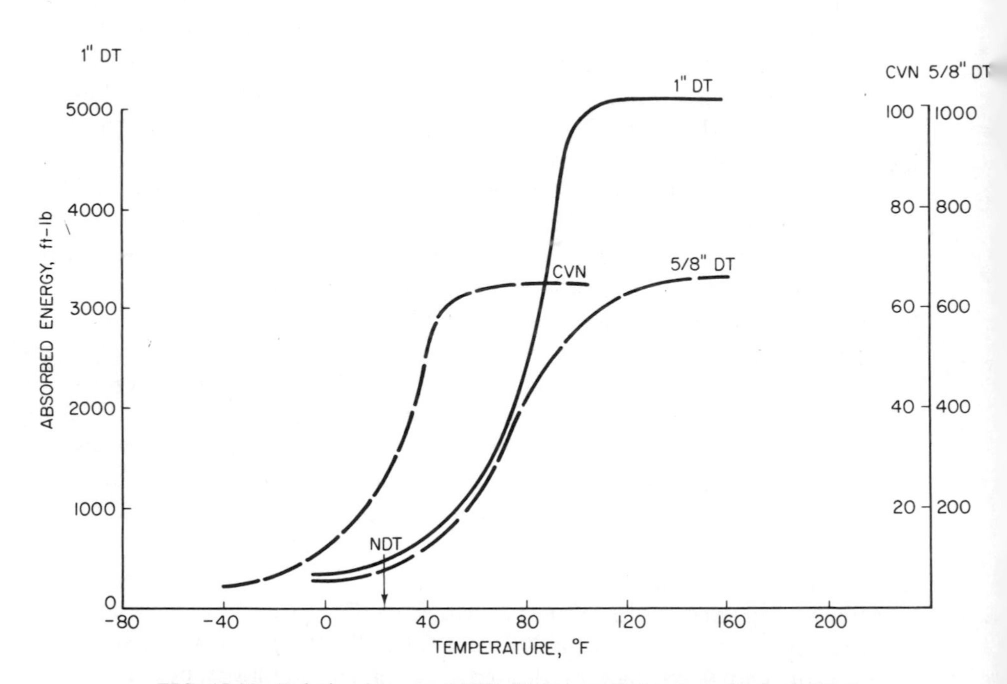

FIG. 15.22. Relation between NDT, CVN, and DT test results for ABS-C steel.

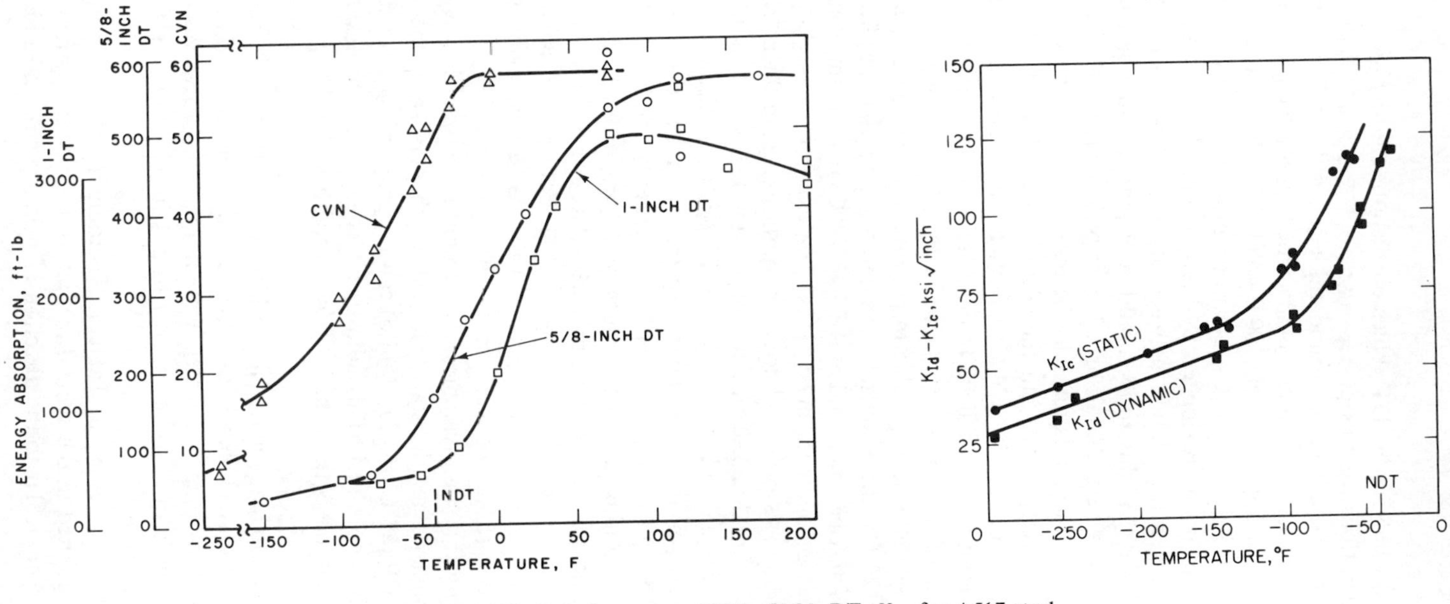

FIG. 15.23. Relation among NDT, CVN, DT, K_{Ic}, for A517 steel.

solutions to establish the DT values corresponding to $K_{Id}/\sigma_{yd} = 0.9$, empirical considerations were used.

A review of available experimental test results indicates that at NDT, where $K_{Id}/\sigma_{yd} \simeq 0.6$, the amount of absorbed energy for $\frac{5}{8}$-in.-thick (15.9-mm) DT specimens is approximately 100 ft-lb (136 J). Thus, at the specified value of $K_{Id}/\sigma_{yd} = 0.9$ at 32°F (0°C), the minimum absorbed energy for the DT specimens can be approximated by 0.9/0.6 times 100, or equal to 150 ft-lb (203 J). The general relation between K_I and energy in the elastic region would indicate that this ratio should be squared. However, in the elastic-plastic region, where the absorbed energy is increasing very rapidly with temperature, a linear relation may be more realistic. The value of 150 ft-lb (203 J) is relatively small, and therefore it was recommended that the DT test be conducted at 75°F (24°C) (room temperature) rather than at 32°F (0°C) because it may be difficult to measure a significant change in resistance to fracture between 0°F (−18°C) and 32°F (0°C). Although from a technical viewpoint it would be preferable to conduct the DT test at *both* 32°F (0°C) and 75°F (24°C), the practical considerations of the specification suggest that the DT test be conducted at +75°F (24°C) (room temperature).

If the test is conducted at 75°F (24°C), the minimum K_{Id}/σ_{yd} ratio should be 1.5 on the basis of a nonlinear extrapolation from 0.9 at 32°F (0°C), as shown in Figure 15.24. Thus, the minimum DT value should be 1.5/0.9 times 150, or equal to 250 ft-lb (339 J). Figure 15.24 also shows a schematic representation of the lower-bound specification curve of *required* values [NDT = 0°F and $K_{Id}/\sigma_{yd} > 1.5$ at 75°F—actually 250 ft-lb (339 J) in a DT test] and the minimum desired values of $K_{Id}/\sigma_{yd} = 0.9$ at 32°F (0°C) compared with possible curves for ship steels that either do or do not meet the criterion. This figure shows that by meeting both of the toughness requirements at 0°F (−18°C) and 75°F (24°C) the desired behavior at 32°F (0°C) ($K_{Id}/\sigma_{yd} \geq 0.9$) should be met.

Assuming that the dynamic yield strength is approximately 20 ksi (138 MN/m²) higher than the static yield strength of a steel, the required DT values at 75°F (24°C) ($K_{Id}/\sigma_{yd} \geq 1.5$) can be proportioned for strength level, as shown in Table 15.1. This adjustment is necessary to ensure that high-strength steels have the same *relative* toughness levels as lower-strength steels.

Thus, the auxiliary material specification in an overall fracture-control plan for welded steel ship hulls was that all steels and weldments used in primary load-carrying plate members in the main-stress regions of ships exhibit the levels of absorbed energy in a $\frac{5}{8}$-in. (15.9-mm) DT specimen as presented in Table 15.3.

The values presented in Table 15.3 should be the minimum values of specimens oriented in the same direction as the primary stress level (notch oriented perpendicular to the direction of primary stress). In most cases, the

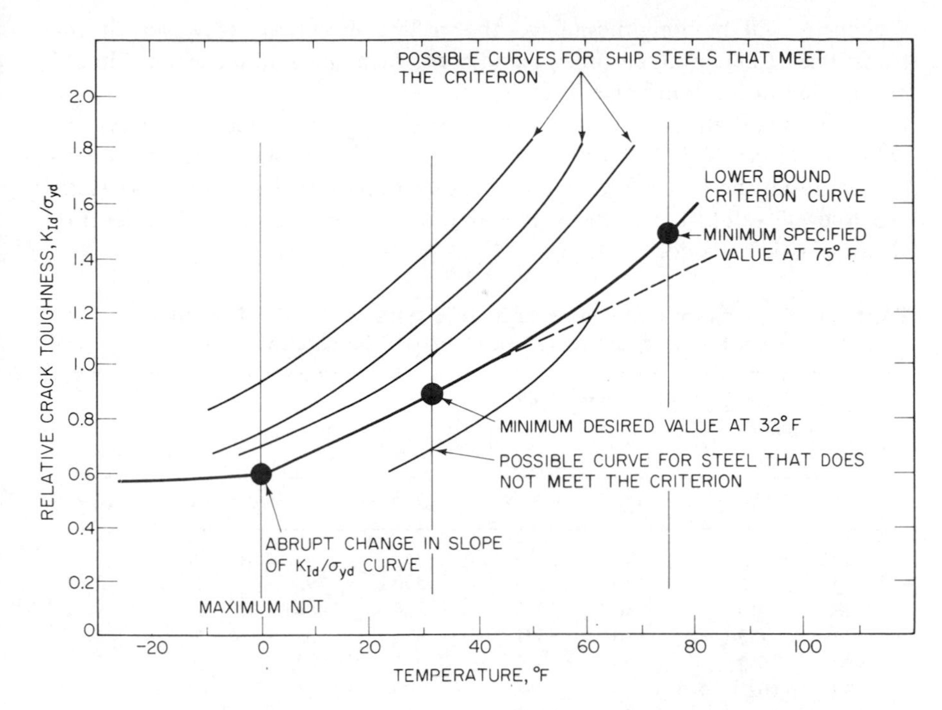

FIG. 15.24. Schematic showing the relation between proposed toughness criterion for members in the main stress region and behavior of actual ship steels.

TABLE 15.3. Dynamic Tear (DT) Requirements at +75°F (24°F) for Steels and Weldments in Main-Stress Regions for Primary Load-Carrying Members* of Ship Hulls

Actual Static Yield Strength, σ_{ys}		*Assumed Dynamic Yield Strength,* σ_{yd}		*Proportionality Factor for Strength Level*	*Absorbed Energy Requirements† for* $\frac{5}{8}$*-in.-thick (15.9 mm) Specimens*	
ksi	*MN/m²*	*ksi*	*MN/m²*		*ft-lb*	*J*
40	276	60	414	(60/60)	250	339
50	345	70	483	(70/60)	290	393
60	414	80	552	(80/60)	335	454
70	483	90	621	(90/60)	375	508
80	552	100	689	(100/60)	415	563
90	621	110	758	(110/60)	460	624
100	689	120	827	(120/60)	500	678

*These members must also meet the requirement of NDT $\leq$ 0°F (−18°C)
†Dynamic elastic-plastic behavior approximating $K_{Id}/\sigma_{yd} \simeq 1.5$.

specimens will be longitudinal to the rolling direction. However, if the transverse stress level becomes significant, then the test specimens should be oriented in the transverse direction.

CVN values (using a standard notch) that are equivalent to a K_{Id}/σ_{yd} value of 0.9 at 32°F (0°C) are presented in Table 15.4. It should be emphasized that the values presented in Tables 15.3 or 15.4 are *not* fully plastic "shelf-level" values but rather values that should ensure the desired level of elastic-plastic behavior.

TABLE 15.4. Equivalent CVN Values for $K_{Id}/\sigma_{yd} \geq 0.9$ at 32°F (0°C) for Primary Load-Carrying Members Using CVN-DT Correlation

Static Yield Strength, σ_{ys}		*Dynamic Yield Strength,* σ_{yd}		*Required DT Value*		*Equivalent CVN Value*	
ksi	*MN/m²*	*ksi*	*MN/m²*	*ft lb*	*J*	*ft lb*	*J*
40	276	60	414	250	339	20	27
50	345	70	483	290	393	24	33
60	414	80	552	335	454	28	38
70	483	90	621	375	508	32	43
80	552	100	689	415	563	36	49
90	621	110	758	460	624	40	54
100	689	120	827	500	678	44	60

The above material specifications will not guarantee the complete absence of brittle fractures in welded ship hulls, although they should be satisfactory for all normal fabrication and loadings. Therefore, a fail-safe philosophy that incorporates properly designed crack arresters fabricated from steels with very high levels of notch toughness should be used in conjunction with the above material requirements, as described in ref. 11.

15.5. Proposed Fracture-Control Plan for Floating Nuclear Plant Platforms

15.5.1. General

The floating nuclear plant (FNP) is a totally integrated nuclear power station mounted on a floating platform and located at an offshore site protected by a breakwater. The platform is essentially a 400-ft-long, 378-ft-wide and 44-ft-deep barge on which a conventional pressurized water reactor, turbine-generator system, and associated facilities are mounted so that the platform becomes a floating foundation for the entire plant, as shown in Figure 15.25. The platform is designed as a large rectangular box

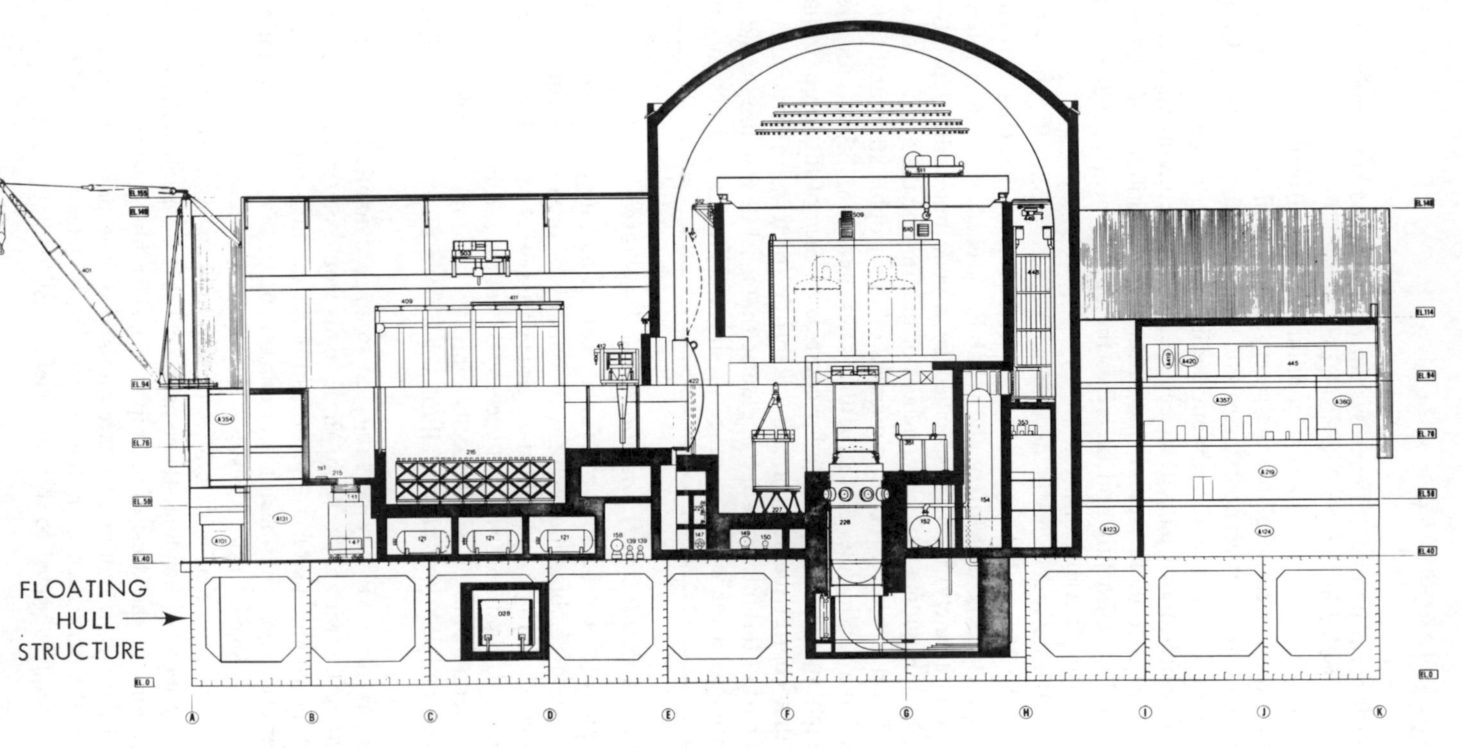

FIG. 15.25. Floating nuclear power plant.

grid utilizing a plate and stiffener system of all-welded steel construction, similar to that used in ship hull construction.

Because the platform hull is a continuous steel structure acting as a foundation for the entire plant, the safety and reliability of the platform structure must be such that no catastrophic brittle fractures occur under any credible service conditions throughout the life of the structure.

This fracture-control plan describes the use of fracture-mechanics concepts to interrelate those factors that affect the susceptibility of a floating platform to brittle fracture and describes the development of that plan for floating nuclear plant platforms[23].

One of the key questions in developing a fracture-control plan is how large must the degree of safety and reliability be for the particular structure in question. The degree of safety and reliability needed (sometimes referred to as factor of safety) is often specified by a code but really depends on many factors, including the consequences of failure. For the floating nuclear plant platform structure, the consequences of catastrophic structural failure are such that a *very* high level of structural safety and reliability was required. Accordingly specific requirements for controls on *each* of the three primary factors that control the susceptibility of brittle fractures in structures are made as part of this fracture-control plan so that a brittle fracture should not occur throughout the expected range of site temperature and design stress conditions.

15.5.2. Loading Conditions and Allowable Stresses

The platform will be analyzed for all known and hypothesized loading conditions and combinations with a probability of occurrence equal to or greater than 10^{-6}–10^{-7}. Allowable hull girder stresses for design base (extreme environmental) conditions will not exceed 90% of the specified yield strength of the material. Stresses will be determined by the finite element method of analysis and the Von-Mises failure theory for combining component stresses.

Although the FNP may be subjected to some dynamic loading during its operating life within the basin these loads would be small compared to the dynamic conditions of a ship operating on the high seas. Historically, the upper deck region has been the dominant source of catastrophic failures in ship hulls. Since tensile stresses are necessary for brittle fractures to occur, this indicates that most cracks in ship hulls have propagated during the hogging mode of bending. It seems logical that the probability of stress concentrations in the deck of a ship is considerably greater than in the bottom shell where discontinuities are minimized. The possibility of critical regions of the deck such as hatch corners, base plates for cargo-handling gear, etc., being subjected to dynamic yield stress loading is also reasonably high. However, this condition is not applicable to the platform. Analyses of the platform structure indicate that the platform hull girder will assume a sagging mode of bending for operating conditions within the basin and that primary stresses in the deck will be compressive except for possibly some localized conditions.

15.5.3. Static Loading Rate

The increase in notch toughness of structural steels with increasing temperature is well known and is the basis for the transition-temperature approach to designing to prevent brittle fracture. Not so widely known is the fact that *loading rate* also has a very significant effect on the behavior of structural steels. Loading rate refers to the time it takes to reach maximum load in a structure and can vary from very slow (strain rates of approximately 10^{-5} in./in./sec) to dynamic (usually impact strain rates of 10–20 in./in./sec). The effect of slow loading rates, compared with impact loading rates, is to shift the fracture-toughness curve to lower temperatures. For low-strength structural steels, such as ABS steels, the shift can be very large, as shown in Figures 15.26 and 15.27 for ABS Class C steel.[24] Note that this behavior occurs regardless of the type of test specimen used and thus is a characteristic of the material, not the test method.

The significance of this behavior is that a particular level of toughness

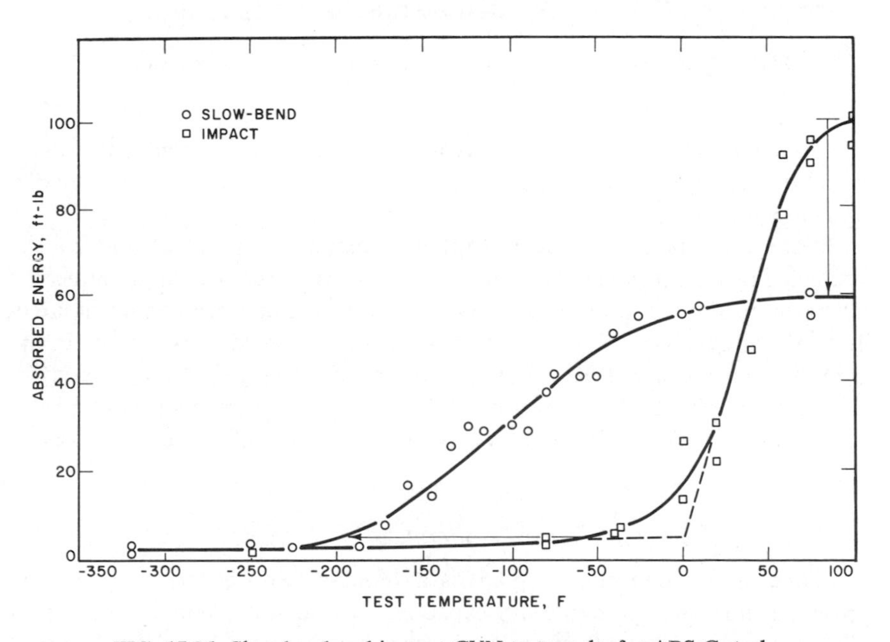

FIG. 15.26. Slow-bend and impact CVN test results for ABS-C steel.

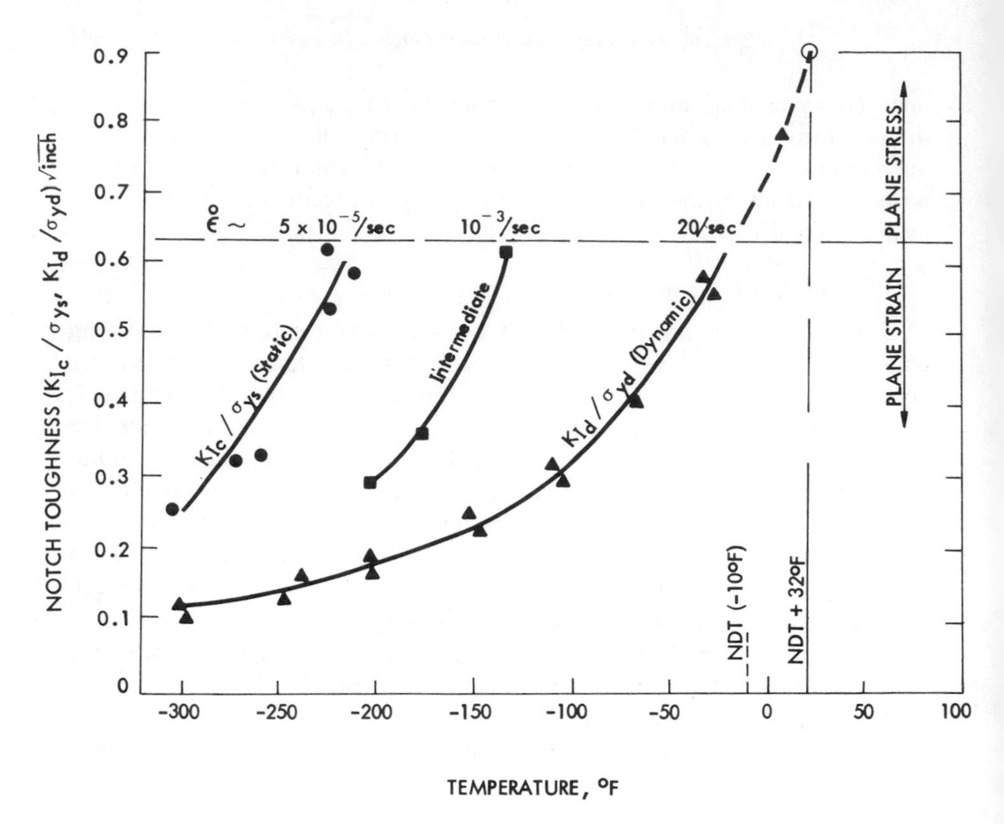

FIG. 15.27. Crack-toughness performance for ABS-C steel, showing effect of loading rate.

(or resistance to fracture) can be reached at much lower temperatures in structures that are loaded statically compared with structures that are loaded dynamically.

Because the floating nuclear plant will remain in one place and will be protected by the breakwater structure, the platform structure will be subjected primarily to static or intermediate loading rates (loading rates approximately 10^{-5}–10^{-3} in./in./sec). Nonetheless, as an added measure of reliability in the overall fracture-control plan, the *dynamic* properties of the hull material as measured by nil-ductility transition (NDT) and impact testing (loading rates approximately 10–20 in./in./sec) were specified with reference to the minimum service temperature.

15.5.4. Minimum Service Temperature

Since the floating nuclear plant is a standardized design, the plant site envelope must include minimum expected air and water temperatures as a site selection parameter. Consequently, material testing is based on the minimum service temperatures of 28.6°F for water and −5°F for air.

The submerged portion of the platform will be subjected to the temperature of the surrounding water. The lowest expected temperature of the seawater would be approximately 30°F. The main deck is completely enclosed, except for a small docking area, by buildings located on the platform. The deck temperature does not vary significantly from the temperature of the enclosed buildings, i.e., 40°F minimum. It approaches the temperature of the outside air only in the event of a prolonged loss of heating, and even then the deck would be in compression.

15.5.5. Dynamic Loading Versus Low Service Temperatures

The floating nuclear plant is exposed to many environmental and meteorological conditions which can occur in numerous combinations. A sound design takes account of the combination of such events or processes which could induce the most damage. From brittle-fracture considerations it is essential to evaluate the simultaneous occurrence of environmental loads such as tornadoes or hurricanes with the minimum expected temperatures associated with these loads.

The probability of simultaneous occurrence of tornadoes or hurricanes with very low temperatures is quite low. Table 15.5 (obtained from Reference 25) shows that the period of frequent tornadoes in the United States is April through August. The same table shows that hurricanes in the western North Atlantic Ocean occur mainly from July to October. Reference 26, which gives temperatures in percentile according to subregions along the eastern seaboard of the United States, indicates that for 99% of the time during the period April to October oversea air temperatures exceed 28°F and water surface temperatures exceed 32°F around Boston and temperatures are much higher southward. Consequently, the assumption of simultaneous occurrence of extreme environmental loads with the specified minimum service temperatures reflects a considerably high degree of conservatism.

15.5.6. Fatigue

Fatigue considerations in specified operating environments depend primarily on the number of load fluctuations of a given structure during its lifetime and the magnitude of the stress changes produced by these fluctuations. Not often recognized is the structural redundancy which limits the consequences of a possible fatigue failure. The more highly redundant a structure is, the less serious will be the consequences of a fatigue failure.

For the projected plant life of 40 years, it was estimated that the plant would have to experience 25,000 cycles/yr of alternating stress magnitude of 6 ksi from a static stress level of 18 ksi to fail in fatigue during its projected life. This is not credible, judging from the full-scale ship measurements reported[27] and especially considering the basin protection.

TABLE 15.5 The Monthly Worldwide Percentage Occurrence of Tropical Cyclones and Frequency Chart Showing the Number of Tornadoes and the Number of Tornado Days Per Month for the United States

Northern Hemisphere	*Jan.*	*Feb.*	*Mar.*	*April*	*May*	*June*	*July*	*Aug.*	*Sept.*	*Oct.*	*Nov.*	*Dec.*	*Period*	*Total*
Eastern North Pacific Ocean	4.4	2.9	2.9	0	2.9	4.4	4.4	19.2	41.3*	11.8	2.9	2.9	1903–40; 1953–60	68
Western North Pacific Ocean	0.8	0.2	0.5	1.1	4.5	6.5	18.9	20.6	20.7*	13.9	8.8	3.2	1884–1955	1,428
Origin in South China Sea	3.6	0	0	1.8	10.7*	8.9	23.2	28.6*	14.3	7.1	1.8	0	1903–40; 1953–60	56
Bay of Bengal	1.3	0	1.6	4.3	6.9	10.9	14.6*	10.7	12.0	16.7*	14.9	6.1	1891–1950	376
Arabian Sea	4.3	0.7	1.4	2.9	11.4	17.9*	2.9	1.4	6.4	24.3*	19.3	7.1	1890–1950	140
Western North Atlantic Ocean	0	0	0.3	0	0.6	4.8	7.0	27.9	36.8*	18.7	3.3	0.6	1886–1961	358
Origin E of 70° W	—	—	—	—	—	1.3	7.0	28.2	41.2*	18.8	3.2	—	1887–1932	155
Origin W of 70° W	—	—	—	—	—	15.8*	4.3	1.4	20.0*	44.1*	14.3	—	1887–1932	70
Southern Hemisphere	*July*	*Aug.*	*Sept.*	*Oct.*	*Nov.*	*Dec.*	*Jan.*	*Feb.*	*Mar.*	*April*	*May*	*June*	*Period*	*Total*
Western South Pacific Ocean	6.2	1.6	4.6	1.6	1.6	3.1	10.7	13.8	35.2*	7.7	7.7	6.2	1903–40; 1953–60	65
E of 160° E	0.4	0.4	0.8	1.6	2.9	13.9	26.9*	19.6	25.3*	6.9	0.8	0.8	1789–1924	245
W of 160° E	4.0	0	2.5	3.0	2.0	8.0	22.0*	17.5	20.5*	8.5	6.5	5.5	1867–1923	200
Eastern South Indian Ocean														
Western Australia	0.4	0.4	0.4	2.5	11.0	25.6*	24.5	25.8*	7.7	1.3	0	0.4	1870–1969	236
Western South Indian Ocean	0	0	0.5	0	2.8	8.9	29.4	31.3*	17.7	7.0	1.9	0.5	1912–1961	214

*Highest values.

15.5.7. Effect of Marine Environment on Fatigue Life

Data on steels which typify the platform steel indicate that this type of steel cyclically loaded at a stress level of 35 ksi will not result in fatigue failure prior to 10^6 cycles. If the steel were cyclically loaded in aerated seawater, the fatigue life may be reduced by the seawater environment. According to Reference 28, if the hull is maintained at a cathodic potential of $-0.79V_{Ag}$/AgCl or greater (more negative) the effects of the seawater are negated and the fatigue limit of the steel in seawater returns to the air value. Thus the fatigue life of the hull material is expected to remain unaffected by the seawater environment as long as the cathodic protection system is operated at the anticipated potential of $-0.8V_{Ag}$/AgCl or greater. Since combined primary and secondary stresses are considerably lower than yield strength for operating basis conditions, crack initiation from fatigue is extremely unlikely and no fatigue problems would be encountered in the primary hull girder plating and welds.

15.5.8. Stress Corrosion

Mild steel is subject to stress-corrosion cracking in nitrates and alkalies; however, low-strength steels including ABS-CS are not subject to stress corrosion in seawater.[29] Stress-corrosion-cracking data for steels similar to the platform steel support this observation. Thus stress corrosion is not considered to be a problem.

15.5.9. Corrosion

The exterior platform plating forming the bottom and sides up to the waterline will be protected by an impressed current cathodic protection system. As long as the external current is maintained, cathodic protection will provide an effective means of reducing the corrosion rate of the submerged portion of the outside hull virtually to zero for an indefinite period. However, as an added measure of conservatism, approximately $\frac{1}{4}$ in. is added to the design thickness of the bottom and side shell plating as a corrosion allowance.

15.5.10. Hull Material Requirements

Traditionally, the fracture characteristics of structural steels have been described in terms of the transition from brittle to ductile behavior as measured by impact tests. This transition in fracture behavior can be related schematically to various fracture states, as shown in Figure 15.28.

For slow or intermediate loading rates, the transition region occurs at lower temperatures than for impact (or dynamic) loading rates. The service loading conditions of the platform structure are expected to be somewhere between static and intermediate, and thus the desired toughness behavior

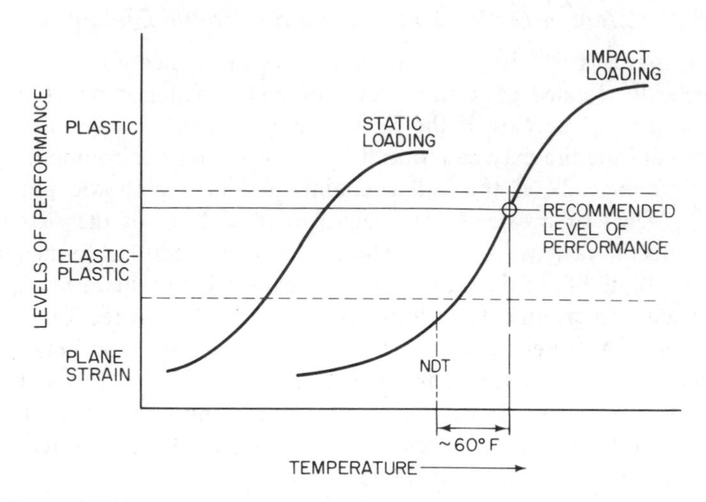

FIG. 15.28. Recommended level of performance for primary load-carrying members in bottom shell structure.

should be analyzed in terms of these loading rates. However, because there may be unusual service loadings at impact rates (accidents, hurricanes, etc.), the desired level of material toughness will be established using impact tests. The use of impact tests to measure notch toughness rather than static tests is a conservative approach for this particular structure but is specified because of the possible consequences of a brittle fracture.

The desired level of performance for the exterior platform plating to prevent brittle-fracture initiation is elastic-plastic behavior approaching fully plastic behavior under *impact* loading, as shown in Figure 15.28. Although several existing ship steels will satisfy this criterion, ABS Class CS steel has been selected as the material for the exterior platform plating. This steel will be required to have a nil-ductility transition (NDT) temperature of approximately 60°F below the minimum service temperature of the bottom shell, which is the region of maximum tensile stress. That is, because the minimum possible service temperature of the bottom shell is +28.6°F, the specified NDT of the ABS Class CS steel will be −30°F.

ABS-Class CS steel is a fine-grained normalized steel with excellent toughness compared with other ship hull steels, as shown in Figure 15.29. In fact, ABS Class CS steel probably has significantly more notch toughness than is required for this application, but it is specified because of the desire for a very high degree of reliability against brittle fracture.

Using Naval Research Laboratory (NRL) terminology, specifying that NDT be approximately 60°F below the minimum service temperature corresponds very closely to the NRL yield criterion (Y.C.) behavior at the

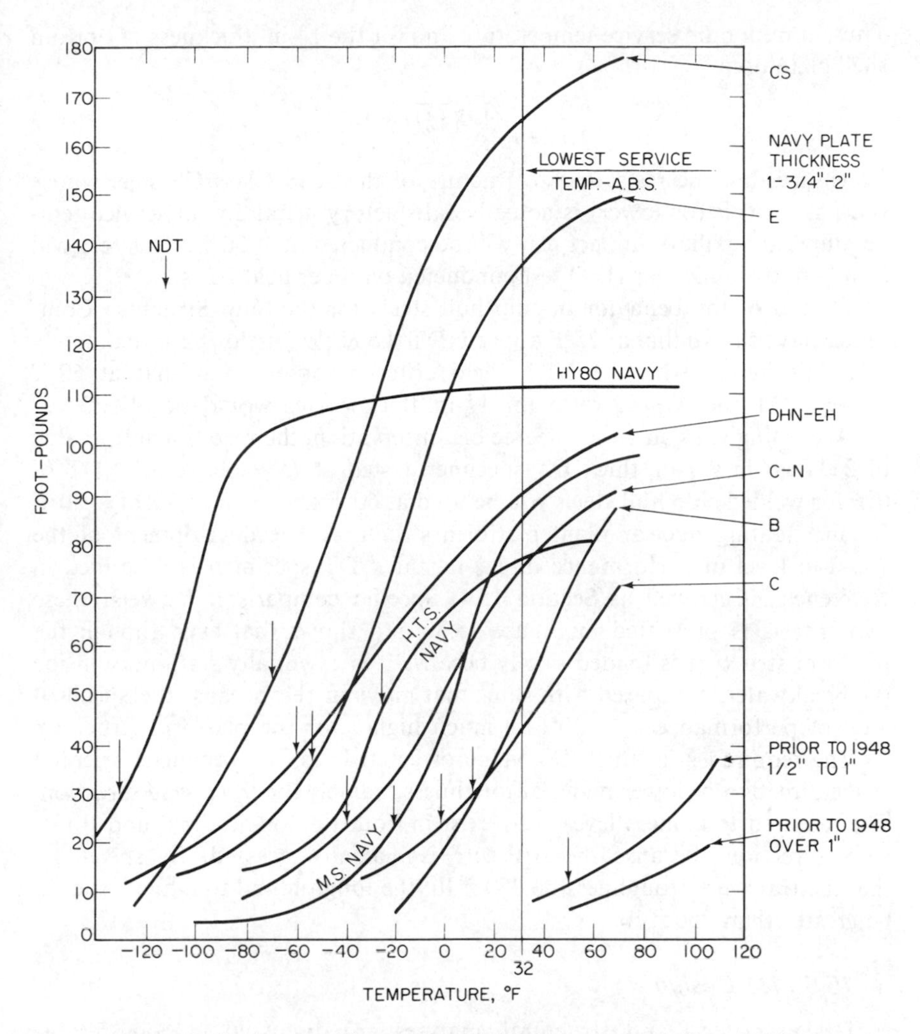

FIG. 15.29. Average Charpy V-notch impact test curves (from ABS).

service temperature. This requirement states that

$$\left(\frac{K_{Id}}{\sigma_{yd}}\right)^2 = 1.0(B)$$

where K_{Id} = critical plane-strain stress intensity factor under conditions of dynamic loading, ksi$\sqrt{\text{in.}}$,

σ_{yd} = dynamic yield strength, ksi (approximated by σ_{ys} static +20–30 ksi),

B = plate thickness, in.

Thus, at minimum service temperature and for the $1\frac{3}{8}$-in. thickness of bottom shell plate used,

$$\frac{K_{Id}}{\sigma_{yd}} = \sqrt{1.0(1\tfrac{3}{8})} \simeq 1.2$$

To ensure that the resistance to fracture of the ABS Class CS steel whose NDT is -30°F (or lower) is actually satisfactory at minimum service temperature an auxiliary impact test will be conducted at $+30$°F. This test will be a $\frac{5}{8}$-in. dynamic tear (DT) test conducted on a per heat basis.

Studies of the behavior of ship hull steels for the Ship Structure Committee have shown that at 75°F above NDT the K_{Id}/σ_{yd} ratio is approximately 1.5. This agrees with the NRL yield criterion observation that at 60°F above NDT the K_{Id}/σ_{yd} ratio for $1\frac{3}{8}$-in.-thick plates would be about 1.2.

Accordingly, as an added degree of conservation, the recommended value of 250 ft-lb in a $\frac{5}{8}$-in.-thick DT specimen tested at 75°F above an NDT of 0°F for welded ship hull steels will be used at 60°F above an NDT of -30°F for the floating nuclear plant platform structure. The development of the 250-ft-lb level of performance in a $\frac{5}{8}$-in.-thick DT specimen is described in Reference 30, as well as Section 15.4. Specific comparison between these two criteria is presented in Figure 15.30 and shows that even though the platform structure is loaded slowly because it is essentially stationary inside the breakwater, compared with ships that move in the oceans, the specified level of performance at $+30$°F is much higher for the platform structure.

The required $\frac{5}{8}$-in.-thick DT value of 250 ft-lb is for specimens oriented in the direction of lower material toughness, namely the *transverse* direction. The required toughness level must exist in both the longitudinal and transverse directions because the structure is biaxially stressed. By specifying that the transverse toughness be 250 ft-lb, the longitudinal toughness should be greater than 250 ft-lb.

15.5.11. Design

Design criteria and structural analyses are discussed in considerable detail in the original document.[23] Emphasis has been placed on design concepts rather than specific code provisions as no structural engineering code has been found to be fully applicable to the platform design. The platform is analyzed for all known and hypothesized loading conditions and combinations with a probability of occurrence equal to or greater than 10^{-6}–10^{-7}. Stresses are determined by the finite element method of analysis and the Von-Mises failure theory for combining component stresses. Allowable hull girder stresses for the most extreme environmental conditions will not exceed 90% of the specified yield strength of the material. Normal design stresses are considerably less.

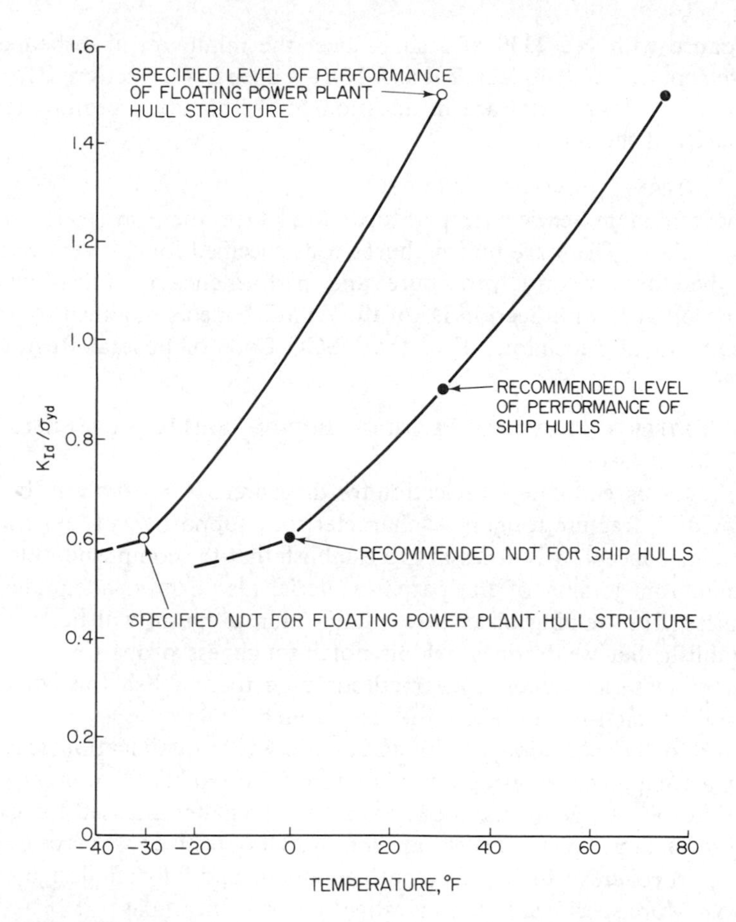

FIG. 15.30. Comparison of levels of performance for FNP hull structure and ship hulls.

15.5.12. Fabrication and Inspection

Welding. Welding procedure and performance qualifications for all welding processes shall be in accordance with Section IX of the ASME code as limited or modified by specific requirements of the USCG Code of Federal Regulations, Title 16, Subchapter F, or as modified herein.

The hull of the floating nuclear plant is categorized into three areas insofar as welding qualifications are concerned:

1. Containment

The portion of the platform which forms the containment boundary is Class MC and has a minimum service temperature (MST) of 30°F. Consequently, the Charpy V-notch impact test will be conducted at 0°F in accordance with NE-2330 and shall meet the minimum absorbed energy requirements of 20 ft-lb per Table I-10.1, Appendix I, Section III, of the ASME code. These tests are in addition to the standard requirements of Section IX of the code.

2. INTERNAL PLATFORM STRUCTURE

The minimum service temperature (MST) for these areas is approximately +30°F. There are no toughness tests specified for this service condition; therefore, welding procedure and performance qualifications shall be in accordance with Section IX of the ASME code as modified by specific requirements of Subchapter F of the USCG Code of Federal Regulations, Title 46.

3. EXTERIOR PLATFORM PLATING—BOTTOM SHELL, SIDE SHELL, AND MAIN DECK

Since the parent material selection for these areas is based on the criterion of providing fracture-toughness characteristics supportive of the fracture-safe design concept, it is necessary to establish that the composite weld zones resulting from joining of the parent material also exhibit adequate notch toughness. The welding procedure qualification program will be structured to establish that weld zones exhibit notch-toughness properties equivalent to parent material values. Test methods used to establish this correlation will be the same for both plate and weld zones.

Prior to the execution of the actual procedure qualification test weldments, a comprehensive program will be made to establish optimum joining procedures for all plant materials, including the materials used for exterior platform plating. For the latter case the overall objective will be to establish welding procedures which satisfy both economic and fail-safe design considerations. More specifically, the investigations will encompass

a. Selection of welding processes.
b. Selection of filler materials.
c. Investigation of the effect of process variables within each welding process.
d. Determination of weld-zone fracture-toughness properties.

With respect to notch-toughness testing the goal will be to use those techniques which utilize a relatively large specimen containing the full weld cross section, with selective placement of the crack-starter notch. The two principal tests used will be the drop weight (DW) test (ASTM E-208) and the dynamic tear (DT) test (MIL-STD-1601). The DT test will be used to determine the DT toughness characteristics of generic weld-zone areas—the results will be depicted as characteristic curves showing performance from upper to lower shelf values. The DW test will be used to establish and confirm nil-ductility transition temperatures for the various welding process variations.

TABLE 15.6 Exterior Platform Plating and Weldments Testing and Acceptance Criteria

Component	*Information*	*Specification*
Parent material		
1. Supplier qualification	a. Chemical & mechanical properties b. Fracture toughness 1. Transverse dynamic tear—two tests each temperature over range of temperature from lower to upper shelf 2. Transverse CVN—three tests each temperature over range of temperature from lower to upper shelf 3. Drop weight tests to establish NDT lower than −30°F	Not Applicable
2. Supplier quality control	Not Applicable	a. Chemical & mechanical properties b. Fracture toughness 1. Two transverse DT tests per heat; minimum 250 ft-lb ($\frac{5}{8}$ in.) at +30°F 2. Two drop weight tests per "plate-as-rolled"; minimum no-break performance at −30°F
Weldments		
3. Prequalification investigation (each welding process)	a. Chemical & mechanical properties b. Fracture toughness 1. Dynamic tear—two tests each temperature over range of temperature from lower to upper shelf in each of three zones: weld, HAZ, and parent material 2. CVN. Three tests each temperature over range of temperature from lower to upper shelf in each of three zones: weld, HAZ, and parent material 3. Drop weight tests to establish *weld metal* NDT lower than −30°F	Not Applicable

TABLE 15.6 (Continued)

Component	*Information*	*Specification*
4. Procedure qualification (each welding process)	Not Applicable	a. Chemical & mechanical properties b. Fracture toughness 1. Two DT tests for each of two locations (weld metal & HAZ) per qualification; minimum 250 ft-lb at +30°F 2. Two drop weight tests (weld metal) per qualification; minimum no-break performance at −30°F
5. Production tests A production toughness test shall be prepared for each 1000 lineal feet deposited by each welding process. Toughness tests shall alternate between the weld metal and heat affected zone; thus, approximately half of all weld production impact tests will be of weld metal and half of heat affected zone material.	Not Applicable	a. Chemical & mechanical properties b. Fracture toughness Weld Metal 1. Two DT specimens per test plate; minimum 250 ft-lb at +30°F. 2. Two drop weight specimens per test plate; no-break performance at −30°F. Heat Affected Zone 1. Two DT specimens per test plate; minimum 250 ft-lbs at +30°F. The notch is to be centered so as to sample, as nearly as practicable, the most critical location for toughness observed in the weld procedure qualification tests.

Details of the prequalification testing are shown in Table 15.6. For comparison of CVN and DT energy values, the preparation, notch orientation, and testing of the two types of specimens will be made identical to eliminate difficult extrapolations in data interpretation. In particular it should be noted that for both DT and CVN specimens from the heat-affected zone (HAZ) the notch will be centered in the HAZ as revealed by macro-etching of each specimen.

Production weld reliability (and conformance with procedure qualification parameters) will be controlled by the inspection and quality-control processes. Full traceability for all platform exterior plating welds will be maintained. This means that all such welds will be fully documented as to parent plate and weld metal chemistry, welding process inspection and surveillance, and other in-process controls. Coupled with the above, the comprehensive nondestructive evaluation (NDE) program specified for the platform plating will assure a very high degree of weld-zone reliability. The stringent control and NDE programs to be used should negate the necessity for on-going production weld testing. However, limited production weld testing will be conducted to demonstrate satisfactory implementation of the welding quality-control program. Production weld testing will consist of two test plates for every thousand lineal feet of welding deposited by each qualified welding process used in fabrication of the platform shell. These welds will be destructively evaluated to the same criteria specified for the welding procedure qualification. Performance to the established criteria will demonstrate that production welding is in accordance with approved welding qualification procedures and that weld-zone properties are equivalent to parent material properties.

15.5.13. Overall Structural Reliability

By using an extremely notch-tough structural steel, namely ABS Class CS,

By specifying that this steel have an NDT of −30°F and an auxiliary impact test requirement at +30°F in the transverse direction,

By controlling the design stress levels and minimizing stress concentration factors,

By establishing controls on welding procedures and using high-quality practices,

By using nondestructive examination of welds to minimize the possibility of small initial flaws,

By use of periodic inspection throughout the life of the plant, and

By numerous other safeguards during the design, fabrication, and life of the structure

the overall structural reliability (safe-life) of the floating nuclear plant platform will be extremely high, and no brittle fractures should occur for all the various design and abnormal loadings considered.

However, as an added degree of safety, a *fail-safe* philosophy using a crack-arrester system will be incorporated in the bottom shell so that, in the extremely unlikely event of a brittle fracture, it will be arrested, thus avoiding catastrophic failure of the structure.

In summary, the structural integrity of the platform incorporating the following material specifications is such that a catastrophic brittle fracture should not occur under any credible service conditions during the life of the floating nuclear plant.

15.5.14. Platform Material Specifications

Materials forming a part of the platform structure shall meet the requirements of the American Bureau of Shipping "Rules for Building and Classing Steel Vessels," 1973, Section 43 and those additional requirements, not included in the ABS Rules, where specified herein.

1. *Internal Platform Structure*
   The following ABS Grades of material in accordance with Section 43 of the ABS Rules are specified as applicable and acceptable for the internal hull structure of the platform, provided that the minimum yield point as defined in Section 43 of the ABS Rules is not less than 32000 psi:

   Rolled Shapes—ABS Grade A
   Flanged Plate Sections—ABS Grade A Cold Flange Quality
   Plating up to and including $\frac{3}{4}$ inch—ABS Grade A
   Plating above $\frac{3}{4}$ inch up to and including 1 inch—ABS Grade B
   Plating above 1 inch up to and including 1-$\frac{3}{8}$ inch—ABS Grade D
   Plating above 1-$\frac{3}{8}$ inch up to and including 2 inch—ABS Grade D Normalized.
2. *Exterior Platform Plating, Bottom Shell, Deck, and Sides*
   ABS Grade CS steel is specified as applicable and acceptable provided that the minimum yield point as defined in Section 43 of the ABS Rules is not less than 35,000 psi and the following testing requirements, not included in the ABS Rules, are adhered to:
   a) Drop weight testing (ASTM E 208 or equivalent) shall be conducted on each heat-treated "plate as rolled" at −30°F. Test shall be no-break performance at the specified test temperature based on two tests for each plate.
   b) An auxiliary $\frac{5}{8}$-inch thick dynamic tear (DT) test will be conducted at 30°F. Testing will be performed in accordance with MIL STD-1601 (ships), "Method for $\frac{5}{8}$-inch Dynamic Tear Testing of Metallic Materials" 8 May 1973 except the specimen orientation shall be the transverse orientation. Two DT specimens shall be tested on a go, no-go basis from each production heat. Both specimens thus tested shall exhibit energy values not less than 250 ft.lbs. at +30°F.

3. *Crack Arrester*
   The crack arrester material utilized in the bottom shell of the platform will be a normalized high-strength steel particularly suited for low temperature applications. A537 or A633 steel is specified as applicable and acceptable provided the minimum tensile strength is not less than 75,000 psi and a high level, full-shelf notch toughness behavior at +30°F is attained. Material toughness qualification (i.e., full shelf behavior at +30°F) will be determined by dynamic tear testing at each source of production. The DT specimens will be oriented in the longitudinal direction.

15.6. Fracture Control and Structural Integrity in Military Aircraft Structures

Fracture control and structural integrity have been an integral part of aircraft structural design since the first aircraft was built. However, the real emphasis on fail-safe aircraft structures came about as a result of the tragic failure of the British Comet jets. After about 5,000 hr of service, a crack appeared in the pressurized cabin skin at a window cutout. The crack was caused by fatigue damage due to the combination of cabin pressure cycles and flight loads. After progressing a very short distance, the crack became unstable and resulted in a sudden failure of the shell. Two Comet aircraft failed in this manner, literally exploding in midair and plummeting to earth. Since that time commercial aircraft manufacturers have become ultraconscious of fail-safe design and build all primary structures in a fail-safe manner.

However, in recent years, there have been premature structural failures of several Air Force aircraft in service and tests, specifically the F-111 and C-5A aircraft, and thus all aspects of structural criteria, analysis, design, and testing have been undergoing intensive review.

As part of the corrective action, fracture mechanics was used extensively to evaluate material characteristics and to quantify proof-testing results along lines similar to those used previously for space-vehicle pressure vessels. In this methodology, the presence of flaws, too small to be reliably detected, is postulated. As an outgrowth of these system problems a fracture-control plan is a requirement in each current major Air Force system and is being incorporated as a requirement for future systems in a structural-integrity-requirement document entitled "Military Standard 1530-Aircraft Structural Integrity Program, Airplane Requirements."

These plans use fracture-mechanics analyses and related material and element testing in an overall plan to preclude catastrophic failures from initial or service-induced flaws that cannot be found by current inspection techniques. The use of fracture-control plans is not limited to Air Force vehicles. Commercial transport aircraft have used similar principles for over a decade. A fracture-control plan will be utilized for the NASA space shuttle and for future Navy aircraft systems.

A fracture-control plan is not a replacement for existing structural design and validation procedures. The fracture-control plan, with the exception of an initial flaw assumption, merely extends and, in some cases, quantifies similar principles that have been applied to the design of aircraft since the beginning of flight.

Recently a special National Materials Advisory Board special committee on *Application of Fracture Prevention Principles to Aircraft* prepared a report[31] on this subject. The significance of that report with respect to fracture control is summarized below.

As a result of the awareness of the fracture problem, increasing emphasis is being placed on developing specific programs, such as those on the F-15 and B-1, to systematize and quantify the fracture-mechanics design and analysis effort for aircraft structures. There is no generally accepted fracture-control program procedure; however, a typical program should address the following elements:

1. Requirements/criteria,
2. Materials,
3. Design and analysis,
4. Testing,
5. Manufacturing (fabrication, quality assurance), and
6. Service life (safety, durability, risk assessment).

No ordering of priorities is intended in this listing. Indeed, a successful fracture-control program will include all these interrelated elements. Each individual element is reviewed briefly in the paragraphs below.

15.6.1. Requirements and Criteria

In the past, no explicit design requirements and criteria pertaining to fracture have been specified contractually. The B-1 program is the first to implement specific requirements at the beginning of a contract. These requirements are based on the essential premises that all critical structures contain flaws of a minimum specified size and that the time for these flaws to propagate to failure must be greater than a specified inspection interval that depends on the structural concept employed and the *in situ* inspectability of the part. These requirements assume that

1. All critical parts and locations within a part can be established.
2. A high degree of confidence exists in the flaw-detection capability.
3. Crack propagation and residual strength can be analyzed within reasonable accuracy.

These assumptions are enumerated to emphasize the significant points and agreements that must be established to permit the designer to use the criteria.

A more general set of requirements also has resulted from the efforts of the Air Force committee established to update the Aircraft Structural Integrity Program (ASIP) document and related structural-design MIL specifications. This set of requirements was developed over several years; it was guided by the Air Force Laboratories at the Aeronautical Systems Division (ASD) at Wright-Patterson Air Force Base and was supported by industry and other government agencies.

This effort is documented in MIL Specification 1530[32] that integrates fracture-mechanics considerations into the total Air Force structural-integrity program. The Navy is also revising its basic aircraft-design specification to include requirements that consider fracture during design and as an integral part of the structural integrity design program.

The calculation of service life of flawed structures requires a more specific definition of the service usage and environment than has been required previously for calculated service life under fatigue; therefore, specific requirements must be developed and stated. Essentially, this definition and the resulting requirements must be derived statistically and can vary widely for different types of aircraft and even within a given series of aircraft. When computing crack propagation, environmental factors such as temperature and the presence of humidity, water, fuels, or other chemicals must be considered, in addition to establishing the load and, subsequently, stress history of an individual part required for the usual fatigue analysis. This environmental consideration is particularly important in structures that employ materials with a high susceptibility to environmental conditions. For example, the crack-propagation rates of high-strength martensitic steels may differ as much as an order of magnitude between dry air and water.

The consistency of requirements within involved government agencies is viewed as a major problem. Also, much additional effort is needed to expand the basic requirements into a usable set of detail-design criteria. To a great extent, practical experience gained in designing to specific requirements, such as in the B-1 program, will determine the development of future criteria. Historically, structural-design specifications have been revised frequently to reflect advancing technology and current experience, and fracture-mechanics requirements probably will evolve in a similar manner.

The criteria for evaluating tests are as important as the design criteria discussed above. Obviously, these test criteria have a significant impact on program aspects, since within the currently understood framework of fracture-mechanics technology, complete validation of the design would involve running innumerable tests to cover all critical areas of each critical part. This testing probably could not be accomplished within the usual schedule constraints placed on research, development, testing, and experimental programs.

Apparently, compromise in this area and some analytical assessments will be necessary instead of complete testing to qualify or certify the design. Careful consideration must be given to the significance of cost, schedule tracks between test and analysis, and the confidence level resulting from compromise in these areas.

15.6.2. Materials

Obviously, the classification and characterization of materials is a necessary element of a fracture-control program. A major problem in material characterization is the standardization of test specimens and development of procedures to obtain valid, directly comparable fracture-mechanics data for the various materials being considered. ASTM Subcommittee E-24 is addressing this subject on a continuing basis. The importance of establishing consistent, directly comparable material data is essential to the materials development program, as well as to design, so that solid benchmarks can be used to measure materials technology development progress.

Basic fracture-mechanics data that are required to perform a fracture analysis must be obtained for all selected materials for use in the design. The required basic data are fracture toughness (K_{Ic} and K_c), crack-propagation data (da/dN), and the stress-corrosion threshold level (K_{Iscc}).

The crack-propagation rate was found to be a function of the stress range and ratio, environment (temperature and chemical), and load frequency and sequencing. Further, an interrelationship depends on the combination of these conditions. Crack-propagation data usually were obtained by running constant-amplitude tests. However, random spectrum loading was observed to cause retardation of crack growth in certain materials by factors of as much as 6 to 10. This observation placed increased emphasis on running random spectrum-type tests. Retardation effects also are influenced by the existing environment, and these environmental/spectrum interactions must be determined. Accordingly, a complete fracture-mechanics data program is very large and indicates the need to develop a consistent and usable data bank.

The original fracture-mechanics efforts were directed to relatively simple structures, such as pressure vessels, and to flaws in plate-type structures. In aircraft, however, fatigue cracks that may occur in service as well as flaws that may exist in the basic material or may be generated as a result of the manufacturing process also must be considered. The service-induced crack that occurs at stress-concentration points, such as bolt holes and machined notches, or results from combat damage presents a more complex situation because the stress state of the material in the immediate vicinity of these areas may be beyond the material's elastic limit and, therefore, beyond the applicability of linear-elastic fracture-mechanics analysis. The fracture-mechanics data test program must include these features of the design.

15.6.3. Design and Analysis

In this discussion, design is defined as the process of creating design concepts and establishing the detail configuration of structural components and elements, including materials selection and supporting and substantiating analysis. As one of the elements of fracture control, the design process involves the application of judgment and the interjection of practical experience. Present refinements of analytical methods, such as finite element analysis and fine grid-stress analysis, have tended to reduce the risk associated with the involved judgment factors and provide better balanced structural configurations.

However, considerable latitude is available in selecting the structural concepts. In the past, when fracture has been considered, these concepts generally have revolved around the question of whether a fail-safe or a safe-life type of structure should be used. The possibilities of including design features that may arrest, or reduce, flaw growth have been considered only recently. In effect, a fail-safe structure does imply crack arrestment, i.e., failure confined to one element of a multielement load path. In addition, an appreciable reduction in strength is generally implied and is tolerable for a short period. Therefore, a distinction exists between a fail-safe design and a structural design that employs crack-arrestment or retardation features to enable a structure to perform without significant degradation of strength or service life even though flaws may propagate up to the point of arrestment.

Generally, the structural portions of the aircraft's airframe employ a large number of discrete elements, such as bulkheads, spars, ribs, and load-carrying covering. In most of these elements, some degree of tolerance to damage does exist whether planned or not. Further, there has been a trend toward including structural redundancy in the design even though particular requirements for fail safety may not be specified. The redundancy is employed particularly for the aerodynamic surfaces where multiple spar configurations are commonly used. The fuselage portion of the structure also usually has a degree of redundancy due to symmetry about the vertical center line. In the past, these factors have been important in preventing catastrophic failures, although failure of individual structural elements did occur.

While fracture-prevention principles must be applied in some degree to all elements of the structure in the design process, obviously emphasis must be placed on those structural parts whose failure would cause catastrophic damage or, in the case of redundant fail-safe structures, would cause a large reduction in residual strength. While in some cases a simple examination of the structural configuration is sufficient to ascertain that failure obviously will be catastrophic, much more elaborate and extensive failure-analysis methods must be applied in the majority of cases.

In the material-selection process, where fracture-mechanics characteristics are concerned, considerations generally are qualitative rather than quantitative. Emphasis is given to other material properties of interest in designing a structure for minimum weight. During the past 10 years, fatigue life was an important consideration in selecting structural materials, but virtually no consideration was given to crack-propagation aspects per se. Unfortunately, a material that has good overall fatigue properties will not necessarily exhibit slow crack-propagation rates. Fracture-prevention consideration must be a factor of equal importance in material selection in the future.

An additional fracture-prevention factor, which is directly a function of design, is the inspectability of the structure in the final assembled configuration. This facet of the fracture-control program is hard to overvalue. The frequent visual inspection of critical components by operating personnel is invaluable in avoiding catastrophic fracture occurrences.

Analytical methods currently used in fracture-control programs are basically deterministic. Occasionally, their complexities are disquieting when the relative accuracy of the empirical data that they manipulate is considered; however, they are the best of a poor lot of alternatives and the only tools available for the time being.

To perform a fracture analysis effectively, all elements of the fracture-control plan must be considered. The basic methodology of computing crack growth and critical crack size is complex due to the many variables that enter into the computation, and the use of high-speed digital computers is required. To limit these problems to a reasonable size, engineering judgment must be exercised in selecting the particular points within structural components and elements that will be analyzed. This judgment must be based on an experienced insight into the total fracture-mechanics problem and on more detailed knowledge of stresses within the part than that required for static-strength assessment or fatigue analysis.

In the past, fracture-mechanics analysis generally dealt with relatively simple crack models, such as the part-through semicircular crack in a simple uniaxial stress field. Usually, the cracks that develop in service involve stress concentrations. A large number of unique stress-concentration conditions, such as bolt holes, machined stiffened runouts, ribs, bosses, lugs, and interference-type fasteners, must be considered in a typical design. More complex test specimens are being employed because either the analytical procedures for solving this problem or the budget to perform analyses, which are technically feasible but very costly, is inadequate. These tests are both time-consuming and costly and still may not be representative of the complete structural situation.

While some fundamental fracture-mechanics equations are based on theory and have been used successfully for the simple flaw models, other equations (i.e., subcritical crack growth) depend entirely on empirically obtained relationships. Current analytic methods are based essentially on these empirically derived data and must be updated continually in the same way as the material-properties data bank. Until adequate theoretical explanations are produced for observed fracture phenomena, and until experimental results based on this theory are reproduced consistently, analytical results must be treated conservatively and, whenever possible, verified by tests. Three types of information are needed to characterize materials and to apply quantitative fracture-mechanics analyses in its present state:

1. Fracture Toughness. Important questions about the characterization of fracture toughness concern

1. The suitability of the K_{Ic} or K_c measurement itself as the best criterion of fracture toughness,
2. The standardization of test specimens and procedures, and
3. The definition of the lower-bound value best suited to the analysis in question.

Considerable visibility has been given recently to efforts (e.g., those of ASTM Committee E-24) to adopt uniform test criteria and data-analysis methods. While much remains to be done, significant progress has been made. The relatively recent upsurge of interest in fracture mechanics as a design tool and the increased uniformity of test methods have resulted in a much better structural materials data base from which to choose more accurate design values, such as the lower bounds for fracture toughness, K_{Ic} and K_c, and the stress-corrosion threshold, K_{Iscc}. Concurrently, research work continues on alternative means of evaluating fracture properties, especially a method theoretically valid in the presence of plastic flow, such as the J integral and the crack-opening displacement (COD) measurement (see Chapter 16).

2. Crack-Growth Measurement. The establishment of flaw- or crack-growth rates (e.g., da/dN) is required for the analysis of structural integrity and prediction of the time (or number of stress cycles) for the crack to develop to the critical size resulting in unstable growth. Commonly, crack-growth rates are obtained in constant amplitude-fatigue tests and presented as a function of the stress-intensity range (ΔK) and stress ratio. Also, such tests are conducted under various environmental conditions, including temperature, chemical, and atmospheric conditions. Crack growth is affected generally by the strong interacting effects of operational load spectra, and most structural materials exhibit a propagation-rate retardation by a factor of as much as 6 to 10. Other testing variables, such as metallurgical and microstructural variations within a specific alloy composition, dissimilar test specimens, discrepancies in stress-intensity correction factors for flaw geometry, and minor environmental changes between experiments, contribute to the scatter in data normally observed. These interacting effects establish operational life estimates that are based on the slow crack-growth interval preceding the initiation of unstable growth.

3. NDE Flaw Detectability Level. The time interval before unstable crack growth (during which slow crack growth proceeds) is usually estimated from the crack-propagation function, da/dN. When estimating a "safe" interval, an initial flaw or crack size (a_0) must be assigned from which propagation starts at time t_0 at rate da/dN. The assumed or measured initial size (a_0) must equal or exceed the minimum flaw size reliably detectable by nondestructive evaluation (NDE). Larger flaws than this minimum will result in premature failures due to their faster growth rates. Thus, the prediction of the time to reach critical flaw size depends primarily on the probability of NDE detection of small flaws. Therefore, the establishment of the lower limit of flaw detectability with demonstrated reliability (probability of detection at a given confidence level) is a *critical* factor in the characterization of materials and structures and must be an integral part of design criteria containing fracture-mechanics requirements.

In the conventional design process, the designer assumes that the handbook values of tensile ultimate stress, tensile yield stress, and modulus of his materials are representative for his particular design. To establish the design stress, the handbook value is divided by a suitable factor of safety that depends on design specifications or experience. Specifications generally require inspection of the part to reveal flaws, defects, or inhomogeneities that will cause material properties to deviate significantly from the handbook values. In this process, the designer recognizes that flaws may occur but uses NDE as a go, no-go quality-control process. If no flaws are detected, the part is passed. If flaws are detected, the part is withdrawn for further examination and possible destructive testing.

The current state-of-the-art in critical aerospace design for a safe-life structure can use a design criterion based on the concepts of fracture mechanics. This criterion permits calculation of the failure stress (σ_f) for a part containing a flaw length ($2a$) by

$$c\sigma_f\sqrt{\pi a} = K_{Ic}$$

where c is a factor associated with the geometry of the flaw and the type of part.

Generally, the initial design assumes that at the outset of its service the part contains no flaw greater than a critical size. Primarily, this assumption (at a given confidence or probability level) is made for unassembled parts or simple monolithic structures, and the role of quality-assurance (QA) personnel is to determine the limits of acceptability of that assumption for any given part. However, during assembly of the structure or subsequent aircraft structural examinations, such as inspection, and after necessary repair, additional assurance must be given that no critical flaws exist in the operating structures. Here, the emphasis is on critical flaws and flaws larger than the limits of detectability assumed in the QA inspection. The inspection for such flaws is a field inspection and usually does not allow complete disassembly of the component parts. The degree of accuracy and coverage of such an inspection is controlled by basic flaw-detection capabilities and the level of inspectability of the assembled part.

The NDE procedure must be quantified so that the NDI (non-destructive inspection) indication of a flaw size ($2a_{\mathrm{NDI}}$), if detected, can be used in the above equation to predict the failure of the part.

The possibility of a design method, based on a nondestructive testing fracture-mechanics (NDT-FM) approach, has been verified.[33] When combined with the NDI indications of flaw size as measured by one or more tests, such as X-ray, penetrant, magnetic particle, and shear-wave ultrasonics, the fracture-mechanics formulation predicted failure loads of flawed 70-75-T6511 aluminum and 4330 V modified steel to within $\pm 10\%$ for almost all flaw sizes examined.

These fracture-mechanics design requirements demand the application of high confidence levels to the fracture-toughness values and an inspection process to ensure the safety of the resulting structure at a minimum weight.

The use of an NDE-FM design criterion for critical aerospace parts places a great burden on the "limits of detectability" of the NDE technique. The exact value of the smallest flaw that can be detected with a high degree of confidence must be determined to predict the maximum flaw size that will be overlooked at a known probability and confidence level. Thus, at a given confidence level, NDE and production quality-control personnel are expected to remove all parts with flaws larger than the assumed critical size.

15.6.4. Testing

The testing of structural elements and components is an essential part of the design process. Usually, development-type testing is conducted in conjunction with the design to establish confidence in the concepts and configurations under consideration and to evaluate alternative designs. In this development-test program, fracture-mechanics considerations now must be integrated into the test planning and given equal weight with other factors being evaluated.

At the present confidence level in the analytical treatment of flaw growth, it is reasonable that demonstration testing will be necessary for the more complex situations existing in critical parts. Undoubtedly, this testing will be done on a component or perhaps even a single-part basis. Since testing usually will have a large impact if the results are disappointing, it must be done with meticulous attention to environment, loading, and physical preparation of the specimen. The interrelatedness of the elements of a fracture-control program is obvious when the various features that must be considered in making a representative test of the actual hardware and its utilization are contemplated.

The qualification test program, which proves the final design and is conducted normally on major components or the complete airframe, will be affected also by the introduction of a fracture-mechanics requirement. The more usual structural qualification programs, such as static and fatigue testing, more or less involve end-point tests to destruction. In the case of fracture-mechanics testing, a large number of failure-type tests must be conducted if all critical elements are to be tested. Extensive testing of this nature would be extremely costly, and it is probable that a complete qualification program, involving all critical elements and critical areas within these elements, never could be conducted practically.

The testing, discussed above, is associated with the static, or slow-loading, strength of the aircraft structure. The requirements of operating in a dynamic environment must be met and validated. Tests that verify analytically determined normal modes of vibration, with certain members failed, must be considered. Requirements of static aeroelastic behavior, e.g., control surface effectivity, and flutter-free operation somehow must be demonstrated. The cost of this testing must be fed back into the design decision-making process for its impact on configuration.

15.6.5. Manufacturing (Fabrication, Quality Assurance)

The total manufacturing process, including inspection, must be addressed if a complete fracture-control program is to be implemented. The correct and precise interpretation of the design rendered into hardware is essential. In the past, a significant number of fracture problems originated in manufacturing operations due to lack of emphasis on and understanding of the significance of minor defects. This situation is compounded when coupled with the general inadequacy of NDE and when dealing with high-strength metallic structures. To accomplish the objective of assuring proper manufacturing and inspection, an intimate relationship is required among the designer, the materials engineer, and the manufacturing and quality-assurance elements of the organization. Also, specific and detailed fabrication and inspection information and direction are required.

This may appear somewhat anomalous when the basic premise of current fracture-prevention programs, such as the B-1, assumes that preexisting cracks are present and the design is tolerant to this condition However, the assumption also is made that flaws are detectable within certain limits with a high degree of reliability, and therefore the penalties that must be incurred in the design obviously are a direct reflection of the quality of fabrication and inspection. Theoretically, if it were possible to fabricate and inspect to standards that would eliminate all flaws above the crack-propagation-threshold level, crack growth would never occur in service from a manufactured-induced defect. However, the size of permissible defects in the areas of stress concentrations are so minute that this situation is never realized and not completely understood at this time due to the plastic strains present.

15.6.6. Service-Life Estimation (Safety, Durability, and Risk Assessment)

The establishment of accurate service-life estimates for airframe structures is among the most important, yet difficult to implement, elements of an effective fracture-control plan. Ideally, such estimates should provide a total characterization of performance with specific reference to

1. Verification of structural integrity (safety),
2. Prediction of the expected service life under anticipated operational usage (durability), and
3. Assessment of the risk associated with such predictions (reliability).

Furthermore, the general methodology, as well as the specific procedures for providing the above characterization, should involve all phases of the structural-development cycle from materials evaluation and selection to the reliability demonstration and acceptance of a fleet of completed structures, including an appropriate inspection and maintenance program keyed to service-life estimates.

Current conventional procedures of service-life estimation produce only a partial characterization and are masked necessarily by empiricism imposed by limitations in the state-of-the-art and experience. The establishment of static-strength margins of safety is accomplished initially by (arbitrarily) selecting the operational limit load as two thirds (or occasionally another fraction) of the ultimate load and minimizing (or disallowing) the exceeding of this limit load during the intended structure lifetime. Generally, structural-safety analysis, whether performed using a safe-life or fail-safe criterion, infers (statistically or otherwise) some measure of confidence that can be associated with the assumed strength margin for a group of structures. The process depends heavily on the experimental results of component elemental development tests and does not follow a well-established or set procedure, although concepts of structural-safety analysis have been proposed for many years.

Conventional procedures for estimating structural fatigue-life expectancy are based on the establishment (through appropriate structural tests) of a life "estimate" to which a somewhat arbitrarily chosen fixed scatter factor (usually ranging from 2 to 5) is applied to reduce this "estimate" to a presumably "safe" service life. The size of this scatter factor usually is based on judgment or a simple probability analysis. By its very nature, this approach deals only with the mean or "average" fatigue life, and the probability of lives shorter than this reduced "safe" operational life in an aircraft fleet is undefined and presumably ignored. Historically, this type of fatigue-life estimation process has been applied principally to cases where only total life to a predefined failure is important and directly measurable in conventional fatigue tests without regard for the instantaneous rate of damage development (e.g., da/dN). The estimation of life that is based on a projection from crack-growth rate introduces new complexities in the otherwise relatively straightforward analytical experimental situation. Further, an experience base for aircraft structural design and durability characterization, using a fracture-mechanics model, is essentially nonexistent.

Contributing to the complexities of using a fracture-mechanics model for fatigue-life prediction are various facts, including the following:

1. Life estimates are subject to error since observed crack-propagation rates (da/dN) at specific values of the stress-range intensity factor (ΔK) show as much scatter as fatigue lives at specific values of the stress amplitude in conventional fatigue tests.
2. On the same metal, a scatter range of at least 1 : 3 in observed crack-propagation rates must be anticipated, while crack-propagation rates in the same class of metal may vary still more.

A similar uncertainty applies to the determination of the critical crack size. The rate of slow crack propagation is not a simple power function of the (constant) stress-range intensity factor (ΔK), as is assumed usually, but deviates significantly from such a function at both ends of the ΔK range. The shape of this function is not known for variable stress-range intensities (ΔK) either without or with constant or variable mean-stress intensity. Further, in view of the severe interactions between high and low intensities that are due mainly to residual stress fields at the crack roots, additional uncertainty arises from attempts at superposition of such rates at variable stress-range intensities. These interactions also may reduce severely or negate the "limiting levels" of ΔK below which crack propagation is nonexistent at a constant ΔK level.

Finally, uncertainty arises because the fracture-mechanics model applies directly only to a monolithic, nonredundant, single-load-path structure containing a preexisting defect (or crack) that may be critical for both ultimate load and fatigue failure. Empirical assumptions are necessary if an analysis is applied to a redundant, multiple-load-path, "damage-tolerant" structure with (or without) preexisting defects. It is on these complexities that considerable work is required, and the success of the work will be predicated on a more accurate determination of operating conditions.

REFERENCES

1. Point Pleasant Bridge, National Transportation Safety Board, "Collapse of U.S. 35 Highway Bridge, Point Pleasant, West Virginia, December 15, 1967," *Report No. NTSB-HAR-71-1*, Washington, D.C., 1971.
2. J. M. Barsom, J. F. Sovak, and S. R. Novak, "AISI Project 168-Toughness Criteria for Structural Steels: Fracture Toughness of A36 Steels," *U.S. Steel Corporation Research Laboratory Report 97.021-001(1)*, May 1, 1972; American Iron And Steel Institute, Washington, D.C.
3. J. M. Barsom, J. F. Sovak, and S. R. Novak, "AISI Project 168-Toughness Criteria for Structural Steels: Fracture Toughness of A572 Steels," *U.S. Steel Corporation Research Laboratory Report 97.021-001(2)*, Dec. 29, 1972; American Iron And Steel Institute, Washington, D.C.
4. G. R. Cudney, "Stress Histories of Highway Bridges," *Journal of the Structural Division, ASCE, 94*, No. ST12, Dec. 1968, pp. 2725–2737.
5. R. B. Madison, "Application of Fracture Mechanics to Bridges," *Fritz Engineering Laboratory Report No. 335.2*, Lehigh University Institute of Research, Bethlehem, Pa., June 1969.
6. Highway Research Board of the NAS-NRC Division of Engineering and Industrial Research, the AASHTO Road Test, Report 4, Bridge Research, Special Report CID, Publication No. 953, National Academy of Science-National Research Council, Washington, D.C., 1962.
7. J. W. Fisher, K. H. Frank, M. A. Hirt, and B. M. McNamee, "Effect of Weldments on the Fatigue Strength of Steel Beams," *NCHRP Report 102*, Washington, D.C., 1970.
8. C. G. Schilling, K. H. Klippstein, J. M. Barsom, S. R. Novak, and G. T. Blake, "Low-Temperature Tests of Simulated Bridge Members," *Journal of the Structural Division, ASCE*, Vol. 101, No. ST1, January 1975.
9. "Standard Specifications for Highway Bridges," 1969, and "Interim Specifications 1971 Developed by the AASHTO Committee on Bridges and Structures," Association of State Highway Officials, Washington, D.C., 1969.
10. S. R. Heller, R. Nielsen, A. R. Lytle, and J. Vasta, "Twenty Years of Research Under the Ship Structure Committee," *Ship Structure Committee Report Serial No. SSC-182*, U.S. Coast Guard Headquarters, Washington, D.C., Dec. 1967.
11. S. T. Rolfe, D. M. Rhea, and B. O. Kuzmanovic, "Fracture Control Guidelines for Welded Steel Ship Hulls," *SSC-244*, Washington, D.C., 1974.
12. D. Hoffman and E. F. Lewis, "Analysis and Interpretation of Full-Scale Data on Midship Bending Stresses of Dry Cargo Ships," *Ship Structure Committee Report Serial No. SSC-196*, U.S. Coast Guard Headquarters, Washington, D.C., 1970.

13. J. J. W. NIBBERING, "Permissible Stresses and Their Limitations," *Ship Structure Committee Report Serial No. SSC-206*, U.S. Coast Guard Headquarters, Washington, D.C., 1970.
14. R. NIELSON, P. Y. CHANG, and L. C. DESCHAMPS, "Computer Design of Longitudinally Framed Ships," *Ship Structure Committee Report Serial No. SSC-225*, U.S. Coast Guard Headquarters, Washington, D.C., 1972.
15. E. R. STENROTH, "Reflections Upon Permissible Longitudinal Stresses in Ships," *Transactions of the The Royal Institution of Naval Architects* (*London*), *109*, No. 2, April 1967.
16. J. HODGSON and G. M. BOYD, "Brittle Fracture in Welded Ships—An Empirical Approach from Recent Experience," *Quarterly Transactions, The Royal Institution of Naval Architects* (*London*), edited by Capt. A. D. Duckworth, R. N., *100*, No. 3, July 1958.
17. W. S. PELLINI, "Design Options for Selection of Fracture Control Procedures in the Modernization of Codes, Rules and Standards," in *Proceedings: Joint United States-Japan Symposium on Application of Pressure Component Codes*, Tokyo, March 13–15, 1973.
18. *Rules for Building and Classing Steel Vessels*, The American Bureau of Shipping, 45 Broad Street, New York, 1976.
19. "Standard Method for Conducting Drop-Weight Test to Determine Nil-Ductility Transition Temperature of Ferritic Steels," *1972 Annual Book of ASTM Standards, Part 31, E-208*, American Society for Testing and Materials, pp. 594–613.
20. A. K. SHOEMAKER and S. T. ROLFE, "Static and Dynamic Low-Temperature K_{Ic} Behavior of Steels," *Transactions of the ASME, Journal of Basic Engineering*, Sept. 1969.
21. "Standard Method of Test for Plane-Strain Fracture Toughness of Metallic Materials," *Annual Book of ASTM Standards, Part 31, E-399*, American Society for Testing and Materials, Philadelphia, pp. 955–974.
22. Method for $\frac{5}{8}$-Inch Dynamic Tear Testing for Metallic Materials, *MIL-STD-1601* (ships) May 8, Washington, D.C., 1973.
23. C. DOTSON and S. T. ROLFE, "Fracture Control Plan for Floating Nuclear Plant Platforms, to be published.
24. A. K. SHOEMAKER, "Notch-Ductility Transition of Structural Steels of Various Yield Strengths," *Transactions of the ASME, Journal of Engineering for Industry*, Paper No. 71-PVP-19, 1971.
25. *Encyclopedia Britannica*, Vols. 9 and 18, pp. 62 and 516, respectively.
26. K. J. SKIPKA and RICHARD GREISSMON, *Atmospheric and Oceanic Temperature Data for Atlantic and Gulf Coastal Regions of the United States*, Smith-Singer Meteorologists Inc., 1972.
27. S. J. THEARLE, "Note on Some Cases of 'Fatigue' in the Steel Material of Steamers," *Transactions of the Institution of Naval Architects*, *55*, Part II, 1913.
28. HERBERT H. UHLIG, *Corrosion and Corrosion Control*, 2nd ed., Wiley, 1971, pp. 150–152.
29. KOICHI MASUBUCHI, *Materials for Ocean Engineering*, M.I.T. Press, Cambridge, Mass., 1970.

30. S. T. Rolfe, "Fracture Mechanics, Fracture Criteria and Fracture Control for Welded Steel Ship Hulls," The Society of Naval Architects and Marine Engineers, Ship Structure Symposium, Washington, D.C., Oct. 6–8, 1975.

31. "Application of Fracture Prevention Principles to Aircraft," *Publication NMAB-320*, National Materials Advisory Board, Washington, D.C., Feb. 1973.

32. Military Standard 1530, "Aircraft Structural Integrity Program, Airplane Requirements," *MIL-STD-1530* (*USAF*), Sept. 1, Dayton, Ohio, 1972.

33. P. F. Packman, H. S. Pearson, J. S. Owens, and G. Young, "Definition of Fatigue Cracks through Non-destructive Testing," *Journal of Materials, ASTM*, Sept. 1969, p. 666.

16

Elastic-Plastic Fracture Mechanics

16.1. Introduction

Almost all low- to medium-strength structural steels that are used in the section sizes of interest for large complex structures such as bridges, ships, pressure vessels, etc., are of insufficient thickness to maintain plane-strain conditions under slow-loading conditions at normal service temperatures. Thus for many structural applications, the linear-elastic analysis used to calculate K_{Ic} values is invalidated by the formation of large plastic zones and elastic-plastic behavior. One approach to the fracture analysis of these materials is to use empirical correlations to approximate K_{Ic} or K_{Id} values, as described in Chapter 6. In addition, considerable effort is being devoted to the development of elastic-plastic fracture-mechanics analyses as an extension of the linear-elastic analyses.

Three of the most promising extensions of linear-elastic fracture mechanics into the elastic-plastic region are the following:

1. Crack-opening displacement (COD).
2. *R*-Curve analysis.
3. *J* Integral.

All three techniques are relatively new, and each technique holds considerable promise for specifying material toughness in terms of an allowable stress or crack size. Considerable research effort is in progress on the development of standardized test methods as well as research and design engineering applications. The engineering significance of each of these techniques is reviewed in the following sections.

16.2. Crack-Opening Displacement (COD)

In 1961, Wells[1] proposed that the fracture behavior in the vicinity of a sharp crack could be characterized by the opening of the notch faces—namely the crack-opening displacement (COD), as shown in Figure 16.1.

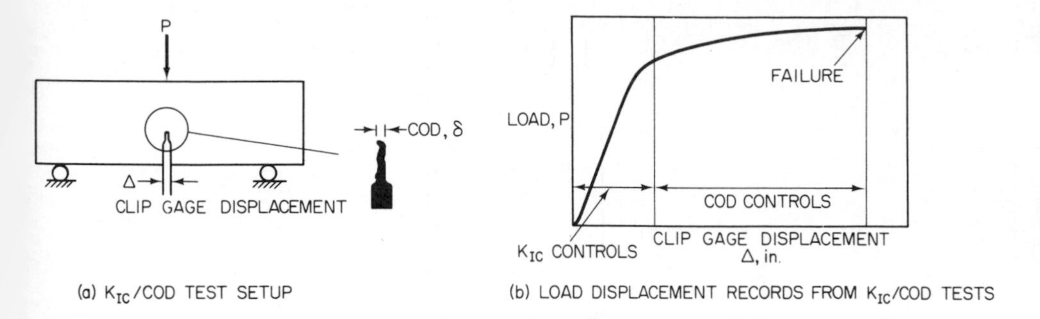

FIG. 16.1. Relation between K_{Ic} and COD test behavior.

Furthermore he showed that the concept of crack-opening displacement was analogous to the concept of critical crack extension force (G_c as described in Chapter 2), and thus the COD values could be related to the plane-strain fracture toughness, K_{Ic}. Because COD measurements can be made even when there is considerable plastic flow ahead of a crack, such as would be expected for elastic-plastic or fully plastic behavior, this technique might still be used to establish critical design stresses or crack sizes in a quantitative manner similar to that of linear-elastic fracture mechanics.

Using a crack-tip plasticity model proposed by Dugdale,[2] referred to as the strip yield model analysis, it is possible to relate the COD to the applied stress and crack length. The strip yield model consists of a through-thickness crack in an infinite plate that is subjected to a tensile stress normal to the plane of the crack (Figure 16.2). The crack is considered to have a length equal to $2a + 2p$. At each end of the crack there is a length p that is subjected to yield-point stresses that tend to close the crack, or, in reality, to prevent

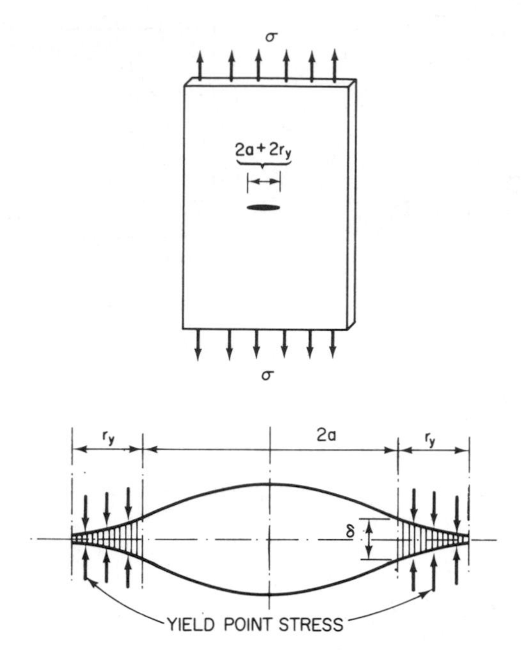

FIG. 16.2. Dugdale strip yield model.

it from opening. Thus the length of the "real" crack would be $2a$. Another way of looking at the behavior of this model is to assume that yield zones of length p spread out from the tip of the real crack, a, as the loading is increased. Thus the displacement at the original crack tip, δ, which is the COD, increases as the real crack length increases or as the applied loading increases. The basic relationship developed by Dugdale is

$$\delta = 8\frac{\sigma_{ys}a}{\pi E}\ln\sec\left(\frac{\pi}{2}\frac{\sigma}{\sigma_{ys}}\right)$$

where σ_{ys} = yield strength of the material, ksi,
$a = \frac{1}{2}$ real crack length, in.,
σ = nominal stress, ksi,
E = modulus of elasticity of the material, ksi.

Using a series expansion for ln sec $[(\pi/2)(\sigma/\sigma_{ys})]$, this expression becomes

$$\delta = \frac{8\sigma_{ys}a}{\pi E}\left[\frac{1}{2}\left(\frac{\pi}{2}\frac{\sigma}{\sigma_{ys}}\right)^2 + \frac{1}{12}\left(\frac{\pi}{2}\frac{\sigma}{\sigma_{ys}}\right)^4 + \frac{1}{45}\left(\frac{\pi}{2}\frac{\sigma}{\sigma_{ys}}\right)^6 + \cdots\right]$$

For nominal stress values less than $\frac{3}{4}\sigma_{ys}$, a reasonable approximation for δ, using only the first term of this series, is

$$\delta = \frac{\pi\sigma^2 a}{E\sigma_{ys}}$$

In Chapter 2, it was shown that, for a through-thickness crack of length $2a$,

$$K_I = \sigma\sqrt{\pi a}$$

or

$$K_I^2 = \sigma^2 \pi a$$

Thus $\delta E \sigma_{ys} = K_I^2$, and since $E = \sigma_{ys}/\epsilon_{ys}$, the following relation exists:

$$\frac{\delta}{\epsilon_{ys}} = \left(\frac{K_I}{\sigma_{ys}}\right)^2$$

Also, the strain energy release rate, G, is equal to

$$G = \frac{\pi\sigma^2 a}{E}$$

$$= \delta \cdot \sigma_{ys}$$

At the onset of crack instability, where K_I reaches K_{Ic} and COD reaches a critical value, δ_c,

$$\frac{\delta_c}{\epsilon_{ys}} = \left(\frac{K_{Ic}}{\sigma_{ys}}\right)^2$$

Because $(K_{Ic}/\sigma_{ys})^2$ can be related to the critical crack size in a particular structure, it is reasonable to assume that the parameter δ_c/ϵ_{ys} can likewise be related to the critical crack size in a particular structure. The advantage of the COD approach is that COD values can be measured throughout the entire plane-strain, elastic-plastic, and fully plastic behavior regions, whereas K_{Ic} values can be measured only in the plane-strain region or approximated in the early portions of the elastic-plastic region.

As with the K_I analysis, the application of the COD approach to engineering structures requires the measurement of a fracture-toughness parameter, δ_c, which is a material property that is a function of temperature, loading rate, specimen thickness, and possibly specimen geometry, i.e., notch acuity, crack length, and overall specimen size.

After considerable study of various methods to measure δ_c, the British Standards Institution has published a "Draft for Development of Methods for Crack Opening Displacement (COD) Testing".[3] Basically the δ_c specimen is a slow-bend test specimen similar to a K_{Ic} slow-bend test specimen (Figure 16.3). Since the δ_c test is regarded as an extension of K_{Ic} testing, the British draft test method is very similar to the ASTM E399 test method for K_{Ic} (Chapter 3). Similar specimen preparation, fatigue-cracking procedures, instrumentation, and test procedures are followed. The displacement gage is similar to the one used in K_{Ic} testing and a continuous load-displacement record is obtained during the test.

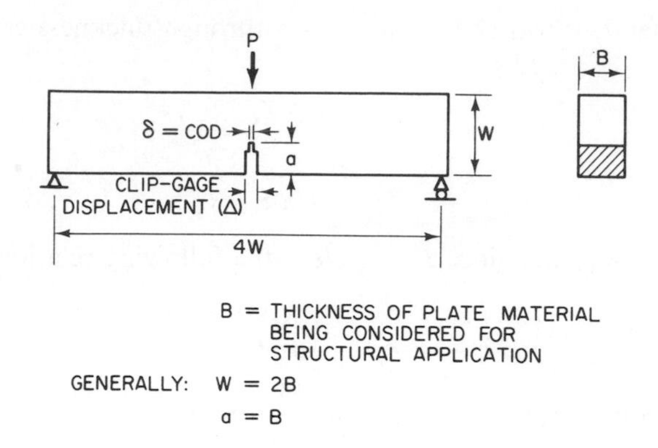

FIG. 16.3. COD test specimen.

Because the COD test is usually conducted on materials that behave in an elastic-plastic manner, linear load-displacement records to failure are rarely obtained. Rather "round-house" or nonlinear load-displacement curves are obtained as shown in Figure 16.4, and as described in the British draft standard.[3] For smooth continuous records in which the applied load rises with increasing displacement to the onset of unstable fracture [Figure 16.4, type a], the critical displacement (Δ_c) should be taken as the total value corresponding to maximum applied load including both elastic and plastic components. If failure occurs close to the linear range, the secant offset procedure described in Chapter 3 should be used to determine whether a valid K_{Ic} measurement can be made.

In some cases, the applied force-displacement curve shows a region of increasing displacement with falling or constant applied force followed by a further region of rising applied force before complete fracture [Figure 16.4, type b]. When this is associated with crack extension as determined by auxiliary measurements, the critical displacement shall be taken as the total

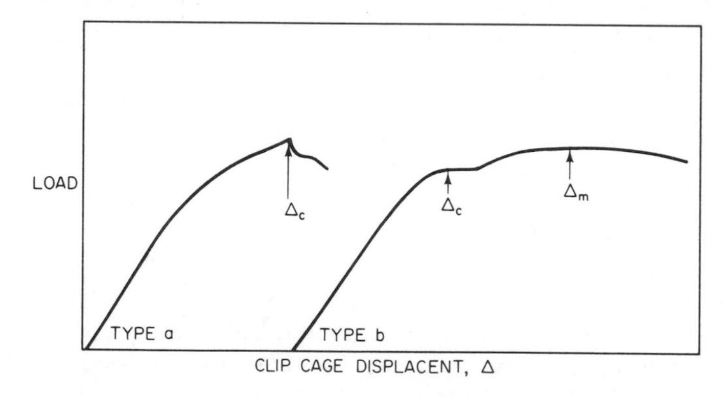

FIG. 16.4. Force displacement records for calculation of COD.

value (Δ_c) corresponding to the applied force at the first instability or curve discontinuity.

Where the applied force-displacement curve goes through a maximum point and increasing displacement is observed with falling applied force, either stable crack growth is occurring or a plastic hinge is forming. The critical displacement required is that value at the point which a specified amount of crack growth commenced as defined in the British standard. If onset of crack growth cannot be established, the COD for crack initiation cannot be measured. For material comparison purposes, a crack-opening displacement (δ_m) calculated from the clip gage displacement (Δ_m) at the first attainment of a maximum load can be used. However, the results may depend on specimen geometry also.

Having obtained the critical value of the clip gage displacement, it is necessary to convert this to the critical crack-tip COD (δ_c) value. The British Draft Standard[3] describes several methods, all of which assume deformation to occur by a hinge mechanism about a center of rotation at a depth of $r(W - a)$ below the crack tip (Figure 16.5).

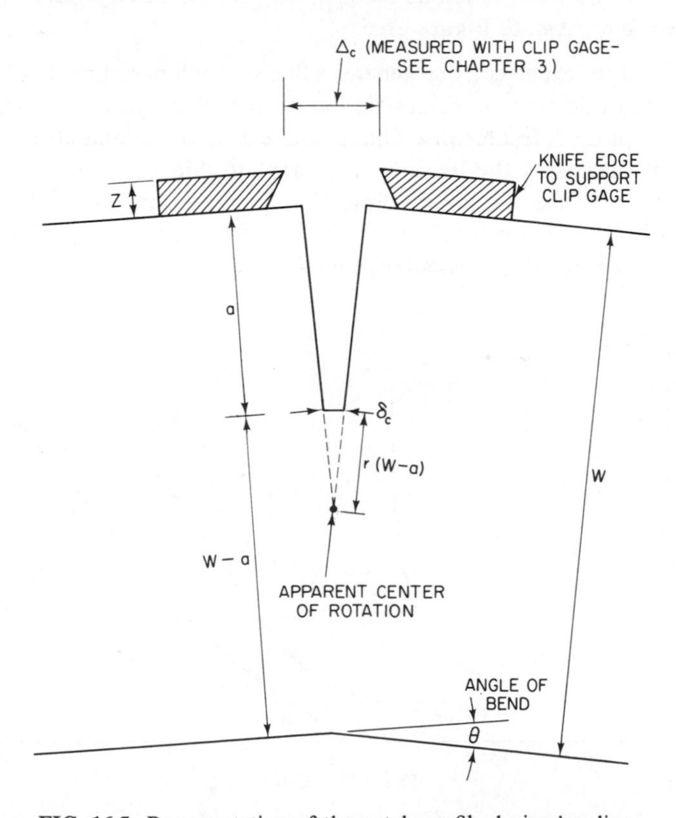

FIG. 16.5. Representation of the notch profile during bending.

16.2.1. Calculation of δ_c

Experimental calibrations using test pieces up to 2 in. thick have shown that, for the COD range of approximately 0.002–0.02 in., or about 0.0625–0.625 mm, the practical range of interest for specimens up 2-inches thick, good approximations to the crack-opening displacement can be obtained using

$$\delta_c = \frac{(W - a)\Delta_c}{W + 2a + 3z}$$

This relationship is derived by assuming deformation to occur by a hinge mechanism about a center of rotation at a depth of $(W - a)/3$ below the crack tip (Figure 16.5). Test results indicate this approximation to the center of rotation is reasonable, although r can be smaller for very small values of Δ_c.

The following measurements, although not essential to the test, will provide useful information and also can be recorded:

1. *Notch root contraction* as a check on the calculated value.
2. *Angle of bend*, θ, Figure 16.5.

As would be expected, the general effects of temperature, loading rate, and specimen thickness on δ_c are similar to the effects of these same conditions on K_{Ic} or CVN test results. That is, increasing the temperature increases δ_c, whereas increasing the loading rate or plate thickness decreases δ_c. An example of this behavior for mild steel[4] is presented in Figure 16.6. These results show that for slow-bend loading the COD value increases very rapidly at about −50°F for this particular steel using specimens 0.394 in. thick. For

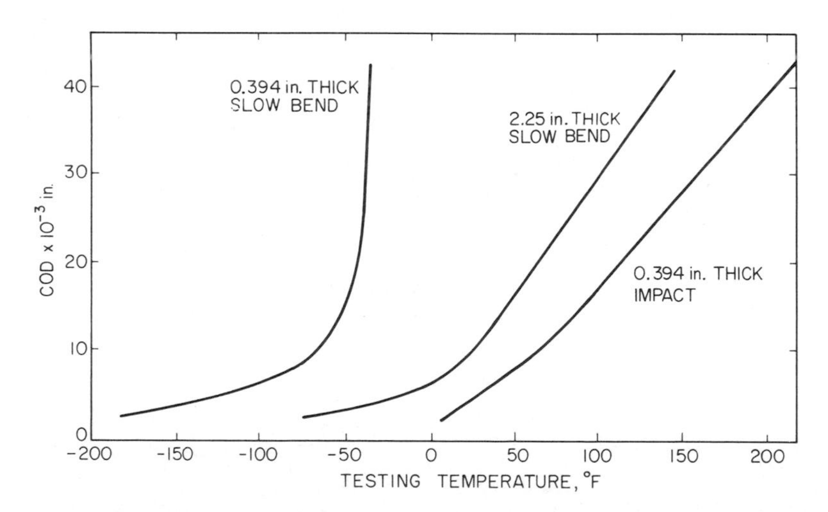

FIG. 16.6. Effect of strain rate and thickness on the transition temperature of mild steel.

thicker test specimens or for test specimens loaded under impact conditions, the transition temperature is higher and is less well defined.

Cotton et al.[5] point out that, as in most fracture-toughness testing, particularly in weldment testing, there is considerable scatter. Test results for a fine-grain control-rolled C-Mn steel are presented in Figures 16.7 and 16.8 and demonstrate this fact. Tests of two-pass submerged arc weld

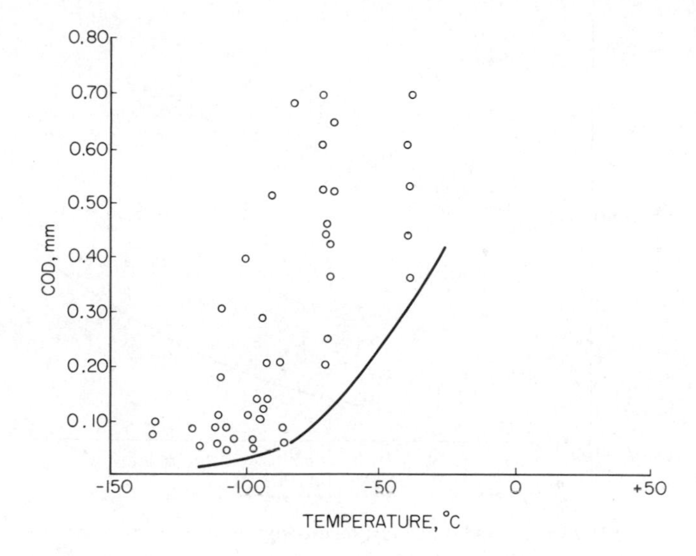

FIG. 16.7. Longitudinal COD transition curve for parent material.

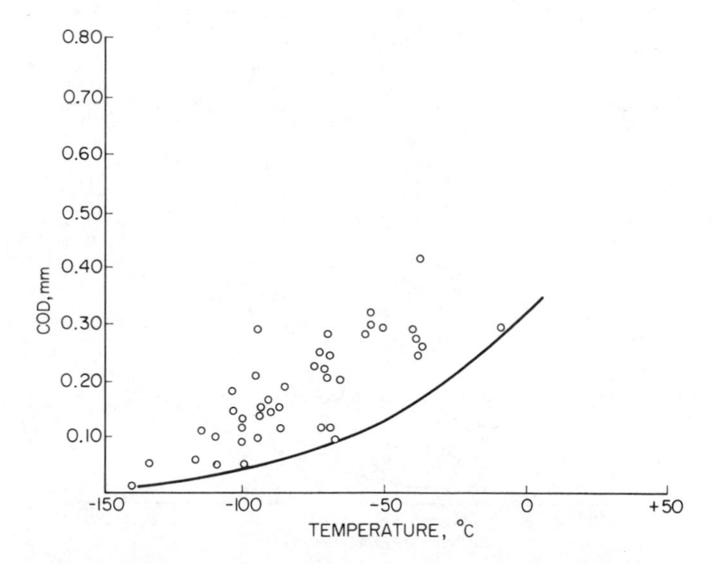

FIG. 16.8. Transverse COD transition curve for parent material.

metal and heat-affected zone material used to weld this same steel also exhibit considerable scatter (Figures 16.9 and 16.10, respectively). For design purposes, lower-bound curves were used as shown in each of the preceding four figures.

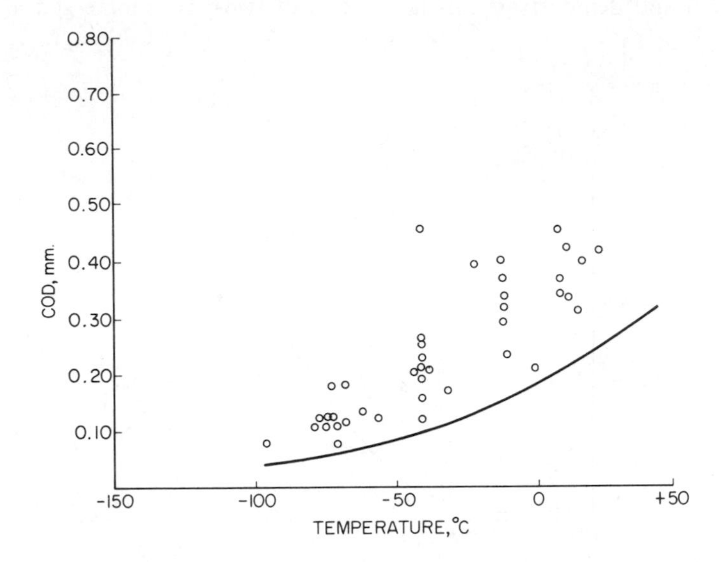

FIG. 16.9. Weld metal test results for sub-arc weldments.

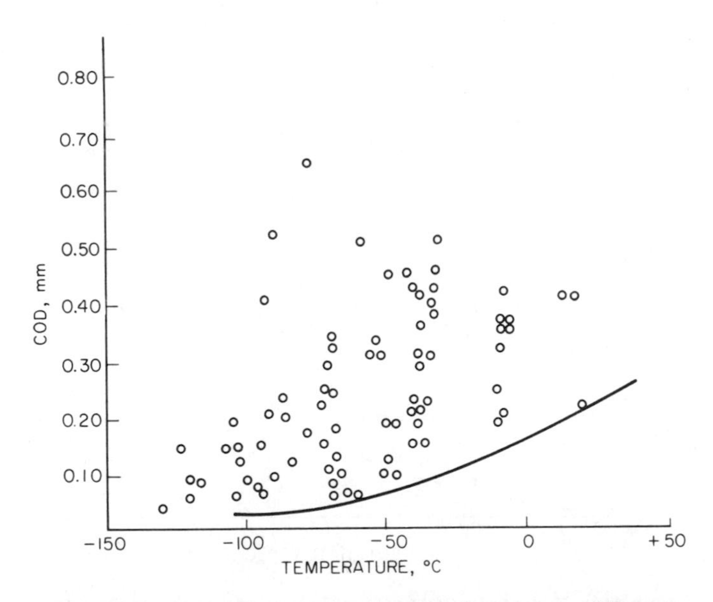

FIG. 16.10. Heat-affected zone test results for sub-arc weldments.

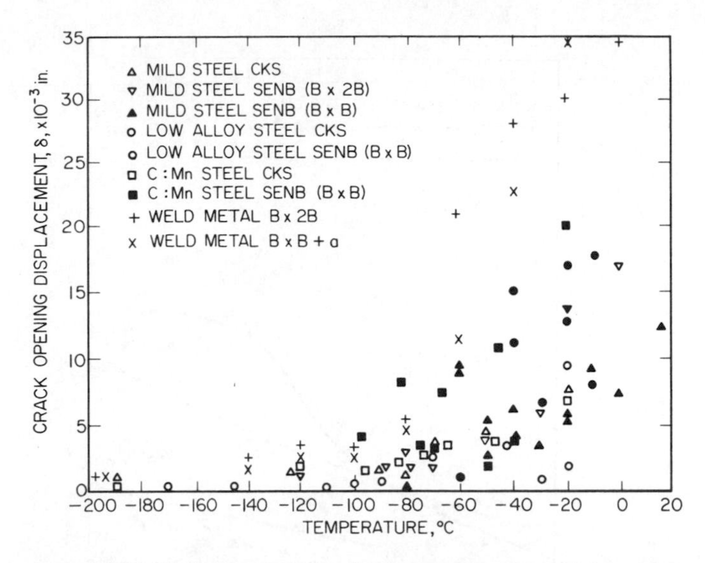

FIG. 16.11. Variation of COD with temperature for various steels.

Additional test results by Egan[6] for a similar C-Mn mild steel and a low-alloy steel are presented in Figure 16.11.

Test results by Nishioka and Iwanaga[7] on HT-50, X-60, X-70, and HT-70 steels (Table 16.1) are presented in Figures 16.12 and 16.13 and demonstrate the effects of loading rate on δ_c. Similar test results by Michida[8] of a structural steel also demonstrate the effect of loading rate on δ_c (Figure 16.14).

TABLE 16.1 Materials Used To Obtain Results of Dynamic COD Test Values Presented in Figures 16.12 and 16.13

Chemical compositions (%)

	Thickness (mm)	*C*	*Si*	*Mn*	*P*	*S*	*Cu*	*Cr*	*Ni*	*Mo*	*V*	*Nb*
HT50	22	0.20	0.33	1.30	0.018	0.015	0.04	0.03	—	—	—	—
X-60	14	0.17	0.46	1.46	0.025	0.016	0.03	—	—	—	0.08	—
X-70	14	0.14	0.41	1.30	0.015	0.005	0.04	0.09	—	—	0.09	0.029
HT70	25	0.12	0.30	0.79	0.014	0.013	0.02	0.43	1.26	0.47	0.03	—

Mechanical properties

	σ_{ys} *(kg/mm²)*	σ_t *(kg/mm²)*	*El. (%)*
HT50	43.9	66.0	31.0
X-60	42.5	64.0	30.0
X-70	54.6	63.3	37.7
HT70	71.0	77.4	26.7

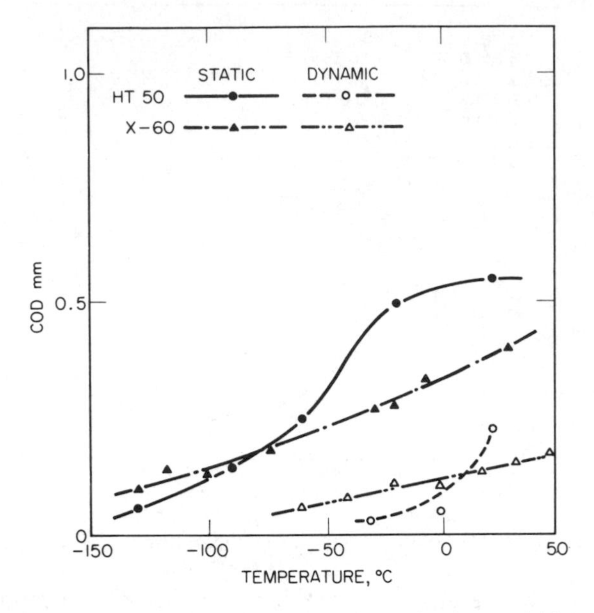

FIG. 16.12. Static and dynamic COD test results from Nishioka and Iwanaga.[7]

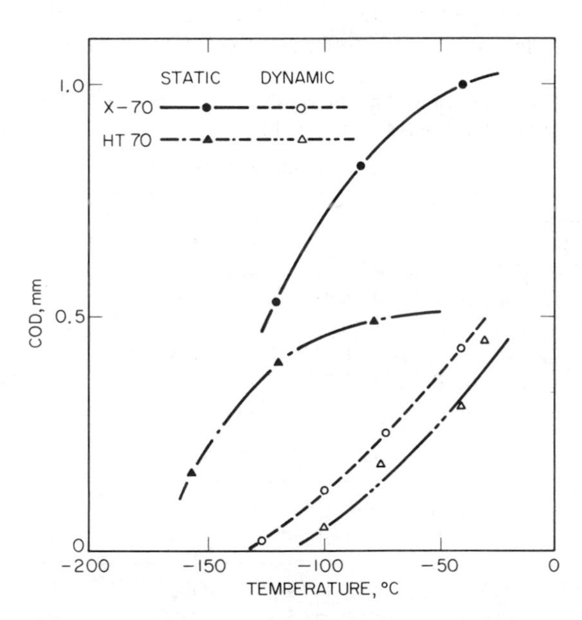

FIG. 16.13. Static and dynamic COD test results from Nishioka and Iwanaga.[7]

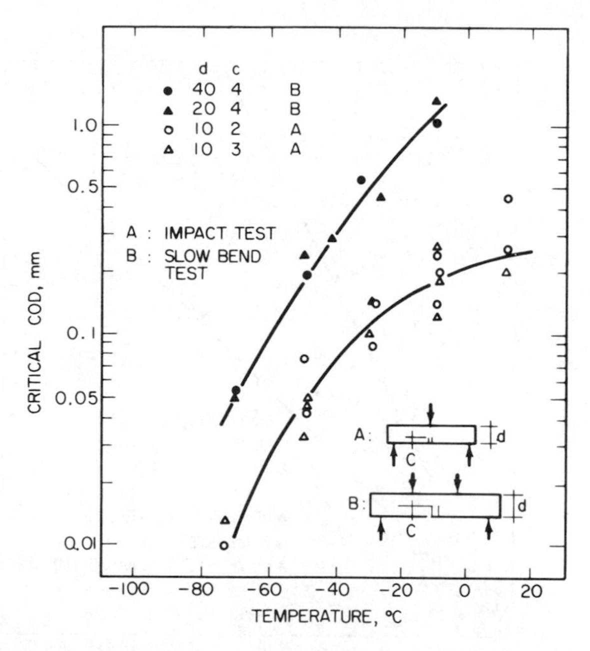

FIG. 16.14. Effect of impact loading on critical COD (Ref. 8).

In summary, δ_c test results are affected by temperature, loading rate, and specimen thickness in the same general way that these variables affect the test results of other types of fracture test results, such as K_{Ic} or CVN.

16.2.2. Interpretation of COD Test Results

The key questions regarding the possible use of δ_c test results in engineering design are, can they be related to actual service conditions, and is the proposed relation between δ_c/σ_{ys} and K_{Ic}/σ_{ys} sufficiently well developed so that δ_c values can be used in engineering design to calculate critical crack sizes for various design stress levels? In this respect, Egan[6] has shown a general relation that varies from a 1:1 relation for elastic behavior to a 2:1 relation for tougher materials (Figure 16.15). Finally, can δ_c concepts be extended into the inelastic region so that meaningful critical crack sizes can be calculated for specific design stress levels?

A design curve, proposed by Burdekin and Dawes,[9] for δ_c, strain, and crack size relationships is shown in Figure 16.16(a). It is based on the strip yielding model discussed above and on considerable experimental work at The Welding Institute in the United Kingdom [Figure 16.16(b)]. This design curve was developed primarily for pressure vessels but can be used for other structural applications as well. It should be emphasized that the design curve [Figure 16.16(a)] is based primarily on the results of actual experimental

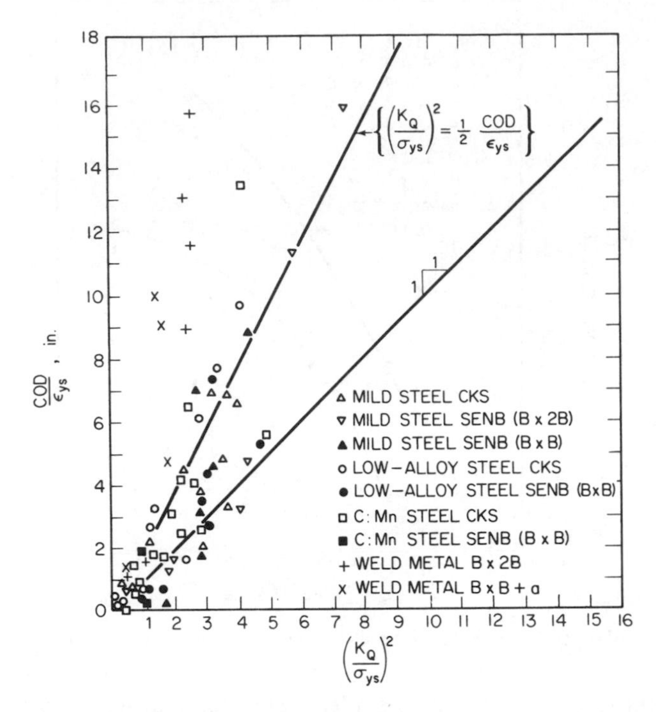

FIG. 16.15. Relationship between $(K_Q/\sigma_y)^2$ and COD/ϵ_y for different materials and specimen designs.

results rather than on theoretical predictions [Figure 16.16(b)]. In fact, Burdekin and Dawes[9] point out that the proposed design curve presented in Figure 16.16(a) "should be checked by further research work on materials with different strain hardening characteristics. The assumptions with regard to alternative interpretations for a_{cr} in the general yielding field should also be checked by experimental work. At the present time, however, they represent practical simplifications which enable much greater use to be made of fracture mechanics techniques and they are consistent with service observations."

Using the relationship for a through-thickness crack, $K_{Ic} = \sigma\sqrt{\pi a}$ (Chapter 2), and for the design stress equal to some percentage of the yield strength, then

$$a_{cr} = \text{constant} \times \left(\frac{K_{Ic}}{\sigma_{ys}}\right)^2$$

Similarly, because $(K_{Ic}/\sigma_{ys})^2$ was shown to be proportional to δ_c/ϵ_{ys},

$$a_{cr} = \text{constant} \times \left(\frac{\delta_c}{\epsilon_{ys}}\right)$$

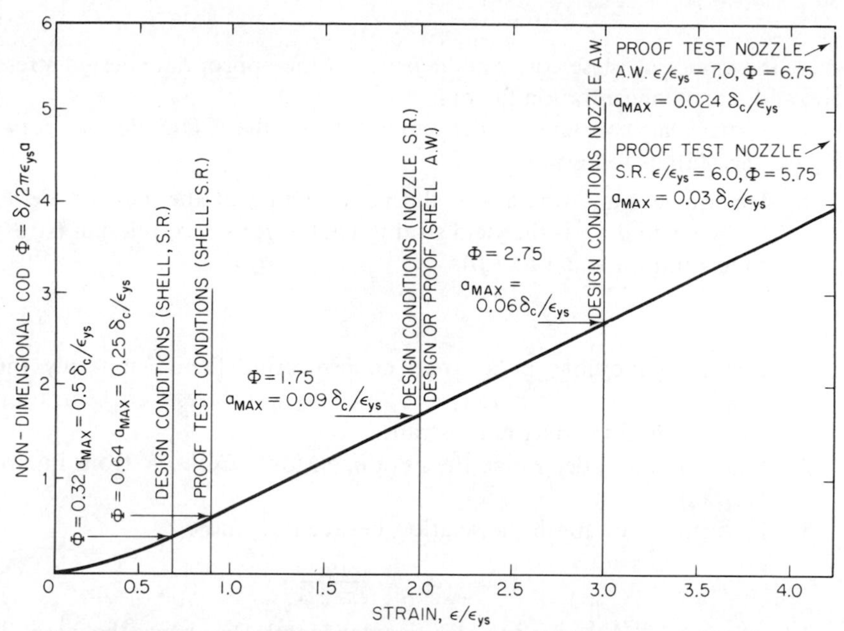

(a) DESIGN CURVE FOR COD, STRAIN, AND CRACK SIZE RELATIONSHIP AS PROPOSED BY BURDEKIN AND DAWES

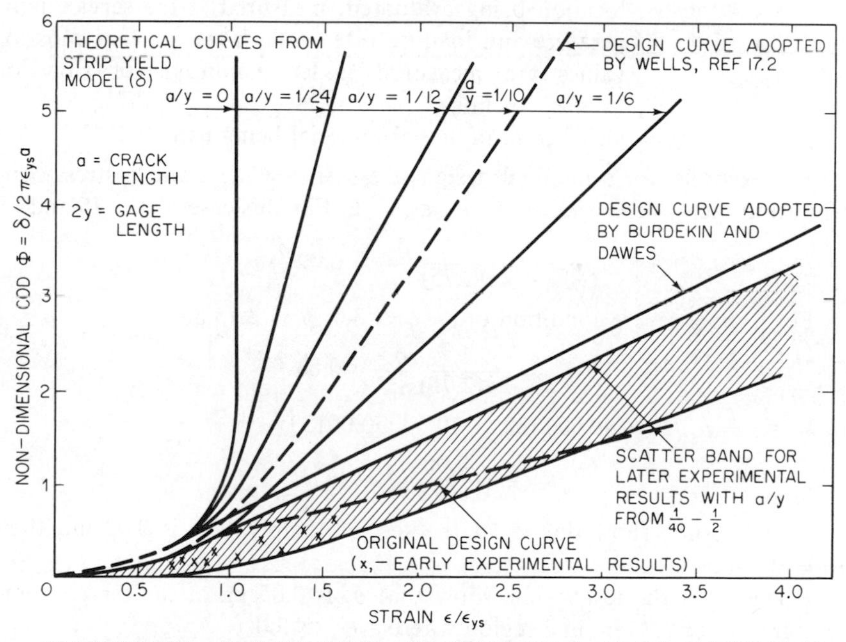

(b) THEORETICAL AND EXPERIMENTAL COD, STRAIN, AND CRACK SIZE RELATIONSHIPS

FIG. 16.16. Theoretical, experimental, and design curve for COD, strain, and crack size relationship.

where the values of these constants depend on the appropriate design stress level and stress concentration factors.

The particular constant is determined from the COD design curve [Figure 16.16(a)] as follows:

1. Determine ϵ/ϵ_{ys}, where ϵ is the actual strain at the location being analyzed, and ϵ_{ys} is the yield strain. That is, for general design (stress concentration equal to 1.0),

$$\epsilon = \frac{\sigma_{\text{des}}}{E}$$

   For the conditions at a stress concentration factor, multiply the nominal design strain ($\epsilon_{\text{des}} = \sigma_{\text{des}}/E$) by the stress concentration factor to find the maximum strain ϵ.
2. Knowing ϵ/ϵ_{ys}, determine the nondimensional COD, ϕ, from Figure 16.16(a).
3. Knowing ϕ, establish the relation between a_{cr} and δ_c:

$$a_{\text{cr}} = \frac{\delta_c}{2\pi\epsilon_{ys}\phi}$$

   where $a_{\text{cr}} = \frac{1}{2}$ through-thickness crack length, $2a$—hence the factor 2,
   $\delta_c =$ critical COD for the actual material being used at the location being evaluated, measured at the service temperature and loading rate (note, however, that most δ_c values are measured under conditions of slow or "static" loading),
   $\epsilon_{ys} =$ yield strain of actual material being used.

For example, for a nominal design stress of $\sigma = \sigma_{ys}/2$ and a stress concentration factor of 4, $\epsilon/\epsilon_{ys} = 4 \cdot \frac{1}{2}\epsilon_{ys}/\epsilon_{ys} = 2$. For this case $\phi = 1.75$ and

$$a_{\text{cr}} = \frac{1}{2\pi(1.75)}\frac{\delta_c}{\epsilon_{ys}} = 0.09\frac{\delta_c}{\epsilon_{ys}}$$

For a more severe condition of $\epsilon/\epsilon_{ys} = 3.0$, $\phi = 2.75$ and

$$a_{\text{cr}} = \frac{1}{2\pi(2.75)}\frac{\delta_c}{\epsilon_{ys}} = 0.06\frac{\delta_c}{\epsilon_{ys}}$$

If $\epsilon_{ys} = \sigma_{ys}/E = 40{,}000/(30 \times 10^6) = 0.00133$,

$$a_{\text{cr}} = \frac{0.06}{0.00133}\delta_c = 45\delta_c$$

If test results show that δ_c at the service temperature is 0.02 in., then $a_{\text{cr}} = 0.9$ in.

However, if the test results show that δ_c is 0.05 in. at the service temperature, $a_{\text{cr}} \simeq 2.25$ in. in a region where $\epsilon/\epsilon_{ys} \simeq 3.0$.

Burdekin and Dawes[9] developed various factors for maximum crack sizes in different regions of a pressure vessel for $\sigma = 0.67\sigma_{ys}$ and $\sigma = 0.87\sigma_{ys}$, as presented in Table 16.2.

TABLE 16.2 Summary of Factors for Maximum Crack Sizes Derived in Paper by Burdekin and Dawes[9]

	Design Stress Level (0.67 Yield)			Proof Test Stress Level (0.87 Yield)			
	*Shell (s.r.)**	*Shell (a.w.)*** *or nozzle (s.r.)*	*Nozzle (a.w.)*	*Shell (s.r.)*	*Shell (a.w.)*	*Nozzle (s.r.)*	*Nozzle (a.w.)*
$\bar{a}_{max} =$	$0.5(\delta/\epsilon_{ys})$ $0.5(K_{Ic}/\sigma_{ys})^2$	$0.09(\delta/\epsilon_{ys})$ $0.09(K_{Ic}/\sigma_{ys})^2$	$0.06(\delta/\epsilon_{ys})$ $0.06(K_{Ic}/\sigma_{ys})^2$	$0.25(\delta/\epsilon_{ys})$ $0.25(K_{Ic}/\sigma_{ys})^2$	$0.09(\delta/\epsilon_{ys})$ $0.09(K_{Ic}/\sigma_{ys})^2$	$0.03(\delta/\epsilon_{ys})$ $0.03(K_{Ic}/\sigma_{ys})^2$	$0.024(\delta/\epsilon_{ys})$ $0.024(K_{Ic}/\sigma_{ys})^2$

*s.r. = stress relieved
**a.w. = as welded

Recently, Cotton et al.[5] have proposed a simplified design method based on Burdekin's method. In their method, a multiplying factor is used to find the limiting crack size, a_{cr}, such that

$$a_{cr} = f \cdot \delta_c$$

where δ_c = critical COD for the material, weld metal, or heat-affected zone material of interest, at the service temperature, and

f = a multiplying factor that varies according to the total strain in the critical region.

Development of the multiplying factor, f, was based on the following observations:

1. Examination of Burdekin's design curve [Figure 16.16(a)] shows that for values of ϵ/ϵ_{ys} greater than 0.89 (i.e., into the inelastic region), a straight-line relation exists between ϕ and ϵ/ϵ_{ys} which is

$$\phi = \frac{\epsilon}{\epsilon_{ys}} - 0.25$$

2. Because $\phi = \delta_c/2\pi\epsilon_{ys} \cdot a_{cr}$, $\delta_c = 2a\pi\epsilon_{ys}[\epsilon/\epsilon_{ys} - 0.25]$ or

$$\delta_c = 2a\pi(\epsilon - 0.25\epsilon_{ys})$$

3. Letting $\beta = 2a$ = total critical crack length (through-thickness crack),

$$\delta_c = \beta\pi(\epsilon - 0.25\epsilon_{ys})$$

4. For $\epsilon_{ys} = 0.002$ ($\sigma_{ys} \simeq 60$ ksi),

$$\delta_c = \beta\pi(\epsilon - 0.0005)$$

This relationship between δ_c and ϵ is plotted in Figure 16.17 for various values of β.

5. By letting $f = 1/[\pi(\epsilon - 0.0005)]$

$$\beta = 2a_{cr} = f \cdot \delta_c$$

which is plotted as a single design curve in Figure 16.18.

Cotton et al.[5] point out that this design approach is not valid for low values of δ_c. However, for very low values of δ_c, elastic conditions control and K_{Ic} values can be used to calculate critical crack sizes. Furthermore, they state that "values of δ_c much less than 0.005 inches should not be relied upon because such low values are greatly affected by the experimental technique and their interpretation is subject to considerable disagreement. Practical observations indicate that it is rarely necessary to accept conditions giving rise to COD of less than 0.005 inches and for the same reason it seems that a strain of less than $\frac{1}{3}$ yield strain has little practical application."

Because the determination of strain at a given location is very difficult,

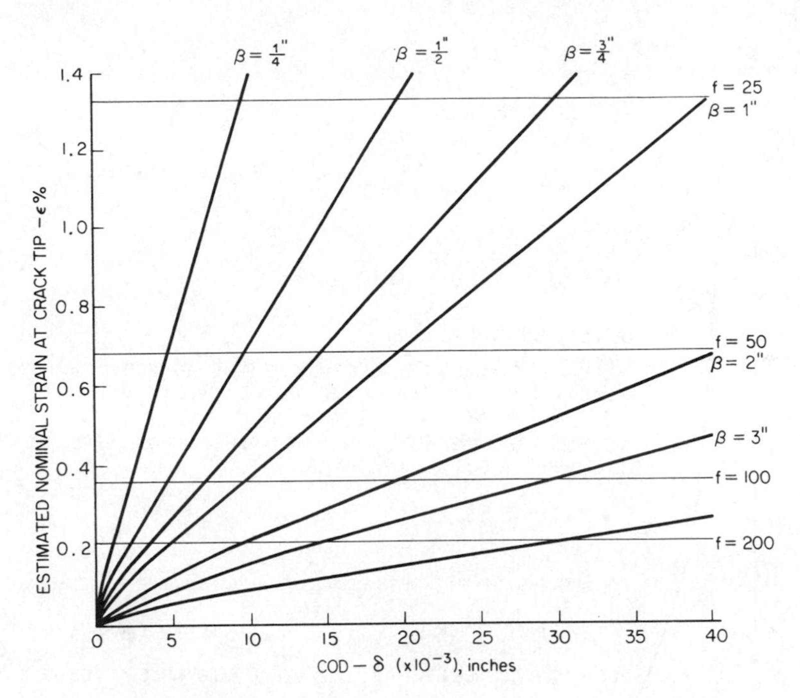

FIG. 16.17. Permissible defect size parameters for values of COD and estimated nominal strain at crack tip.

the following values for the multiplying factor, *f*, have been suggested by Cotton et al.[5] for pressure vessels:

	As Welded Material		*Stress-Relieved Material*	
Location of Defect	ϵ, *in./in.*	*f* (*factor*)	ϵ, *in./in.*	*f* (*factor*)
Cracks along pressure vessel axis (longitudinal)	0.004	90	0.002	200
Cracks transverse to pressuse vessel axis (circumferential)	0.003	115	0.001	700
Cracks around nozzles and other points of strain intensification including indeterminant areas	0.014	25	0.012	27

Note: In the case of "as-welded" structures a strain of 0.002 in./in. has been added to allow for residual stress.

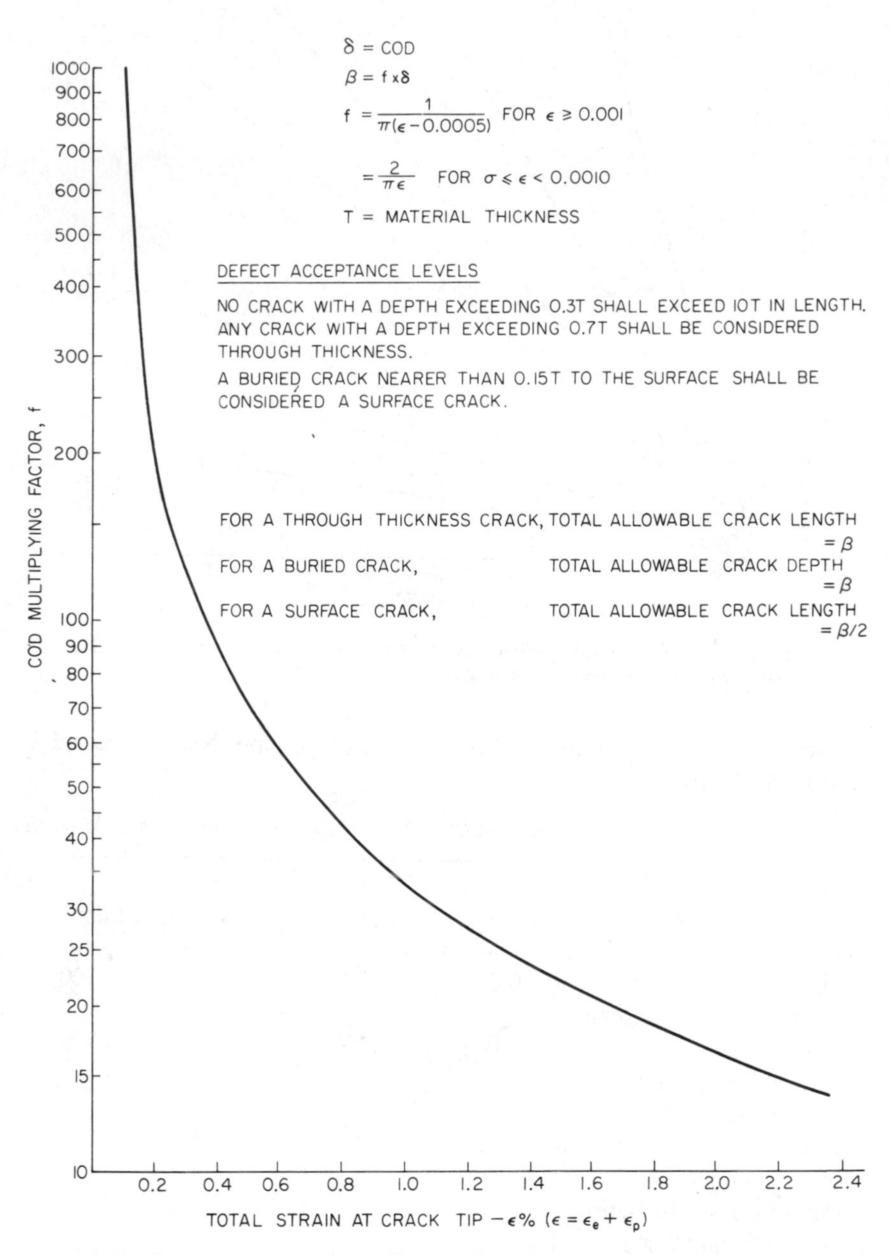

FIG. 16.18. Relationship between COD multiplying factor and total strain at crack tip used in Cotton's[5] analysis.

Cotton et al.[5] suggest that for most cases the need for extreme accuracy in COD values is not great and therefore the design values that can be obtained from Figure 16.18 are sufficiently accurate for most engineering applications. Table 16.3 lists the proposed multiplying factors for various

TABLE 16.3 Proposed Multiplying Factors for Various Types of Construction (Reference 5)

	Construction	*Seam*	*Crack Location*	*Multiplying Factor, f*
A	Oil storage tanks (as welded)	Horizontal	Longitudinal	500
	Pressure vessels & pipes (stress-relieved)	Circumferential	Longitudinal	500
		Longitudinal	Transverse	500
B	Pressure vessels & pipes (stress-relieved)	Circumferential	Transverse	200
		Longitudinal	Longitudinal	200
C	Oil storage tanks (as welded)	Horizontal	Transverse	100
		Vertical	Longitudinal	100
	Pressure vessels & pipes (as welded)	All cracks except those in category D		100
D	Oil storage tanks, pressure vessels, & pipes (as welded & stress-relieved)	Cracks around nozzles and points of high stress or in locations of indeterminate design.		25

other types of vessel construction, such as oil storage tanks. As the degree of severity of possible cracks increases (moving from condition A to D), the multiplying factor decreases. Thus to obtain the same critical crack size for the more severe conditions as for the less severe cases, a tougher material with a higher value of δ_c must be used.

In summary, δ_c test results obviously can be used to compare the relative fracture-toughness behavior of different structural materials, weld metals, or heat-affected zones in the same manner as CVN, DT, or precracked CVN test specimens are used because all these tests can be conducted throughout the full transition-temperature region. Figure 16.19 shows a COD-CVN ft-lb correlation for a quenched and tempered $2\frac{1}{4}$Cr-1Mo steel and weldment, indicating that a general relation between these two tests may exist, although the scatter is considerable.

The obvious potential advantage of the COD technique, compared with other testing techniques such as K_{Ic}, K_{Id}, CVN, etc., is that it may be possible

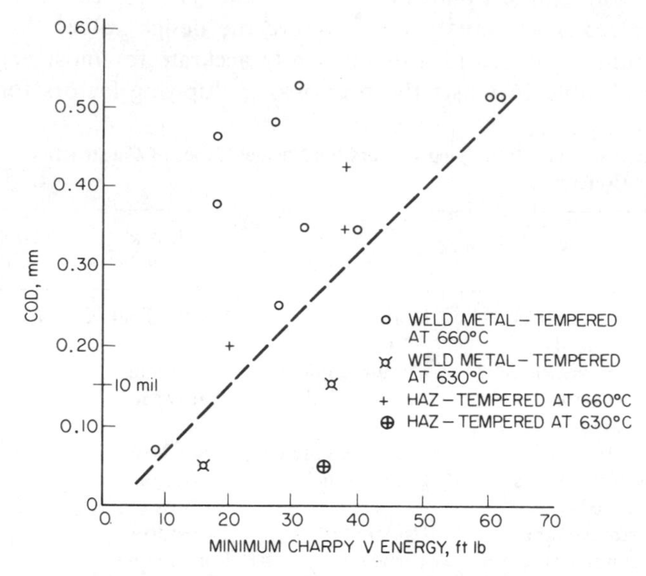

FIG. 16.19. COD/Charpy V energy correlation for Q and T $2\frac{1}{4}$ Cr-1Mo steel weldments.

to relate δ_c measurements to critical crack sizes in regions of elastic-plastic or fully plastic behavior.

Hood[4] has studied three examples of the use of COD measurements to predict service behavior and concluded that there is a wide variation in the ability of the COD test to predict the failure stress of flawed pressure vessels. Some of the inaccuracy was attributed to local material variation, bulging of the vessel walls around the crack tip, and inherent scatter. However, the major difference in predicted and actual test results was attributed to the occurrence of stable crack extension prior to instability. Because stable crack extension occurs primarily in the elastic-plastic or fully plastic region (ductile fracture), and it is these regions that COD measurements were really developed for, the effect of stable crack extension on COD measurements needs to be more clearly understood. Fearnehough et al.[10] have studied the role of stable ductile crack growth in the failure of structures and concluded that cracking in full-size burst tests can be predicted by COD test results on small specimens, providing the initiation of cracking in the vessels is carefully detected. That is, although failure occurred at a considerably later stage than crack initiation (in some cases the crack growth was 36% of the initial crack size), the initiation COD test results were very useful in predicting failure stresses.

The development of the R-curve concept, which accounts for stable crack extension, offers considerable promise for solving this problem.

16.3. *R*-Curve Analysis

As described in Chapters 2, 3, and 4, K_{Ic} is the critical stress-intensity factor under conditions of plane strain ($\epsilon_z = 0$) with attendant small-scale crack-tip plasticity. Conversely K_c is the critical stress intensity factor under plane-stress conditions ($\sigma_z = 0$) with attendant large-scale crack-tip plasticity. Thus the behavior represented by K_c is the opposite of that represented by K_{Ic}, that is, negligible rather than complete through-thickness elastic constraint (stress) at fracture. K_c values are generally 2–10 times larger than K_{Ic} and vary not only with temperature (T) and strain rate ($\dot{\epsilon}$), as does K_{Ic}, but with plate thickness (B) as well. Furthermore, for fixed conditions of temperature, strain rate, and plate thickness (T, $\dot{\epsilon}$, and B), the K_c values may also vary with initial crack length, a_0.

The operating temperatures, rates of loading, and thickness of most structural materials used in actual structures are generally such that plane-stress rather than plane-strain conditions actually exist in service. Accordingly, considerable effort has been devoted toward plane-stress fracture-toughness evaluations using an *R*-curve or resistance curve analysis as one of several extensions of linear-elastic fracture mechanics into elastic-plastic fracture mechanics.

R-curves characterize the resistance to fracture of materials during incremental slow-stable crack extension. They provide a record of the toughness development as a crack is driven stably under increasing crack-driving forces, i.e., loads. They are dependent upon specimen thickness, as well as test temperature and loading rate. An *R*-curve is a plot of crack growth resistance in a material as a function of actual or effective crack extension. K_R is the crack growth resistance expressed in units corresponding to K, namely ksi$\sqrt{\text{in.}}$ K_c is the plane stress fracture toughness and is equal to the value of K_R at a particular instability condition determined during an *R*-curve test.

The *R*-curve describes the variation in K_R with crack length, a_0. It consists of a plot of K_R versus Δa, where K_R represents the driving force required to produce stable crack extension (Δa) prior to unstable crack growth at K_c. The K_c value that results for a given crack length, a_0, is the value associated with the point of tangency between the line representing the applied load and the *R*-curve itself (Figure 16.20).

The dashed lines in Figure 16.20 represent the variation in K_I with crack length, a, for constant load P_1, P_2, or P_3, where $P_3 > P_2 > P_1$. That is, for a given load level and increasing crack length, a, K_I will increase because $K_I = f(P, \sqrt{a})$.

The solid lines represent the increase in K_R with increasing load and increasing crack length for two different initial crack lengths. The two points of tangency where $K_R = K_c$ for $a_0 = a_1$ and $a_0 = a_2$ represent points of instability, or the critical plane-stress-intensity factor, K_c, at the particular

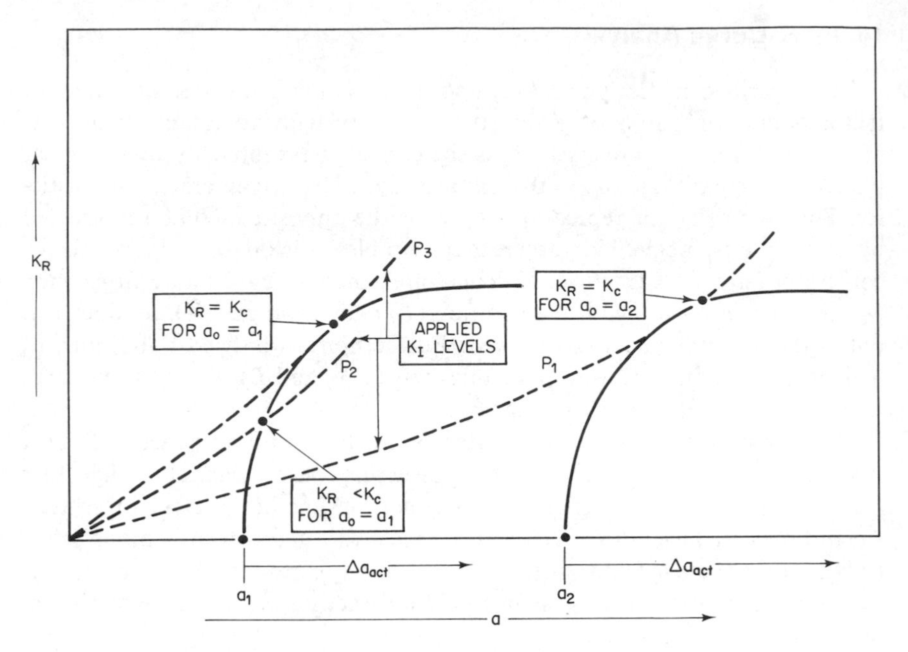

FIG. 16.20. Basic principle of *R*-curves for use in determining K_c under different conditions of initial crack length, a_0 (Ref. 15).

crack length and, of course, for the given conditions of temperature, loading rate, and plate thickness. The K_R value is always calculated by using the effective crack length, a_{eff}, and is plotted against the actual crack extension, a_{act}, that takes place physically in the material during the test.

Under plane-strain conditions, the fracture toughness of a material depends on only two variables [$K_{Ic} = f(T \text{ and } \dot{\epsilon})$], whereas under plane-stress conditions, the fracture toughness depends on four variables [$K_c = f(T, \dot{\epsilon}, B, \text{ and } a_0)$]. For given test and material conditions (T, $\dot{\epsilon}$, and B), a K_c value merely represents a singular point on an *R*-curve. On the other hand, an *R*-curve describes the complete variation of K_c with changes in initial crack length, a_0. As such, a single *R*-curve is a highly efficient method of fracture characterization since it is equivalent to a large number (10–15) of direct K_c tests conducted with various (initial) crack lengths, a_0. Thus, the *R*-curve is the most general characterization of plane-stress-fracture behavior and depends on only three variables [*R*-curve $= f(T, \dot{\epsilon}, \text{ and } B)$].

An ASTM special technical publication describing the state-of-the-art of *R*-curve testing has recently been published.[11] Of particular interest as part of this book is a paper by Heyer, which presents a literature survey of *R*-curve testing, including some noteworthy historical aspects. Novak[12] has published an extensive study of the *R*-curve behavior, and some of his results are described in this chapter as an example of the use of *R*-curves. Recently, ASTM has developed a "Recommended Practice for *R*-Curve Determination."[13]

In elementary terms, *R*-curves can be determined by using either of two experimental methods: *load control* or *displacement control.* The load-control technique can be used to obtain only that portion of the *R*-curve up to the K_c value (where complete unstable fracture occurs), whereas the displacement-control technique can be used to obtain the entire *R*-curve and therefore offers a fundamental advantage. The equivalence of the two techniques for determining K_c has been demonstrated by the work of Heyer and McCabe,[14,15] the originators of the displacement-control technique. However, their demonstration of equivalence for the two test techniques was restricted to very high-strength steels and aluminum alloys, where the principles of linear-elastic fracture mechanics (LEMF) are directly applicable as a result of limited crack-tip plasticity.

The evaluation of *R*-curves for relatively low-strength, high-toughness alloys is more complex. Because such materials exhibit large-scale crack-tip plasticity (r_y) at fracture, relative to the test-specimen in-plane dimensions (W and a), LEFM principles cannot be applied directly. As a consequence, a nonlinear, elastic-plastic approach is required. In this elastic-plastic approach, the crack-opening displacement (δ) at the physical crack tip is measured and used in calculating the equivalent elastic K value.* This nonlinear approach is based on theoretical considerations advanced earlier by Wells[16] and reviewed more recently by Wells[17] and Irwin.[18] This elastic-plastic crack model is designated the crack-opening-stretch (COS) method, where δ and COS are equivalent terms. The application of the COS analysis method to *R*-curve testing has been developed to an advanced degree by McCabe and Heyer.[19] Furthermore, this method can be used with either the load-control or displacement-control test procedures.

The general procedure for conducting *R*-curve tests is described briefly in this chapter. It is a complex testing procedure and is not standardized, although ASTM has developed a recommended practice.[13] For specific details of the different testing techniques used, the reader is referred to References 11–19.

Generally the thickness of the test specimens, B, is equal to the plate thickness being considered for actual service usage. The other dimensions, W and H where H is one-half the height of the specimen and generally is equal to 0.6 W (Figure 16.21), are made considerably larger. This is done so that the plastic-zone size ahead of the crack tip, r_y, is small with respect to the

*The equivalent elastic K value is the analog K value that would be measured under elastic conditions for which LEFM principles can be used directly when specimens of the same thickness, B, but much larger planar dimensions, W and a, are tested.

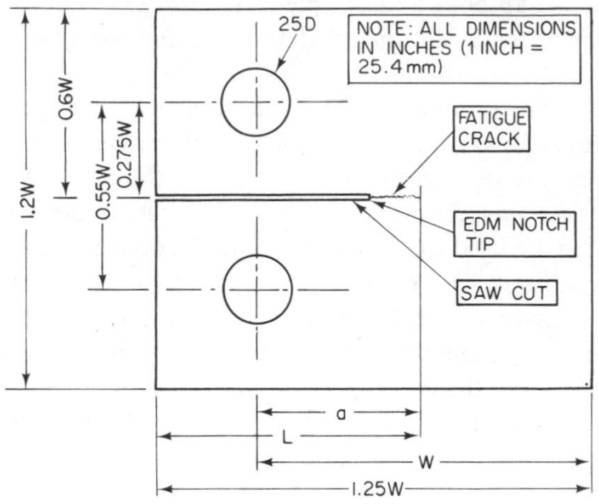

(a) CT SPECIMEN (H/W = 0.600) USED FOR LOAD-CONTROL TESTS. TYPES 2T AND 4T

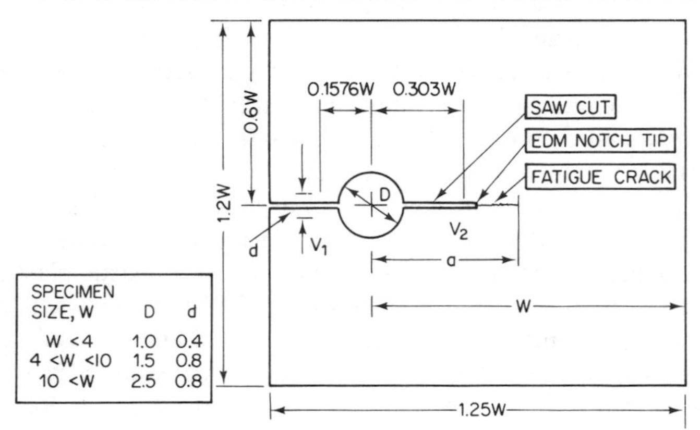

(b) CT SPECIMEN (H/W = 0.600) USED FOR DISPLACEMENT-CONTROL TESTS. TYPES 4C AND 7C

FIG. 16.21. CT specimens used for load-control and displacement-control tests (Ref. 12).

remaining ligament length, $W - a$, Figure 16.22. In load control, the specimen is loaded in a large tension machine with loading pins [Figure 16.21(a)]. For displacement control, special wedges are pushed into a hole on the crack line to wedge the crack surfaces apart [Figure 16.21(b)]. The overall testing setup for displacement control is shown in Figure 16.23, where this particular test specimen is about 2 × 2 ft. As stated, the displacement-control method offers a primary advantage in being able to obtain the entire R-curve for many values of crack length, whereas the load-control technique can be used only to obtain that portion of the R-curve up to the K_c value, where complete unstable fracture occurs. However, the advantage of the displacement-control technique is partially offset by the necessity for new or unique loading

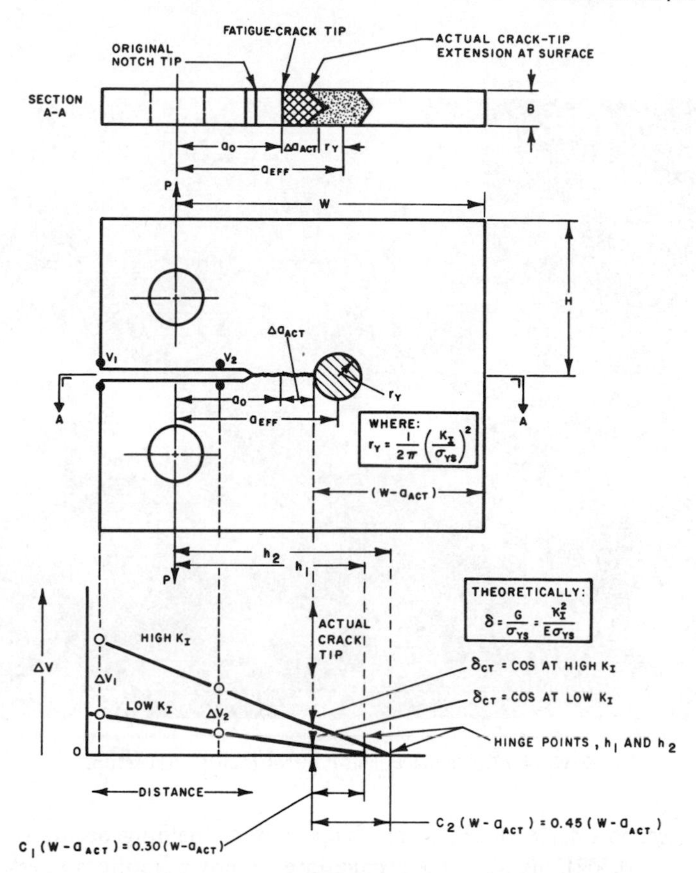

FIG. 16.22. Schematic of procedure for measuring δ_{CT} or COS at the actual crack tip (a_{ACT}) relative to applied load level (K_I) under plane-stress conditions.

facilities and sophisticated instrumentation (Figure 16.23), whereas with the load-control method a conventional tension machine may be used in conjunction with relatively simple measuring devices (clip gage and vernier micrometer).

After fabrication of the test specimens, they are fatigue-cracked and then tested at various testing temperatures. Loading is essentially monotonic, with displacements being measured at various points, such as V_1 and V_2 (Figure 16.22).

For linear-elastic behavior, i.e., no back surface yielding,

$$K_R = \frac{P \cdot f(a/W)}{B \cdot W^{1/2}}$$

where $f(a/W)$ is a specific function of the relative crack length and a is the effective crack length as described in Figure 16.22.

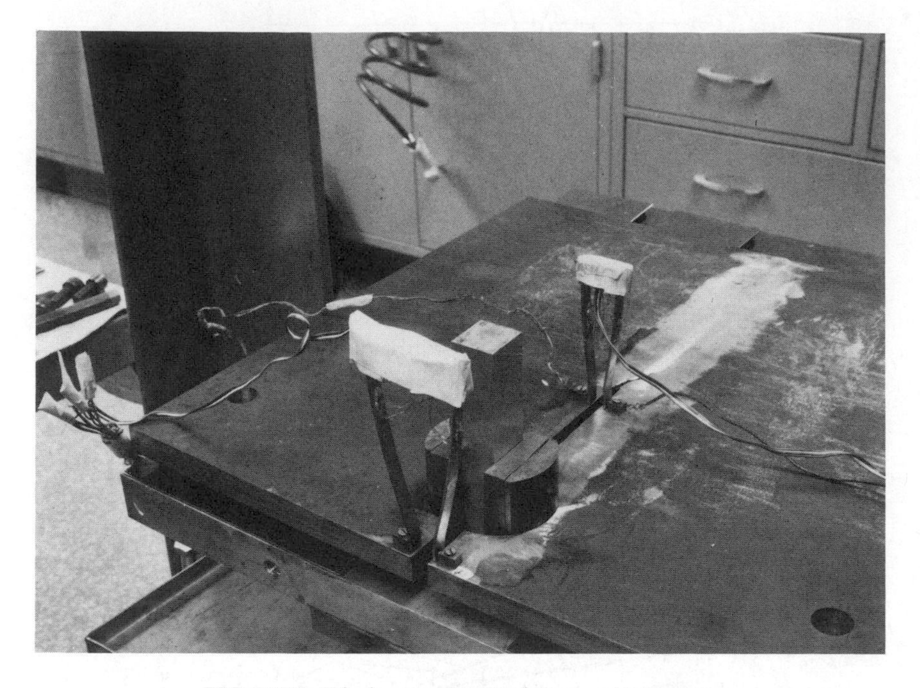

FIG. 16.23. Displacement-control *R*-curve test setup.

At larger values of load, where back surface yielding occurs (defined as $\epsilon_{BS} > \epsilon_{ys} + 0.002$), the K_R value is calculated at any point by first establishing the hinge point, h_1 or h_2, using the relationship

$$h_1 \text{ or } h_2 = \frac{0.1576W + 0.303W(V_{1\,\text{abs}}/V_{2\,\text{abs}})}{(V_{1\,\text{abs}}/V_{2\,\text{abs}}) - 1}$$

where $(V_{1\,\text{abs}}/V_{2\,\text{abs}})$ is measured just prior to unloading at a particular crack length.[13] The value of h is then used to caluclate the COS (or δ) value at the tip of the actual crack length (a_{act}) (Figure 16.22) by using the relation

$$\delta = V_{2\,\text{abs}}\left(\frac{h - a}{h - 0.303W}\right)$$

The value of δ is used, in turn, to calculate the equivalent elastic-stress-intensity value, K_R, by the relation[16-18]

$$K_R = (E \cdot \sigma_{ys} \cdot \delta)^{1/2}$$

where E and σ_{ys} are the modulus of elasticity and the yield strength of the material being tested, respectively. For materials with high levels of toughness this analysis becomes increasingly less accurate. A more complete discussion of this particular technique is found in Reference 12.

To illustrate the types of results that might be expected, results of an *R*-curve investigation for a relatively low-strength high-toughness structural steel (A572) by Novak[12] are described. Novak used two types of compact tension (CT) test specimens (Figure 16.21), namely load-control and displacement-control specimens, in his investigation.

Typical *R*-curves for various-sized specimens of A572 steel tested at +72°F are presented in Figure 16.24. Test results at +40°F and −40°F for this same steel are presented in Figures 16.25 and 16.26, respectively, and show that the effect of temperature on *R*-curve behavior is similar to the effect obtained using other types of fracture-toughness tests such as K_{Ic} or CVN impact test specimens; i.e., the values decrease significantly at lower temperatures.

The results in Figures 16.24 through 16.26 show, collectively, that a rapid increase in plane-stress crack tolerance occurs with increasing temperature. Evidence for this can be seen in both the increasing amounts of stable crack extension preceding fracture, Δa_c, and the increasing K_c values that occur with increasing temperature. The specific variation in K_c values with temperature for all the 1.5-in.-thick specimens tested is presented in a

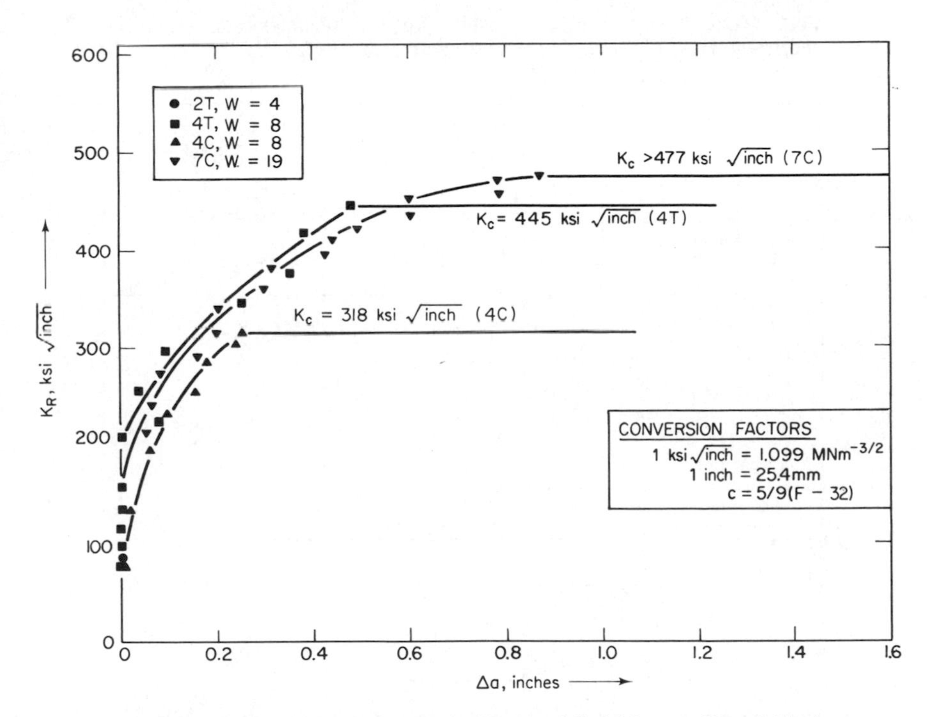

FIG. 16.24. *R*-curve and K_c results for full-thickness ($B = 1.5$ in.) specimens of A572 Grade 50 steel tested at +72°F (Ref. 12).

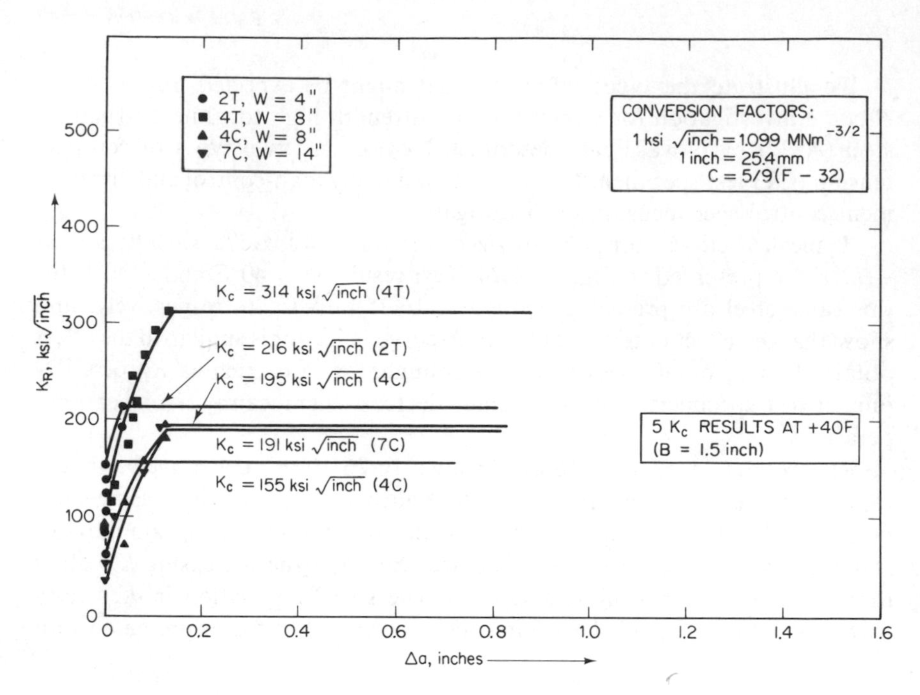

FIG. 16.25. *R*-curve and K_c results for full-thickness ($B = 1.5$ in.) specimens of A572 Grade 50 steel tested at +40° (Ref. 12).

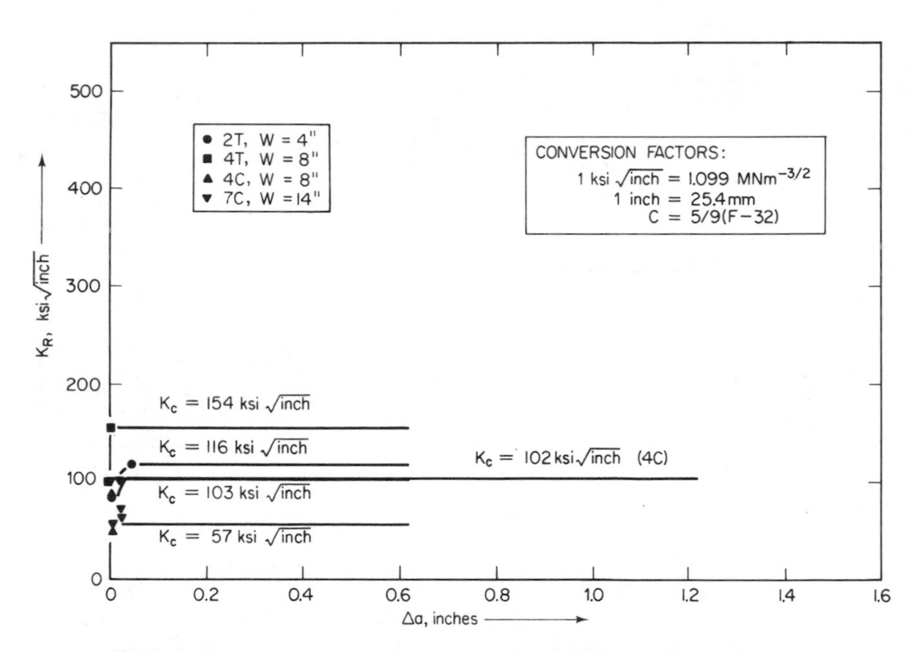

FIG. 16.26. *R*-curve and K_c results for full-thickness ($B = 1.5$ in.) specimens of A572 Grade 50 steel tested at −40°F (Ref. 12).

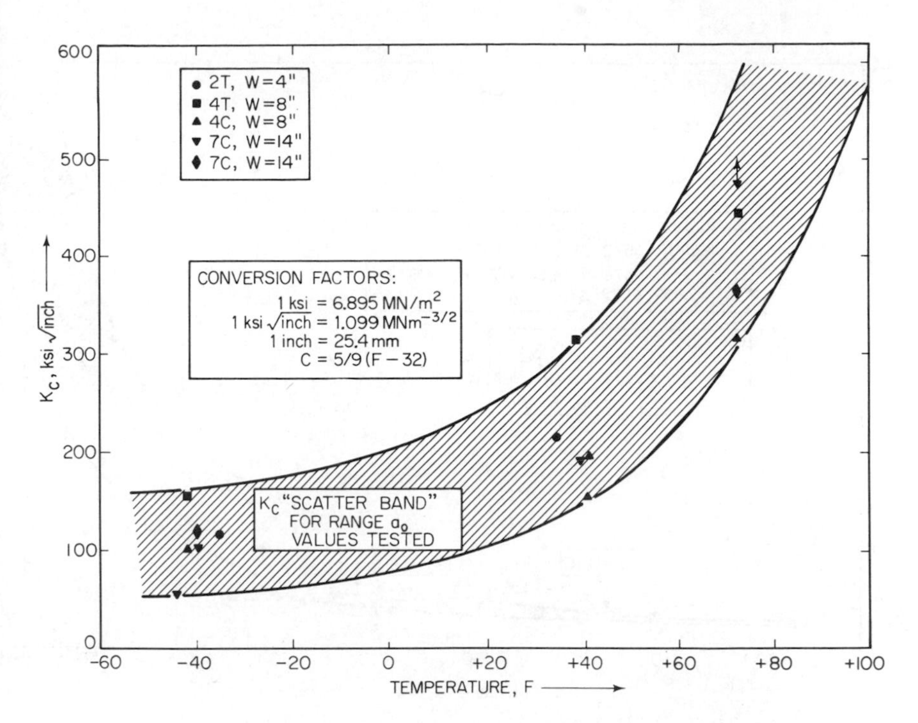

FIG. 16.27. Summary of K_c results for full-thickness ($B = 1.5$ in.) specimens of A572 Grade 50 steel and A572 steel processed to 62-ksi strength level (Ref. 12).

summary plot (Figure 16.27). This figure shows that the K_c transition is quite steep at temperatures above 0°F (−18°C).

Static K_{Ic} and K_c test results for this same steel are compared in Figure 16.28 and show that at the higher temperatures the plane-stress K_c transition behavior is extremely large, much larger than the plane-strain transition in K_{Ic} described in Chapter 4. This behavior would be expected as K_c values extend well into the elastic-plastic region, whereas K_{Ic} values are limited to plane-strain behavior. The significance of the K_c values obtained from an R-curve analysis is in the calculation of the critical flaw size (a_{cr}) required to cause fracture instability under the same material and test conditions (T, $\dot{\epsilon}$, and B) used to measure the specific K_c value. The specific a_{cr} value is further related to the level of the design stress, σ_D, relative to σ_{ys} for a given specimen or structural geometry. A normalized plot showing the general relationship of a_{cr} to such design parameters for a large center-cracked tension (CCT) specimen subjected to uniform tension is presented in Figure 16.29. Because of the normalized basis of the plot, Figure 16.29 can be used to calculate a_{cr} values for a CCT specimen of any material (i.e., with any yield strength σ_{ys}) for which valid fracture-mechanics results (K_{Ic}, K_{Id}, K_c, K_{Iscc})

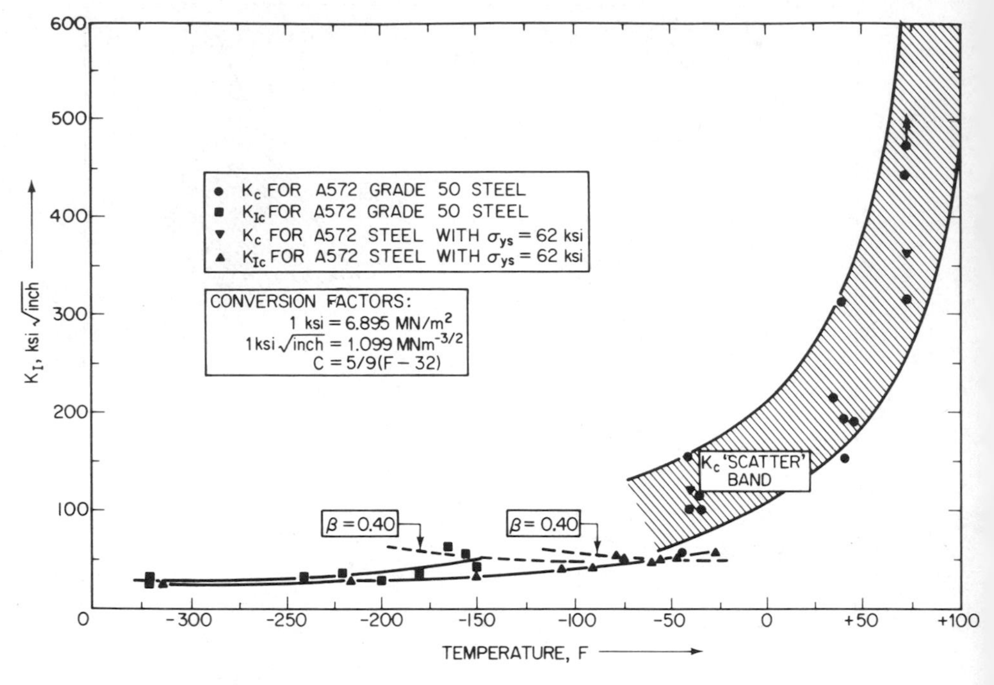

FIG. 16.28. Summary comparison of K_c and K_{Ic} behavior obtained from 1.5-in. thick plates of A572 steel (Ref. 12).

are available under the loading rate, temperature, and state-of-stress of interest.

An example of the applicability of the a_{cr} values can be given in terms of a typical structural member, such as a large H-beam (girder) with typical thicknesses for both the flange and the web on the order of $\frac{1}{2}$–$1\frac{1}{2}$ in. Specifically, the a_{cr} values cited would have application for through-thickness cracks located in the web of such a beam, where plane-stress conditions would exist and where the crack is surrounded by a thin plate having large in-plane dimensions. However, the same a_{cr} values would have no application for partial-thickness cracks (PTC) emanating from the top surface of the tension flange of the beam (such as would occur at the base of a cover plate due to fatigue).

These same stress-state (plane-stress) and structural (large planar dimensions) requirements are necessary for the interpretation of essentially *all* R-curve measurements, since such measurements intrinsically deal with materials exhibiting high levels of crack tolerance. In turn, high levels of crack tolerance ($K_c/\sigma_{ys} > 1$) under plane-stress conditions imply the existence of either very large critical flaws (a_{cr}) under low levels of elastic stress ($\sigma_D \leq \frac{1}{2}\sigma_{ys}$) (Figure 16.29) or high K_c levels that translate, for short cracks (a), into large values of the corresponding critical crack-tip plastic zone, r_y

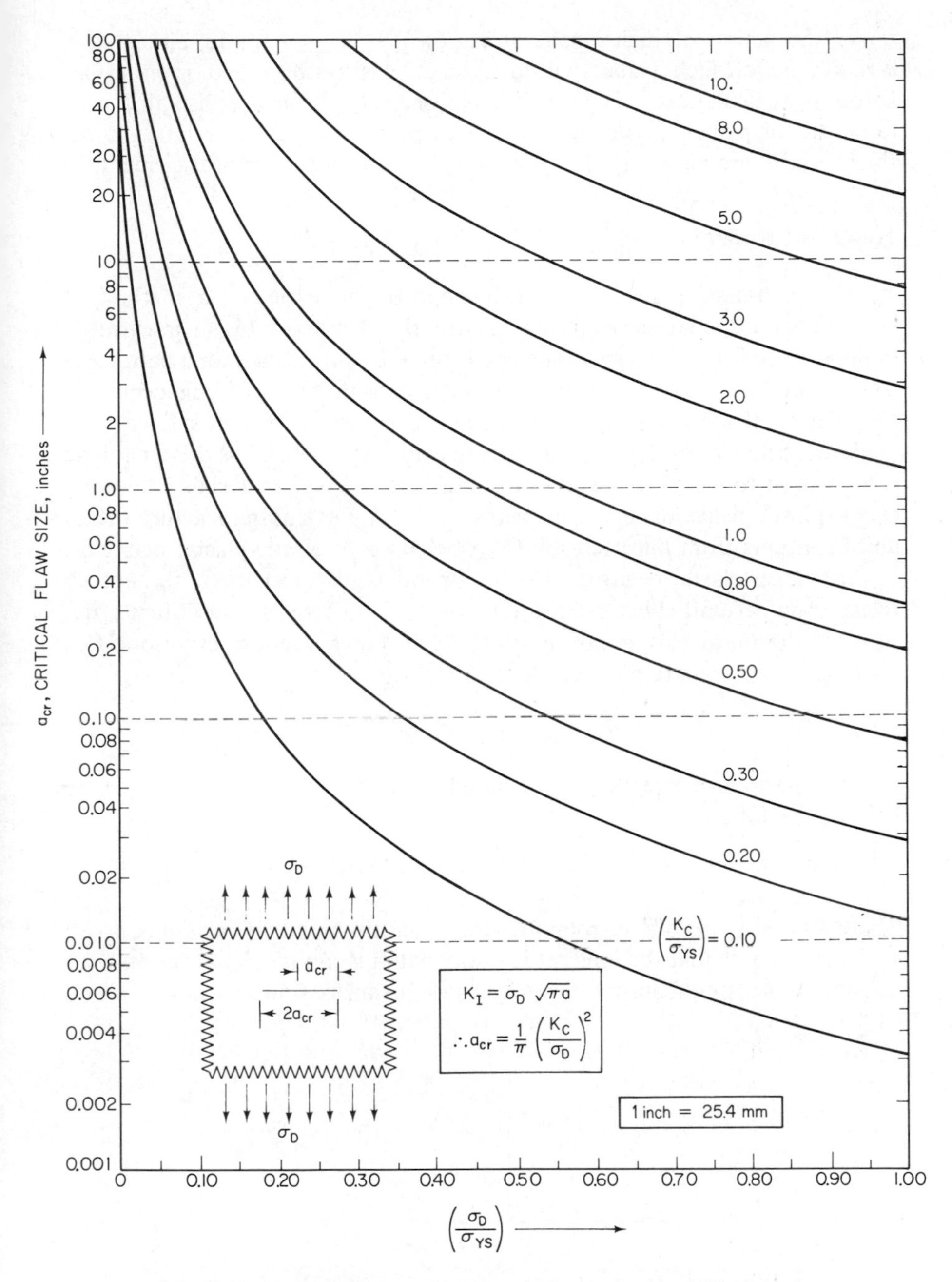

FIG. 16.29. Critical-flaw-size (a_{cr}) requirements for the initiation of a critical event (K_{Icr}) for cracks contained in an infinite center-cracked tension (CCI) specimen and subjected to a uniform tension stress (σ_D) applied remotely (Ref. 12).

under the action of high elastic stress ($\frac{1}{2}\sigma_{ys} \leq \sigma_D \leq \sigma_{ys}$). In either case, *containment* of such values within a large elastic-stress field is *necessary* before a_{cr} calculations can be valid. Accordingly, to accomplish this containment for plane-stress conditions, large planar dimensions relative to the thickness, B, are necessary for either a specimen or a structural element.

16.4. *J* Integral

The path-independent J integral proposed by Rice[20] is a method of characterizing the stress-strain field at the tip of a crack by an integration path taken sufficiently far from the crack tip to be substituted for a path close to the crack-tip region. Thus, even though considerable yielding occurs in the vicinity of the crack tip, if the region away from the crack tip can be analyzed, behavior of the crack-tip region can be inferred. Thus this technique can be used to estimate the fracture characteristics of materials exhibiting elastic-plastic behavior and is a means of directly extending fracture-mechanics concepts from linear-elastic (K_{Ic}) behavior to elastic-plastic behavior.

For linear-elastic behavior, the J integral is identical to G, the energy release rate per unit crack extension (Chapter 2). Therefore a J failure criterion for the linear-elastic case is identical to the K_{Ic} failure criterion. For linear-elastic plane-strain conditions,

$$J_{Ic} = G_{Ic} = \frac{(1 - \nu^2)K_{Ic}^2}{E}$$

The energy line integral, J, is defined for either elastic or elastic-plastic behavior as follows:

$$J = \int_R W\,dy - T\left(\frac{\partial U}{\partial x}\right) dx$$

where R = any contour surrounding the crack tip as shown in Figure 16.30 (note that the integral is evaluated in a counterclockwise manner starting from the lower flat notch surface and continuing along

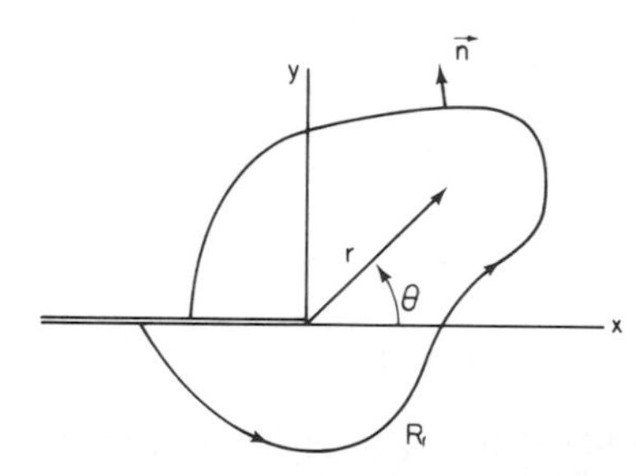

FIG. 16.30. Crack-tip coordinate system and arbitrary line integral contour.

an arbitrary path R to the upper flat surface),

$W =$ the strain energy density $= \int_0^{\epsilon} \sigma \, d\epsilon$,

$T =$ the traction vector defined according to the outward normal n along R, $T_i = \sigma_{ij} n_j$,

$U =$ displacement vector,

$s =$ arc length along R.

For any linear-elastic or elastic-plastic material treated by deformation theory of plasticity, Rice[20] has proven path independence of the J integral.

From a more physical viewpoint J may be interpreted as the potential energy difference between two identically loaded bodies having neighboring crack sizes, or

$$J = -\frac{1}{B}\frac{\partial U}{\partial a}$$

This definition is shown schematically in Figure 16.31, where the shaded area is $\partial U = JB \, da$. Note that Δ is measured in the load line.

Considerable work on developing the J integral as an analytical tool for elastic-plastic crack-tip field analysis has been performed by Begley and Landes[21,22] using a compliance technique. In this method several specimens

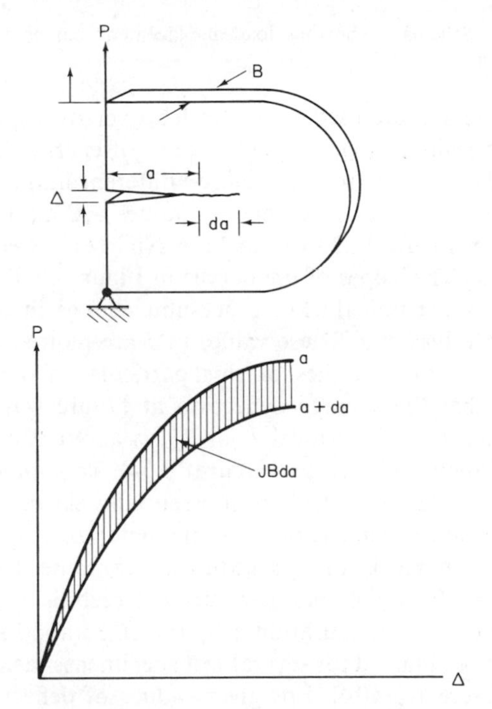

FIG. 16.31. Interpretation of J integral.

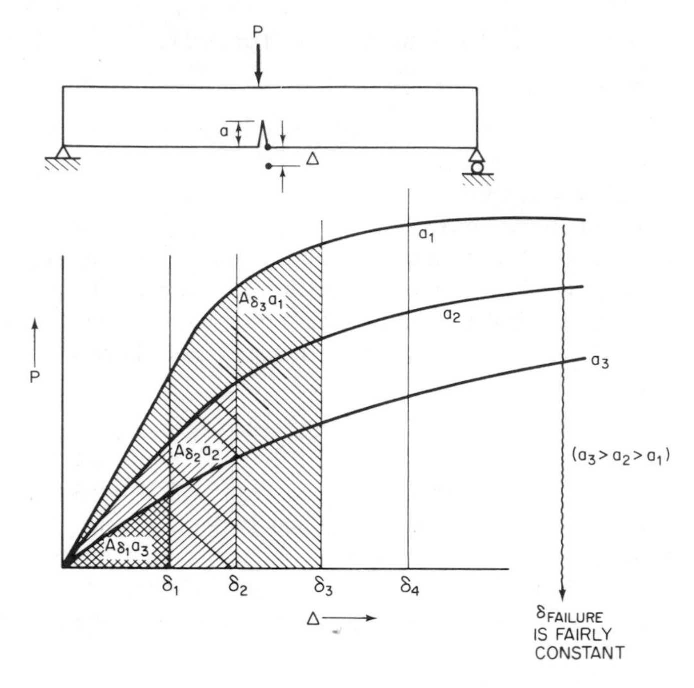

FIG. 16.32. Schematic showing load-displacement curves for various crack depths.

of varying crack length are used to obtain load versus displacement curves, as shown schematically in Figure 16.32. Values of energy per unit thickness (area under the P-Δ curve) are obtained for different initial crack lengths at various values of deflection. These energy values are plotted against crack length as shown in Figure 16.33 for an 1196 NiCrMoV steel having a yield strength of 135 ksi. The slopes of the curves in Figure 16.33 are the changes in potential energy per unit thickness per unit change in crack length and thus are equal to values of J. These values of J are plotted versus deflection in Figure 16.34. In their studies of this particular material, Begley and Landes[21] found that the average deflection at failure was about 0.024 in. (Figure 16.32), and thus the critical J, or J_{Ic}, is about 950 in.-lb/in.^{2} or an equivalent K_{Ic} of about 180 ksi$\sqrt{\text{in.}}$ (Figure 16.34). This value agreed reasonably well with G_{Ic} values obtained from essentially elastic failures of 8-in.-thick compact-tension specimens of the same material.

Bucci et al.[23] summarize this procedure to compute the J integral as a function of δ from a family of load-displacement records as follows: "Given a typical test specimen configuration [Figure 16.35(a)], load-displacement ($P - \Delta$) records are obtained for several test specimens, each having different crack lengths [Figure 16.35(b)]. For given values of deflection, Δ, the area under each load-displacement record may be interpreted as pseudo-potential

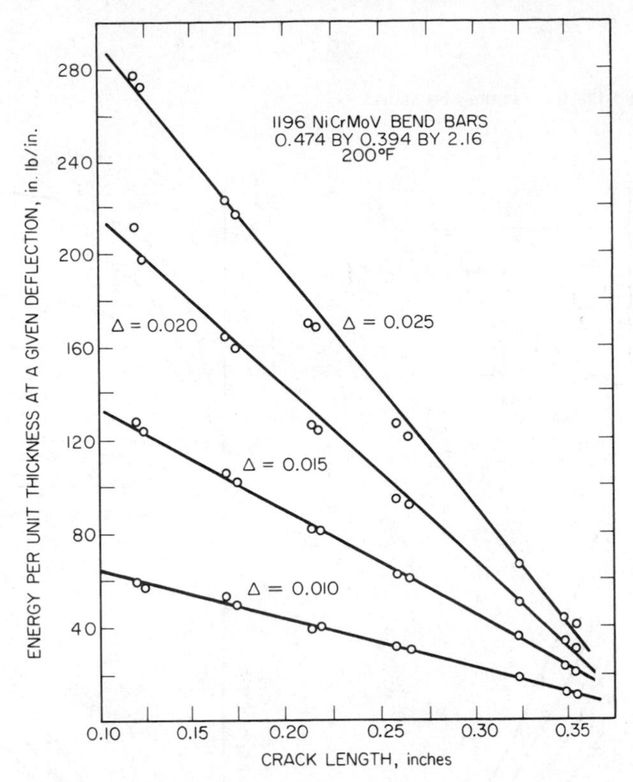

FIG. 16.33. Energy absorbed at a given deflection versus crack length, NiCrMoV steel bars (Ref. 21).

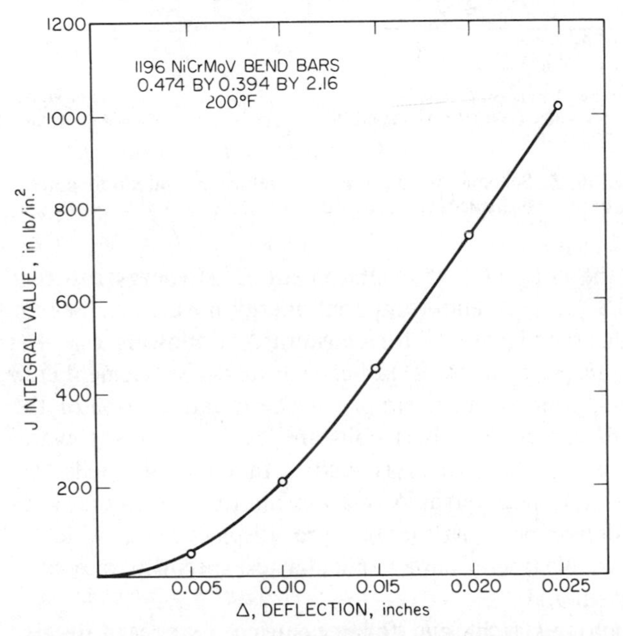

FIG. 16.34. *J* Value as a function of deflection, NiCrMoV bend bars (Ref. 21).

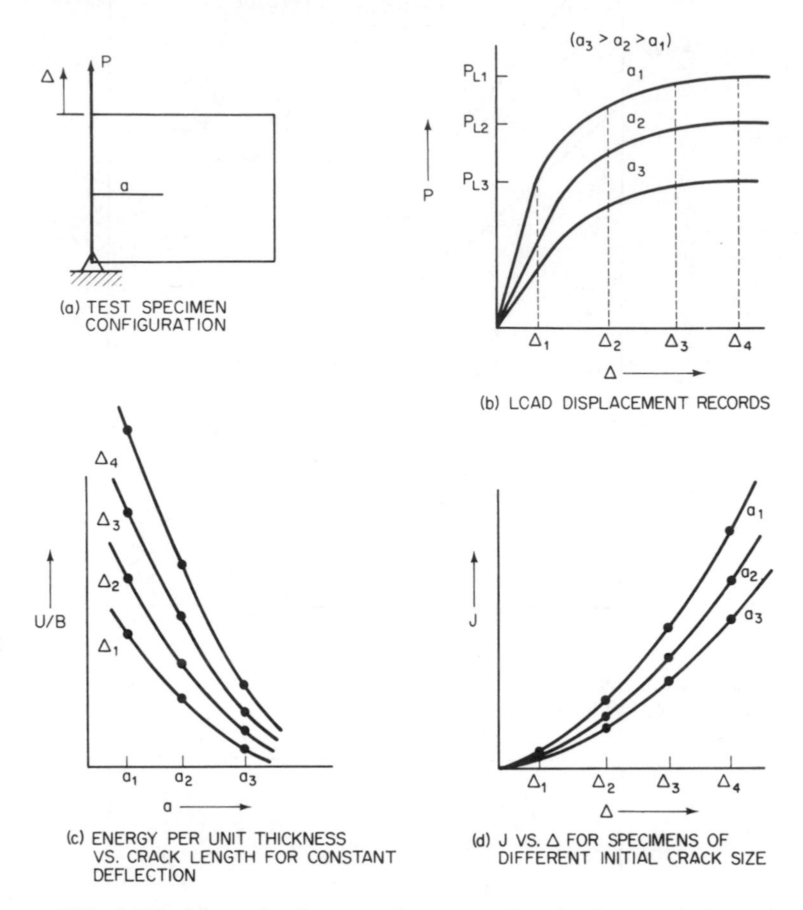

FIG. 16.35. Schematic diagram of J versus δ evaluation employing a family of load-displacement records (Ref. 23).

energy of the body at that displacement. This energy can then be plotted [Figure 16.35(c)] as pseudo-potential energy normalized per unit thickness, U/B, versus crack length, a, for constant Δ. Following $J = -(1/B)(\partial U/\partial a)$, J may be interpreted as the area between load-displacement curves of neighboring crack sizes, or more simply as the negative slope of the U/B versus crack-length curves, for given constant Δ. This permits evaluation of a J versus Δ relationship which is also a function of crack size, a, [Figure 16.35(d)]. Given the J versus Δ relationship for a given crack size, an experimentally determined critical fracture displacement, which characterizes onset of unstable fracture in a particular test specimen, may be used to determine a critical J, J_c."

Although this technique for measuring J appears to give reasonable results, methods for measuring J from a single specimen design (i.e. same initial crack size) would be much more desirable, and such methods were developed by Bucci et al.[23] as well as by Rice et al.[24]

The method developed by Rice et al.[24] for measuring J on deeply notched, bend-type specimens appeared to offer the simplest single specimen design technique for measuring J and was further developed by Landes and Begley.[25]

In this technique, several bend-type specimens, bend bars, or compact-tension specimens with a deep crack ($a/W \geq 0.6$) are loaded to the displacements of interest, and J is determined as a function of displacement from

$$J = \frac{2A}{B \cdot b}$$

where A = area under the load-displacement curve up to the various displacements of interest,
B = specimen thickness,
b = uncracked ligament (for a bend bar, $W - a$).

Critical values of J were labeled J_{Ic} and were taken when crack extension was first encountered. One of the critical questions in establishing a standard J_{Ic} testing procedure is the question of where to take the measurement point for J_{Ic}. Accordingly Landes and Begley[25] tested a 44-ksi yield strength A216 steel to develop a tentative J_{Ic} test method. Preliminary results on the A216 steel are presented in Figure 16.36, which shows that J_{Ic} values (converted to K_{Ic} values) agreed very well with K_{Ic} values obtained using specimens up to 12 in. thick.

Based on their experience with J_{Ic} testing, Landes and Begley[25] recommended the following procedure for a tentative J_{Ic} test method:

Specimens

1. A bend-type specimen is needed. This includes 3- or 4-point bend bars, compact-tension specimens, or WOL specimens.
2. The specimen should be deeply notched with $a/W \geq 0.6$.
3. All specimens should be precracked in fatigue with maximum loads less than half of the expected specimen limit load and $K_{max}/E <$ 0.002 in.$^{1/2}$.
4. Four to six specimens with identical crack lengths should be prepared.

Testing Equipment

1. Testing machine with a load monitor.
2. A load-point-displacement monitor.
3. An X-Y recorder for plotting load versus load-point displacement.
4. A crack-tip-marking facility.
5. A crack extension measurement facility. (This can often be as simple as a scale and a magnifying glass.)

Testing Procedure

1. Load each specimen to different displacement values using displacement control if possible [Figure 16.37(a)]. (It is desirable to load one

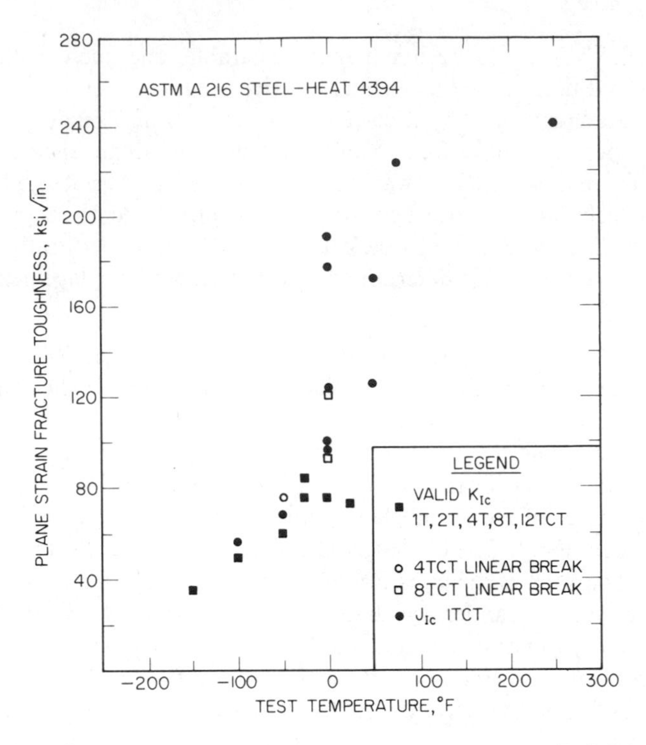

FIG. 16.36. Fracture toughness versus test temperature (Ref. 25).

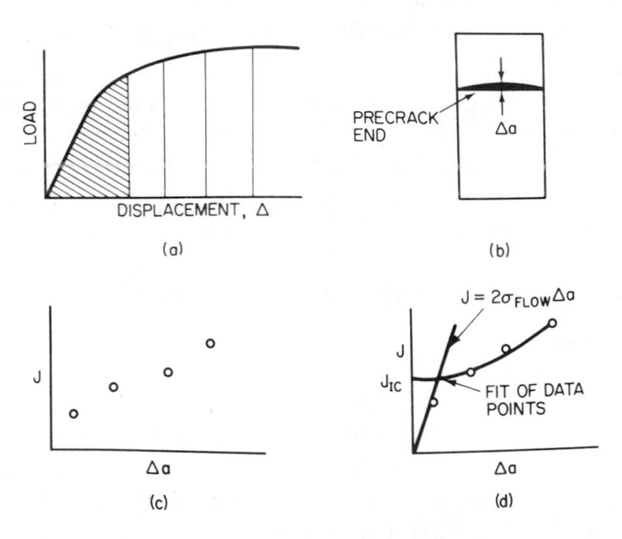

FIG. 16.37. Procedure for J measurement: (a) Load specimens to various displacements, (b) measure crack extension, (c) calculate J for each specimen and plot versus Δa, and (d) construct two curves for J_{Ic} measurement (Ref. 25).

specimen to a point where no actual crack extension has occurred and the others to different amounts of crack extension. It may be difficult to determine this prior to testing; however, if specimens are loaded individually and each fracture surface is examined before the next specimen is loaded, this can easily be done.)

2. Unload each specimen and mark the crack. (Heat tinting is an easy way to mark the crack for steel. For other materials a dye penetrant or a fatigue mark could be used.)
3. Pull the specimen apart and measure crack extension. The crack extension should be measured at its maximum point and taken to include all crack extension from the fatigue precrack to the end of the mark [Figure 16.37(b)].

Data Analysis

1. Calculate J values from the load versus load-point-displacement record using $J = 2A/bB$, where A is the area under the curve up to the point of unloading [Figure 16.37(a)], b is the uncracked ligament measured from the end of the fatigue crack, and B is specimen thickness. (*Note:* In measuring the area under the curve, do not subtract out the area due to elastic unloading. The right-hand side of the area measured should be a vertical line at the unloading displacement.)
2. Plot J versus crack extension [Figure 16.37(c)].
3. Construct the line $J = 2\sigma_{\text{flow}}\,\Delta a$ [Figure 16.37(d)]. (σ_{flow} can be taken halfway between yield and ultimate stresses.)
4. Draw the best fit line to the J versus crack extension points [Figure 16.37(d)]. Include only the points where actual crack extension has occurred. Where crack extension appears only as a stretch zone the point should fall along the line $J = 2\sigma_{\text{flow}}\,\Delta a$.
5. Mark J_{Ic} at the intersection of the lines constructed in steps 3 and 4 [Figure 16.37(d)].

Specimen Size Analysis

1. Calculate J/σ_{flow}.
2. Compare specimen dimensions a, B, b with J/σ_{flow} as follows

$$\alpha = \frac{(a, B, \text{ or } b)}{J/\sigma_{\text{flow}}}$$

3. α should be greater than 25 for a valid specimen size.

This procedure is recommended as a possible standard method for J_{Ic} testing. However, it is likely that individual steps will have to be altered, expanded, or clarified. This can be done only as more test results are acquired.

In summary, the J_{Ic} extension of linear-elastic fracture mechanics into the elastic-plastic region of behavior appears to be very promising. However, testing procedures are not standardized and probably will not be until the late 1970s. Furthermore, the lack of service experience for this technique limits its present usefulness to the design engineer. Yet this approach potentially offers considerable promise in application of fracture-mechanics concepts to the elastic-plastic failure analysis of large complex structures.

REFERENCES

1. A. A. WELLS, "Unstable Crack Propagation in Metals—Cleavage and Fast Fracture," *Cranfield Crack Propagation Symposium, 1*, Sept. 1961, p. 210.
2. D. S. DUGDALE, "Yielding of Steel Sheets Containing Slits," *Journal of the Mechanics and Physics of Solids, 8*, 1960, p. 100.
3. "Methods for Crack Opening Displacement (COD) Testing," *British Standards Institution Draft for Development 19*, London 1972.
4. J. E. HOOD, "Fracture Initiation in Tough Materials," Conference of Metallurgists, CIMM, Montreal, P.Q. Aug. 31, 1971.
5. H. E. COTTON, J. B. DENHAM, and K. R. DOBLE, *How To Conduct COD Tests and Utilize the Results in Practice*, to be published.
6. G. R. EGAN, "Compatibility of Linear Elastic (K_{Ic}) and General Yielding (COD) Fracture Mechanics," *Engineering Fracture Mechanics, 5*, 1973, pp. 167–185.
7. K. NISHIOKA and H. IWANAGA, "Some Results of Dynamic COD-Test," *Significance of Defects in Welded Structures*, University of Tokyo Press, Tokyo, 1973.
8. S. MICHIDA, "COD Approach to Brittle Fracture Initiation in Welded Structures," *Significance of Defects in Welded Structures*, University of Tokyo Press, Tokyo, 1973.
9. F. M. BURDEKIN and M. G. DAWES, "Practical Use of Linear Elastic and Yielding Fracture Mechanics with Particular Reference to Pressure Vessels," *Practical Application of Fracture Mechanics to Pressure-Vessel Technology*, Institute of Mechanical Engineering, London May 1971.
10. G. D. FEARNEHOUGH, G. M. LEES, J. M. LOWES, and R. T. WEINER, "The Role of Stable Ductile Crack Growth in the Failure of Structures," *Practical Application of Fracture Mechanics to Pressure-Vessel Technology*, Institution of Mechanical Engineers, London, May 3–5, 1971.
11. "Fracture Toughness Evaluation by R-Curve Methods," *ASTM STP 527*, American Society for Testing and Materials, Philadelphia, 1973.
12. S. R. NOVAK, "Resistance to Plane-Stress Fracture (*R*-Curve Behavior) of A572 Structural Steel," *ASTM STP 591*, American Society for Testing and Materials, Philadelphia, 1976.
13. "Recommended Practice for R-Curve Determination," *ASTM Annual Book of Standards,* Part 10, Philadelphia, 1975.
14. R. H. HEYER and D. E. MCCABE, "Plane-Stress Fracture Toughness Testing Using a Crack-Line-Loaded Specimen," *Engineering Fracture Mechanics, 4*, 1972, p. 393.

15. R. H. Heyer and D. E. McCabe, "Crack Growth Resistance in Plane-Stress Fracture Testing," *Engineering Fracture Mechanics*, *4*, 1972, pp. 413–430.

16. A. A. Wells, "Notched Bar Tests, Fracture Mechanics, and the Brittle Strengths of Welded Structure," *British Welding Journal*, *12*, No. 2, 1962.

17. A. A. Wells, "Crack Opening Displacements from Elastic-Plastic Analyses of Externally Notched Tension Bars," *Engineering Fracture Mechanics*, *1*, No. 3, April 1969, pp. 399–410.

18. G. R. Irwin, "Linear Fracture Mechanics, Fracture Transition, and Fracture Control," *Engineering Fracture Mechanics*, *1*, No. 2, Aug. 1968, pp. 241–257.

19. D. E. McCabe and R. H. Heyer, "*R*-Curve Determination Using a Crack-Line-Wedge-Loaded (CLWL) Specimen," *ASTM STP 527*, American Society for Testing and Materials, Philadelphia, 1973, pp. 17–35.

20. J. R. Rice, "A Path Independent Integral and the Approximate Analysis of Strain Concentration by Notches and Cracks," *Journal of Applied Mechanics, Transactions ASME*, *35*, June 1968.

21. J. A. Begley and J. D. Landes, "The *J* Integral as a Fracture Criterion," *ASTM STP 514*, American Society for Testing and Materials, Philadelphia 1972, pp. 1–20.

22. J. D. Landes and J. A. Begley, "The Effect of Specimen Geometry on J_{Ic}," *ASTM STP 514*, American Society for Testing and Materials, Philadelphia, 1972, pp. 24–39.

23. R. J. Bucci, P. C. Paris, J. C. Landes, and J. D. Rice, "*J* Integral Estimation Procedures," *ASTM STP 514*, American Society for Testing and Materials, Philadelphia, 1972, pp. 40–69.

24. J. R. Rice, P. C. Paris, and J. G. Merkle, "Some Further Results on *J*-Integral Analysis and Estimates," *ASTM STP 536*, American Society for Testing and Materials, Philadelphia, 1973, pp. 231–245.

25. J. D. Landes and J. A. Begley, "Test Results from *J*-Integral Studies: An Attempt to Establish a J_{Ic} Testing Procedure," *ASTM STP 560*, American Society for Testing and Materials, Philadelphia, 1974, pp. 170–186.

Index

C A P U S
BO STORE

08/2 4

UTEX 31.45
NTEX 39.95
SUB 71.40
TAX 2.86
TOT 74.26
CHE 74.26
CHN 0.00

TEM

00 09:02
CA BOOKS